# Differential Equations

# Differential Equations

DAVID LOMEN / JAMES MARK
*University of Arizona*

PRENTICE HALL, Englewood Cliffs, New Jersey 07632

*Library of Congress Cataloging-in-Publication Data*

Lomen, David, (date)
  Differential equations.

  Includes index.
  1. Differential equations.   I. Mark, James, (date)
II. Title
QA371.L64   1988         513.3'5        87-14454
ISBN O-13-211558-1       01

Editorial/production supervision and
   interior design: Maria McColligan
Cover design: Karen Stephens
Manufacturing buyer: Paula Benevento

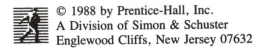 © 1988 by Prentice-Hall, Inc.
A Division of Simon & Schuster
Englewood Cliffs, New Jersey 07632

Printed in the United States of America

10   9   8   7   6   5   4   3   2   1

ISBN   0-13-211558-1     01

Prentice-Hall International (UK) Limited, *London*
Prentice-Hall of Australia Pry. Limited, *Sydney*
Prentice-Hall Canada Inc., *Toronto*
Prentice-Hall Hispanoamericana, S.A., *Mexico*
Prentice-Hall of India Private Limited, *New Delhi*
Prentice-Hall of Japan, Inc., *Tokyo*
Simon & Schuster Asia Pte. Ltd., *Singapore*
Editora Prentice-Hall do Brasil, Ltda., *Rio de Janeiro*

# Contents

# Preface

This text is intended to be used in an introductory one semester course in methods of solution and applications of ordinary differential equations, or as a two semester course including topics from Laplace transforms, systems of autonomous differential equations, Fourier series, and partial differential equations. Users of this text, usually from Engineering, Applied Science, Physics, and Mathematics as well as other disciplines, are expected to have completed the usual calculus sequence required of scientists and engineers. A glance at the table of contents shows that there is more material than is possible to cover in one or even two semesters. This is deliberate and provides for flexibility on the part of the instructor in structuring the course.

Throughout the text the author's present the material in a manner that is mathematically correct but *stresses clarity*. The fundamental existence and uniqueness theorems are stated, in all rigor, but not proven. Examples have sufficient detail for the students to understand each example and not be puzzled by the details missing between two steps in their solution.

The following chapter by chapter summary gives an idea of the book's content. A generous collection of applications appears at appropriate places, both in the text and the exercises.

Chapter 1 establishes terminology and answers basic questions such as what is an ordinary differential equation? What do we mean by a solution to an O.D.E.? Under what conditions are we guaranteed a solution? Differential operators are also covered in this chapter.

Chapter 2 is a study of first order equations which are exact, separable, or reducible to one of these types.

Chapter 3 is devoted to the first order linear equation. The general solution is found and the response of the solution to various forms of input is discussed. The ideas of steady state and transient parts of the solution are introduced and the principle of superposition is utilized.

Chapter 4 is a study of linear equations of order higher than one. It is here that the method of undetermined coefficients is presented in a way that makes memorization of numerous families of functions unnecessary. The method of variation of parameters is presented and the Cauchy-Euler equation is solved. Boundary value problems are solved, introducing eigenvalues and eigenfunctions.

Chapter 5 shows how systems of ordinary differential equations arise, how one equation may be written as a system of equations and presents solution techniques for systems of equations. A section reviewing properties of matrices is included.

Chapter 6 is a detailed treatment of series solutions beginning with Taylor series solutions obtained from the differential equation itself and including a clear and careful treatment of the method of Frobenius. The method of undetermined coefficients is revisited and the solutions to Legendre's and Bessel's equation are developed in detail.

Chapter 7 is an introduction to Laplace transforms in which the existence of the transform is treated in some detail. Transform properties are demonstrated in a series of examples and the convolution theorem is stated and used. Solutions to differential and integral equations and systems of equations are demonstrated by example.

Chapter 8 is a brief introduction to numerical methods, BASIC language programs (unsophisticated) are included in some of the examples. This chapter is purposely limited as we feel this material should be presented in depth in a separate course.

Chapter 9 contains an introduction to qualitative methods of analyzing systems of autonomous differential equations. The comparison method and the direct method are used on nonlinear systems. Periodic solutions and limit cycles are briefly treated.

Chapter 10 starts with a general discussion of orthogonal functions. Fourier series are then considered as a special case. Half range expansions are discussed as well as Sturm-Liouville boundary value problems.

Chapter 11 covers general solutions of partial differential equations and the method of separation of variables is used on the heat equation, the wave equation, and Laplace's equation. Careful attention is given to nonhomogeneous boundary value problems.

Chapters 5, 6, 7, 8, 9, and 10 are independent of each other and may be covered in any desired order.

The authors express their deep appreciation to the following reviewers for their comments and valuable suggestions: Donald Friedlen, Georgia Institute of Technology; John A. Hildebrant, Louisiana State University; Marvin Mundt, Valparaiso University; Hiram Paley, University of Illinois-Urbana; David B. Surowski, Kansas State University; David Lovelock, University of Arizona; Juan A. Gatica, The University of Iowa; John W. Petro, Western Michigan University; Bruce Conrad, Temple University; Thomas W. Rishel, Cornell University; Terry Herdman, Virginia Polytechnic Institute and State University; Joachim Jasiulek, Case Western University, and David Wend, Montana State University.

This book has been classroom tested at the University of Arizona and Northern Arizona University. We thank the students who have read this material for their comments and skill in finding misprints. The valued assistance of the secretarial staff in the mathematics department at the University of Arizona is also appreciated.

D.L.   J.M.

# Differential Equations

# CHAPTER 1

# Basic Concepts of Differential Equations

The purpose of this chapter is to establish some terminology and to gain some familiarity with the basic notions of what an ordinary differential equation (ODE) is, what we mean by a solution to one, and what the solution represents.

## 1.1 ORIGINS, DEFINITIONS, AND CLASSIFICATION OF ODEs

Ever since the ideas of calculus were developed by Newton, Leibniz, and others in the seventeenth century, people have been using differential equations to describe many phenomena which touch our lives. The differential equation is, in fact, the most common mathematical tool used for the precise formulation of the laws of nature and any other phenomena described by a relationship between a function and its derivative.

In this text, first order derivatives will be noted by $dy/dx$, $y'(x)$, $y'$, and $Dy$. The first two clearly show that $y$ is considered to be the dependent variable and $x$ the independent variable. When $y'$ and $Dy$ are used, or primes and $D$'s in front of other letters, the context will determine the independent variable. Second derivatives are written as $d^2y/dx^2$, $y''(x)$, $y''$, and $D^2y$. $d^ny/dx^n$, $y^{(n)}(x)$, $y^{(n)}$, and $D^ny$ are expressions for $n$th derivatives.

We now give a general definition of a differential equation. While at first reading it may appear abstract, it is the form best suited for the statements of some theorems to follow.

**Definition 1.1.** By an *ordinary differential equation of order n* we will mean an equation of the form

$$G(x, y, y', y'', \ldots, y^{(n)}) = 0, \tag{1.1}$$

1

which expresses a functional relationship between the independent variable, $x$, the dependent variable, $y$, and the derivatives of $y$ through the $n$th order.

As an example, we consider the equation

$$y' - 3x^2 = 0. \tag{1.2}$$

Equation (1.2) is of the form (1.1) where $n = 1$, that is, only the first derivative appears. The order of (1.2) is 1 and we say that (1.2) is a *first order ordinary differential equation*.

As another example,

$$x^2y'' - 3x(y')^3 + 4y - \sin x = 0 \tag{1.3}$$

is also an ordinary differential equation. Since the order of the highest derivative appearing is 2, (1.3) is a *second order ordinary differential equation*.

In the first of these examples we have

$$G(x, y, y') = y' - 3x^2$$

and there is no explicit dependence upon the dependent variable $y$.

In the second example we have

$$G(x, y, y', y'') = x^2y'' - 3x(y')^3 + 4y - \sin x.$$

Note that the derivatives appearing in the definition and in the examples are all total derivatives. No partial derivatives occur in an ordinary differential equation. Equations which contain partial rather than total derivatives are called *partial differential equations* and will be treated in Chapter 11.

A *linear differential equation of order $n$* is one that may be written as

$$a_n(x)y^{(n)} + a_{n-1}(x)y^{(n-1)} + \cdots + a_0(x)y = Q(x), \tag{1.4}$$

with $a_n(x) \neq 0$. Notice that no products of terms involving $y$ or its derivatives occur in (1.4) and no transcendental functions of $y$ occur such as in (1.5):

$$(x + \cos y)y' + y + 4x^3 = 0. \tag{1.5}$$

Equation (1.2) is linear, while equations (1.3) and (1.5) are not, and hence (1.3) and (1.5) are called *nonlinear*. Linear equations form a subject of study unto themselves and are taken up in great detail in later chapters.

The following are all linear ordinary differential equations of various orders.

$$y' + kx = 0, \quad k \text{ a constant},$$

$$y'' + 4y' + 4 = 0,$$

$$D^2y - 6Dy + 9y = e^x \csc^3 x,$$

$$y^{(4)} + 2xy = 4x^2.$$

Note that in the last equation the order of differentiation is indicated by a superscript Arabic numeral in parentheses. This will be done to prevent a large number of primes being used.

**EXAMPLE 1.1**    If the rate of change of a dependent variable is proportional to the dependent variable, we have a differential equation

$$\frac{dy}{dt} = \alpha y,$$

where $\alpha$ is the proportionality constant. If $y$ represents some positive quantity, $y$ will be an increasing function of $t$ for $\alpha > 0$ and a decreasing function of $t$ for $\alpha < 0$. The growth of a bacteria culture is an example for $\alpha > 0$, and the decay of a radioactive substance is an example for $\alpha < 0$.

**EXAMPLE 1.2**    Suppose we have a pendulum consisting of an object with mass $m$ at the end of a thin wire of length $l$ as shown in Figure 1.1. If $y$ denotes the angle between the wire and the vertical axis, $t$ is time, and $g$ is a constant giving the acceleration due to gravity, we have the motion of the pendulum governed by the differential equation

$$ml\frac{d^2y}{dt^2} + mg \sin y = 0.$$

For oscillations where $y$ is small, the term $\sin y$ in this differential equation is often replaced by the first term in its Taylor series, namely $y$, and gives a linear approximation for this nonlinear differential equation.

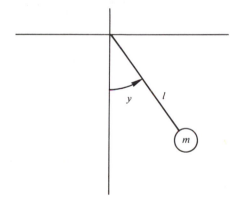

**Figure 1.1**   Simple pendulum.

**EXAMPLE 1.3**    The use of pacemakers to ensure regular heartbeats has led to the development of a system of differential equations to describe the heartbeat cycle. If $y$ denotes the length of a typical heart muscle fiber, $z$ denotes an electrical control variable, and $t$ represents time, we have

$$a\frac{dy}{dt} = -y^3 + by - z,$$

$$\frac{dz}{dt} = (y - y_0) + (y_0 - y_1)h(z).$$

Here $a$, $b$, $y_0$, and $y_1$ are constants which are determined by the heart muscle, and $h(z)$ is a function which typifies the pacemaker. This is an example of a coupled pair of ordinary differential equations. They are called *coupled* because the $y$ of the first equation appears in the second one and the $z$ of the second equation appears in the first.

## 1.2 SOLUTIONS

Having defined an ordinary differential equation, we must now explain what we mean by a solution, so that we have a criterion to use to tell us when a solution has been found. An explicit solution to an ordinary differential equation is a function, given by $y = f(x)$, defined and possessing derivatives up to order $n$ for $x$ in some interval $I$, which identically satisfies the equation (1.1).

An implicit solution to an ordinary differential equation is any relationship between $x$ and $y$, say $g(x, y) = 0$, which through implicit differentiation may be reduced to the original differential equation. Sometimes it is a simple matter to determine an explicit solution from an implicit solution, while sometimes it seems impossible. The word *solution* in this text may refer to either an explicit solution or an implicit solution. The words *explicit* or *implicit* will be added only when needed for emphasis. The form of the equation giving us the solution will determine whether it is an explicit or implicit solution.

We already know how to find solutions to certain simple ordinary differential equations from things we have learned in the study of the calculus. As an example, consider (1.2), which we write as

$$\frac{dy}{dx} - 3x^2 = 0.$$

Solving for $dy/dx$, we have

$$\frac{dy}{dx} = 3x^2. \tag{1.6}$$

If we rewrite (1.6) in an equivalent form using differentials, we have

$$dy = 3x^2 \, dx. \tag{1.7}$$

If we now integrate both sides of (1.7), we get

$$\int dy = \int 3x^2 \, dx$$

or

$$y = x^3 + C. \tag{1.8}$$

Then $f(x) = x^3 + C$ is defined and has a derivative for all $x$ on the interval $I = (-\infty, +\infty)$ which, since $n = 1$, is all that is required. If we substitute $y = f(x)$ directly into (1.2), we see that (1.2) is satisfied identically. Therefore, $y = f(x)$ satisfies all the criteria to be a solution, and is in fact an explicit solution.

Note that (1.8) contains an arbitrary constant, namely, the integration constant arising from the direct integration of both sides of (1.7). Thus the solution to (1.2) is in reality a one-parameter family of functions, a different member of the family being obtained for each different choice of the arbitrary constant $C$. This state of affairs will persist for higher order linear equations as well and we shall see that the solution to an $n$th order linear ordinary differential equation will contain $n$ arbitrary constants and will thus be an $n$-parameter family of functions, each of whose members is specified by a choice of the $n$ *integration constants*. We will say that such a solution is the *general solution* to the differential equation. Thus (1.8) is the general solution to (1.2). Specific members of this general solution are obtained by choosing specific values of $C$. For the example above, $C = 1$ yields a solution

$$f_1(x) = x^3 + 1,$$

while a different choice, say $C = \pi/3$, yields a different solution,

$$f_2(x) = x^3 + \frac{\pi}{3}.$$

We may graph the members of this family, labeling each curve by the corresponding choice of the constant $C$ in (1.8) as shown in Figure 1.2. A solution curve is defined as the graph of a solution of a differential equation, so we may say that Figure 1.2 contains three solution curves.

Notice that the technique used in solving (1.2) resulted automatically in an explicit solution. If we differentiate the relation

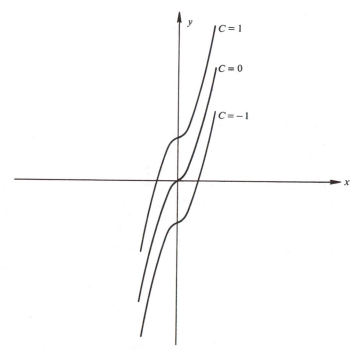

**Figure 1.2**   Examples of the family of functions $f(x) = x^3 + C$.

$$xy + \sin y + x^4 + C = 0$$

with respect to $x$, we obtain

$$xy' + y + y' \cos y + 4x^3 = 0,$$

which is simply (1.5). Thus we have an implicit solution of (1.5) that contains an arbitrary constant.

## 1.3 THE INITIAL VALUE PROBLEM: EXISTENCE AND UNIQUENESS THEOREM

Consider the differential equation

$$\frac{d^2y}{dx^2} = \sin x. \tag{1.9}$$

Letting $v = dy/dx$, we see that (1.9) is equivalent to

$$\frac{dv}{dx} = \sin x$$

and integration gives

$$v = -\cos x + C_1. \tag{1.10}$$

But $v = dy/dx$, so (1.10) is just

$$\frac{dy}{dx} = -\cos x + C_1$$

and again by integration we have

$$y = -\sin x + C_1 x + C_2 \tag{1.11}$$

as the general solution to (1.9).

We may now require that the graph of a solution from (1.11) pass through a given point $(x_0, y_0)$ in the plane, with a given direction specified by assigning a value to the derivative $dy/dx$ at $x = x_0$. That is, for a specific value of $x$, we specify values for $y$ and for $dy/dx$. These are called *initial values*. For the differential equation (1.9), a statement of initial values might be

$$y(0) = 3, \qquad y'(0) = 1. \tag{1.12}$$

These initial values determine the integration constants in the following way. Substituting values for $x_0$ and $dy/dx$ at $x_0$ from (1.12) into (1.10) gives

$$1 = -\cos(0) + C_1 \tag{1.13}$$

or

$$C_1 = 2. \tag{1.14}$$

If we now use (1.14) and (1.12) in (1.11), we have

$$3 = -\sin(0) + 0 + C_2$$

or

$$C_2 = 3. \tag{1.15}$$

Substitution of (1.14) and (1.15) into (1.11) yields the solution given by

$$y = f(x) = -\sin x + 2x + 3,$$

which satisfies the initial conditions (1.12).

The general $n$th order ODE has the form

$$G(x, y, y', \ldots, y^{(n)}) = 0.$$

We are interested in the situation where this equation may be solved for the highest derivative, $y^{(n)}$,

$$y^{(n)} = F(x, y, y', \ldots, y^{(n-1)}) \tag{1.16}$$

and initial conditions are specified by giving values of $y, y', \ldots, y^{(n-1)}$ at some $x = x_0$.

The following theorem tells us when this equation, subject to given initial conditions, has exactly one solution.

**Theorem 1.1.**    Let $F(x, y, y', \ldots, y^{(n-1)})$ be defined and continuous when

$$|x - x_0| < h, |y - y_0| < h, \ldots, |y^{(n-1)} - y_0^{(n-1)}| < h, \qquad h > 0,$$

and have continuous first partial derivatives with respect to $y, y', \ldots, y^{(n-1)}$. Then there exists a $\delta > 0$ such that a unique solution to the equation

$$y^{(n)} = F(x, y, y', \ldots, y^{(n-1)})$$

is defined in some interval $|x - x_0| < \delta$ and satisfies the given initial conditions

$$y(x_0) = y_0, y'(x_0) = y_0', y''(x_0) = y_0'', \ldots, y^{(n-1)}(x_0) = y_0^{(n-1)}.$$

Points for which any of the partial derivatives referred to in Theorem 1.1 fail to be continuous are called *singular points* and must be dealt with separately.

As an example of an application of Theorem 1.1, consider the equation

$$y'' - 4y' + 4y = 4e^{2x} \sin x.$$

Solving for the highest derivative, we have

$$y'' = F(x, y, y') = 4e^{2x} \sin x + 4y' - 4y.$$

Now $F$ is everywhere continuous as a function of $x$, $y$, and $y'$. Furthermore, the partial derivatives

$$\frac{\partial F}{\partial y} = -4$$

$$\frac{\partial F}{\partial y'} = 4$$

are also continuous for all $x$, $y$, and $y'$. Thus Theorem 1.1 guarantees a unique solution satisfying given initial conditions for all $x$. The solution may be found, and techniques for doing so are presented later. For the benefit of the curious, it is

$$y = (C_1 + C_2 x - \sin x)e^{2x}.$$

On the other hand, the equation

$$y'' - xy^{1/2} = 0$$

has $y'' = F(x, y, y') = xy^{1/2}$. Here $\partial F/\partial y = \frac{1}{2}xy^{-1/2}$ and fails to be continuous at $y = 0$. Thus Theorem 1.1 fails to apply and the case $y = 0$ must be examined by itself. As it turns out, $y = 0$ *is* a solution since it satisfies the equation identically for all $x$, but it is a singular solution, that is, a solution at singular points in the $xy$-plane, namely, the entire $x$-axis, and is in no way guaranteed by Theorem 1.1.

## EXERCISES

1. Give the order for the following differential equations and determine if each is linear or nonlinear.

   (a) $\dfrac{d^2y}{dx^2} + w^2 y = 6x^2$.

   (b) $x^2 y'' + (y')^2 - \sin x^2 = 0$.

   (c) $(y''')^2 + xy + 13e^x = 0$.

   (d) $x^2 y'' + 3xy' + y = 0$.

   (e) $D^4 y + 3D^2 y + 4(y)^2 = 0$.

   (f) $D^2 y + 2Dy + \cos y = 0$.

2. Show that the functions in braces are solutions of the differential equations that follow.

   (a) $\{e^x, e^{3x}\}$,         $y'' - 4y' + 3y = 0$.

   (b) $\{e^x, xe^x\}$,         $y'' - 2y' + y = 0$.

   (c) $\{\sin 4x, \cos 4x\}$,     $y'' + 16y = 0$.

   (d) $\{1, x, x^2\}$,         $y''' = 0$.

   (e) $\{x^2, x^{-1}\}$,         $x^2 y'' - 2y = 0, x > 0$.

   (f) $\{e^x, e^{-x}, \sin x, \cos x\}$,     $D^4 y - y = 0$.

3. For what values of $r$ is $e^{rx}$ a solution of the following differential equations?

   (a) $y'' - y' - 6y = 0$.

   (b) $D^2 y - 6Dy + 9y = 0$.

   (c) $(D^2 + D - 12)y = 0$.

   (d) $y'' - 3y' - 10y = 0$.

   (e) $y^{(4)} - 16y = 0$.

   (f) $y^{(4)} - 3y'' - 4y = 0$.

   (g) $(D^3 - 8)y = 0$.

   (h) $9D^2 y + 6Dy + y = 0$.

4. For what value of $r$ is the given function a solution of the given differential equation?

   (a) $x^r$,         $(xD - 3)y = 0$.

   (b) $rx + x^2$,     $xy' - 2y + x = 0$.

   (c) $1 - e^{-rx}$,     $y' + 32y - 32 = 0$.

   (d) $(x - r)^2 e^x$,   $[(x - 2)D - x]y = 0$.

   (e) $x^r e^x$,       $(D^2 - 2D + 1)y = 0$.

   (f) $\exp(1 - re^x)$,   $(D + e^x)y = 0$.

5. Show that the following functions are solutions of the given differential equations for any choice of the constant $C$.

   (a) $Cx^4 - 1$,       $xy' - 4y = 4$.

   (b) $1 - Ce^{-2x}$,     $(D + 2)y = 2$.

   (c) $C(x - 2)^2 e^x$,   $(x - 2)y' - xy = 0$.

   (d) $C \exp(1 - e^x)$,   $y' + e^x y = 0$.

**6.** For what value of $C$ does the graph of a solution in each part of exercise 5 pass through the point $(0, -1)$?

**7.** A linear combination of $f_1(x)$ and $f_2(x)$ is $C_1 f_1(x) + C_2 f_2(x)$, where $C_1$ and $C_2$ are constants. Find linear combinations of the functions in exercise 2 which satisfy
(a) $y(0) = 0$, $y'(0) = 2$.
(b) $y(0) = 0$, $y'(0) = 3$.
(c) $y(0) = 9$, $y'(0) = 0$.
(d) $y(1) = 2$, $y'(1) = 1$, $y''(1) = 0$.
(e) $y(1) = 2$, $y'(1) = -1$.
(f) $y(0) = 0$, $y'(0) = 0$, $y''(0) = 0$, $y^{(3)} = 4$.

**8.** Which of the following initial value problems satisfy the conditions of Theorem 1.1?
(a) $x^2 y'' + 3xy' + (\sin x)y = e^x$, $y(1) = y'(1) = 3$.
(b) $2y'' + 3xy' + 5y = \cos x$, $y(0) = y'(0) = 2$.
(c) $(1 - x)y' = \sin(1 - x)$, $y(1) = 3$.
(d) $(3 - x)y'' + xy = 6y^3$, $y(0) = 2$.
(e) $xyy' + 3e^x y + 7x = 4$, $y(1) = 0$.

**9.** Show that the following are implicit solutions of the given differential equations for any choice of the constant $C$.

(a) $x^2 + y^2 = C$, $\qquad y' = -\dfrac{x}{y}$.

(b) $\sin xy + x^3 + xy^2 = C$, $\quad (x \cos xy + 2xy)y' = -3x^2 - y^2 - y \cos xy$.

(c) $xe^{-xy} + x^2 y = C$, $\qquad (x^2 - x^2 e^{-xy})y' = xye^{-xy} - e^{-xy} - 2xy$.

(d) $\ln |e^C y| + e^{x^2} = y^3$, $\qquad \left(-3y^2 + \dfrac{1}{y}\right)y' + 2xe^{x^2} = 0$.

(e) $xy + (Cy)^{-1} = 3x$, $\qquad C \neq 0$, $(3x - 2xy)y' = y^2 - 3y$.

## 1.4 THE BOUNDARY VALUE PROBLEM

In Section 1.3 we defined an initial value problem as one where the dependent variable, $y$, and possibly some of the derivatives of $y$, were specified for a given value of the independent variable, $x$. On the other hand, a *boundary value problem* prescribes the dependent variable (and possibly some of its derivatives) at *two* or more values of the independent variable. The following examples illustrate the difference between an initial value problem and a boundary value problem.

**EXAMPLE 1.4**    The differential equation describing a freely falling body under the influence of gravity (assuming no resistance) is

$$\frac{d^2 y}{dt^2} = -g, \tag{1.17}$$

with $y$ a vertical distance, $t$ time, and $g$ the gravitational constant. If we prescribe the initial conditions

$$y(0) = y_0, \qquad y'(0) = v_0 \tag{1.18}$$

we specify the initial height in some reference frame and the initial velocity. However, if we are only interested in values of $t$ in the interval $0 \leq t \leq 4$ and require that

$$y(0) = y_0, \qquad y(4) = y_1 \qquad\qquad (1.19)$$

(i.e., specify the height at two different times), we have a boundary value problem. Conditions such as (1.19), where the dependent variable, $y$, is specified at the two endpoints, or boundaries, of the domain (0 and 4) are called *boundary conditions*.

Notice that one integration of (1.17) results in

$$\frac{dy}{dt} = -gt + C_1.$$

With an initial value problem [conditions of (1.18)] we can solve for $C_1$ directly in terms of a given quantity, $v_0$, but with a boundary value problem [as with condition (1.19)] we usually cannot. This situation is indicative of the fact that some of the techniques used in satisfying initial conditions do not always apply to boundary value problems. In this example, however, one more integration and applying the other initial condition gives the solution as

$$y = -\frac{gt^2}{2} + v_0 t + y_0.$$

The solution with boundary conditions (1.19) is

$$y = -\frac{gt^2}{2} + \left(\frac{y_1 - y_0}{4} + 2g\right)t + y_0.$$

**EXAMPLE 1.5**   The buckling of a long shaft or column under an axial load is governed by the differential equation

$$\frac{d^4 y}{dx^4} + \lambda \frac{d^2 y}{dx^2} = 0$$

with $\lambda$ a positive constant. If the end of the column given by $x = 0$ is built in and the end given by $x = l$ is hinged, the associated boundary conditions may be taken as

$$y(0) = 0, \qquad y(l) = 0,$$
$$y'(0) = 0, \qquad y''(l) = 0.$$

Boundary value problems are covered in some detail in Section 4.5, and again in Chapters 7, 10, and 11.

## EXERCISES

**1.** Consider the boundary value problem

$$\frac{d^2 y}{dx^2} + \omega^2 y = 0, \qquad \omega \neq 0,$$

$$y'(0) = 0, \qquad y'(a) = 0.$$

(a) Show that sin $\omega x$ and cos $\omega x$ are solutions of this differential equation.

(b) For what choices of $A$ and $B$ will a solution of this differential equation, $y = A \cos \omega x + B \sin \omega x$, also satisfy the boundary conditions, $y'(0) = 0$, $y'(a) = 0$, if $a = \pi/\omega$?

(c) Show that $y = A \cos \omega x + B \sin \omega x$ can satisfy $y'(0) = 0$, $y'(a) = 0$ only if $a = n\pi/\omega$, $n$ a nonzero integer.

2. Show that the only solution of

$$\frac{d^2y}{dx^2} + \omega^2 y = 0, \qquad \omega \neq 0,$$

$$y'(0) = 0, \qquad y'(2) = 0,$$

of the form $y = A \cos \omega x + B \sin \omega x$ is $y = 0$.

3. Show that the only solution of

$$\frac{d^2y}{dx^2} = 0,$$

$$y(0) = 0, \qquad y'(a) = 0,$$

of the form $y = A + Bx$ is $y = 0$.

4. Find values of $A$ and $B$ so that $y = A + Bx$ satisfies the boundary value problem

$$\frac{d^2y}{dx^2} = 0, \qquad y'(0) = 0, \quad y'(a) = 0.$$

5. The differential equation in Example 1.5 is

$$\frac{d^4y}{dx^4} + \lambda \frac{d^2y}{dx^2} = 0.$$

(a) Find values of $r$ such that $x^r$ is a solution of this differential equation.

(b) Find values of $r$ such that cos $rx$ is a solution of this differential equation.

(c) Find values of $C_1$, $C_2$, $C_3$, and $C_4$ such that

$$y = C_1 + C_2 x + C_3 \cos(\sqrt{\lambda}\, x) + C_4 \sin(\sqrt{\lambda}\, x)$$

satisfies

$$y(0) = y''(0) = 0.$$

(d) Show that requiring the solution from part (c) to satisfy

$$y(\ell) = y''(\ell) = 0$$

leads to the equations

$$C_2\ell + C_4 \sin(\sqrt{\lambda}\, \ell) = 0,$$
$$C_4\lambda \sin(\sqrt{\lambda}\, \ell) = 0.$$

(e) Show that solutions of the equations in part (d) exist for the situations: $C_2 = C_4 = 0$, or $C_2 = 0$, $C_4$ arbitrary if $\sqrt{\lambda}\, \ell = n\pi$, $n = 1, 2, 3, \ldots$.

## 1.5 DIRECTION FIELDS AND THE METHOD OF ISOCLINES

Even though this text will develop many techniques for solving various types of differential equations, there will be times when a closed-form solution will not exist. For such cases we can still obtain an idea of what the family of *solution curves* looks like by using graphical techniques.

We start by considering a first order differential equation of the form

$$\frac{dy}{dx} = F(x, y). \tag{1.20}$$

If $(x_0, y_0)$ is a point on the solution curve of (1.20), $F(x_0, y_0)$ gives the value of the slope of the tangent line to the curve at $(x_0, y_0)$. If a short line segment with this slope is drawn through $(x_0, y_0)$, the tangent line to the solution curve at $(x_0, y_0)$ will coincide with this short line segment. If this procedure is followed for many points, a so called *direction field* of (1.20) is obtained. Solution curves (sometimes called *integral curves*) are now drawn so that the tangent lines to the curve coincide (or are consistent) with this direction field. To illustrate these ideas, we consider a simple example.

**EXAMPLE 1.6**   Find a direction field for

$$\frac{dy}{dx} = \frac{-x}{y}, \qquad y > 0. \tag{1.21}$$

Table 1.1 gives the slope at several points, and short, heavy line segments are plotted on Figure 1.3 to indicate the proper slope at these points. Additional short line segments, with slopes given by (1.21), are plotted on Figure 1.3 to give a direction field.

**TABLE 1.1**   VALUES OF $x$, $y$, AND $dy/dx$ FOR $dy/dx = -x/y$

| $x$ | $y$ | $\dfrac{dy}{dx}$ |
| --- | --- | --- |
| 1 | 1 | $-1$ |
| $-1$ | 1 | 1 |
| 2 | 2 | $-1$ |
| $\frac{1}{2}$ | 2 | $-\frac{1}{4}$ |
| 2 | $\frac{1}{2}$ | $-4$ |
| $-\frac{1}{2}$ | 2 | $\frac{1}{4}$ |
| $-2$ | $\frac{1}{2}$ | 4 |

It should be apparent that semicircles centered at the origin (the solid curves on Figure 1.4) will have their tangent lines consistent with this direction field. This can be seen analytically by implicitly differentiating

$$y^2 + x^2 = C \tag{1.22}$$

with respect to $x$ to obtain

$$2y\frac{dy}{dx} + 2x = 0. \tag{1.23}$$

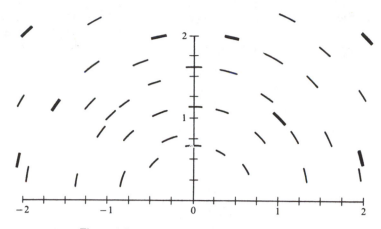

**Figure 1.3**   Directional field for $dy/dx = -x/y$.

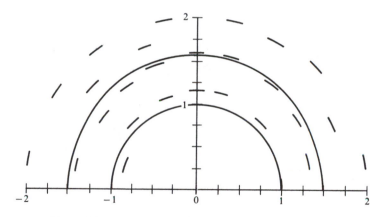

**Figure 1.4**   Solution curves (integral curves) for $dy/dx = -x/y$.

Rearranging (1.23) gives (1.21), our original differential equation. It is probably apparent that computing lots of slopes of line segments can be a very time consuming process. However, this is an easy task for a computer and efficient routines have been developed to construct direction fields.

We can also obtain direction fields quickly by using the *method of isoclines,* defined as follows. If we consider values of $x$ and $y$ where the right-hand side of (1.20) is a constant,

$$F(x, y) = k, \tag{1.24}$$

we obtain a curve in the $xy$-plane along which the solution curve has the same slope. The curve given by (1.24) is called an *isocline* (same inclination) of (1.20) since all solution curves have slope $k$ as they cross this curve. In Example 1.6, isoclines are given by

$$-\frac{x}{y} = k \quad \text{or} \quad y = \left(-\frac{1}{k}\right)x. \tag{1.25}$$

Thus the solution curve will have slope $k$ at points where it crosses the isocline given by (1.25). As an aid in determining the direction, we plot isoclines and draw small line segments with slope $k$. Doing this for several values of $k$ gives a systematic method of determining the direction field. In Figure 1.5, isoclines of

$$\frac{dy}{dx} = \frac{-x}{y} \tag{1.26}$$

are given for $k = 0, \pm 0.5, \pm 1,$ and $\pm 2$ [see (1.25)], together with two solution curves.

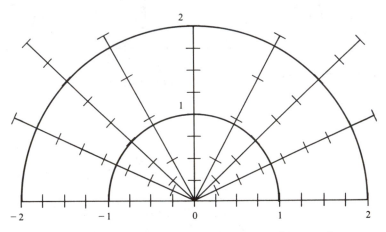

**Figure 1.5**   Isoclines and integral curves for $dy/dx = -x/y$.

**EXAMPLE 1.7**   Find a direction field and sketch solution curves for the differential equation

$$\frac{dy}{dx} = \frac{xy}{1 + x^2}. \tag{1.27}$$

Drawing short segments with the proper slope at points $(x, y)$, $x = -10, -9,$ $\ldots, 9, 10, y = -10, -9, \ldots, 9, 10$ results in the direction field shown in Figure 1.6. Note that all horizontal tangents occur on the two axes. Note also that $y = 0$ satisfies the differential equation and is a solution curve as well as an isocline. Since the differential equation (1.27), together with an arbitrary initial conditon, $y(x_0) = y_0$, satisfies the hypothesis of Theorem 1.1, a *unique* solution curve passes through $y(x_0) = y_0$. These solution curves may never cross. Since $y = 0$ is the solution curve for initial condition $y(0) = 0$, no solution curve may cross the $x$-axis. That is, if $y_0 > 0$, the solution curve remains above the $x$-axis, and if $y_0 < 0$, the solution curve is always below the $x$-axis. A few solution curves are drawn consistent with the direction field of Figure 1.6 and shown in Figure 1.7.

Straightforward differentiation shows that the function

$$y = A(1 + x^2)^{1/2}, \tag{1.28}$$

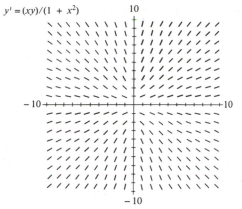

**Figure 1.6**   Direction field for $dy/dx = xy/(1 + x^2)$.

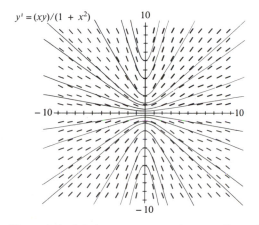

**Figure 1.7**   Solution curves for $dy/dx = xy/(1 + x^2)$.

where

$$A = y_0(1 + x_0^2)^{-1/2}, \tag{1.29}$$

satisfies (1.27) with initial condition $y(x_0) = y_0$. From this form of the solution we also see that the sign of $y$ is determined by the sign of $y_0$.

## 1.6 PICARD'S ITERATION METHOD

In Section 1.3 we presented, without proof, a theorem regarding the existence and uniqueness of the solution to an $n$th order ordinary differential equation. The theorem for a first order equation is as follows.

**Theorem 1.2.**   Let $F(x, y)$ and $\partial F/\partial y$ be defined and continuous when

$$|x - x_0| < h, \qquad |y - y_0| < h. \tag{1.30}$$

Then there exists a $\delta > 0$ such that a unique solution to

$$y' = F(x, y) \tag{1.31}$$

is defined in the interval $|x - x_0| < \delta$ and satisfies the given initial condition

$$y(x_0) = y_0. \tag{1.32}$$

This theorem may be proved by means of the method of successive approximation developed by Picard. The proof needs ideas from advanced calculus which are not familiar to most users of this text. Even so, we present an outline of this iterative method since it introduces a technique which you may use later in a different context.

To use Picard's method, we let $y(x)$ be the solution of (1.31) and (1.32). Then we integrate both sides of (1.31) to obtain

$$\int_{x_0}^{x} y'(t) \, dt = y(x) - y(x_0) = \int_{x_0}^{x} F(t, y(t)) \, dt,$$

where we use $t$ as a dummy variable of integration to avoid confusion with $x$ as the upper limit of the integral. We rearrange this equation as

$$y(x) = y_0 + \int_{x_0}^{x} F(t, y(t)) \, dt \tag{1.33}$$

and notice that (1.32) is satisfied since $\int_{x_0}^{x_0} F(t, y(t)) \, dt = 0$. Picard's idea was to obtain an approximate solution by assuming that the $y$ in $F(t, y)$ equaled the known value $y_0$ for all $t$ covered in the integration. This gives the first approximation, $y_1(x)$, as

$$y_1(x) = y_0 + \int_{x_0}^{x} F(t, y_0) \, dt. \tag{1.34}$$

Once $y_1(x)$ is completely determined, we obtain a second approximation, $y_2(x)$, by assuming $y = y_1(x)$ for the range of integration of the integral in (1.33). This gives

$$y_2(x) = y_0 + \int_{x_0}^{x} F(t, y_1(t)) \, dt, \tag{1.35}$$

In this manner we obtain a sequence of functions $y_1(x), y_2(x), y_3(x), \ldots, y_n(x)$ defined by

$$y_k(x) = y_0 + \int_{x_0}^{x} F(t, y_{k-1}(t)) \, dt, \qquad k = 1, 2, 3, \ldots. \tag{1.36}$$

The repetitive use of (1.36) is called the *method of successive approximation* or *Picard's iteration method*.

The difficult part of proving the theorem is to show that such a technique will always converge. For some specific differential equations, the convergence can easily be shown.

**EXAMPLE 1.8**   Solve the initial value problem

$$y' = 2xy, \qquad y(0) = 3 \tag{1.37}$$

using Picard's iteration method. [Note that $F(x, y) = 2xy$ and $\partial F/\partial y = 2x$ are both continuous for all values of $x$ and $y$.]

Our first step is to write (1.37) as

$$y(x) = 3 + \int_0^x 2ty(t)\, dt$$

and compute the approximate solutions

$$y_1(x) = 3 + \int_0^x 2t(3)\, dt = 3 + 3x^2,$$

$$y_2(x) = 3 + \int_0^x 2t(3 + 3t^2)\, dt$$

$$= 3 + 3\left(x^2 + \frac{x^4}{2}\right),$$

$$y_3(x) = 3 + 3\int_0^x 2t\left(1 + t^2 + \frac{t^4}{2}\right) dt$$

$$= 3 + 3\left(x^2 + \frac{x^4}{2} + \frac{x^6}{6}\right),$$

$$y_4(x) = 3 + 3\int_0^x 2t\left(1 + t^2 + \frac{t^4}{2} + \frac{t^6}{6}\right) dt$$

$$= 3 + 3\left(x^2 + \frac{x^4}{2} + \frac{x^6}{6} + \frac{x^8}{24}\right).$$

By induction we can show that

$$y_n(x) = 3\left(1 + x^2 + \frac{x^4}{2!} + \frac{x^6}{3!} + \cdots + \frac{x^{2n}}{n!}\right). \tag{1.38}$$

The right-hand side of (1.38) is the $n$th partial sum of the infinite series

$$3\sum_{k=0}^{\infty} \frac{x^{2k}}{k!}, \tag{1.39}$$

which converges for all $x$ (using the ratio test). Since

$$e^t = \sum_{k=0}^{\infty} \frac{t^k}{k!},$$

we may write the solution of (1.37) as

$$y = 3e^{x^2}.$$

In practice the integrations are not always so easy, nor is the proof that the resulting series converges.

Notice that once we have a simple form for a solution, we may check to see if it satisfies the original differential equation and initial condition by direct substitution.

Thus the theorem for existence and uniqueness is especially needed for situations where the form of the solution is too complicated to check easily or where we need to seek an approximate solution.

## EXERCISES

**1. (a)** Give the equation of the isoclines corresponding to

$$\frac{dy}{dx} = 3x^2.$$

**(b)** What is the equation of the isocline where all solutions have zero slope?

**(c)** Draw isoclines for $k = 0, 1, 3$, and $12$.

**(d)** Draw the integral curve that passes through $(0, -1)$.

**(e)** Show that $y = x^3 - 1$ passes through $(0, -1)$ and satisfies $dy/dx = 3x^2$.

**2. (a)** Give the equation of the isoclines corresponding to

$$y\frac{dy}{dx} = x.$$

**(b)** What is the equation of the isocline where all solutions have zero slope?

**(c)** Draw isoclines for $k = 0, \pm 0.5, \pm 1$, and $\pm 2$ and add short line segments on the isocline with the proper slope.

**(d)** Draw the integral curve which passes through the point $x = 0, y = 1$.

**(e)** Show that $y^2 - x^2 = 1$ passes through $(0, 1)$ and satisfies $y(dy/dx) = x$.

**3. (a)** Give the equation of the isoclines corresponding to

$$\frac{dy}{dx} = x^2 + y^2.$$

**(b)** When will the solution to this differential equation have a horizontal tangent?

**(c)** Draw isoclines for $k = \frac{1}{4}, 1, 4$, and $9$ and add small line segments on the isocline with the proper slope.

**(d)** Draw the integral curve that passes through $(-1, -1)$.

Use the method of isoclines to sketch integral curves for the following.

**4.** $\dfrac{dy}{dx} = 1 + y^2.$

**5.** $\dfrac{dy}{dx} = \sqrt{x^2 + y^2}.$

**6.** $\dfrac{dy}{dx} = -2xy.$

Use the Picard method of approximation on the following initial value problems.

**7.** $\dfrac{dy}{dx} = 1 + y^2, \quad y(0) = 1(n = 3).$

**8.** $\dfrac{dy}{dx} = -2xy, \quad y(0) = 1.$ Show that your sequence goes to $y = e^{-x^2}$ in the limit as $n \to \infty$.

## 1.7 DIFFERENTIAL OPERATORS

Some of the methods we will develop for solving differential equations will be easier to understand if we consider the derivative as a linear operator.

**Definition 1.2.**    $L$ is a *linear operator* if

$$L[af + bg] = aLf + bLg \tag{1.40}$$

for all constants $a$ and $b$ and functions $f$ and $g$ for which $Lf$ and $Lg$ are defined.

Clearly, if $f$ and $g$ are differentiable functions of $x$ and $L = D = d/dx$ is the derivative operator, (1.40) just gives a well-known property of derivatives.

We formally define an $n$th order differential operator, $D^n$, by

$$D^1 = D,$$

$$D^2 = D[D^1],$$

$$D^3 = D[D^2],$$

$$\vdots$$

$$D^n = D[D^{n-1}].$$

The fact that $D^n$ is also a linear operator may easily be seen using (1.40) with $L$ replaced by $D^n$. Furthermore, since $D^n$ and $D^m$ are linear differential operators, then $a_n D^n + a_m D^m$, where $a_n$ and $a_m$ are constants, is a linear operator. This is shown using the properties of the derivative as

$$(a_n D^n + a_m D^m)(af + bg) = a_n D^n(af + bg) + a_m D^m(af + bg)$$

$$= a_n D^n af + a_n D^n bg + a_m D^m af + a_m D^m bg$$

$$= a a_n D^n f + a a_m D^m f + b a_n D^n g + b a_m D^m g$$

$$= a(a_n D^n + a_m D^m)f + b(a_n D^n + a_m D^m)g.$$

Formally, these operators look like polynomials in $D$ and may be manipulated in similar ways. For example, operators may be multiplied

$$(D - 1)(D - 2) = D^2 - 3D + 2$$

just like polynomials, and

$$[(D - 1)(D - 2)]y = (D^2 - 3D + 2)y. \tag{1.41}$$

As it turns out, expansions like those in (1.41) may be reversed and operators may be factored like polynomials also, thus

$$(D^2 + 2D + 1)y = (D + 1)(D + 1)y = (D + 1)^2 y. \tag{1.42}$$

We must be careful, however, to interpret $(D + 1)^2$ in (1.42) as $(D + 1)(D + 1)$, that is, as the repeated application of the single operator $(D + 1)$. The symbolic operator $(D + a)^n$ then means

$$\underbrace{(D + a)(D + a) \cdots (D + a)}_{n \text{ factors}}.$$

It is also possible to form differential operators of the type

$$L = a_n(x)D^n + a_{n-1}(x)D^{n-1} + \cdots + a_1(x)D + a_0(x) \qquad (1.43)$$

in which the coefficients of $D^j$ are functions of $x$. We must exercise some caution in the multiplication of these operators. For example, let $L_1 = xD + 1$ and $L_2 = xD - 1$. Then we define the product by

$$(L_1 L_2)y = L_1(L_2 y).$$

In this case

$$
\begin{aligned}
(L_1 L_2)y &= (xD + 1)[(xD - 1)y] \\
&= (xD + 1)(xDy - y) \qquad\qquad (1.44) \\
&= xD(xDy - y) + (xDy - y).
\end{aligned}
$$

In forming the first term on the right-hand side we must apply the product rule to the term $xD(xDy)$. This gives

$$xD(xDy) = xDy + x^2 D^2 y. \qquad (1.45)$$

Using (1.45), the right-hand side of (1.44) becomes

$$
\begin{aligned}
xDy + x^2 D^2 y - xDy + xDy - y &= x^2 D^2 y + xDy - y \\
&= (x^2 D^2 + xD - 1)y.
\end{aligned}
\qquad (1.46)
$$

On the other hand, incorrectly treating the operators as polynomials yields

$$(xD + 1)(xD - 1) = x^2 D^2 - 1$$

and the result is different from that in (1.46). Thus operators in (1.43) may be treated as polynomials (in that they commute) only when the coefficients $a_i$ are real or complex numbers.

## EXERCISES

1. Evaluate the following expressions.
   (a) $(D + 2)e^{-x}$
   (b) $(D + 1)e^{-x}$
   (c) $(D + 1)^2 e^{-x}$
   (d) $(D - 1)e^{-x}$
   (e) $(D + 1)(D + x)x^2$
   (f) $(D + x)(D + 1)x^2$
   (g) $(D^2 + w^2)\sin \gamma x$
   (h) $(D^2 - w^2)e^{\gamma x}$
   (i) $(aD^2 + bD + c)e^{\gamma x}$
   (j) $(ax^2 D^2 + bxD + c)x^n$

2. Factor the following operator equations.
   (a) $(D^2 + D - 6)y = 0.$
   (b) $(D^2 + 6D + 9)y = 0.$
   (c) $(D^2 - D - 12)y = 0.$
   (d) $(D^2 + 3D - 10)y = 0.$
   (e) $(D^4 - 16)y = 0.$
   (f) $(D^4 + 3D^2 - 4)y = 0.$
   (g) $(D^3 - 3D^2 - 4D + 12)y = 0.$
   (h) $(D^3 + 8)y = 0.$
   (i) $(9D^2 - 6D + 1)y = 0.$
   (j) $(2D^2 + 4D - 6)y = 0.$

3. Prove that the operators $L_1 = aD + b$ and $L_2 = cD + e$ ($a$, $b$, $c$, and $e$ are constants) commute.

4. For what values of $a$, $b$, $c$, and $e$ do the operators $L_1 = aD + bx$ and $L_2 = cxD + e$ commute?

5. Prove that
   (a) $De^{rx} = re^{rx}$.
   (b) $D^2e^{rx} = r^2e^{rx}$.
   (c) $D^ne^{rx} = r^ne^{rx}$.

6. Prove that
   (a) $xDx^m = mx^m$.
   (b) $x^2D^2x^m = m(m-1)x^m$.
   (c) $x^nD^nx^m = \dfrac{m!}{(m-n)!}x^m$, $n \le m$.

7. Prove that $L = a_jx^jD^j$ is a linear operator for any nonnegative integer $j$.

8. Show that the functions listed below are solutions of the corresponding differential equation in exercise 2.
   (a) $e^{-3x}$, $e^{2x}$                        (b) $e^{-3x}$, $xe^{-3x}$
   (c) $e^{4x}$, $e^{-3x}$                        (d) $e^{-5x}$, $e^{2x}$
   (e) $e^{2x}$, $e^{-2x}$, $\sin 2x$, $\cos 2x$      (f) $e^x$, $e^{-x}$, $\sin 2x$, $\cos 2x$
   (g) $e^{3x}$, $e^{2x}$, $e^{-2x}$                (h) $e^{-2x}$, $e^x \cos(\sqrt{3}x)$, $e^x \sin(\sqrt{3}x)$
   (i) $e^{x/3}$, $xe^{x/3}$                       (j) $e^{-3x}$, $e^x$

9. Show that $e^{-x}$, $e^x$, $xe^x$, $e^{-x}\cos x$, and $e^{-x}\sin x$ all satisfy $Ly = 0$, where $L$ is the operator $L = D^5 + D^4 - D^3 - 3D^2 + 2$.

10. For what values of $n$ is $x^n$ $(x > 0)$ a solution of the following differential equations.
    (a) $(x^2D^2 - 3xD + 3)y = 0$.          (b) $(3x^2D^2 - 5xD)y = 0$.
    (c) $(x^2D^2 - 2xD + 2)y = 0$.          (d) $(4x^2D^2 + 8xD + 1)y = 0$.

11. Show that if $y_1$ satisfies $Ty = f$, where $T$ is a linear operator and $y_2$ satisfies $Ty = 0$, then $y = y_1 + \alpha y_2$ satisfies $Ty = f$ for all choices of the constant $\alpha$.

12. Show that if $y_0$ satisfies $Ty = f_0$ and $y_1$ satisfies $Ty = f_1$, $y_0 + y_1$ satisfies $Ty = f_0 + f_1$.

13. If $L_1 = \sum_{i=0}^n a_iD^i$ and $L_2 = \sum_{j=0}^m b_jD^j$, where $a_i$, $i = 0, 1, 2, \ldots, n$ and $b_j$, $j = 0, 1, 2, \ldots, m$ are constants:
    (a) Prove that $L_1L_2 = L_2L_1$ for $m = n = 2$.
    (b) Prove that $L_1L_2 = L_2L_1$ for arbitrary $m$ and $n$.

## REVIEW EXERCISES

1. Give the order for the following differential equations and determine if each is linear or nonlinear.
   (a) $\dfrac{d^4y}{dx^4} + \lambda\dfrac{d^2y}{dx^2} = 0$.          (b) $y^4\dfrac{dy}{dx} + 3y = 4$.
   (c) $D^2y + 2y^2 = 4e^x$.          (d) $D^3y + 4y = e^x$.

2. For what values of $r$ is $e^{rx}$ a solution of the following differential equations?
   (a) $(D^2 + D - 6)y = 0$.          (b) $(D^2 + 3D - 10)y = 0$.
   (c) $y'' - 4y' + 4y = 0$.          (d) $y''' - y'' - 12y' = 0$.

3. Which of the following initial value problems satisfy the conditions of Theorem 1.1?
   (a) $(1 + x)y' + xy = 0$,   $y(0) = 1$.
   (b) $(1 + x)y' + y = 0$,   $y(-1) = 2$.
   (c) $xy' - 4y = 0$,   $y(0) = -1$.

4. Show that $y^3 - x^3 = 6$ is an implicit solution of $y' = x^2/y^2$.

5. Find solutions of $y'' = 0$ which satisfy $y(0) = 1$, $y(1) = 3$.

6. Show that the only solution of the boundary value problem

$$y'' - 4y = 0 \qquad y(0) = 0, \quad y(2) = 0$$

of the form $y = C_1 e^{2x} + C_2 e^{-2x}$ requires that $C_1 = C_2 = 0$.

7. Show that $C \sin \pi x$ satisfies the boundary value problem

$$y'' + \pi^2 y = 0, \qquad y(0) = y(1) = 0,$$

for any choice of the constant $C$.

8. **(a)** Give the equation of the isoclines corresponding to

$$\frac{dy}{dx} = y + x.$$

    **(b)** What is the equation of the isocline where all solutions have zero slope?

    **(c)** Draw isoclines for $k = 0$, $\pm 1$, $\pm 2$.

    **(d)** Draw the solution curve that passes through $(0, 0)$.

    **(e)** Show that $y = e^x - x - 1$ passes through $(0, 0)$ and satisfies $dy/dx = y + x$.

9. Use the Picard method of approximation on the following initial value problems.

    **(a)** $\dfrac{dy}{dx} = x^2 + y^2$,   $y(0) = 2$ (stop with $n = 3$).

    **(b)** $\dfrac{dy}{dx} = -y + 2$,   $y(0) = 1$.

10. Show that in the limit as $n \to \infty$, the Picard approximation in review exercise 9(b) converges to the solution $2 - e^{-x}$.

# CHAPTER 2

# First Order Differential Equations

The solutions of first order differential equations of the form

$$\frac{dy}{dx} = F(x)$$

may be readily obtained by direct integration as

$$y = \int F(x)\, dx + C$$

for any F which is integrable. However, solutions of

$$\frac{dy}{dx} = F(x, y)$$

are not always so easily obtained.

In this chapter we will discover that it is possible to obtain solutions when $F(x, y)$ is such that the equation is either separable or exact. Once techniques for solving these two types of equations are mastered, we will spend three sections showing how some other differential equations may be reduced to one of these types.

## 2.1 SEPARABLE EQUATIONS AND HOMOGENEOUS FUNCTIONS

The general form for an ordinary differential equation of first order is taken as

$$N(x, y)\frac{dy}{dx} + M(x, y) = 0. \qquad (2.1)$$

We may rearrange this equation in a number of ways. The way we choose depends on the forms of the functions $N(x, y)$ and $M(x, y)$. For example, for all points for which $N(x, y) \neq 0$, we may write

$$\frac{dy}{dx} = -\frac{M(x, y)}{N(x, y)};$$ 
(2.2)

and if we define a new function $F(x, y)$ by

$$F(x, y) = -\frac{M(x, y)}{N(x, y)},$$

(2.2) becomes

$$\frac{dy}{dx} = F(x, y),$$ 
(2.3)

which has the form required by Theorem 1.1. Thus (2.3) has a unique solution passing through a given point $(x_0, y_0)$ if $F(x, y)$ and $\partial F/\partial y$ are continuous on appropriate intervals. The form of the equation given in (2.3), then, is useful for determining where the equation is solvable.

Other forms, easily obtained from (2.1), are

$$M(x, y)\, dx + N(x, y)\, dy = 0$$

and

$$N(x, y) + M(x, y)\frac{dx}{dy} = 0,$$

the second of which is used to find $x$ as a function of $y$. The first will appear in our study of exact equations.

A differential equation is separable if it may be "separated" into two terms, one containing only $x$ and the other containing only $y$. Thus if (2.1) has the form

$$P_1(x)P_2(y)\, dx + Q_1(x)Q_2(y)\, dy = 0,$$ 
(2.4)

division by $P_2(y)Q_1(x)$ gives

$$\frac{P_1(x)}{Q_1(x)}\, dx + \frac{Q_2(y)}{P_2(y)}\, dy = 0$$ 
(2.5)

and the functions of $x$ have been "separated" from the functions of $y$. The general solution of this separable differential equation is found by integration as

$$\int \frac{P_1(x)}{Q_1(x)}\, dx + \int \frac{Q_2(y)}{P_2(y)}\, dy = C.$$ 
(2.6)

We have already seen an elementary example of a separable equation in

$$\frac{dy}{dx} - 3x^2 = 0.$$

Here we have

$$dy - 3x^2\, dx = 0.$$

so integration gives the solution as

$$y(x) = x^3 + C.$$

Another example of a separable equation is

$$\frac{1}{2x}\frac{dy}{dx} - y(x^2 - 1)^{-1} = 0. \tag{2.7}$$

Rearrangement brings (2.7) to the form

$$\frac{dy}{y} = \frac{2x\,dx}{x^2 - 1}, \quad y \neq 0, x \neq \pm 1,$$

and integration of both sides yields

$$\ln|y| = \ln|x^2 - 1| + C. \tag{2.8}$$

Applying the exponential function to both sides of (2.8), we obtain

$$e^{\ln|y(x)|} = e^{\ln|x^2-1|+C}$$

$$= e^C e^{\ln|x^2-1|}$$

or

$$y(x) = k(x^2 - 1) \tag{2.9}$$

as the general solution to (2.7). [Note that k can be either a positive or negative constant. This convention allows us to remove the absolute value bars. Also note that we assumed that $y \neq 0$ in rearranging the equation. However, $y = 0$ is a solution and it may be obtained by setting $k = 0$ in (2.9).]

The only difficulty that might arise with separable equations is in the actual integration after the variables have been separated. Even if one of the integrals cannot be determined, we may proceed to determine the solution in a useful form. To illustrate this situation, consider the equation

$$\frac{dy}{dx} = 3x^2 e^{y^2}, \tag{2.10}$$

with the added condition $y = y_0$ when $x = x_0$. Because $3x^2 e^{y^2}$ is continuous and possesses a continuous partial derivative with respect to $y$, namely $6x^2 y e^{y^2}$, we are guaranteed a solution by Theorem 1.1.

Separation of the variables in (2.10) yields

$$e^{-y^2}\,dy = 3x^2\,dx,$$

so that integration of both sides gives

$$\int e^{-y^2}\,dy = \int 3x^2\,dx = x^3 + C. \tag{2.11}$$

The integral appearing on the left-hand side of (2.11) is not one which may be expressed in closed form by elementary techniques. Two methods will now be given to show you how to determine $C$ when closed-form integrations are not possible.

*Method 1.* If we write the integral on the left-hand side of (2.11) as

$$\int_{yo}^{y} e^{-t^2}\, dt,$$

it will have the value of 0 when $y = y_0$. Thus the constant of integration must be $-x_0^3$ and (2.11) can be written as

$$x^3 = x_0^3 + \int_{y0}^{y} e^{-t^2}\, dt. \tag{2.12}$$

If $x_0$ and $y_0$ are specific numbers, an approximation to the solution is obtained by specifying $y$, using any form of numerical integration (Simpson's rule, trapezoidal rule, etc.) for approximating the value of the definite integral occurring on the right-hand side of (2.12), and calculating $x$ from (2.12).

*Method 2.* Here we will evaluate the integral in (2.11) by using the Taylor series representation for $e^t$, namely,

$$e^t = \sum_{k=0}^{\infty} \frac{t^k}{k!}.$$

Replacing $t$ by $-y^2$ yields

$$e^{-y^2} = \sum_{k=0}^{\infty} (-1)^k \frac{y^{2k}}{k!}. \tag{2.13}$$

Since (2.13) is a convergent power series with an infinite radius of convergence, it may be integrated term by term and the result will be a convergent power series. Thus (2.11) becomes

$$\int \sum_{k=0}^{\infty} (-1)^k \frac{y^{2k}}{k!}\, dy = x^3 + C$$

and

$$\sum_{k=0}^{\infty} \frac{(-1)^k\, y^{2k+1}}{k!(2k+1)} = x^3 + C$$

is the general solution of (2.10). If $x = x_0$ when $y = y_0$,

$$C = \sum_{k=0}^{\infty} \frac{(-1)^k y_0^{2k+1}}{k!(2k+1)} - x_0^3.$$

Sometimes equations which do not appear to be separable may be brought to the form of a separable equation after a change of variable. Consider the equation

$$\frac{dy}{dx} = \frac{x^3 y}{x^4 + y^4}, \quad (x, y) \neq (0, 0). \tag{2.14}$$

Dividing the numerator and denominator of (2.14) by $x^4$ yields

$$\frac{dy}{dx} = \frac{y/x}{1 + (y/x)^4}. \tag{2.15}$$

We now make the change of variable

$$v = \frac{y}{x},$$

(2.16)

so that

$$y = xv$$

and differentiation with respect to $x$ gives

$$\frac{dy}{dx} = x\frac{dv}{dx} + v.$$

(2.17)

Substitution of (2.16) and (2.17) into (2.15) yields the equation

$$x\frac{dv}{dx} + v = \frac{v}{1 + v^4},$$

from which

$$x\frac{dv}{dx} = \frac{v}{1 + v^4} - v = -\frac{v^5}{1 + v^4}.$$

(2.18)

Equation (2.18) may now be put into the form

$$\frac{1 + v^4}{v^5}\, dv = -\frac{dx}{x}$$

(2.19)

in which the integration may be carried out to yield

$$\left(\frac{-1}{4}\right)v^{-4} + \ln|v| = -\ln|x| + C$$

or

$$4\ln|xv| = v^{-4} + 4C.$$

Thus $(xv)^4 = ke^{v^{-4}}$, with $k$ an arbitrary constant, is a solution of (2.19). A return to the original variables yields

$$y^4 = ke^{x^4/y^4}$$

as a solution to (2.14). Note that this is an implicit solution.

The example above is a particular case of the situation in which the form of the first order equation is

$$\frac{dy}{dx} = F\left(\frac{y}{x}\right).$$

(2.20)

The substitution $v = y/x$, as before, brings (2.20) to the form

$$x\frac{dv}{dx} + v = F(v),$$

from which

$$\frac{dv}{F(v) - v} = \frac{dx}{x} \tag{2.21}$$

is recognized as a separable differential equation.

A criterion for determining when a differential equation may be brought to the form (2.20) may be given using the ideas of homogeneous functions.

**Definition 2.1.**    A function $M(x, y)$ is said to be *homogeneous of degree n* if

$$M(tx, ty) = t^n M(x, y),$$

where $t$ is arbitrary but such that $(tx, ty)$ is in the domain of $M$.

**Theorem 2.1.**    Let

$$\frac{dy}{dx} = \frac{-M(x, y)}{N(x, y)}, \tag{2.22}$$

where $M$ and $N$ are homogeneous of degree $n$. Then (2.22) may be written as a separable differential equation by an appropriate change of variables.

*Proof.*    Since $M$ and $N$ are homogeneous of degree $n$, we have

$$M(tx, ty) = t^n M(x, y)$$

or

$$M(x, y) = t^{-n} M(tx, ty).$$

Since this is true for arbitrary values of $t$, it is true if we let $t = 1/x$. This gives

$$M(x, y) = x^n M\left(1, \frac{y}{x}\right)$$

and similarly,

$$N(x, y) = x^n N\left(1, \frac{y}{x}\right).$$

Thus

$$\frac{M(x, y)}{N(x, y)} = \frac{x^n M(1, y/x)}{x^n N(1, y/x)}$$

$$= \frac{M(1, y/x)}{N(1, y/x)}. \tag{2.23}$$

Substituting (2.23) into (2.22) yields

$$\frac{dy}{dx} = \frac{-M(1, y/x)}{N(1, y/x)} = F\left(\frac{y}{x}\right).$$

Since this equation was shown earlier to be transformable to a separable differential equation by a change of dependent variable, the theorem is proved.

**EXAMPLE 2.1**    Consider the differential equation

$$(x^2 - xy + y^2)\frac{dy}{dx} + y^2 = 0$$

or                                                                                                        (2.24)

$$\frac{dy}{dx} = \frac{-y^2}{x^2 - xy + y^2}.$$

Here $M(x, y) = y^2$, so

$$M(tx, ty) = (ty)^2 = t^2 M(x, y).$$

Similarly,

$$N(tx, ty) = (tx)^2 - (tx)(ty) + (ty)^2$$
$$= t^2(x^2 - xy + y^2) = t^2 N(x, y).$$

Thus both $M(x, y)$ and $N(x, y)$ are homogeneous of degree 2 and the transformation $v = y/x$ changes (2.24) to the separable differential equation [see (2.21)]

$$\frac{1 - v + v^2}{v + v^3}\frac{dv}{dx} = -\frac{1}{x}.$$

Although the integral of the left-hand side of this equation may be obtained using partial fractions, let us show that a simpler method of integration results from a different change of variable. If we consider $x$ as the dependent variable and $y$ as the independent variable, we may write (2.24) as

$$\frac{dx}{dy} = \frac{x^2 - xy + y^2}{-y^2}$$                                           (2.25)

and make the change of variable

$$u = \frac{x}{y} \quad \text{or} \quad x = uy.$$

Now $u$ is the new dependent variable, so differentiation with respect to the variable $y$ gives

$$\frac{dx}{dy} = u + \frac{du}{dy}y$$

and (2.25) becomes

$$u + \frac{du}{dy}y = -u^2 + u - 1$$

or

$$\left(\frac{1}{1 + u^2}\right)\frac{du}{dy} = -\frac{1}{y}.$$                        (2.26)

Integration yields

$$\tan^{-1} u = -\ln|y| + C$$

or, in terms of the original variables,

$$\tan^{-1}\frac{x}{y} = -\ln|y| + C.$$

This example shows that if the substitution $v = y/x$ gives a differential equation that is difficult to integrate, try the substitution $x = uy$ instead.

## EXERCISES

1. Find the solution of the following differential equations using the method of separation of variables.

(a) $x^2 \dfrac{dy}{dx} = y^2$.

(b) $e^{-x} \dfrac{dy}{dx} - \sec y = 0$.

(c) $\dfrac{y}{x} dy - \sin x^2 \, dx = 0$.

(d) $\dfrac{dy}{dx} = \dfrac{2x + xy^2}{4y + x^2 y}$.

(e) $x^2 \dfrac{dy}{dx} + 4 + y^2 = 0$.

(f) $(y^2 + 1) dy + y \tan x \, dx = 0$.

(g) $\dfrac{dy}{dx} = e^{x+y}$.

(h) $y \dfrac{dy}{dx} = e^{x+y}$.

(i) $y \, dy - (1 + y)\cos^2 x \, dx = 0$.

(j) $\dfrac{dy}{dx} = \dfrac{1 + y^2}{1 + x^2}$.

2. Solve the following differential equations subject to the given condition.

(a) $x \dfrac{dy}{dx} = y, \quad y(e) = 1$.

(b) $\dfrac{dy}{dx} + y^2 \cos x = 0, \quad y(0) = \frac{1}{2}$.

(c) $3e^{x^2} dy + y^{-2} \, dx = 0, \quad y(1) = 2$.

(d) $y \dfrac{dy}{dx} = \sqrt{4 - y^2}, \quad y(8) = 1$.

(e) $\sqrt{1 - x^2} \dfrac{dy}{dx} + y^3 = 0, \quad y(1) = 1$.

(f) $2(2 + y) dy + y(1 - x^2) dx = 0, \quad y(0) = -1$.

(g) $3y^2 \dfrac{dy}{dx} = \sin x^2, \quad y(0) = 3$.

(h) $(4x^2 + x - 1) \dfrac{dy}{dx} + (8x + 1)(y + 2) = 0, \quad y(1) = 2$.

(i) $y \sin x \, dx + (y^2 + 1)e^{\cos x} \, dy = 0, \quad y\left(\dfrac{\pi}{2}\right) = 1$.

3. Check to see if the following functions are homogeneous. If so, give the degree.

(a) $f(x, y) = x^2 + 3xy - x^3(x + y)$.

(b) $f(x, y) = 3x^2 + y^2 - \dfrac{x^3}{x + y}$.

(c) $f(x, y) = x \sin\dfrac{x}{y} + \dfrac{x^2}{y}$.

(d) $g(x, y) = \sqrt{x^4 + 4y^2x^2 + x^2} + x^3$.

(e) $g(x, y) = \cos\dfrac{x + 4y}{x}$.

(f) $g(x, y) = x \ln y + ye^x$.

(g) $f(x, y) = \ln x - \ln y + e^{x/y}$.

4. Solve the following differential equations after making an appropriate change of variable.

(a) $(x^2 + y^2)\, dx + 2xy\, dy = 0$,   $y(1) = 2$.

(b) $y\, dx - (x - \sqrt{x^2 + y^2})\, dy = 0$,   $y(1) = 0$.

(c) $x\dfrac{dy}{dx} - y\left(\ln\dfrac{y}{x} + 1\right) = 0$.

(d) $\dfrac{dy}{dx} = \dfrac{x - y}{x + y}$,   $y(0) = 1$.

(e) $(x^2 + y^2)\, dx + 3xy\, dy = 0$.

(f) $\dfrac{dy}{dx} = \dfrac{x + y \cos(y/x)}{x \cos(y/x)}$

(g) $(x + y)\, dx + (x + 2y)\, dy = 0$,   $y(1) = 1$.

5. (a) Show that $f(x, y) = x^3 + 2xy^2 + y^3$ satisfies

$$x\frac{\partial f}{\partial x} + y\frac{\partial f}{\partial y} = 3f.$$

(b) Show that if $f(x, y)$ is a homogeneous function of degree $n$,

$$x\frac{\partial f}{\partial x} + y\frac{\partial f}{\partial y} = nf.$$

(This is known as Euler's theorem for homogeneous functions.)

6. A differential equation which arises in study of a chemical of concentration $x$ that is undergoing both first and second order reactions is

$$\frac{dx}{dt} = Ax - Bx^2.$$

(a) Find the solution of this differential equation which satisfies the initial condition $x(0) = x_0$. (Note what happens if $B = 0$.)

(b) Repeat part (a) if $A = 0$.

## 2.2 THE EXACT EQUATION

We return now to the equation

$$N(x, y)\frac{dy}{dx} + M(x, y) = 0, \tag{2.1}$$

which, as before, may be brought to the differential form

$$M(x, y)\, dx + N(x, y)\, dy = 0. \tag{2.27}$$

This equation is said to be *exact* if there exists a function $u(x, y)$ for which

$$\frac{\partial u(x, y)}{\partial x} = M(x, y), \qquad \frac{\partial u(x, y)}{\partial y} = N(x, y), \tag{2.28}$$

for in this case substitution of (2.28) into (2.27) yields

$$\frac{\partial u}{\partial x} dx + \frac{\partial u}{\partial y} dy = 0. \tag{2.29}$$

Since the left-hand side of (2.29) is just the total (or exact) differential of $u(x, y)$, (2.29) becomes

$$d[u(x, y)] = 0. \tag{2.30}$$

The general solution of (2.30) is then given by

$$u(x, y) = C. \tag{2.31}$$

Curves along which a function such as $u(x, y)$ takes on constant values are called *level curves* of the function. Thus, if (2.1) is exact, its solutions are given by the level curves of $u(x, y)$.

Suppose that we have put (2.1) into the form (2.27). How do we decide whether or not it is exact? The answer lies in the equality of the second mixed partial derivatives of $u(x, y)$. If the conditions (2.28) are satisfied and the function $u(x, y)$ has continuous mixed partial derivatives, we must also have

$$\frac{\partial^2 u}{\partial y \, \partial x} = \frac{\partial M}{\partial y}, \qquad \frac{\partial^2 u}{\partial x \, \partial y} = \frac{\partial N}{\partial x},$$

but since $\partial^2 u / \partial y \, \partial x = \partial^2 u / \partial x \, \partial y$, it follows that

$$\frac{\partial M}{\partial y} = \frac{\partial N}{\partial x}. \tag{2.32}$$

Equation (2.32) is then a necessary condition for (2.27) to be exact. It may be shown that the condition (2.32) is also sufficient, provided that $M$ and $N$ have continuous first partial derivatives.

Consider the equation

$$(y + 3x^3 y^2) \frac{dy}{dx} + 3x^2 y^3 - 5x^4 = 0 \tag{2.33}$$

or, in an equivalent form,

$$(y + 3x^3 y^2) \, dy + (3x^2 y^3 - 5x^4) \, dx = 0. \tag{2.34}$$

Here

$$M(x, y) = 3x^2 y^3 - 5x^4 \tag{2.35}$$

and

$$N(x, y) = y + 3x^3 y^2.$$

Taking appropriate partial derivatives gives

$$\frac{\partial M}{\partial y} = 9x^2y^2$$

and

$$\frac{\partial N}{\partial x} = 9x^2y^2.$$

Since $\partial M/\partial y = \partial N/\partial x$, (2.34) is exact.

The question now is: How do we construct the general solution? Well, since the equation is exact there exists a function, $u(x, y)$, such that

$$du = \frac{\partial u}{\partial x}\, dx + \frac{\partial u}{\partial y}\, dy = 0.$$

Thus

$$\frac{\partial u}{\partial x}\, dx = M(x, y)\, dx \tag{2.36}$$

and the integration of both sides of (2.36) gives

$$\int \frac{\partial u}{\partial x}\, dx = \int M(x, y)\, dx. \tag{2.37}$$

The integration on the left-hand side of (2.37) just yields $u(x, y)$. The integration of the right-hand side proceeds as follows. We substitute the expression for $M(x, y)$ from (2.35) into (2.37) and integrate, holding $y$ constant,

$$u(x, y) = \int (3x^2y^3 - 5x^4)\, dx = x^3y^3 - x^5 + f(y). \tag{2.38}$$

The integration "constant" is here taken as an arbitrary function $f(y)$ of $y$. This is because the integration was with respect to $x$, with $y$ considered a fixed parameter. [Notice that (2.38) identically satisfies $\partial u/\partial x = M(x, y)$ for any choice of $f(y)$, since $\partial f(y)/\partial x = 0$.] We now apply the condition

$$\frac{\partial u}{\partial y} = N(x, y)$$

to (2.38). Taking the partial derivative of (2.38) with respect to $y$ and setting the result equal to $N(x, y)$ gives

$$3x^3y^2 + \frac{df}{dy} = y + 3x^3y^2.$$

This implies

$$\frac{df}{dy} = y \tag{2.39}$$

and integration gives $f(y) = y^2/2 + k$, $k$ a constant. Equation (2.38) then becomes

$$u(x, y) = x^3y^3 - x^5 + \frac{y^2}{2} + k \tag{2.40}$$

and the general solution is

$$u(x, y) = C_1$$

or

$$x^3y^3 - x^5 + \frac{y^2}{2} = C,$$

where the sum of two arbitrary constants has been replaced by one arbitrary constant. In future examples, we will not include the arbitrary constant $k$ if the next step equates our result to an arbitrary constant $C$. [Note that you could have arrived at this same result by starting with

$$\int \frac{\partial u}{\partial y}\, dy = \int N(x, y)\, dy,$$

holding $x$ constant.]

This procedure of solving an exact equation will be given as a three-step process.

**Step 1.**   Write the differential equation in the form

$$M(x, y)\, dx + N(x, y)\, dy = 0$$

and check to make sure that

$$\frac{\partial M}{\partial y} = \frac{\partial N}{\partial x}.$$

**Step 2.**   Evaluate

$$u(x, y) = \int M(x, y)\, dx \qquad \left[\text{or } u(x, y) = \int N(x, y)\, dy\right]$$

(treating $x$ and $y$ as independent variables in the integration process.)

**Step 3.**   Evaluate the arbitrary function $f(y)$ [or $g(x)$] that occurs in step 2 and write your solution as

$$u(x, y) = C.$$

**EXAMPLE 2.2**   Putting the differential equation

$$(y^2 - x^2 \sin xy) \frac{dy}{dx} = xy \sin xy - \cos xy - e^{2x} \tag{2.41}$$

in the form $M(x, y)\, dx + N(x, y)\, dy = 0$ requires that

$$M(x, y) = -xy \sin xy + \cos xy + e^{2x} \tag{2.42}$$

and

$$N(x, y) = y^2 - x^2 \sin xy. \qquad (2.43)$$

Since

$$\frac{\partial M}{\partial y} = \frac{\partial N}{\partial x} = -2x \sin xy - x^2 y \cos xy,$$

the differential equation

$$(-xy \sin xy + \cos xy + e^{2x}) \, dx + (y^2 - x^2 \sin xy) \, dy = 0 \quad (2.44)$$

is exact. We now have a choice of evaluating

$$\int \frac{\partial u}{\partial x} \, dx = \int M(x, y) \, dx \qquad \text{or} \qquad \int \frac{\partial u}{\partial y} \, dy = \int N(x, y) \, dy.$$

Since $\int M(x, y) \, dx$ will require integration by parts, we try

$$\int \frac{\partial u}{\partial y} \, dy = \int N(x, y) \, dy = \int (y^2 - x^2 \sin xy) \, dy \qquad (2.45)$$

instead. From (2.45) we have

$$u(x, y) = \frac{y^3}{3} + x \cos xy + g(x), \qquad (2.46)$$

where the "constant of integration" is now an arbitrary function of $x$. [Note that $\partial g(x)/\partial y = 0$.] The function $u(x, y)$ from (2.46) must satisfy

$$\frac{\partial u}{\partial x} = M(x, y)$$

so $\cos xy - xy \sin xy + g'(x) = -xy \sin xy + \cos xy + e^{2x}$. Thus

$$g'(x) = e^{2x}, \qquad g(x) = \frac{e^{2x}}{2}$$

and the general solution of (2.41) is

$$\frac{y^3}{3} + x \cos xy + \frac{e^{2x}}{2} = C.$$

## EXERCISES

1. Check each of the following differential equations for exactness and solve those that are exact.
   (a) $(x^2 - y^2) \, dx - 2xy \, dy = 0$.
   (b) $(x \cos y - y) \, dx + (x \sin y + x) \, dy = 0$.
   (c) $(e^x + y - 1) \, dx + (3e^y + x - 7) \, dy = 0$.
   (d) $(y \cos xy + 3y - 1) \, dx + (x \cos xy + 3x) \, dy = 0$.

**(e)** $\left(4x^3y^3 + \dfrac{x^2}{4}\right) dx + (3x^4y^2 - 16) \, dy = 0.$

**(f)** $(3x^2y + e^y) \, dx + (x^3 + xe^y + \sin y) \, dy = 0.$

**(g)** $[x \cos(x + y) + \sin(x + y)] \, dx + [x \cos(x + y) + y] \, dy = 0.$

**(h)** $\left(xy^3 - \dfrac{1}{xy}\right) dx + \left(3x^2y^2 - \dfrac{1}{y} + x^2\right) dy = 0.$

**(i)** $(1 + y \cos xy) \, dx + (x \cos xy + 2) \, dy = 0.$

**(j)** $(2yx + x^3) \, dx + (y + x^2) \, dy = 0.$

**(k)** $3x^2 \ln y \, dx + x^3 y^{-1} \, dy = 0.$

2. Find values of the constant $r$ such that the following differential equations are exact.

   **(a)** $2y \, dx + (y^2 - rx) \, dy = 0.$

   **(b)** $\sin y \, dx + (x^r \cos y + y^3) \, dy = 0.$

   **(c)** $(y^4 + 2rxy) \, dx + (4xy^3 + rx^2) \, dy = 0.$

   **(d)** $(3x^2 - 3y^r + rx) \, dx + (3xy^{-2} - 3r + 7) \, dy = 0.$

## 2.3 INTEGRATING FACTORS

Sometimes we have an equation of the form (2.1) which is not exact but which may be made exact by multiplication by some function of $x$ and $y$. If we call this function $\rho(x, y)$, then it may turn out that

$$\rho(x, y)M(x, y) + \rho(x, y)N(x, y) \frac{dy}{dx} = 0 \qquad (2.47)$$

is exact. Such a function, $\rho(x, y)$, is called an *integrating factor*.

We consider the following special cases.

*1.* $\rho(x, y) = x^m y^n$.   One possible type of integrating factor would be of the form $x^m y^n$. As an example, consider the equation

$$(1 - xy)y' + y^2 + 3xy^3 = 0. \qquad (2.48)$$

Here

$$M(x, y) = y^2 + 3xy^3,$$

$$N(x, y) = 1 - xy,$$

$$\frac{\partial M}{\partial y} = 2y + 9xy^2,$$

$$\frac{\partial N}{\partial x} = -y,$$

and this equation is *not* exact. We try to obtain an exact equation by multiplying (2.48) by $x^m y^n$:

$$(x^m y^n - x^{m+1}y^{n+1})y' + x^m y^{n+2} + 3x^{m+1}y^{n+3} = 0. \qquad (2.49)$$

Now we have

$$M(x, y) = x^m y^{n+2} + 3x^{m+1} y^{n+3}$$

and

$$N(x, y) = x^m y^n - x^{m+1} y^{n+1}$$

so that

$$\frac{\partial M}{\partial y} = (n + 2)x^m y^{n+1} + (n + 3)3x^{m+1} y^{n+2}$$

and

$$\frac{\partial N}{\partial x} = mx^{m-1} y^n - (m + 1)x^m y^{n+1}.$$

If the two functions $\partial M / \partial y$ and $\partial N / \partial x$ are to be equal, we must have

$$(n + 2)x^m y^{n+1} + 3(n + 3)x^{m+1} y^{n+2} = mx^{m-1} y^n - (m + 1)x^m y^{n+1}$$

or                                                                                                    (2.50)

$$[(n + 2) + (m + 1)]x^m y^{n+1} + 3(n + 3)x^{m+1} y^{n+2} - mx^{m-1} y^n = 0.$$

Since $x^m y^{n+1}$, $x^{m+1} y^{n+2}$, and $x^{m-1} y^n$ are linearly independent functions, we may set the coefficients of each of these terms to zero, giving

$$(n + 2) + (m + 1) = 0,$$
$$3(n + 3) = 0,$$                                                                              (2.51)
$$m = 0.$$

The second and third of equations (2.51) tell us that

$$m = 0,$$
$$n = -3.$$

Substituting these into the first equation in (2.51), we get

$$(-3 + 2) + (0 + 1) = 0$$

so all three equations are satisfied. Thus we have an integrating factor of the form

$$\rho(x, y) = x^0 y^{-3} = y^{-3}.$$                                                          (2.52)

Multiplying (2.48) by $y^{-3}$ gives

$$(y^{-3} - xy^{-2})y' + y^{-1} + 3x = 0.$$                                                     (2.53)

Here

$$M(x, y) = y^{-1} + 3x,$$

and

$$N(x, y) = y^{-3} - xy^{-2},$$

so

$$\frac{\partial M}{\partial y} = \frac{\partial N}{\partial x} = -y^{-2},$$

and (2.53) is indeed exact.

Solving the equation

$$\frac{\partial u}{\partial x} = M(x, y) = y^{-1} + 3x$$

gives

$$u(x, y) = xy^{-1} + \frac{3x^2}{2} + f(y), \qquad (2.54)$$

while setting $\partial u/\partial y = N(x, y)$ yields

$$-xy^{-2} + f'(y) = y^{-3} - xy^{-2}.$$

Thus

$$f'(y) = y^{-3}$$

and integration gives

$$f(y) = -\frac{y^{-2}}{2}.$$

Finally, we have the general solution of (2.53) [and (2.48)] given by

$$xy^{-1} + \frac{3x^2}{2} - \frac{1}{2y^2} = C. \qquad (2.55)$$

**2. $\rho(x, y) = \rho(x)$.**   If the integrating factor of

$$M(x, y)\, dx + N(x, y)\, dy = 0 \qquad (2.56)$$

is to be a function only of $x$, the equation

$$\rho(x)M(x, y)\, dx + \rho(x)N(x, y)\, dy = 0 \qquad (2.57)$$

must be exact. The condition for exactness gives

$$\frac{\partial}{\partial x}[\rho(x)N(x, y)] = \frac{\partial}{\partial y}[\rho(x)M(x, y)]. \qquad (2.58)$$

Differentiating yields

$$\rho'(x)N(x, y) + \rho(x)\frac{\partial N}{\partial x} = \rho(x)\frac{\partial M}{\partial y} \qquad (2.59)$$

and rearranging gives

$$\frac{\rho'(x)}{\rho(x)} = \frac{1}{N(x, y)}\left(\frac{\partial M}{\partial y} - \frac{\partial N}{\partial x}\right). \qquad (2.60)$$

Notice that the left-hand side of (2.60) is a function of $x$ only, while the right-hand side may contain both $x$ and $y$. If this equation is to make sense, the right-hand side must be a function only of $x$; that is, if the integrating factor is to be a function only of $x$, the expression

$$\frac{1}{N(x, y)}\left[\frac{\partial M(x, y)}{\partial y} - \frac{\partial N(x, y)}{\partial x}\right]$$

must be independent of $y$. To see if this is the case, we need to compute the expression above and see if the $y$ dependence cancels out. Such is the case in the next example.

**EXAMPLE 2.3**    In the linear differential equation

$$\frac{dy}{dx} + p(x)y = q(x) \qquad (\text{or } [p(x)y - q(x)]\, dx + dy = 0) \qquad (2.61)$$

we have that

$$M(x, y) = p(x)y - q(x), \qquad N(x, y) = 1$$

so

$$\frac{1}{N(x, y)}\left[\frac{\partial M(x, y)}{\partial y} - \frac{\partial N(x, y)}{\partial x}\right] = p(x)$$

and is a function only of $x$. Thus $p(x)$ can be determined from (2.60) as

$$\ln |\rho(x)| = \int p(x)\, dx + k \qquad (2.62)$$

or

$$\rho(x) = \pm e^{k} e^{\int p(x)\, dx}.$$

Since we are only looking for *one* function, $\rho(x)$, which is an integrating factor, we usually take the plus sign and let $k = 0$, to obtain

$$\rho(x) = e^{\int p(x)\, dx}. \qquad (2.63)$$

(See exercises 26, 27, and 28 at the end of this section.)
    To solve

$$\frac{dy}{dx} + 2xy = e^{-x^2} \qquad (2.64)$$

we first calculate an integrating factor $\rho(x)$ as [see (2.63)]

$$\rho(x) = e^{\int 2x\, dx} = e^{x^2}.$$

Thus

$$e^{x^2}\frac{dy}{dx} + e^{x^2}2xy = (e^{x^2})(e^{-x^2})$$

is an exact equation. If we write it as

$$\frac{d}{dx}(e^{x^2}y) = 1,$$

we can integrate and have the general solution of (2.64) in the form

$$e^{x^2}y = x + C$$

or

$$y = (x + C)e^{-x^2}.$$

Since linear first order equations are extremely important, Chapter 3 is devoted entirely to them.

**EXAMPLE 2.4**    The differential equation

$$(e^{-x} + \sin y)\, dx + \cos y \, dy = 0 \qquad (2.65)$$

is not exact, since

$$\frac{\partial M}{\partial y} = \frac{\partial}{\partial y}(e^{-x} + \sin y) = \cos y,$$

while

$$\frac{\partial N}{\partial x} = 0.$$

However,

$$\frac{1}{N(x, y)}\left[\frac{\partial M(x, y)}{\partial y} - \frac{\partial N(x, y)}{\partial x}\right] = \frac{\cos y - 0}{\cos y} = 1$$

is in fact independent of $y$, so a solution of

$$\frac{\rho'(x)}{\rho(x)} = \frac{1}{N}\left(\frac{\partial M}{\partial y} - \frac{\partial N}{\partial x}\right) = 1 \qquad (2.66)$$

gives an integrating factor which is a function of $x$ only. Obtaining the integrating factor as $\rho(x) = e^x$ from the solution of (2.66) and multiplying (2.65) by $\rho(x)$ gives the exact differential equation

$$(1 + e^x \sin y)\, dx + e^x \cos y \, dy = 0. \qquad (2.67)$$

[Note that

$$\frac{\partial}{\partial y}(1 + e^x \sin y) = \frac{\partial}{\partial x}(e^x \cos y).]$$

The solution of (2.67) is readily obtained as

$$x + e^x \sin y = C. \qquad (2.68)$$

**3. $\rho(x, y) = \rho(y)$.**    If the integrating factor of

$$M(x, y)\, dx + N(x, y)\, dy = 0 \qquad (2.69)$$

is to be a function only of $y$, the equation

$$\rho(y)M(x, y)\,dx + \rho(y)N(x, y)\,dy = 0 \tag{2.70}$$

must be exact. The condition for exactness is

$$\frac{\partial}{\partial x}[\rho(y)N(x, y)] = \frac{\partial}{\partial y}[\rho(y)M(x, y)]. \tag{2.71}$$

Carrying out the differentiation in (2.71) gives

$$\rho(y)\frac{\partial N(x, y)}{\partial x} = \rho'(y)M(x, y) + \rho(y)\frac{\partial M(x, y)}{\partial y} \tag{2.72}$$

and rearranging (2.72) gives

$$\frac{\rho'(y)}{\rho(y)} = \frac{1}{M(x, y)}\left(\frac{\partial N}{\partial x} - \frac{\partial M}{\partial y}\right). \tag{2.73}$$

Thus if the integrating factor is to be a function only of $y$, the right-hand side of (2.73) must be independent of $x$.

**EXAMPLE 2.5**    The differential equation

$$2x\,dx + [(x^2 + 1)\cot y - 1]\,dy = 0 \tag{2.74}$$

is not exact since

$$\frac{\partial M}{\partial y} = 0$$

while $\partial N/\partial x = 2x\cot y$. However,

$$\frac{1}{M(x, y)}\left(\frac{\partial N}{\partial x} - \frac{\partial M}{\partial y}\right) = \frac{1}{2x}(2x\cot y) = \cot y$$

is independent of $x$, so a solution of

$$\frac{\rho'(y)}{\rho(y)} = \cot y = \frac{\cos y}{\sin y} \tag{2.75}$$

is an integrating factor of (2.74). Solving (2.75) gives

$$\ln|\rho(y)| = \ln|\sin y| + C$$

and we choose $C = 0$ so that

$$\rho(y) = \sin y.$$

Multiplying (2.74) by this integrating factor gives

$$2x\sin y\,dx + [(x^2 + 1)\cos y - \sin y]\,dy = 0. \tag{2.76}$$

Equation (2.76) is an exact differential equation, since

$$\frac{\partial M}{\partial y} = \frac{\partial N}{\partial x} = 2x\cos y$$

and has a general solution given by

$$(x^2 + 1) \sin y + \cos y = C. \tag{2.77}$$

## EXERCISES

Solve the following differential equations after finding an integrating factor in the form $x^m y^n$.

**1.** $y\,dx + (2x - y^2)\,dy = 0.$

**2.** $-y\,dx + (2x + y)\,dy = 0.$

**3.** $x\dfrac{dy}{dx} + 2xy^2 - y = 0.$

**4.** $xy^2\dfrac{dy}{dx} - x^3 - y^3 = 0.$

**5.** $y^2\,dx + x^2\,dy = 0.$

**6.** $x^3y\,dx - (x^4 + y^4)\,dy = 0.$

**7.** $2x^2y\dfrac{dy}{dx} + xy^2 + x^{-1} = 0.$

**8.** $(xy^2 - 2x^2y)\dfrac{dy}{dx} + 2xy^2 - y^3 = 0.$

**9.** $2xy\,dx + (y^2 - x^2 + y)\,dy = 0.$

**10.** $(2x^2y + y^2)\,dx + (2x^3 - xy)\,dy = 0.$

Find an appropriate integrating factor and solve the following linear differential equations. (More exercises concerning linear differential equations are given in Chapter 3.)

**11.** $\dfrac{dy}{dx} + xy = 3x.$

**12.** $\dfrac{dy}{dx} - \dfrac{1}{2x}y = 2.$

**13.** $\dfrac{dy}{dx} - 2y = xe^{2x}.$

**14.** $\dfrac{dy}{dx} + \dfrac{4x}{x^2 + 1}y = 3x.$

**15.** $\dfrac{dy}{dx} + \dfrac{1}{x \ln x}y = 3x^2.$

Find a solution of the following differential equations after first obtaining an appropriate integrating factor.

**16.** $e^y\dfrac{dy}{dx} + (1 + x)e^{-y} = 0.$

**17.** $(x - \sec y \ln x)\dfrac{dy}{dx} + \tan y - \dfrac{y}{x}\sec y = 0.$

**18.** $(y^2 - yx \sin y)\,dx + (xy \ln x - x^2y \cos y)\,dy = 0.$

**19.** $y\,dx + (2y - x)\,dy = 0.$

**20.** $(2y + 3xy^2)\,dx + (3x + 4yx^2)\,dy = 0.$

**21.** $2xy\,dx + (3x^2 + 4y)\,dy = 0.$

**22.** $x(y + x)\dfrac{dy}{dx} + 2x + y - 1 = 0.$

**23.** $x(y + x + 1)\dfrac{dy}{dx} + y^2 + 3xy + 2y = 0.$

**24.** $(2xy + 4x^3)\,dx + (x^2 + x^2y + x^4)\,dy = 0.$

**25.** Find at least three integrating factors of the form $x^m y^n$ for the differential equation

$$x^2\dfrac{dy}{dx} + xy = 0.$$

**26.** Show that if $\rho(x, y)$ is an integrating factor, so is $C\rho(x, y)$, where $C$ is any nonzero constant.

**27.** Show that $\rho(x) = x^2$ is an integrating factor for

$$\frac{dy}{dx} + \frac{2}{x}y = \frac{1}{x^3 + 2}.$$

**28.** Show that $\rho(x) = 3x^2$ is also an integrating factor of the differential equation in exercise 27, and find its solution.

## 2.4 CHANGES OF VARIABLES

We have already seen that differential equations of the form

$$\frac{dy}{dx} = F\left(\frac{y}{x}\right)$$

can be put in a separable form with the transformation $v = y/x$ (or $u = x/y$). In this section we discover other types of first order differential equations which may be transformed into exact or linear differential equations with an appropriate change of variables.

### 2.4.1 Bernoulli's Equation

Bernoulli's equation is a differential equation of the form

$$\frac{dy}{dx} + p(x)y = q(x)y^n, \qquad n \neq 0, 1. \tag{2.78}$$

(Note that $n = 0$ and $n = 1$ correspond to linear differential equations.)
    The change of dependent variable from $y$ to $u$ by the equation

$$u = y^{1-n} \tag{2.79}$$

transforms (2.78) to a linear differential equation. To find this linear equation, differentiate (2.79) with respect to $x$ and obtain

$$\frac{du}{dx} = (1 - n)y^{-n}\frac{dy}{dx}. \tag{2.80}$$

If we multiply (2.78) by $y^{-n}$ and use (2.79) and (2.80), we can transform (2.78) into

$$\left(\frac{1}{1 - n}\right)\frac{du}{dx} + p(x)u = q(x). \tag{2.81}$$

This is obviously a first order linear equation which may be solved using an integrating factor. This solution will be in terms of $u$ and $x$, so another transformation using (2.79) is needed to obtain a solution of (2.78) in terms of $y$ and $x$.

**EXAMPLE 2.6**    In the Bernoulli equation

$$\frac{dy}{dx} - \frac{1}{x}y = -xy^2 \tag{2.82}$$

we see that $n = 2$. Making the change of variable

$$u = y^{1-2} = y^{-1}$$

gives

$$\frac{du}{dx} = -y^{-2}\frac{dy}{dx}$$

and transforms (2.82) to

$$-\frac{du}{dx} - \frac{1}{x}u = -x$$

or

$$\frac{du}{dx} + \frac{1}{x}u = x. \tag{2.83}$$

This linear equation has $x$ as an integrating factor, so multiplying (2.83) by $x$ gives the exact equation

$$x\frac{du}{dx} + u = x^2$$

or

$$\frac{d}{dx}(xu) = x^2. \tag{2.84}$$

Integration gives

$$xu = \frac{x^3}{3} + C$$

or

$$u = \frac{x^2}{3} + \frac{C}{x}, \qquad x \neq 0.$$

Returning to our original variable $y$ using $u = y^{-1}$ results in a solution of our original differential equation as

$$y = \frac{1}{x^2/3 + C/x} = \frac{3x}{x^3 + 3C}.$$

Notice that even though we could have written the exact form of the solution of a general Bernoulli equation [using an integrating factor in (2.81)] in terms of integrals, we did not. The three important things to remember from this section are

1. The form of a Bernoulli equation, $y' + p(x)y = q(x)y^n$.
2. The transformation $u = y^{1-n}$.
3. The method of solution for the resulting linear equation.

## EXERCISES

Solving the following differential equations.

**1.** $y' + \sqrt{x}y = \dfrac{2}{3}\sqrt{\dfrac{x}{y}}.$

**2.** $y' + \dfrac{y}{x} = \dfrac{1}{x^3 y^3}.$

**3.** $x^2\dfrac{dy}{dx} + y^2 - xy = 0.$

**4.** $x\dfrac{dy}{dx} - (1 + x)y - y^2 = 0.$

**5.** $y' + \dfrac{2y}{x} = -x^9 y^5.$

**6.** $y' + y = 2x^2 y^2.$

**7.** $yy' + xy^2 - x = 0.$

**8.** $2yy' + y^2 \sin x - \sin x = 0.$

**9.** $x\dfrac{dy}{dx} - \dfrac{y}{2 \ln x} - y^2 = 0.$

**10.** $\dfrac{dy}{dx} - y + xe^{-2x}y^3 = 0.$

**11.** A differential equation of the form

$$\frac{dy}{dx} = a_0(x) + a_1(x)y + a_2(x)y^2$$

with $a_0(x)$ and $a_2(x)$ not identically zero is called a Riccati differential equation. [If $a_0(x)$ were identically zero, this equation would be a Bernoulli differential equation.] Note that this equation is not linear, so if $y_1(x)$ is a solution, $Cy_1(x)$, with $C$ an arbitrary constant, will not necessarily be a solution.

(a) Show that if we know one solution, $y_1(x)$, of a Riccati differential equation, then the change of variable

$$y = y_1(x) - \frac{1}{u}$$

results in the following linear differential equation for $u$:

$$u' + [a_1(x) + 2a_2(x)y_1(x)]u = a_2(x).$$

(b) To find this first solution of a Ricatti differential equation, we often try simple functions like $ax^b$ or $ae^{bx}$. Find values of $a$ and $b$ such that $ax^b$ is a solution of

$$\frac{dy}{dx} = 1 + x + 2x^2 \cos x - (1 + 4x \cos x)y + 2(\cos x)y^2.$$

(c) Make the change of variable given in part (a) to obtain the differential equation

$$u' - u = 2 \cos x.$$

(d) Solve the differential equation in part (c) and show that the general solution to our original differential equation is

$$y = x - [\sin x - \cos x + C_1 e^x]^{-1},$$

where $C_1$ is an arbitrary constant.

**12.** Solve the following Riccati differential equations.

(a) $\dfrac{dy}{dx} = 5 + \left(\dfrac{4}{x}\right)y + \left(\dfrac{1}{4x^2}\right)y^2.$

(b) $\dfrac{dy}{dx} = (1 - y)\left(\dfrac{1}{x} - \dfrac{1}{10} + \dfrac{y}{10}\right).$

(c) $\dfrac{dy}{dx} = 6x^{-2} - 2y^2$.

(d) $\dfrac{dy}{dx} = -1 + 2y - y^2$.

(e) $\dfrac{dy}{dx} = 3e^{4x} + 2y - 12y^2$.

(f) $\dfrac{dy}{dx} = e^{2x} + [1 + (5/2)e^x]y + y^2$.

(g) $\dfrac{dy}{dx} = 4x^{-2} - \dfrac{y}{x} - y^2$.

**13.** Show that if $y$ is a solution of a Riccati differential equation (in the form from exercise 11), then $u = 1/y$ is a solution of the differential equation

$$u' = -a_2(x) - a_1(x)u - a_0(x)u^2.$$

### 2.4.2 Linear Coefficients

Another first order differential equation that can be solved after a change of variables is one where both $M(x, y)$ and $N(x, y)$ are linear functions of $x$ and $y$.

We write such an equation as

$$(a_1 x + b_1 y + c_1)\,dx + (a_2 x + b_2 y + c_2)\,dy = 0. \tag{2.85}$$

Notice the special cases listed below.

| Condition | Type of equation |
|---|---|
| $b_2 = 0$ | Linear |
| $b_1 = a_2$ | Exact |
| $c_1 = c_2 = 0$ | $M(x, y)$ and $N(x, y)$ are both homogeneous of degree 1, so the transformation $v = y/x$ makes it separable |

Since $b_2 = 0$ is already classified above, we assume $b_2 \neq 0$ and make the change of variables

$$u = x + d_1, \qquad v = a_2 x + b_2 y + c_2,$$

$$(x = u - d_1,\ y = [v - a_2 u + a_2 d_1 - c_2]/b_2)$$

in (2.85) and obtain

$$\left[\left(a_1 - \frac{a_2 b_1}{b_2}\right)u + \left(\frac{b_1 - a_2}{b_2}\right)v + \frac{b_1}{b_2}(a_2 d_1 - c_2) - a_1 d_1 + c_1\right]du + \frac{v}{b_2}dv = 0. \tag{2.86}$$

If $a_1 - (a_2 b_1/b_2) = 0$, (2.86) is a separable equation regardless of the choice of $d_1$, so we can choose $d_1$ to be any number we wish (usually 0). If $a_1 - (a_2 b_1/b_2) \neq 0$, choose $d_1$ so that

$$\frac{b_1}{b_2}(a_2 d_1 - c_2) - a_1 d_1 + c_1 = 0. \tag{2.87}$$

For this choice of $d_1$, since (2.86) has coefficients of $du$ and $dv$ that are homogeneous of degree 1, a change of variables $w = u/v$ will reduce it to a separable equation. Thus for all possible values of the constants $a_1$, $b_1$, $c_1$, $a_2$, $b_2$ and $c_2$, (2.85) is solvable.

**EXAMPLE 2.7**    To solve

$$(y + 9) \, dx + (2x + y + 3) \, dy = 0 \qquad (2.88)$$

we make the change of variables

$$u = x + d_1, \qquad v = 2x + y + 3$$

or

$$x = u - d_1, \qquad y = v - 2(u - d_1) - 3.$$

With this change of variables, $dx = du$ and $dy = dv - 2 \, du$, so (2.88) becomes

$$(v - 2u + 2d_1 + 6) \, du + v(dv - 2 \, du) = 0$$

or

$$(-v - 2u + 2d_1 + 6) \, du + v \, dv = 0. \qquad (2.89)$$

Choosing $d_1 = -3$ makes both the coefficients of $du$ and $dv$ homogeneous functions of degree 1. Thus we use the change of variables $w = u/v$, so $du = w \, dv + v \, dw$ and (2.89) becomes

$$(-1 - 2w)(w \, dv + v \, dw) + dv = 0$$

or

$$(-2w^2 - w + 1) \, dv + (-1 - 2w)v \, dw = 0. \qquad (2.90)$$

Equation (2.90) is a separable equation that can be integrated using partial fractions:

$$\frac{dv}{v} = \frac{(1 + 2w) \, dw}{(-2w + 1)(w + 1)} = \left( \frac{4/3}{-2w + 1} + \frac{-1/3}{w + 1} \right) dw.$$

Thus

$$\ln |v| = -\frac{2}{3} \ln |-2w + 1| - \frac{1}{3} \ln |w + 1| + \ln |C|$$

$$= \ln \left| \frac{C}{(-2w + 1)^{2/3}(w + 1)^{1/3}} \right|.$$

If $C$ is allowed both positive and negative values, this gives

$$v = \frac{C}{(-2w + 1)^{2/3}(w + 1)^{1/3}}$$

or

$$v^3(-2w + 1)^2(w + 1) = C^3. \qquad (2.91)$$

We now must back-substitute ($w = u/v$, $u = x - 3$, $v = 2x + y + 3$) to obtain the solution, (2.91), in terms of our original variables as

$$v^3\left(-2\frac{u}{v} + 1\right)^2\left(\frac{u}{v} + 1\right) = C^3$$

or

$$(-2u + v)^2(u + v) = C^3$$

and finally,

$$(y + 9)^2(3x + y) = C^3.$$

## EXERCISES

Solve the following differential equations.

1. $(x + 2y - 5)\, dx + (4 - 2x - y)\, dy = 0.$
2. $(6x - y + 5)\, dx - (4x - y + 3)\, dy = 0.$
3. $(2x + 3y)\, dx + (3x - y - 11)\, dy = 0.$
4. $(x - y - 2)\, dx + (2 - x - 2y)\, dy = 0.$
5. $(x + 2y + 3)\, dx + (x + 7)\, dy = 0.$
6. $(x - 2y - 1)\, dx - (4x - 9y - 4)\, dy = 0.$
7. $(2x - 3y + 5)\, dx + (y - 1)\, dy = 0.$
8. $(4x + y - 2)\, dx + (3x + y - 2)\, dy = 0.$
9. $(5x + 4y - 4)\, dx + (4x + 5y - 5)\, dy = 0.$
10. $(9x + 3y - 12)\, dx - (3x - y - 2)\, dy = 0.$

## 2.5 EQUATIONS REDUCIBLE TO FIRST ORDER

*1. The equation $F(y'', y', x) = 0$.* If the dependent variable, $y$, is missing in the second order differential equation

$$F(y'', y', x) = 0, \tag{2.92}$$

the change of variables

$$w = y' = \frac{dy}{dx}$$

reduces it to a first order equation in $w$, namely

$$F(w', w, x) = 0. \tag{2.93}$$

To illustrate this technique, consider the differential equation

$$\frac{d^2y}{dx^2} + 2\frac{dy}{dx} - 1 = 0. \tag{2.94}$$

If we let

$$w = \frac{dy}{dx} \tag{2.95}$$

(2.94) may be written as the separable first order equation

$$\frac{dw}{dx} = 1 - 2w \quad \text{or} \quad \frac{dw}{1 - 2w} = dx.$$

Integration gives

$$1 - 2w = C_1 e^{-2x}. \tag{2.96}$$

Since $w = dy/dx$, we may express (2.96) in the form

$$\frac{dy}{dx} = \frac{1}{2}(1 - C_1 e^{-2x})$$

and integrate to obtain the general solution of (2.94) as

$$y = \frac{1}{2}x + \frac{C_1}{4}e^{-2x} + C_2. \tag{2.97}$$

**2. The equation $F(y'', y', y) = 0$.**    If the independent variable is missing in the second order differential equation

$$F(y'', y', y) = 0, \tag{2.98}$$

we may again use the substitution

$$w = y' = \frac{dy}{dx}.$$

However, in this case we must calculate $y''$ in a different manner to eliminate $x$ as a variable. Here we use the chain rule to write

$$y'' = \frac{d}{dx}\left(\frac{dy}{dx}\right) = \frac{d}{dx}(w) = \frac{dw}{dy}\frac{dy}{dx} = w\frac{dw}{dy}.$$

Substitution of $w = y'$ and the foregoing expression for $y''$ into (2.98) gives

$$F\left(w\frac{dw}{dy}, w, y\right) = 0,$$

which is a first order differential equation in $w$ with $y$ as the independent variable.

**EXAMPLE 2.8**

$$\frac{d^2y}{dx^2} + 3y^2\left(\frac{dy}{dx}\right)^2 = 0, \tag{2.99}$$

$$y(0) = 1, \quad y'(0) = \frac{3}{e}.$$

Setting

$$w = \frac{dy}{dx} \qquad \text{and} \qquad \frac{d^2y}{dx^2} = w\frac{dw}{dy}$$

gives

$$w\frac{dw}{dy} + 3y^2w^2 = 0.$$

This is a separable differential equation, $dw/w = -3y^2 \, dy$, with solution

$$\ln|w| = -y^3 + C. \tag{2.100}$$

In order to integrate (2.100) to obtain $y$, we first put it in the equivalent form

$$|w| = \left|\frac{dy}{dx}\right| = e^C e^{-y^3}. \tag{2.101}$$

Since $y'(0) = 3/e$, $e^C = 3$ and we can drop the absolute value sign in (2.101) and integrate to obtain

$$\int e^{y^3} \, dy = 3x + C_1. \tag{2.102}$$

Even though the integral of $e^{y^3}$ is not given in terms of elementary functions, we can still solve for $C_1$ using the condition $y(0) = 1$ as the lower limit of a definite integral; that is, write (2.102) as

$$\int_1^y e^{t^3} \, dt = 3x + C_1.$$

Since $y = 1$ when $x = 0$, $C_1 = 0$ and we have the solution in the implicit form

$$\int_1^y e^{t^3} \, dt = 3x$$

(see Section 2.1). A numerical evaluation of this integral will give the value of $x$ for any desired value of $y$.

*3. Reduction of order.*    Sometimes, one solution of a differential equation is obvious, or is obtained by trial and error. In this case the reduction of order technique will result in a new differential equation of degree 1 less than the original one. We illustrate this procedure for equations of second order.

If one nonzero solution of a second order differential equation

$$a_2(x)\frac{d^2y}{dx^2} + a_1(x)\frac{dy}{dx} + a_0(x)y = 0 \tag{2.103}$$

is known, a change of variable will put (2.103) in a form [equation (2.92)] which we can always solve. Since it is convenient to use operator notation, we rewrite (2.103) as

$$Ty = [a_2(x)D^2 + a_1(x)D + a_0(x)]y = 0. \tag{2.104}$$

If $y_1(x)$ satisfies (2.104), we make a change of dependent variable in the form

$$y(x) = v(x)y_1(x),$$

calculate

$$Dy = (Dv)y_1 + vDy_1$$

and                                                                                         (2.105)

$$D^2y = (D^2v)y_1 + 2DvDy_1 + vD^2y_1,$$

and substitute into (2.104) to obtain

$$Ty = a_2[(D^2v)y_1 + 2DvDy_1 + vD^2y_1] + a_1[(Dv)y_1 + vDy_1] + a_0vy_1$$

$$= (a_2D^2v + a_1Dv)y_1 + 2a_2DvDy_1 + vTy_1 = 0. \qquad (2.106)$$

Since $y_1(x)$ satisfies (2.104), $Ty_1 = 0$ and (2.106) can be reduced to a first order equation by letting

$$w = Dv.$$

If we make this substitution and rearrange (2.106), we obtain the first order equation

$$a_2Dw + \left(a_1 + 2a_2\frac{Dy_1}{y_1}\right)w = 0. \qquad (2.107)$$

This equation is separable and may be solved as follows. Separating $w$ and $y_1$ dependencies, we have

$$\frac{a_2(x)Dw + a_1(x)w}{w} + 2a_2(x)\frac{Dy_1}{y_1} = 0.$$

Thus

$$\frac{Dw}{w} + \frac{a_1(x)}{a_2(x)} + 2\frac{Dy_1}{y_1} = 0$$

and

$$\ln|w| + \int\frac{a_1(x)}{a_2(x)}\,dx + 2\ln|y_1| = \ln|C_1| \qquad (2.108)$$

with $C_1$ an arbitrary constant. Solving (2.108) for $w = Dv = dv(x)/dx$,

$$w = \frac{dv(x)}{dx} = \frac{C_1}{y_1^2(x)}\exp\left[-\int\frac{a_1(x)}{a_2(x)}dx\right],$$

and integrating once more gives

$$v(x) = C_1\int\frac{\exp\{-\int[a_1(x)/a_2(x)]\,dx\}}{y_1^2(x)}\,dx + C_2.$$

Since our original variable was $y$, using $y = vy_1$ gives the general solution of (2.104) as

$$y(x) = C_1y_1(x)\int\frac{\exp\{-\int[a_1(x)/a_2(x)]\,dx\}}{y_1^2(x)}\,dx + C_2y_1(x). \qquad (2.109)$$

The point to be emphasized with this development is that a change of variables transforms a second order equation (2.104) to a first order equation (2.107). This reduction of 1 in order is true in general and results in transforming an $n$th order equation to one of order $n - 1$.

**EXAMPLE 2.9**　If $y_1(x) = \sin x/\sqrt{x}$, $0 < x < \pi$, satisfies the differential equation

$$4x^2 \frac{d^2y}{dx^2} + 4x \frac{dy}{dx} + (4x^2 - 1)y = 0, \qquad (2.110)$$

use (2.109) to find a second solution. For this equation

$$a_2(x) = 4x^2, \qquad a_1(x) = 4x, \qquad a_0(x) = 4x^2 - 1,$$

$$\int \frac{a_1(x)}{a_2(x)} dx = \int \frac{4x}{4x^2} dx = \ln|x| = \ln x, \qquad \text{since } 0 < x < \pi,$$

and

$$\int \frac{\exp\{-\int [a_1(x)/a_2(x)] \, dx\}}{y_1^2(x)} dx = \int \frac{x^{-1}}{x^{-1} \sin^2 x} dx$$

$$= \int \csc^2 x \, dx = -\cot x.$$

Thus the second solution is given by

$$y_2(x) = \frac{\sin x}{\sqrt{x}} (-\cot x) = \frac{-\cos x}{\sqrt{x}}.$$

# EXERCISES

Solve the following differential equations.

1. $xy'' + y' = 0$.
2. $2xy'' + (y')^2 - 1 = 0$.
3. $yy'' + (y')^2 = 0$.
4. $xy'' + y' + x = 0$.
5. $xy'' - y' = (y')^3$.
6. $y'' + (y')^2 = 1$.
7. $x^2y'' + (y')^2 - 2xy' = 0$.
8. $yy'' + (y')^2 + 4 = 0$.
9. $(y^2 + 1)y'' - 2y(y')^2 = 0$.
10. $y'' + 2y - 2y^3 = 0$.
11. Show that $y = e^{2x}$ satisfies

$$(D^2 - 4D + 4)y = 0$$

and use reduction of order to find a second solution.

12. Show that $y = x^2$ satisfies

$$(x^2D^2 + xD - 4)y = 0$$

and use reduction of order to find a second solution.

**13.** Show that $y = x$ satisfies

$$x^2y'' - xy' + y = 0$$

and use reduction of order to find a second solution.

**14.** Show that $y = e^x$ satisfies

$$xy'' - 2(x + 1)y' + (x + 2)y = 0$$

for $x > 0$ and use reduction of order to find a second solution.

**15.** Show that for $0 < x < \pi$, $y = x \sin x$ satisfies

$$x^2y'' - 2xy' + (x^2 + 2)y = 0$$

and use reduction of order to find a second solution.

**16.** Show that $y = x$ satisfies

$$x^2y'' - x(x + 2)y' + (x + 2)y = 0$$

and use reduction of order to find a second solution.

**17.** If a projectile of mass $m$ is fired vertically from the earth, the distance from the surface of the earth, $x$, is governed by the differential equation

$$m\frac{d^2x}{dt^2} = -\frac{mgR^2}{x^2},$$

where $g$ is the gravitational constant and $R$ is the radius of the earth

**(a)** Letting

$$\frac{dx}{dt} = v \quad \text{and} \quad \frac{d^2x}{dt^2} = v\frac{dv}{dx}$$

solve the foregoing differential equation for $v$, obtaining

$$v = \left[v_0^2 + 2gR^2\left(\frac{1}{x} - \frac{1}{R}\right)\right]^{1/2},$$

where $v_0$ is the initial velocity.

**(b)** Show that if $v_0^2 - 2gR > 0$, the velocity will always be nonzero. (This critical value of $v_0$, $v_0 = \sqrt{2gR}$, is called the *escape velocity* of the earth.)

**18.** The differential equation

$$\frac{d^2y}{dx^2} = g(y)$$

is a special case of (2.98) so the substitution $w = y'$ could be used to solve this equation. Use the following (quicker) method to find the solution. Multiply both sides of the differential equation by $dy/dx$ and integrate to find that

$$\left(\frac{dy}{dx}\right)^2 = 2G(y) + C,$$

where $G(y)$ is any antiderivative of $g(y)$. Integrate one more to find that

$$\int \frac{dy}{\sqrt{2G(y) + C}} = x + K,$$

where $K$ is also an arbitrary constant.

**19.** Let a cable of uniformly distributed weight be suspended between two supports (see Figure 2.1). If we put the origin of our coordinate system at the lowest point of the cable and the coordinates of any point on the cable are $(x, y)$, the differential equation relating $x$ and $y$ is

$$a\frac{d^2y}{dx^2} = \left[1 + \left(\frac{dy}{dx}\right)^2\right]^{1/2}.$$

The constant $a$ in this equation is the ratio of the tension in the cable at the lowest point to the density of the cable. Show that the solution of this differential equation, subject to the conditions $y(0) = a$, $y'(0) = 0$, may be expressed as $y = a \cosh(x/a)$. (The graph of this equation is called a "catenary," which is the Latin word for "chain.")

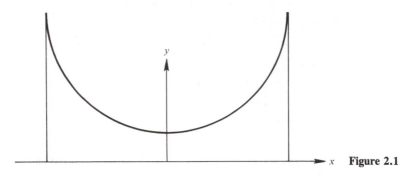

**Figure 2.1**

## 2.6 APPLICATIONS

### 2.6.1 Mixture Problems

Determining the concentration of solute in a container or pollutant in an enclosed region are the two examples included in this section to illustrate mixture problems. To derive a differential equation which describes this process, we let $x$ be a function which represents the amount of substance in a given container at time $t$, and assume that the instantaneous rate of change of $x$ with respect to $t$ is given by

$$\frac{dx}{dt} = \begin{pmatrix} \text{rate at which the} \\ \text{substance is added} \\ \text{to the container} \end{pmatrix} - \begin{pmatrix} \text{rate at which the} \\ \text{substance is withdrawn} \\ \text{from the container} \end{pmatrix}. \qquad (2.111)$$

This equation is often called an *equation of continuity* or a *conservation equation*.

**EXAMPLE 2.10**    Let a 200-gallon container of pure water have a salt solution of concentration 4 pounds per gallon added to the container at a rate of 2 gallons per minute. If the well-stirred mixture is drained from the container at the rate of 2 gallons per minute, find the number of pounds of salt in the container as a function of time.

We first note that since solution is being added and drained at the same rate, the number of gallons in the container remains constant. If $x$ represents the number of pounds of salt in the container at time $t$, the concentration of salt is

$x/200$. Thus the rate at which salt is leaving the container is $(x/200)(2)$ and from the continuity equation we have

$$\frac{dx}{dt} = (4)(2) - \left(\frac{x}{200}\right)(2)$$

$$= 8 - \frac{x}{100}. \qquad (2.112)$$

Since there is no salt in the container at time $t = 0$, the initial condition is

$$x(0) = 0.$$

Equation (2.112) is both linear and separable, and if we write it as

$$\frac{dx}{800 - x} = \frac{dt}{100}$$

and integrate, we obtain

$$-\ln|800 - x| = \frac{t}{100} + C,$$

or

$$800 - x = Ke^{-t/100}.$$

Since $x(0) = 0$, $K = 800$ and the amount of salt at time $t$ is given by

$$x = 800(1 - e^{-t/100})$$

and the concentration is given by $x/200 = 4(1 - e^{-t/100})$. Notice that in the limit as $t \to \infty$, the concentration of salt in the container approaches that of the incoming solution.

**EXAMPLE 2.11**    A conference room with volume 2000 cubic meters contains air with 0.002% carbon monoxide. At time $t = 0$, the ventilation system starts blowing in air which contains 3% carbon monoxide (by volume). If the ventilation system inputs (and extracts) air at a rate of 0.2 cubic meters per minute, how long before the air in the room contains 0.015% carbon monoxide?

If $x$ represents the volume (cubic meters) of carbon monoxide at time $t$, using the conservation equation gives

$$\frac{dx}{dt} = (\text{inflow}) - (\text{outflow})$$

$$= (0.03)(0.2) - \left(\frac{x}{2000}\right)(0.2)$$

$$= 0.006 - 0.0001x. \qquad (2.113)$$

The initial condition is

$$x(0) = (0.00002)(2000) = 0.04.$$

Separating variables in (2.113) and integrating gives

$$-\ln |0.006 - 0.0001x| = 0.0001t + C,$$

or

$$x = 60 - Ke^{-0.0001t}.$$

Choosing $K$ so that $x(0) = 0.04$ yields $K = 59.96$ and

$$x = 60 - 59.96e^{-0.0001t}.$$

To determine how many minutes it takes for the air to contain 0.015% carbon monoxide, set

$$60 - 59.96e^{-0.0001t} = (0.00015)(2000) = 0.3.$$

Thus

$$e^{-0.0001t} = \frac{59.70}{59.96},$$

or

$$t = -10,000 \ln \frac{59.7}{59.96} \approx 43.5 \text{ minutes.}$$

### 2.6.2 Orthogonal Trajectories

Orthogonal trajectories occur naturally in many physical situations, including wave propagation, planar flow of an incompressible fluid, and the interpretation of weather maps. In the latter situation, isobars are curves connecting places with equal barometric pressure, while their orthogonal trajectories represent paths for air movement from places with high pressure to those with low pressure.

A family of straight lines with slope 3 is given by

$$y = 3x + b, \tag{2.114}$$

where $b$ is an arbitrary constant (the $y$-intercept). The family of straight lines

$$y = -\frac{1}{3}x + C \tag{2.115}$$

intersects each line satisfying (2.114) at right angles. (Recall that lines with slopes $m_1$ and $m_2$ are orthogonal if $m_1 = -1/m_2$.) The two families of lines are *orthogonal trajectories* of each other (Figure 2.2), since whenever two elements of these two sets intersect, they do so at right angles. This definition also holds for two families of curves if at any point of intersection, their tangent lines are orthogonal (perpendicular).

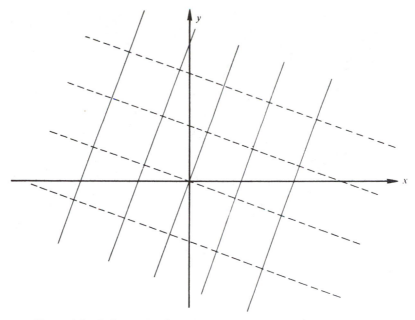

**Figure 2.2** Orthogonal trajectories $y = 3x + b$; $y = -\frac{1}{3}x + C$.

**EXAMPLE 2.12** Find the orthogonal trajectories of the family of parabolas $y = bx^2$.

Since the condition for a family of curves to be orthogonal trajectories requires a condition on slopes of tangent lines, we seek a differential equation (giving this slope, $dy/dx$) satisfied by all these parabolas. One way of doing this is to solve for $b$ and differentiate the resulting equation implicitly. Thus

$$x^{-2}y = b$$

and

$$-2x^{-3}y + x^{-2}\frac{dy}{dx} = 0 \qquad \text{or} \qquad \frac{dy}{dx} = \frac{2y}{x}.$$

The orthogonal trajectories to these parabolas must therefore satisfy the differential equation

$$\frac{dy}{dx} = -\frac{x}{2y},$$

where $-x/2y$ is the negative reciprocal of $2y/x$. This is a separable differential equation with solution

$$\frac{y^2}{2} = -\frac{x^2}{4} + C. \tag{2.116}$$

Thus the family of ellipses given by (2.116) are the orthogonal trajectories of the parabolas $y = bx^2$. Curves for $b = -1, 1, -2$, and 2 and $C = 1$ and 2 are shown in Figure 2.3.

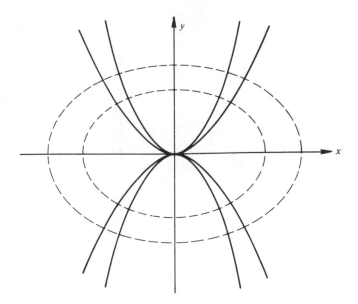

**Figure 2.3**   Orthogonal trajectories $y = bx^2$; $y^2/2 + x^2/4 = C$.

**EXAMPLE 2.13**    Find the orthogonal trajectories to the family of circles

$$x^2 + y^2 = 2by$$

with $b$ an arbitrary constant.

Rather than solving for $b$ and then differentiating, we will now differentiate implicitly and then eliminate $b$ from the resulting differential equation. Thus we have

$$2x + 2y\frac{dy}{dx} = 2b\,\frac{dy}{dx}$$

$$= \left(\frac{x^2 + y^2}{y}\right)\frac{dy}{dx},$$

or

$$\frac{dy}{dx} = \frac{-2xy}{y^2 - x^2}$$

as the differential equation satisfied by the family of circles. The differential equation satisfied by the orthogonal trajectories of these circles is

$$\frac{dy}{dx} = \frac{y^2 - x^2}{2xy}.$$

This equation is not linear, separable, or exact, but it is homogeneous as well as a type of Bernoulli equation. We will solve the equation by changing the dependent variable from $y$ to $v$ using the transformation

$$y = xv, \qquad \frac{dy}{dx} = x\frac{dv}{dx} + v.$$

This gives

$$x\frac{dv}{dx} + v = \frac{1}{2}\left(v - \frac{1}{v}\right).$$

This is a separable equation, so rearranging gives

$$-\frac{dx}{x} = \frac{dv}{1/(2v) + v/2} = \frac{2v}{v^2 + 1}\, dv$$

and integration gives

$$x(v^2 + 1) = C.$$

In terms of the original variables, we obtain the orthogonal trajectories as circles given by

$$y^2 + x^2 = Cx. \qquad\qquad (2.117)$$

In Figure 2.4 the curves $x^2 + y^2 = 2by$ are shown as solid lines for $b = \pm 1$ and $\pm 2$ and their orthogonal trajectories from (2.117) are shown as dashed lines for $C = \pm 4$ and $\pm 2$.

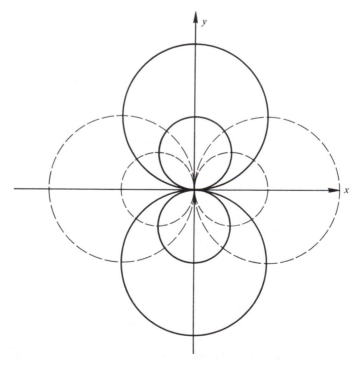

**Figure 2.4**   Orthogonal trajectories $x^2 + y^2 = 2by$; $y^2 + x^2 = Cx$.

### 2.6.3 Growth and Decay

The differential equation

$$\frac{dy}{dt} = ky \tag{2.118}$$

states that the rate of change of $y$ with respect to $t$ is proportional to the value of $y$. Many different phenomena may be modeled by such an equation. For example, if $y$ represents the amount of a substance that is disintegrating through radioactive decay, the constant $k$ is negative. (Since $y > 0$, this says that $y$ is a decreasing function, $dy/dt < 0$.)

**EXAMPLE 2.14**    Radium disintegrates according to (2.118). If the half-life of radium is 1600 years, what percentage radium will remain in a given sample after 800 years?

The solution of (2.118) is

$$y = y_0 e^{kt},$$

where $y_0$ is the amount of radium at $t = 0$. If the half-life of radium is 1600 years, half of the original amount of radium exists after 1600 years, that is,

$$\frac{y_0}{2} = y_0 e^{1600k}.$$

Solving for $k$ gives

$$2^{-1} = e^{1600k} \quad \text{or} \quad k = -\frac{\ln 2}{1600}.$$

If this value of $k$ is used in the original solution, we may write

$$y = y_0 e^{-(t \ln 2)/1600} = y_0 (e^{\ln 2})^{-t/1600} = y_0 2^{-t/1600}.$$

Since

$$\frac{y(800)}{y_0} = \frac{y_0 2^{-800/1600}}{y_0} = \frac{1}{\sqrt{2}} \approx 0.707,$$

after 800 years about 70.7% of the original amount of radium would remain.

The fact that radioactive materials decay at known rates was used by Willard Libby to determine the approximate age of plant and animal remains, such as wood, bone, and so on. The half-life of radioactive carbon (called carbon 14 or $^{14}C$) is about 5730 years. Since the ratio of $^{14}C$ to ordinary carbon in the atmosphere is constant, this ratio also applies to animal and plant tissue as long as they are alive. However, when the animal or plant dies, the $^{14}C$ starts to decay through radioactivity, while the ordinary carbon does not disappear. Thus, knowing the ratio of $^{14}C$ to ordinary carbon can determine an approximation to the year of death. For his work, Libby received a Nobel prize in chemistry in 1960.

This method of dating objects, usually called *carbon-14 dating*, has been used in such applications as dating the Dead Sea scrolls and dating the remains of ancient ruins. Similar observations for a radioactive isotope of lead have been used to distinguish between authentic and forged oil paintings.

**EXAMPLE 2.15**    Scientists discover that 15% of the original $^{14}C$ remains in a wooden archeological specimen. Use a half-life of 5730 years to determine the approximate age of the specimen. (Assume that this age approximates the time when the wood ceased being part of a living tree.)

Since $y = y_0 e^{kt}$ is the solution of (2.118), the value of $k$ for any given half-life can be determined as follows. If $T$ is the time when half the sample remains,

$$\frac{y_0}{2} = y_0 e^{kT}$$

and taking the natural logarithm of both sides of this equation yields

$$kT = \ln \frac{1}{2} = -\ln 2.$$

Thus $k = -\ln 2/T$, $T$ is the half-life, and

$$y = y_0 e^{-t \ln 2/T} = y_0 (e^{\ln 2})^{-t/T} = y_0 2^{-t/T}. \tag{2.119}$$

Equation (2.119) is the formula for determining the present amount of radioactive material with a half-life of $T$ and an initial amount of material $y_0$. For our example

$$0.15 y_0 = y_0 2^{-t/5730}$$

or

$$t = -5730 \frac{\ln 0.15}{\ln 2} \approx 15,683 \text{ years.}$$

If the value of $k$ in (2.118) is positive, $y$ describes an increasing function ($ky > 0$). The next example treats this situation.

**EXAMPLE 2.16**    A colony of bacteria increases at a rate proportional to the amount present. If the number of bacteria doubles in 1 hour, how long does it take for the colony to attain four times its initial size?

The governing differential equation

$$\frac{dy}{dt} = ky, \qquad k > 0$$

has a solution in the form

$$y = y_0 e^{kt},$$

where $y_0$ gives the initial size of the colony. In 1 hour the size of the colony is $2y_0$, so

$$2y_0 = y_0 e^k$$

or

$$k = \ln 2.$$

Thus

$$y = y_0 e^{t \ln 2} = y_0 2^t$$

gives the size of the colony as a function of time.

The time for the colony to grow to four times its size is given by the solution of

$$4y_0 = y_0 2^t$$

or

$$t = \frac{\ln 4}{\ln 2} = \frac{2 \ln 2}{\ln 2} = 2 \text{ hours.}$$

### 2.6.4 Population Dynamics

Example 2.16 predicts that the size of the bacteria colony is proportional to $2^t$, where $t$ is time. Since $2^t$ is unbounded as $t \to \infty$, this model obviously is not valid for all time. To develop a more realistic model of population growth, the facts of limited food supply, overcrowding, disease, and so on, must be taken into account. In fact, let us look at the population of a species in terms of a continuity equation. Common ways for a population to grow are through birth (which is proportional to the present population) and immigration. Decreases in the population are death (proportional to the present population) and emigration. Also a term, $cy^2$, is added to account for the fact that death rates increase for large populations (overcrowding, lack of food, etc.). Thus

$$\frac{dy}{dt} = \left( \begin{matrix} \text{additions to} \\ \text{the population} \end{matrix} \right) - \left( \begin{matrix} \text{subtractions from} \\ \text{the population} \end{matrix} \right)$$

$$= (By + I) - (Dy + E + cy^2) \qquad (2.120)$$

$$= ay + b - cy^2$$

where $a = B - D = $ births $-$ deaths (per unit population),

$b = I - E = $ immigration $-$ emigration,

$c$ accounts for competition or inhibition of large populations.

Note that we could think of the right hand side of (2.120) as the first three terms in a Taylor series of a function. If $c = 0$, the integrating factor for $(dy/dt) - ay = b$ is $e^{-at}$ and the solution is given by integrating

$$\frac{d}{dt}(e^{-at}y) = be^{-at}$$

as

$$e^{-at}y = -\frac{b}{a}e^{-at} + C$$

or

$$y = -\frac{b}{a} + Ce^{at}.$$

If the size of the population at $t = t_0$ is $y_0$, then

$$y_0 = -\frac{b}{a} + Ce^{at_0} \quad \text{and} \quad C = \left(y_0 + \frac{b}{a}\right)e^{-at_0}.$$

Thus

$$y = -\frac{b}{a} + \left(y_0 + \frac{b}{a}\right)e^{a(t-t_0)}$$

gives the population at any time $t > t_0$. If $b = 0$, we have the separable equation

$$\frac{dy}{dt} = ay - cy^2, \tag{2.121}$$

which has been used to successfully predict human population for short periods of time. The solution of (2.121) subject to $y(0) = y_0$ is

$$y = \frac{ay_0}{cy_0 + (a - cy_0)e^{-at}}$$

(see exercise 10). Equation (2.121) is known as the differential equation of logistics. It describes many diverse physical situations, such as dissemination of technological innovations, spread of infectious disease, rumor spreading, and kinetics of auto-catalytic reactions. A graph of the solution of the logistics equation for $y_0 < a/c$ may be an S-shaped curve as shown in Figure 2.5. In contrast to the simple growth model, $dy/dt = ay$, the solution of (2.121) is bounded as $t \to \infty$.

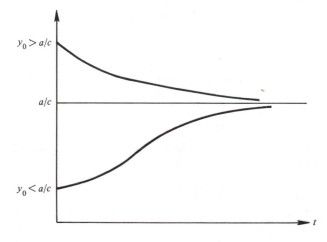

Figure 2.5

## EXERCISES

1. Let a 200-gallon container of pure water have a salt solution of concentration 3 pounds per gallon added to the container at a rate of 4 gallons per minute.
   (a) If the well-stirred mixture is drained from the container at a rate of 4 gallons per minute, find the number of pounds of salt in the container as a function of time.
   (b) How many minutes does it take for the concentration in the container to reach 2 pounds per gallon?
   (c) What does the concentration in the container approach for large values of time?

2. (a) and (b) Solve exercise 1(a) and (b) if the rate at which the well-stirred mixture is drained is changed to 5 gallons per minute.
   (c) For how many minutes is the solution in part (a) valid?

3. A container is filled with 10 gallons of water which initially contains 5 pounds of salt. A salt solution of concentration 3 pounds per gallon is pumped into the container at a rate of 2 gallons per minute and the well-stirred mixture drains at this same rate. How much salt remains in the container after 15 minutes?

4. An open container has 5 pounds of impurities dissolved in 150 liters of water. Pure water is pumped into the container at a rate of 3 liters per minute and the well-stirred mixture is drained at a rate of 2 liters per minute.
   (a) How many pounds of impurities remain after 10 minutes?
   (b) If the container holds a maximum of 200 liters, what is the concentration of impurities just before the solution overflows?

5. A conference room contains 3000 cubic meters of air which is free of carbon monoxide. The ventilation system blows in air, free of carbon monoxide, at a rate of 0.3 cubic meters per minute and extracts it at the same rate. If at $t = 0$, people in the room start smoking and add carbon monoxide to the room at a rate of 0.02 cubic meters per second, how long before the air in the room contains 0.015% carbon monoxide?

6. Find the orthogonal trajectories of the families of curves given by
   (a) $y = -2x + b$.                (b) $y = bx + 4$.
   (c) $y^2 = bx$.                   (d) $x^2 - b^2y^2 = 16$.
   (e) $x^2 + b^2y^2 = 16$.          (f) $y = b/x$.
   (g) $y = be^x$.                   (h) $y = be^{-x}$.
   (i) $y = \ln(x^3 + b)$.

7. Oblique (or isogonal) trajectories occur when one family of curves intersects another family of curves at some angle $\theta$, $\theta \neq \pi/2$. If one family is given by the solution of

$$\frac{dy}{dx} = f(x, y),$$

   the other family, which intersects these curves at an angle of $\theta$, is given by

$$\frac{dy}{dx} = \frac{f(x, y) + \tan \theta}{1 - f(x, y) \tan \theta}.$$

   (a) Derive this formula. (*Hint:* Use the identity for the tangent of the sum of two angles.)
   (b) Find the family of lines that intersect

$$y = x + b$$

   at an angle of $\pi/3$.

(c) Find the family of curves that intersect the family of hyperbolas

$$y^2 = 2x^2 + b$$

at an angle of $\tan^{-1}(\frac{1}{2})$.

(d) Find the family of curves that intersect the circle

$$x^2 + y^2 = b^2$$

at an angle of $\pi/4$.

(e) Find the family of curves that intersect the parabolas

$$y^2 = bx$$

at an angle of $\pi/4$.

8. In 1800 the world's population was approximately 1 billion, while in 1900 it was 1.7 billion. Use an appropriate solution (2.118) to estimate the world's population in the year 2000.

9. A culture of bacteria doubles its size every 6 hours. How many hours will it take to be five times its original size?

10. Obtain the solution of (2.121) subject to $y(0) = y_0$.

11. At time $t = 0$ a bacterial culture has $N_0$ bacteria. One hour later the population has grown by 25%. How long will it take the population to double?

12. After administration of a dose, the concentration of a drug in one's body decreases by 50% in 10 hours. How long before the concentration of the drug is 10% of its original amount?

13. Of 10,000 companies, 100 have adopted a new development at time $t = 0$. If the differential equation that describes the number, $y(t)$, of companies that have adopted the innovation satisfies the differential equation

$$y' = 0.005y(10{,}000 - y),$$

find the number of companies that can be expected to adopt the innovation after 10 years.

14. A differential equation which models the growth of a bacterial colony situated on top of a nutrient-containing medium is

$$\frac{d^2y}{dx^2} = q(x) = \begin{cases} A[1 - (1 + ky)^{-1}], & 0 < y \le 1 \\ A[1 - (1 + k)^{-1}], & 1 \le y < y_0 \end{cases}.$$

Solve this differential equation in the two regions $0 < y < 1$ and $1 < y < y_0$ separately, subject to the conditions

$$y(0) = y_0, \qquad y(1) = 0, \quad y'(1) = 0,$$

if $k = 1$ and $A = 10$.

# REVIEW EXERCISES

Solve the following differential equations.

1. $(y^2 - 1)\, dx + (2xy + \tan y)\, dy = 0$.
2. $(x - 3)(y + 1)\, dx + (xy^2 + 2xy + x)\, dy = 0$.
3. $(y + x(1 - x)^2)\, dx + (1 - x)\, dy = 0$.

**4.** $(y + x^2y^4) \, dx + 3x \, dy = 0$.

**5.** $(x + xy^2) \, dx + (1 + x^2)^3 \, dy = 0$.

**6.** $(3y^2 + 2x) \, dx - (3y^2 - 6xy) \, dy = 0$.

**7.** $(4x^3y^2 + y \cos x) \, dx + (2x^4y + \sin x) \, dy = 0$.

**8.** $(y^2 + y - 1) \, dx + (2y + 1)\sqrt{1 - x^2} \, dy = 0$.

**9.** $(2x^2y - 2y + 1) \, dx - x \, dy = 0$.

**10.** $2xy \, dx - (3x^2 + y^2) \, dy = 0$.

**11.** $(y^3 + x^2) \, dx - xy^2 \, dy = 0$.

**12.** $(xy^2 - y) \, dx - dy = 0$.

**13.** $(xy + y + 2x + 2) \, dx - (1 - x^2)(y + 2)^{-1} \, dy = 0$.

**14.** $e^x(1 + y)\dfrac{dy}{dx} + (1 + e^x)(y^2 + y - 2) = 0$.

**15.** $(xe^{xy} + 6xy^2 - y^{-1} \ln y)\dfrac{dy}{dx} + ye^{xy} + 2y^3 + 3 = 0$.

**16.** $\dfrac{dy}{dx} + \dfrac{5x^3y^3 - 2y}{2x} = 0$.

**17.** $\dfrac{dy}{dx} + 2xy + 4x = 0$.

**18.** $3xy\dfrac{dy}{dx} + x^2 + y^2 = 0$.

**19.** $y\dfrac{dy}{dx} + y^2 + y \sin x = 0$.

**20.** $(x \cosh y)\dfrac{dy}{dx} - \left(\dfrac{\ln x}{x} - \sinh x + 4\right) = 0$.

**21.** Observations on the growth of animal tumors indicate that they obey the differential equation

$$\frac{dy}{dt} = -ky \ln\left(\frac{y}{A}\right),$$

where $k$ and $A$ are positive constants. This differential equation is sometimes called the Gompertz growth law.

**(a)** Solve this differential equation subject to the initial condition $y(0) = y_0$.

**(b)** Show that if $y_0 < A$, the solution is an increasing function for all time and find relations between $y_0$ and $A$ such that the graph of $y$ versus $t$ will not have an inflection point.

# CHAPTER 3

# Linear First Order Differential Equations

In this chapter a procedure is given for obtaining the general solution of a first order linear differential equation. The response of the equation to various forms of input is discussed and the ideas of steady-state and transient parts of solutions are brought out. Applications in population dynamics and engineering are given.

## 3.1 THE GENERAL SOLUTION

Since the general form for the linear equation of order $n$ is [from (1.5) with $a_n(x) \neq 0$]

$$a_n(x) \frac{d^n y}{dx^n} + a_{n-1}(x) \frac{d^{n-1}y}{dx^{n-1}} + \cdots + a_1(x) \frac{dy}{dx} + a_0(x)y = Q(x), \qquad (3.1)$$

we may obtain the general form for the linear equation of order 1, the first order linear equation, by setting $n = 1$ in (3.1). Doing so yields

$$a_1(x) \frac{dy}{dx} + a_0(x)y = Q(x). \qquad (3.2)$$

Now $a_1(x)$, in (3.2), is not zero, so we may divide both sides of the equation by it and obtain the equation in the form

$$\frac{dy}{dx} + \frac{a_0(x)}{a_1(x)}y = \frac{Q(x)}{a_1(x)}. \qquad (3.3)$$

If we now let

$$p(x) = \frac{a_0(x)}{a_1(x)}, \qquad q(x) = \frac{Q(x)}{a_1(x)},$$

we may write (3.3) in the form

$$\frac{dy}{dx} + p(x)y = q(x). \tag{3.4}$$

We will say that (3.4) is the first order linear equation in standard form, and it is this equation that we will use to obtain solutions. In Section 2.3 we discovered that

$$\rho(x) = \pm e^k \exp\left[\int p(x)\, dx\right] \tag{3.5}$$

was an integrating factor of (3.4). Thus if we multiply (3.4) by this $\rho(x)$, we obtain the exact differential equation

$$\rho(x)\frac{dy}{dx} + \rho(x)p(x)y = \rho(x)q(x). \tag{3.6}$$

Combining the two terms on this left side of (3.6) gives

$$\frac{d}{dx}[\rho(x)y] = \rho(x)q(x). \tag{3.7}$$

[Note that from (3.5)

$$\frac{d}{dx}\rho(x) = \rho(x)p(x).$$

This is why the left-hand side of (3.6) may be written as the derivative of a product.] If we integrate both sides of (3.7) with respect to $x$, we obtain

$$\rho(x)y = \int \rho(x)q(x)\, dx + C,$$

or

$$y = \frac{1}{\rho(x)} \int \rho(x)q(x)\, dx + \frac{C}{\rho(x)}. \tag{3.8}$$

Substituting the form of $\rho(x)$ from (3.5) gives

$$y = \frac{1}{\pm e^k e^{\int p(x)\, dx}} \int \pm e^k e^{\int p(x)\, dx} q(x)\, dx + \frac{C}{\pm e^k e^{\int p(x)\, dx}}. \tag{3.9}$$

Notice that the choice of the $\pm$ sign and the value of $k$ is irrelevant since these make no difference in the first term above and may be absorbed into the integration constant $C$ in the second term. Thus we may take $k = 0$ and the plus sign in (3.5) and write the general solution of (3.4) as (3.8) with an integrating factor $\rho(x)$ given by

$$\rho(x) = e^{\int p(x)\, dx}. \tag{3.10}$$

From the way in which this solution has been constructed, we obtain the following three-step procedure for solving first order linear equations:

**Step 1.**    Put the equation in the form

$$\frac{dy}{dx} + p(x)y = q(x) \tag{3.11}$$

and find an integrating factor, $\rho(x)$, from

$$\rho(x) = e^{\int p(x)\,dx}. \tag{3.12}$$

**Step 2.**    Multiply both sides of the differential equation (3.11) by $\rho(x)$, noting that the left side will be a total derivative,

$$\rho(x)\frac{dy}{dx} + \rho(x)p(x)y = \frac{d}{dx}[\rho(x)y] = \rho(x)q(x). \tag{3.13}$$

[Note computing $d\rho(x)/dx$ to see if it equals $\rho(x)p(x)$ is a way you may check your integration in finding the integrating factor.]

**Step 3.**    Integrate to obtain

$$\rho(x)y = \int \rho(x)q(x)\,dx + C \tag{3.14}$$

and determine the solution by dividing by $\rho(x)$ to obtain

$$y = \frac{1}{\rho(x)} \int \rho(x)q(x)\,dx + \frac{C}{\rho(x)}. \tag{3.15}$$

As an example, consider the equation

$$x\frac{dy}{dx} + 2y = 4x^2. \tag{3.16}$$

Applying the three steps given above:

**Step 1.**    Divide by $x$ to put (3.16) in the form

$$\frac{dy}{dx} + \frac{2}{x}y = 4x. \tag{3.17}$$

Here $p(x) = 2/x$, $q(x) = 4x$, and

$$\rho(x) = e^{\int (2/x)\,dx} = e^{2\ln|x|} = e^{\ln x^2} = x^2.$$

**Step 2.**    Multiply (3.17) by $x^2$ to obtain

$$x^2\frac{dy}{dx} + 2xy = \frac{d}{dx}(x^2y) = x^2(4x) = 4x^3.$$

**Step 3.**    Integrate to obtain

$$x^2y = \int 4x^3\,dx + C = x^4 + C$$

and divide to obtain the solution as

$$y = \frac{x^4}{x^2} + \frac{C}{x^2} = x^2 + Cx^{-2}. \tag{3.18}$$

As a check, the right-hand side of (3.18) may be substituted for $y$ in (3.16).

As a second example, consider the equation

$$(x^2 + 1)\frac{dy}{dx} - 2xy = x^2 + 1. \tag{3.19}$$

**Step 1.**  Divide by $x^2 + 1$ to get

$$\frac{dy}{dx} - \frac{2x}{x^2 + 1}y = 1, \tag{3.20}$$

so $p(x) = -2x/(x^2 + 1)$, $q(x) = 1$, and

$$\rho(x) = \exp\left(-\int \frac{2x}{x^2 + 1} dx\right) = \exp[-\ln(x^2 + 1)] = \frac{1}{x^2 + 1}.$$

**Step 2.**  Multiply (3.20) by $1/(x^2 + 1)$ to obtain

$$\frac{1}{x^2 + 1}\frac{dy}{dx} - \frac{2x}{(x^2 + 1)^2}y = \frac{d}{dx}\left(\frac{y}{x^2 + 1}\right) = \frac{1}{x^2 + 1}.$$

**Step 3.**  Integrate to obtain

$$\frac{y}{x^2 + 1} = \int \frac{1}{x^2 + 1} dx + C = \tan^{-1} x + C.$$

Thus

$$y = (x^2 + 1) \tan^{-1} x + C(x^2 + 1). \tag{3.21}$$

Again, we may check by substituting from (3.21) into (3.19).

As a special case we may look at the case in which $q(x) \equiv 0$ in the first order linear equation. Then (3.4) becomes

$$\frac{dy}{dx} + p(x)y = 0 \tag{3.22}$$

and the general solution from (3.15) is just

$$y = \frac{C}{\rho(x)} \tag{3.23}$$

where $\rho(x)$ is obtained as in the procedure outlined above.

## EXERCISES

Find the general solution of the following differential equations.

**1.** $\dfrac{dy}{dx} + y = x.$

**2.** $\dfrac{dy}{dx} + \dfrac{1}{x+1}y = \dfrac{\cos x}{x+1}.$

**3.** $\dfrac{dy}{dx} + y = e^x.$

**4.** $\dfrac{1}{x}\dfrac{dy}{dx} + 2y = 2x^2.$

**5.** $\cos x\dfrac{dy}{dx} + \sin x\, y = 3\sin x\cos^2 x.$

**6.** $\dfrac{dy}{dx} - 2xy = 1.$

**7.** $x\dfrac{dy}{dx} - y = x\cos x.$

**8.** $y' - 6y = e^x.$

**9.** $y' - \dfrac{2}{x^2}y = \dfrac{1}{x^2}.$

**10.** $y' + 2xy = 2x.$

**11.** $y' + (2 + x^{-1})y = 2e^{-2x}.$

**12.** $(x^2 + 1)y' + 4xy = x.$

**13.** $y' + 3y = 3x^2e^{-3x}.$

**14.** $x\dfrac{dy}{dx} + \dfrac{2x+1}{x+1}y = x - 1.$

**15.** $x^4y' + 2x^3y = 1.$

**16.** $(x^2 + 1)y' + 4xy = 3x.$

## 3.2 RESPONSE TO SPECIAL FORCING FUNCTIONS

The equation

$$a_1(t)\frac{dx}{dt} + a_0(t)x(t) = f(t) \tag{3.24}$$

is a first order linear equation in which the independent variable is $t$ and the dependent variable is $x$. Here $t$ may be thought of as time, $x$ some physical quantity of interest such as position, temperature, or electric current, and $f(t)$ is called the *forcing function*.

A simple example of equation (3.24) has

$$a_0(t) = 1,$$

$$a_1(t) = a, \qquad a > 0, \quad \text{constant.}$$

Thus we will look at the equation

$$a\frac{dx}{dt} + x = f(t) \tag{3.25}$$

or

$$\frac{dx}{dt} + \frac{1}{a}x = \frac{1}{a}f(t) \tag{3.26}$$

so that $p(t) = 1/a$ and $q(t) = (1/a)f(t)$.

The integrating factor, $\rho(t)$, for this equation is just

$$\rho(t) = e^{\int (1/a)\,dt} = e^{t/a}. \tag{3.27}$$

We now consider the solutions of (3.26) for various forms of $f(t)$.

**1. $f(t) \equiv 0$.** Here we have

$$\frac{dx}{dt} + \frac{1}{a}x = 0,$$

which is a separable equation and may actually be solved by methods of Chapter 2. Nevertheless, an application of the three-step procedure given in Section 3.1 yields the solution

$$x = \frac{C}{\rho(t)} = Ce^{-t/a}, \tag{3.28}$$

where $C$ is a constant. Solutions of the form (3.28) are said to decay exponentially since in the limit as $t \to \infty$ we have

$$\lim_{t \to \infty} x = C \lim_{t \to \infty} e^{-t/a} = 0.$$

Thus no matter what $C$ we choose, each solution approaches 0 as $t$ increases. The constant $a$ is called the *time constant* for the system and governs how fast the solution decays. For small values of $a$ the decay is rapid and for large values of $a$ the decay is slow (see Figure 3.1).

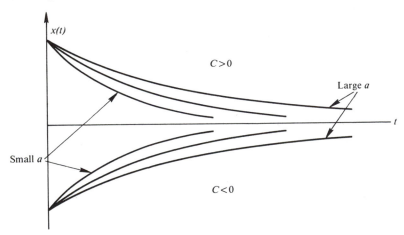

**Figure 3.1**　Exponential decay with $a > 0$.

If we denote the initial value of $x(t)$ (at $t = 0$) by $x_0$, we have

$$x = x_0 e^{-t/a}. \tag{3.29}$$

For different initial values $x_0$ the starting point is different but all solutions decay in time. This state of affairs is shown in Figure 3.2.

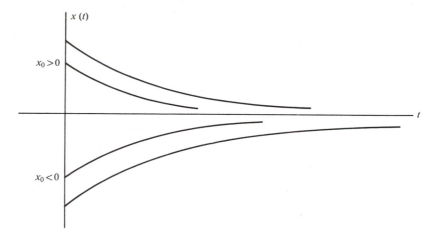

**Figure 3.2**   Response for various initial values.

**2. $f(t) = f_0$, a constant.**    In this case

$$\frac{dx}{dt} + \frac{1}{a}x = \frac{f_0}{a}.$$

The integrating factor is still given by (3.12), so that

$$x = e^{-t/a} \int e^{t/a}\frac{f_0}{a} \, dt + Ce^{-t/a}. \tag{3.30}$$

Now

$$\int e^{t/a}\frac{f_0}{a} \, dt = \frac{f_0}{a} \int e^{t/a} \, dt = f_0 e^{t/a}$$

and (3.30) becomes

$$x = f_0 + Ce^{-t/a}. \tag{3.31}$$

The solution (3.31) has two pieces. One is just the forcing function $f_0$ itself and the other is the exponentially decaying solution of case 1. Thus these solutions decay in time also, but instead of going to zero with increasing time, they approach the value $f_0$. This piece, $f_0$, is said to be the steady-state part of the solution and the exponential piece is said to be the transient part of the solution. Note that $f_0 = 0$ recovers case 1.

Figure 3.3 shows the response of the system to constant input.

**3. $f(t) = kt$, where $k$ is a constant.**    Here (3.26) becomes

$$\frac{dx}{dt} + \frac{1}{a}x = \frac{k}{a}t \tag{3.32}$$

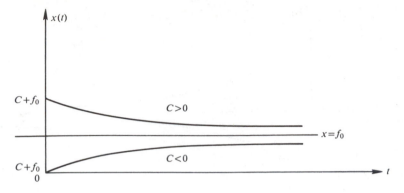

**Figure 3.3** Response to constant input.

and the solution is

$$x = e^{-t/a} \int \frac{k}{a} t e^{t/a} \, dt + C e^{-t/a}. \tag{3.33}$$

Now

$$\int t e^{t/a} \, dt = a t e^{t/a} - \int a e^{t/a} \, dt$$

$$= a t e^{t/a} - a^2 e^{t/a},$$

the integration being done by parts, so that (3.33) becomes

$$x = k(t - a) + C e^{-t/a}. \tag{3.34}$$

Again, we see that the solution has two parts. If we write

$$x(t) = x_1(t) + x_2(t),$$

we see that

$$x_1(t) = k(t - a) \tag{3.35}$$

and

$$x_2(t) = C e^{-t/a}. \tag{3.36}$$

The function in (3.36) represents the exponentially decaying part of the solution, while that in (3.35) represents the steady-state part.

A comparison of $x_1(t)$ to $f(t) = kt$ shows that $x_1(t)$ is also linear, with the same slope, but has its $t$ intercept shifted by an amount $a$. Thus the solutions tend to the steady state, $x_1(t) = k(t - a)$, as $t$ increases as shown in Figure 3.4.

**4. $f(t) = A_i \sin \omega t$.**    Here $A_i$ denotes an initial, or input, amplitude and $\omega$ an initial, or input, frequency. The differential equation becomes

$$a \frac{dx}{dt} + x = A_i \sin \omega t \tag{3.37}$$

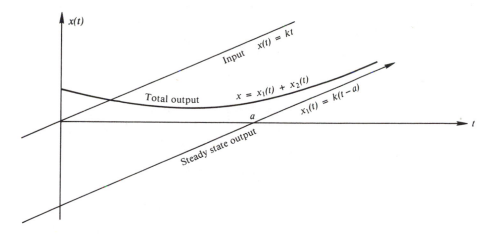

**Figure 3.4**   Response to linear input.

or, in standard form,

$$\frac{dx}{dt} + \frac{1}{a}x = \frac{A_i}{a} \sin \omega t,$$

with

$$p(t) = \frac{1}{a}, \qquad q(t) = \frac{A_i}{a} \sin \omega t.$$

Using the three-step procedure yields

$$\rho(t) = e^{t/a}$$

and

$$x(t) = e^{-t/a} \int \frac{1}{a} e^{t/a} A_i \sin \omega t \, dt + C e^{-t/a}. \tag{3.38}$$

Integration by parts yields

$$x(t) = \frac{A_i}{1 + a^2 \omega^2} (\sin \omega t - a\omega \cos \omega t) + C e^{-t/a}. \tag{3.39}$$

The trigonometric parts of (3.39) may be rewritten in a more suggestive form in the following way:

$$\frac{A_i}{1 + a^2 \omega^2} (\sin \omega t - a\omega \cos \omega t)$$

$$= \frac{A_i}{\sqrt{1 + a^2 \omega^2}} \left( \frac{1}{\sqrt{1 + a^2 \omega^2}} \sin \omega t - \frac{a\omega}{\sqrt{1 + a^2 \omega^2}} \cos \omega t \right). \tag{3.40}$$

We now note that

$$0 < \frac{1}{\sqrt{1 + a^2 \omega^2}} \leq 1$$

and

$$0 \leq \frac{|a\omega|}{\sqrt{1 + a^2\omega^2}} \leq 1$$

and

$$\left(\frac{1}{\sqrt{1 + a^2\omega^2}}\right)^2 + \left(\frac{a\omega}{\sqrt{1 + a^2\omega^2}}\right)^2 = 1,$$

so that

$$\frac{1}{\sqrt{1 + a^2\omega^2}} \quad \text{and} \quad \frac{a\omega}{\sqrt{1 + a^2\omega^2}}$$

have the same properties as the trigonometric functions cosine and sine. That is, they are bounded by 1 and the sum of their squares equals 1. Therefore, we may say that there exists an angle $\phi$ such that

$$\cos \phi = \frac{1}{\sqrt{1 + a^2\omega^2}}, \quad \sin \phi = \frac{a\omega}{\sqrt{1 + a^2\omega^2}}. \tag{3.41}$$

Substituting (3.41) into (3.40) yields

$$\frac{A_i}{1 + a^2\omega^2}(\sin \omega t - a\omega \cos \omega t) = \frac{A_i}{\sqrt{1 + a^2\omega^2}}(\sin \omega t \cos \phi - \cos \omega t \sin \phi)$$

$$= \frac{A_i}{\sqrt{1 + a^2\omega^2}} \sin(\omega t - \phi).$$

Thus, if we call $A_i/\sqrt{1 + a^2\omega^2} = A_0$ the output amplitude, we may write the solution (3.39) as

$$x(t) = A_0 \sin(\omega t - \phi) + Ce^{-t/a}. \tag{3.42}$$

Again the solution has a steady-state part and a transient part. The steady state again mimics the input, but $A_0 < A_i$ and $\phi$ acts as a phase shift so that input and output are out of phase by an amount $\phi/\omega$. The response of the steady state to sinusoidal input is shown in Figure 3.5.

**5. Step functions as inputs.**   In some cases, particularly in electrical engineering, it is useful to be able to solve the equation for input functions which have jump discontinuities at a finite number of points in their domains. Typical of such functions is the unit step function

$$u_1(t) = \begin{cases} 0, & 0 \leq t < 1 \\ 1, & t \geq 1, \end{cases} \tag{3.43}$$

which might be thought of as a direct current of 1 volt which is switched on at time $t = 1$.

The equation in standard form

$$\frac{dx}{dt} + \frac{x}{a} = \frac{f(t)}{a} \tag{3.26}$$

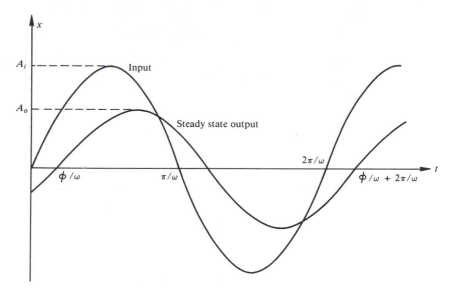

**Figure 3.5**  Response to sinusoidal input (steady state).

may then be solved in each of the regions $0 \leq t < 1$, $t \geq 1$ separately and the appropriate solution used in the appropriate region.

For the specific $f$ given by (3.43) we have

$$\frac{dx}{dt} + \frac{x}{a} = \begin{cases} 0, & 0 \leq t < 1 \\ \dfrac{1}{a}, & t \geq 1 \end{cases}.$$

Again, the integrating factor is found to be

$$\rho(x) = e^{t/a}$$

and the solutions from (3.28) and (3.31) are found to be

$$x = C_1 e^{-t/a}, \qquad 0 \leq t < 1, \tag{3.44a}$$

$$x = 1 + C_2 e^{-t/a}, \quad t \geq 1. \tag{3.44b}$$

We may write the general solution in one equation as

$$x = G_1(t)e^{-t/a} + u_1(t), \tag{3.45}$$

where

$$G_1(t) = \begin{cases} C_1, & 0 \leq t < 1 \\ C_2, & t \geq 1, \end{cases} \qquad u_1(t) = \begin{cases} 0, & 0 \leq t < 1 \\ 1, & t \geq 1 \end{cases} \tag{3.46}$$

**EXAMPLE 3.1**    If the initial value of $x$ is $A$ [i.e., $x(0) = A$], from (3.44a) we have

$$A = C_1.$$

If the solution is continuous at $t = 1$, $\lim_{t \to 1^-} x = \lim_{t \to 1^+} x$ and

$$Ae^{-1/a} = 1 + C_2 e^{-1/a}.$$

Solving this last equation for $C_2$ gives

$$C_2 = A - e^{1/a}.$$

This procedure may be repeated for as many jumps as the function may have. In fact, the input function need not be constant between jumps. The function given by

$$f(t) = \begin{cases} 0, & 0 \le t < 1 \\ t, & 1 \le t < 2, \\ 2, & t \ge 2 \end{cases}$$

as shown in Figure 3.6, may be treated exactly in the same way. Thus in the regions given we have

$$a\frac{dx}{dt} + x = \begin{cases} 0, & 0 \le t < 1 \\ t, & 1 \le t < 2, \\ 2, & t \ge 2 \end{cases}$$

each of which represents a case that has already been solved. Again we may write the solution in the forms typified by (3.45) and (3.46).

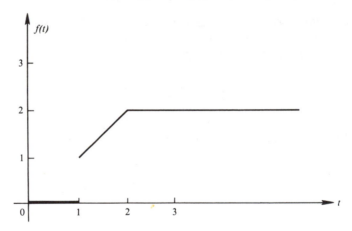

**Figure 3.6**

**6. The principle of superposition.**    We return to (3.2) and consider the case in which

$$Q(x) = b_1(x) + b_2(x) \tag{3.47}$$

and ask how the solution would change. The answer is obtained easily by rewriting (3.2) in operator form,

$$[a_1(x)D_x + a_0(x)]y = Q(x), \tag{3.48}$$

and noting that the polynomial operator on the left-hand side of (3.48) is a linear operator. Hence, if we denote by $y_1(x)$ and $y_2(x)$ functions such that

$$[a_1(x)D_x + a_0(x)]y_1(x) = b_1(x)$$
$$[a_1(x)D_x + a_0(x)]y_2(x) = b_2(x) \tag{3.49}$$

and add the two equations in (3.49), we will obtain

$$[a_1(x)D_x + a_0(x)][y_1(x) + y_2(x)] = b_1(x) + b_2(x), \tag{3.50}$$

because the operator is linear.

Equation (3.50) is telling us the following: If $y_1(x)$ is a solution to (3.48) for $Q(x) = b_1(x)$ and $y_2(x)$ is a solution to (3.48) for $Q(x) = b_2(x)$, their sum $y_1(x) + y_2(x)$ is a solution for $Q(x) = b_1(x) + b_2(x)$. This is known as the *principle of super-position*. We stress that this is a consequence of the linearity of the polynomial operator in (3.48).

The utility of the superposition principle is that it permits us to solve complicated problems in terms of simpler ones. For example, the equation

$$(D + 1)x = e^{-t} + 2 \sin t + t$$

may be solved by first solving each of the equations

$$(D + 1)x_1 = e^{-t},$$

$$(D + 1)x_2 = 2 \sin t,$$

$$(D + 1)x_3 = t,$$

and adding the results to get

$$x = x_1(t) + x_2(t) + x_3(t).$$

**7. The effect of negative a.**    Everything done so far has been done under the assumption that $a > 0$. If $a < 0$, the solution is obtained in the same manner, but (3.27) becomes

$$p(t) = e^{-t/|a|} \tag{3.51}$$

and the general solution becomes

$$x = e^{t/|a|} \int e^{-t/|a|} q(t) \, dt + C e^{t/|a|}.$$

Here the absolute value of the term $C e^{t/|a|}$ increases without bound as $t$ does. The general solution is unbounded and is said to grow exponentially in time.

## EXERCISES

Solve the following differential equations.

**1.** $\dfrac{dx}{dt} + x = 0.$               **2.** $\dfrac{dx}{dt} + x = 3.$

**3.** $\dfrac{dx}{dt} + x = t.$               **4.** $\dfrac{dx}{dt} + x = \sin t.$

**5.** $\dfrac{dx}{dt} + x = 3 + t + \sin t.$     **6.** $\dfrac{dx}{dt} + x = 3 \sin t + \sin 5t.$

**7.** $\dfrac{dx}{dt} - 3x = 6, \quad x(0) = 1.$   **8.** $\dfrac{dx}{dt} + tx = 0.$

**9.** $\dfrac{dx}{dt} - 7x = 14t, \quad x(0) = 0.$      **10.** $\dfrac{dx}{dt} + \dfrac{2}{t}x = t, \quad x(1) = 1.$

**11.** $\dfrac{dx}{dt} + 2tx = t, \quad x(0) = 2.$      **12.** $\dfrac{dx}{dt} - x = \sin 2t, \quad x(0) = 0.$

**13.** $\dfrac{dx}{dt} + x = e^{it}.$               **14.** $\dfrac{dx}{dt} + x = \cos t.$

**15.** $\dfrac{dx}{dt} + x = e^t.$

Solve the equation $(dx/dt) + x = f(t)$, where

**16.** $f(t) = \begin{cases} 0, & 0 \le t < 1 \\ 1, & t \ge 1 \end{cases}, \quad x(0) = 0.$    **17.** $f(t) = \begin{cases} 2, & 0 \le t < 1 \\ 0, & t \ge 1 \end{cases}, \quad x(0) = 0.$

**18.** $f(t) = \begin{cases} 5, & 0 \le t < 10 \\ 1, & t \ge 10 \end{cases}, \quad x(0) = 6.$    **19.** $f(t) = \begin{cases} e^{-t}, & 0 \le t < 2 \\ e^{-2}, & t \ge 2 \end{cases}, \quad x(0) = x_0.$

**20.** $f(t)$ is the function associated with Figure 3.6.

For the equation $(dx/dt) + x = f(t)$ use the principle of superposition to find the output when the input is

**21.** $2t + 1.$                      **22.** $t + \cos 3t.$

**22.** $6 \cos t + 5 \sin 2t.$          **24.** $2t^2 + t + 3.$

**25.** $5 + 2e^t.$                  **26.** $2 \sin\left(t + \dfrac{\pi}{6}\right).$

## 3.3 FURTHER APPLICATIONS

### 3.3.1 Newton's Law of Cooling

If we let $x(t)$ denote the temperature of a body at time $t$ and assume that the body is immersed in a heat reservoir at temperature $T$, Newton's law of cooling tells us that the rate of change of the temperature of the body is proportional to the difference, $T - x(t)$, of the temperature of the reservoir and that of the body. The differential equation expressing this law of cooling is

$$\frac{dx}{dt} = k(T - x), \quad k > 0. \tag{3.52}$$

This equation may be written as

$$\frac{1}{k}\frac{dx}{dt} + x = T$$

and is first order linear with the constant $T$ as a forcing function. The quantity $k^{-1}$ plays the role of $a$ and the solution, given by (3.31), is

$$x(t) = T + Ce^{-kt}. \tag{3.53}$$

The constant $C$ is related to the initial temperature difference and is found, by setting $t = 0$ in (3.53), to be

$$C = -(T - x(0)) = -(T - x_0),$$

so that

$$x(t) = T - (T - x_0)e^{-kt}. \tag{3.54}$$

The steady-state part of the solution (3.54) is just the reservoir temperature $T$ and the transient part is

$$-(T - x_0)e^{-kt},$$

so that as time increases the body temperature, $x(t)$, approaches that of the reservoir.

### 3.3.2 Motion in the Presence of Velocity-Dependent Friction

Let a body of mass $m$ be moving under the influence of some force $F(t)$ with a frictional force present which is proportional to the velocity. From Newton's second law we have

$$m\frac{d^2x}{dt^2} = F(t) - k\frac{dx}{dt}, \tag{3.55}$$

where $-k(dx/dt)$ represents the frictional force. Since the dependent variable, $x$, is not present explicitly in (3.55), this equation may be turned into a first order linear equation by writing

$$\frac{dx}{dt} = v(t), \quad \frac{d^2x}{dt^2} = \frac{dv}{dt}. \tag{3.56}$$

Substitution of (3.56) into (3.55) yields

$$m\frac{dv}{dt} = F(t) - kv(t),$$

or

$$m\frac{dv}{dt} + kv(t) = F(t). \tag{3.57}$$

Equation (3.57) is a first order linear equation for the velocity as a function of time. Suppose the force $F(t)$ is constant, say $F_0$, then (3.57) may be written as

$$\frac{dv}{dt} + \frac{k}{m}v = \frac{F_0}{m}, \tag{3.58}$$

which is a first order linear equation in standard form with

$$p(t) = \frac{k}{m}, \qquad q(t) = \frac{F_0}{m}.$$

The three-step procedure of Section 3.1 then yields the solution

$$v(t) = \frac{F_0}{k} + C_1 e^{-kt/m}, \tag{3.59}$$

where $C_1$ is an integration constant. Using the first of equations (3.56) again,

$$\frac{dx}{dt} = \frac{F_0}{k} + C_1 e^{-kt/m}, \tag{3.60}$$

and (3.60) may be integrated directly to obtain

$$x(t) = \frac{F_0}{k}t - \frac{mC_1}{k}e^{-kt/m} + C_2. \tag{3.61}$$

### 3.3.3 Simple Electrical Circuits

Let the simple electrical circuit of Figure 3.7 be given where $i$ denotes the current due to $E(t)$, the electromotive force (EMF), and $R$ and $C$ denote the resistance and capacitance, respectively. If we denote the charge on the capacitor at any time $t$ by $q(t)$, we have the circuit described by the differential equation

$$R\frac{dq}{dt} + \frac{1}{C}q = E(t).$$

Dividing by $R$ puts the equation in standard form:

$$\frac{dq}{dt} + \frac{1}{RC}q = \frac{E(t)}{R}.$$

As a specific example we might have

$$E(t) = E_0 \sin \omega t,$$

so that

$$\frac{dq}{dt} + \frac{1}{RC}q = \frac{E_0}{R} \sin \omega t. \tag{3.62}$$

Equation (3.62) may be solved by the method of Section 3.1

The only new item in the electrical circuit of Figure 3.8 is an inductor with inductance $L$. The differential equation which describes the current flow in this circuit is

$$L\frac{di}{dt} + Ri = E(t). \tag{3.63}$$

This is also a linear differential equation with the EMF, $E(t)$, as the forcing function. Exercises pertaining to (3.62) and (3.63) are given below. Additional exercises concerning electric circuits occur in Chapters 4, 5, and 7.

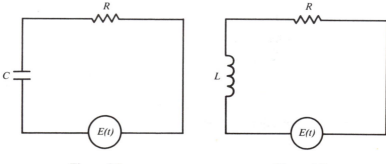

**Figure 3.7**                                          **Figure 3.8**

## EXERCISES

1. A body at temperature 72°F is taken outdoors where the temperature is 20°F. After 5 minutes the temperature of the body is 55°F. How long will it take the body to reach a temperature of 32°F?

2. A body of temperature $T$ is placed in a room held at a constant temperature of 68°F. After 10 minutes the temperature of the body is 50°F and after 15 minutes it is 55°F. What was the original temperature of the body?

3. A simple $RL$ circuit has an EMF of 6 volts, a resistance of 60 ohms, and an inductance of 1 henry. The current is switched on at $t = 0$. Find $i(t)$.

4. Let the circuit of exercise 3 have an EMF in volts given by $E(t) = \sin 2t$. If $i(0) = 6$ amperes, find $i(t)$.

5. An $RC$ circuit has an EMF given by $E(t) = 5$ volts, a resistance of 10 ohms, a capacitance of $10^{-2}$ farad, and an *initial charge* on the capacitor of 5 coulombs. The circuit is closed at $t = 0$. Find the transient and steady-state charges.

6. Let the circuit of exercise 5 have $E(t) = 2 \sin t$ and $q(0) = 0$. Find the resulting $q(t)$.

7. Note that a body falling under the influence of gravity in the presence of velocity-dependent friction is governed by (3.58), where $F_0 = mg$. Use the results of Section 3.3.2 to prove that the limiting velocity (velocity after a long time) of such a falling body is $mg/k$.

8. A 200-pound parachutist opens his parachute when he is falling at 100 feet per second.
   (a) If $F_0 = mg$, $k = 20$, and $g = 32$ in (3.58), find the velocity of the parachutist as a function of time after the parachute opens. (Recall that $m = $ weight$/g$.)
   (b) At what time will his velocity be 50 feet per second?
   (c) What is his terminal velocity?

9. Repeat exercise 8 for his 100-pound friend.

10. If the two parachutists in exercises 8 and 9 jump from opposite sides of an airplane at the same time, which will hit the ground first? [Assume that their fall is governed by (3.58), with $k = 20$ for both parachutists.]

11. A 4-pound rock is thrown vertically with an initial velocity of 64 feet per second.
    (a) Set up an initial value problem for the subsequent velocity (ignore frictional forces).
    (b) Solve this initial value problem and find the value of $t$ at which the rock reaches its maximum height.
    (c) What is the maximum height if $x(0) = 4$?

12. A barrel falls from a boat over a very deep portion of the ocean. The increase in resistance from this viscous water (which is colder and more dense at greater depths) is accounted for by replacing the $k$ in (3.58) by $k_0(3 - [2/(t + 1)])$.
    (a) If also $F_0 = mg$ in (3.58), solve for the resulting velocity assuming $v(0) = v_0$.
    (b) Is there a limiting velocity? If so, what is it?

## REVIEW EXERCISES

Find the general solution of the following differential equations.

1. $y' + (\sin x)y = \sin x$.

2. $\dfrac{dy}{dx} + e^x y = 0$.

3. $\dfrac{dy}{dx} + y = e^{-x} + 3$.

**4.** $5y' + y = 5e^{6x/5}$.

**5.** $y' + 2xy = 4x$.

**6.** $x^2y' - 4xy = x^6e^{2x} + 7$.

**7.** Solve the following initial value problems

(a) $\dfrac{dx}{dt} + 2tx = 6t, \quad x(0) = 5$.

(b) $\dfrac{dx}{dt} + (\cos t)x = \cos t, \quad x(0) = 7$.

**8.** Solve the differential equation $(dx/dt) + 2x = f(t)$, where

(a) $f(t) = \begin{cases} 1, & 0 < t < 2 \\ 0, & t > 2 \end{cases}, \quad x(0) = 4$.

(b) $f(t) = \begin{cases} 0, & 0 < t < 3 \\ 2, & t > 3 \end{cases}, \quad x(0) = 0$.

**9.** A 4-pound rock is thrown vertically with an initial velocity of 64 feet per second.
   (a) Set up an initial value problem for the subsequent velocity if it is governed by (3.58) with $F_0 = mg$, $k = 20$, and $g = 32$ feet per second squared.
   (b) Solve this initial value problem and find the value of $t$ for which the rock reaches its maximum height.
   (c) What is this height if $x(0) = 3$?

**10.** An $RC$ circuit has an EMF given by $E(t) = 6$ volts, a resistance of 100 ohms, a capacitance of $\frac{1}{100}$ farad, and an initial charge on the capacitor of 4 coulombs. The circuit is closed at $t = 0$. Find the transient and steady-state charges.

**11.** The resistance in an $RC$ circuit does not remain a constant when subjected to a current, but varies so that the constant $RC$ in the differential equation is given by $(100 + a \cos t)^{-1}$, with $a$ being a number less than 1. Solve the differential equation if $E(t) = 0$ and $q(0) = 3$. Observe what happens for large times.

**12.** A simple mathematical model of the memorization process is given by the differential equation

$$\frac{dx}{dt} = a(m - x) - bx,$$

where $m$ is the total amount of material to be memorized and $a$ and $b$ are positive constants. $x(t)$ represents the amount of material memorized at time $t$. The first term in this equation should remind you of Newton's law of cooling and the term $-bx$ accounts for the fact that the more you know, the more rapidly you forget.
   (a) Solve this differential equation subject to the initial condition, $x(0) = 0$.
   (b) Evaluate the limiting value of $x(t)$ as $t$ approaches infinity for the cases $b = 0$ and $b \neq 0$ and interpret the results.

# CHAPTER 4

# Higher Order Linear Differential Equations

We have defined the linear ordinary differential equation of $n$th order to be an equation of the form

$$a_n(x)y^{(n)}(x) + a_{n-1}(x)y^{(n-1)}(x) + \cdots + a_0(x)y(x) = Q(x). \qquad (4.1)$$

Our object now is to develop techniques for solving (4.1). Since we will need to determine whether or not a set of functions is linearly independent, we start this chapter with some definitions and examples.

## 4.1 LINEAR INDEPENDENCE OF A SET OF FUNCTIONS

**Definition 4.1.**  The set of functions $f_i$, $i = 1, \ldots, n$, is said to be *linearly independent* for $x$ in some internal $I$ if the only way a linear combination of these $n$ functions may equal zero for all $x \in I$,

$$\sum_{i=1}^{n} c_i f_i(x) = 0, \qquad \text{for all } x \in I, \qquad (4.2)$$

is for $c_i = 0$, $i = 1, \ldots, n$.

Thus $f_1, f_2, \ldots, f_n$ are *linearly dependent* if (4.2) is true for all $x \in I$ and not all of the $c_i = 0$. (Note that in using these definitions the same set of constants must be used for all $x \in I$.)

We obtain an easy method to test for independence of a set of $n$ functions which have $n - 1$ continuous derivatives in the following way. Equation (4.2) when written out is just

$$c_1 f_1 + c_2 f_2 + \cdots + c_n f_n = 0. \qquad (4.3)$$

**85**

Now if (4.3) holds, and the $f_i$, $i = 1, 2, 3, \ldots, n$, have continuous derivatives up to order $n - 1$ for $x \in I$, so do the equations which result from differentiating it, namely,

$$c_1 f_1' + c_2 f_2' + \cdots + c_n f_n' = 0,$$
$$c_1 f_1'' + c_2 f_2'' + \cdots + c_n f_n'' = 0,$$
$$\vdots \qquad \qquad \qquad \qquad (4.4)$$
$$c_1 f_1^{(n-1)} + c_2 f_2^{(n-1)} + \cdots + c_n f_n^{(n-1)} = 0.$$

For a fixed value of $x$, the $f_i^{(j)}$, $i = 1, 2, 3, \ldots, n$, $j = 0, 1, 2, \ldots, n - 1$ are known and (4.3) and (4.4) are $n$ equation in the $n$ unknowns, $c_1, c_2, c_3, \ldots, c_n$. If the functions $f_i$, $i = 1, 2, 3, \ldots, n$, are linearly independent, then we must have

$$c_i = 0, \quad i = 1, 2, 3, \ldots, n$$

as the unique solution to the system (4.3) and (4.4). But this will be the case if at some point $x \in I$ the determinant

$$W[f_1, f_2, \ldots, f_n] = \begin{vmatrix} f_1 & f_2 & \cdots & f_n \\ f_1' & f_2' & \cdots & f_n' \\ \vdots & \vdots & & \vdots \\ f_1^{(n-1)} & f_2^{(n-1)} & \cdots & f_n^{(n-1)} \end{vmatrix} \qquad (4.5)$$

is not zero (see corollary A.1 in the Appendix). The determinant $W$ in (4.5) is called the *Wronskian* of the functions $f_i$, $i = 1, 2, 3, \ldots, n$, and we may state the following theorem.

**Theorem 4.1.**    If the Wronskian of a set of functions, defined on some interval $I$, is different from zero at some point $x \in I$, the functions are linearly independent.

**EXAMPLE 4.1**    Let $f_1(x) = \sin x$ and $f_2(x) = \cos x$. Then

$$W[f_1, f_2] = \begin{vmatrix} \sin x & \cos x \\ \cos x & -\sin x \end{vmatrix} = -1, \qquad \text{for all } x$$

and $\sin x$ and $\cos x$ are linearly independent.

**EXAMPLE 4.2**    Let $f_1(x) = x^2$ and $f_2(x) = 3x^2$. Then

$$W[f_1, f_2] = \begin{vmatrix} x^2 & 3x^2 \\ 2x & 6x \end{vmatrix} = 6x^3 - 6x^3 = 0$$

for all $x$.

Since $W[x^2, 3x^2] = 0$ we cannot conclude anything from Theorem 4.1. However, since

$$-3(x^2) + 1(3x^2) = 0$$

for all $x$, $x^2$ and $3x^2$ are linearly dependent. In fact, $W[f_1, f_2]$ can equal zero even if $f_1$ and $f_2$ are linearly independent. Consider the functions

$$f_1(x) = \begin{cases} 0, & x < 0 \\ x^2, & x \geq 0 \end{cases} \qquad f_2(x) = \begin{cases} -x^2, & x < 0 \\ 0, & x \geq 0 \end{cases}, \qquad (4.6)$$

which are continuous and have continuous derivatives. The Wronskian,

$$W[f_1, f_2] = \begin{cases} \begin{vmatrix} 0 & -x^2 \\ 0 & -2x \end{vmatrix}, & \text{if } x < 0 \\[4mm] \begin{vmatrix} x^2 & 0 \\ 2x & 0 \end{vmatrix}, & \text{if } x \geq 0 \end{cases}$$

vanishes identically on R. Now consider a linear combination of $f_1$, and $f_2$ set equal to zero,

$$c_1 f_1 + c_2 f_2 = 0,$$

that is,

$$\begin{cases} c_1(0) + c_2(-x^2) = 0, & x < 0 \\ c_1(x^2) + c_2(0) = 0, & x \geq 0, \end{cases}$$

for all $x$. This requires $c_1 = c_2 = 0$ and $f_1$ and $f_2$ are linearly independent for $-\infty < x < \infty$. (Note that $f_1$ and $f_2$ in this example are linearly dependent on the interval $-\infty < x < 0$, or $0 < x < \infty$. This shows that the interval for $x$ affects the linear independence or dependence of a set of functions.)

**EXAMPLE 4.3**    Let $f_1(x) = e^{rx}$ and $f_2(x) = xe^{rx}$. Then

$$W[e^{rx}, xe^{rx}] = \begin{vmatrix} e^{rx} & xe^{rx} \\ re^{rx} & (rx + 1)e^{rx} \end{vmatrix} = e^{2rx}$$

and $e^{rx}$ and $xe^{rx}$ are linearly independent for all values of $r$.

**EXAMPLE 4.4**    Let $f_1(x) = 1$, $f_2(x) = x$, $f_3(x) = x^2$, and $f_4(x) = x^3$. Then

$$W[1, x, x^2, x^3] = \begin{vmatrix} 1 & x & x^2 & x^3 \\ 0 & 1 & 2x & 3x^2 \\ 0 & 0 & 2 & 6x \\ 0 & 0 & 0 & 6 \end{vmatrix} = 12$$

and $1$, $x$, $x^2$, and $x^3$ are linearly independent.

## EXERCISES

Are the following sets of functions linearly independent or dependent? (Compute the Wronskian first.)

**1.** $3$, $\sin x$, $\cos x$                          **2.** $3$, $\sin^2 x$, $\cos^2 x$

**3.** $1, 1 + x, x + 3$

**5.** $e^x, e^{-x}$

**7.** $\sin x, \sin^2 x$

**9.** $e^{rx}, xe^{rx}, x^2 e^{rx}$

**11.** The determinant

**4.** $1 - x, 1 + x, x^2$

**6.** $\sin x, x \sin x$

**8.** $e^x, xe^x, x^2 e^x$

**10.** $e^x, e^{-x}, e^{2x}$

$$\begin{vmatrix} 1 & 1 & 1 & \cdots & 1 \\ r_1 & r_2 & r_3 & \cdots & r_n \\ r_1^2 & r_2^2 & r_3^2 & \cdots & r_n^2 \\ \cdot & \cdot & \cdot & & \cdot \\ \cdot & \cdot & \cdot & & \cdot \\ \cdot & \cdot & \cdot & & \cdot \\ r_1^{n-1} & r_2^{n-1} & r_3^{n-1} & \cdots & r_n^{n-1} \end{vmatrix}$$

is called *Vandermonde's determinant*.

**(a)** Show that

$$\begin{vmatrix} 1 & 1 \\ r_1 & r_2 \end{vmatrix} = r_2 - r_1.$$

**(b)** Show that

$$\begin{vmatrix} 1 & 1 & 1 \\ r_1 & r_2 & r_3 \\ r_1^2 & r_2^2 & r_3^2 \end{vmatrix} = (r_2 - r_1)[(r_3 - r_1)(r_3 - r_2)].$$

**(c)** Show that

$$\begin{vmatrix} 1 & 1 & 1 & 1 \\ r_1 & r_2 & r_3 & r_4 \\ r_1^2 & r_2^2 & r_3^2 & r_4^2 \\ r_1^3 & r_2^3 & r_3^3 & r_4^3 \end{vmatrix} = (r_2 - r_1)[(r_3 - r_1)(r_3 - r_2)][(r_4 - r_1)(r_4 - r_2)(r_4 - r_3)].$$

Using induction, we may prove that for any positive integer $n$, this determinant equals

$$(r_2 - r_1)[(r_3 - r_1)(r_3 - r_2)][(r_4 - r_1)(r_4 - r_2)(r_4 - r_3)][\cdots]$$
$$[(r_n - r_1)(r_n - r_2)(r_n - r_3)\cdots(r_n - r_{n-1})].$$

**12.** Use the results in exercise 11 to prove the following sets of functions are linearly independent.

**(a)** $e^x, e^{2x}, e^{4x}$

**(b)** $e^x, e^{2x}, e^{3x}, e^{4x}, e^{5x}$

**(c)** $e^x, e^{2x}, e^{3x}, \ldots, e^{nx}$

**13.** If the constants $c_1, c_2, c_3, \ldots, c_n$ in a linear combination of $f_1, f_2, f_3, \ldots, f_n$ are allowed to be complex numbers, are the following sets of functions linearly independent?

**(a)** $e^{ix}, e^{-ix}; (i^2 = -1)$    $e^{ix} = \cos x + i \sin x$

**(b)** $e^{ix}, \cos x$

**(c)** $e^{ix}, e^{-ix}, \sin x$

## 4.2 THE HOMOGENEOUS EQUATION WITH CONSTANT COEFFICIENTS

We consider the special case of (4.1), in which each of the coefficients $a_i(x)$, $i = 0, \ldots, n$, is a constant and $Q(x) \equiv 0$:

$$a_n y^{(n)}(x) + a_{n-1} y^{(n-1)}(x) + \cdots + a_0 y(x) = 0. \qquad (4.7)$$

In operator notation, (4.7) has the form

$$Ty = 0,$$

with $T$ the linear differential operator

$$T = a_n D^n + a_{n-1} D^{n-1} + \cdots + a_1 D + a_0 = \sum_{j=0}^{n} a_j D^j.$$

We now state the main theorem regarding solutions of (4.7).

**Theorem 4.2.** Given the linear differential equation (4.7) with the $a_0$, $a_1, \ldots, a_n$ specified constants. If $a_n \neq 0$ and the set of functions, $\{y_1(x), y_2(x), \ldots, y_n(x)\}$, is a linearly independent set of solutions of (4.7), then all solutions of (4.7) may be written as

$$y = C_1 y_1 + C_2 y_2 + \cdots + C_n y_n. \qquad (4.8)$$

[Equation (4.8) is called the general solution of (4.7).]

The proof of this theorem is straightforward and will be explored in the exercises in Section 4.2.2. The proof for the case $n = 2$ will now be given.

For $n = 2$, since $a_2 \neq 0$, we may write $Ty = 0$ as

$$y'' = -\frac{1}{a_2}(a_0 y + a_1 y'). \qquad (4.9)$$

Note that this is the form of the differential equation needed in Theorem 1.1. Note also that the partial derivatives of the right-hand side of this equation with respect to $y$ as well as $y'$ are constants and therefore continuous for all $x$. Thus the hypotheses of Theorem 1.1 are fulfilled and there is a unique solution of the differential equation above which satisfies the initial conditions

$$y(x_0) = y_0, \qquad y'(x_0) = y_0'.$$

Now take a linear combination of two solutions of (4.7) as

$$y = C_1 y_1(x) + C_2 y_2(x).$$

Since $y_1(x)$ and $y_2(x)$ both satisfy our linear homogeneous differential equation, so will $y$. $C_1$ and $C_2$ are now determined so that the initial conditions are satisfied. This gives the two equations

$$C_1 y_1(x_0) + C_2 y_2(x_0) = y_0,$$

$$C_1 y_1'(x_0) + C_2 y_2'(x_0) = y_0'.$$

From Cramer's rule we know that we have a unique solution of these two algebraic equations if the determinant of the coefficients

$$\begin{vmatrix} y_1(x_0) & y_2(x_0) \\ y_1'(x_0) & y_2'(x_0) \end{vmatrix}$$

is not zero. However, this determinant is simply the Wronskian of $y_1$ and $y_2$ and we know that the Wronskian of two linearly independent solutions of

$$a_2 y'' + a_1 y' + a_0 y = 0$$

will never be zero (see exercise 24). Thus $C_1$ and $C_2$ are uniquely determined and we have a unique solution of the differential equation subject to our initial conditions. Since $x_0$, $y_0$, and $y_0'$ may be chosen arbitrarily, we see that all solutions of $Ty = 0$ may be written as a linear combination of the linearly independent functions $y_1(x)$ and $y_2(x)$ and the proof is complete.

For arbitrary values of $n$, Theorem 4.2 shows that the general solution of (4.7) will consist of a linear combination of $n$ linearly independent solutions, $y_j(x)$, $j = 1$, $2, \ldots, n$, of (4.7), that is,

$$y = \sum_{j=1}^{n} C_j y_j(x). \tag{4.8}$$

Our task now is to discover how to determine these $n$ linearly independent functions. The case $n = 1$, where the differential equation is

$$(a_1 D + a_0)y = 0, \tag{4.10}$$

has already been discussed in Chapter 3. To summarize these results we may say that

$$y = Ce^{rx} \tag{4.11}$$

is the general solution of (4.10) if $r$ is a solution of the linear equation

$$a_1 r + a_0 = 0. \tag{4.12}$$

Equation (4.12) is called the *characteristic equation* associated with the first order differential equation (4.10).

### 4.2.1 The Second Order Equation

For the case $n = 2$, the differential equation becomes

$$(a_2 D^2 + a_1 D + a_0)y = 0 \tag{4.13}$$

and we seek solutions of this new differential equation of the same form, (4.11), as we found for $n = 1$. Since

$$Dy = D(Ce^{rx}) = Cre^{rx} = ry$$

and

$$D^2 y = D(Dy) = D(ry) = rDy = r^2 y,$$

$y = Ce^{rx}$ satisfies (4.13) if

$$a_2r^2y + a_1ry + a_0y = (a_2r^2 + a_1r + a_0)y = 0.$$

Thus if we are to have a nonzero solution, $r$ must satisfy the algebraic equation

$$a_2r^2 + a_1r + a_0 = 0. \tag{4.14}$$

This is the *characteristic equation* associated with the differential equation (4.13). Notice that for both cases, $n = 1$ and $n = 2$, the characteristic equation may be obtained from the differential operator by replacing $D^n$ by $r^n$, $n = 1, 2$, and setting the resulting polynomial in $r$ equal to zero. [Compare (4.10) and (4.12), (4.13), and (4.14).]

In solving (4.14) we find that there are three cases to consider: two real distinct, one real repeated, and two imaginary roots. Each of these cases will be considered in turn.

**1. Two real distinct roots.**    If $r_1$ and $r_2$ $(r_1 \neq r_2)$ satisfy (4.14), $e^{r_1x}$ and $e^{r_2x}$ will be solutions of (4.13). Since the Wronskian

$$W[e^{r_1x}, e^{r_2x}] = \begin{vmatrix} e^{r_1x} & e^{r_2x} \\ r_1e^{r_1x} & r_2e^{r_2x} \end{vmatrix} = (r_2 - r_1)e^{(r_1+r_2)x}$$

is not zero for $r_2 \neq r_1$, $e^{r_1x}$ and $e^{r_2x}$ are linearly independent and the general solution of (4.13) has the form

$$y = C_1e^{r_1x} + C_2e^{r_2x},$$

where $r_1$ and $r_2$ are distinct real solutions of the characteristic equation (4.14).

**EXAMPLE 4.5**    Let the differential equation be

$$(D^2 + D - 6)y = 0. \tag{4.15}$$

The characteristic equation associated with (4.15) is

$$r^2 + r - 6 = 0,$$

with real distinct roots given by

$$r_1 = -3, \qquad r_2 = 2.$$

Thus the general solution to (4.15) is

$$y(x) = C_1e^{-3x} + C_2e^{2x}.$$

**2. One real repeated root.**    The quadratic formula gives the solutions of (4.14) as

$$r = \frac{-a_1 \pm \sqrt{a_1^2 - 4a_2a_0}}{2a_2}.$$

If $a_1^2 - 4a_2a_0 = 0$, we have $r_1 = -a_1/(2a_2)$ as a repeated root of the characteristic equation and thus have obtained only one linearly independent solution, $e^{r_1x}$, of

(4.13). To obtain a second solution, we could use the reduction-of-order technique from Section 2.5 (this will be explored in the exercises). However, since we know that we must have two linearly independent solutions, if we come up with a second solution of (4.13) independent of $e^{r_1 x}$, our task is complete. If we consider

$$y_2(x) = xe^{r_1 x} \tag{4.16}$$

and compute

$$Dy_2 = e^{r_1 x} + r_1 xe^{r_1 x},$$

$$D^2 y_2 = 2r_1 e^{r_1 x} + r_1^2 xe^{r_1 x},$$

we see that

$$(a_2 D^2 + a_1 D + a_0)y_2 = (a_2 r_1^2 + a_1 r_1 + a_0)xe^{r_1 x} + (2a_2 r_1 + a_1)e^{r_1 x}. \tag{4.17}$$

The coefficient of $xe^{r_1 x}$ equals zero since $r_1$ satisfies the characteristic equation, and the coefficient of $e^{r_1 x}$ equals zero since $r_1 = -a_1/(2a_2)$. Thus $y_2$ satisfies the original differential equation, and since the Wronskian of $y_1$ and $y_2$,

$$W[y_1, y_2] = \begin{vmatrix} e^{r_1 x} & xe^{r_1 x} \\ r_1 e^{r_1 x} & (xr_1 + 1)e^{r_1 x} \end{vmatrix} = e^{2r_1 x},$$

is never 0, $y_1$ and $y_2$ are linearly independent functions. Thus for a repeated root, $r_1$, of the characteristic equation, the general solution of (4.13) is

$$y = (C_1 + C_2 x)e^{r_1 x}. \tag{4.18}$$

**EXAMPLE 4.6**   Consider the differential equation

$$(D^2 + 2D + 1)y = 0, \tag{4.19}$$

with characteristic equation

$$r^2 + 2r + 1 = (r + 1)^2 = 0.$$

Since $r = -1$ is a double root, the general solution to (4.19) is given by

$$y(x) = (C_1 + C_2 x)e^{-x}.$$

*3. Two complex roots.*   When we solve the characteristic equation (4.14) as

$$r = \frac{-a_1 \pm \sqrt{a_1^2 - 4a_0 a_2}}{2a_2} \tag{4.20}$$

and the discriminant, $a_1^2 - 4a_0 a_2$, is negative, we see that the two roots are complex and occur as complex conjugates. Writing $r_1 = \alpha + i\beta$, $r_2 = \alpha - i\beta$, where

$$\alpha = \frac{-a_1}{2a_2}, \qquad \beta = \frac{\sqrt{4a_0 a_2 - a_1^2}}{2a_2},$$

we see that the corresponding functions

$$y_1(x) = e^{(\alpha + i\beta)x} \qquad \text{and} \qquad y_2(x) = e^{(\alpha - i\beta)x}$$

are linearly independent. (The Wronskian, $W[y_1, y_2] = -2i\beta e^{2\alpha x}$, is never zero.)

Thus the general solution of (4.13) is

$$y = C_1 e^{(\alpha + i\beta)x} + C_2 e^{(\alpha - i\beta)x}, \tag{4.21}$$

where $r_1 = \alpha + i\beta$ and $r_2 = \alpha - i\beta$ are complex roots of the characteristic equation.

If we use (4.21) to solve a problem where the dependent variable, $y$, takes on only real values, we will discover that the constants $C_1$ and $C_2$ will often be complex numbers. To avoid this complication, we will expand (4.21) using Euler's identity,

$$e^{i\beta x} = \cos \beta x + i \sin \beta x. \tag{4.22}$$

The equation becomes

$$y = e^{\alpha x}[C_1(\cos \beta x + i \sin \beta x) + C_2(\cos \beta x - i \sin \beta x)]$$
$$= (C_1 + C_2)e^{\alpha x} \cos \beta x + i(C_1 - C_2)e^{\alpha x} \sin \beta x.$$

Since $C_1$ and $C_2$ were arbitrary constants, choose two new arbitrary constants $C_3$ and $C_4$ by $C_3 = C_1 + C_2$, $C_4 = i(C_1 - C_2)$, and write the general solution as

$$y = C_3 e^{\alpha x} \cos \beta x + C_4 e^{\alpha x} \sin \beta x. \tag{4.23}$$

The requirement that the solution be real means that $C_1 + C_2$ is real and $C_1 - C_2$ is pure imaginary. This will be the case if we choose $C_1$ and $C_2$ to be complex conjugates. Either (4.21) or (4.23) is an acceptable form of the general solution for the case of two complex roots.

### EXAMPLE 4.7

$$(D^2 + 2D + 5)y = 0.$$

The characteristic equation associated with this differential equation,

$$r^2 + 2r + 5 = 0,$$

has the complex roots

$$r_1 = -1 + 2i, \qquad r_2 = -1 - 2i.$$

Thus the general real solution is given by

$$y(x) = C_1 e^{(-1+2i)x} + C_2 e^{(-1-2i)x},$$

where $C_1$ and $C_2$ are complex constants, or

$$y(x) = C_3 e^{-x} \cos 2x + C_4 e^{-x} \sin 2x,$$

where $C_3$ and $C_4$ are real constants.

We now summarize the results of this section. To find the general solution of the second order differential equation with constant coefficients

$$(a_2 D^2 + a_1 D + a_0)y = 0,$$

use the following three steps:

**Step 1.** Write the characteristic equation

$$a_2 r^2 + a_1 r + a_0 = 0.$$

**Step 2.** Solve the characteristic equation and call the solutions $r_1$ and $r_2$. There are three possibilities:
  (a) $r_1$ and $r_2$ are real with $r_1 \neq r_2$.
  (b) $r_1 = r_2$.
  (c) $r_1 = \alpha + i\beta, r_2 = \alpha - i\beta, \beta \neq 0.$

**Step 3.** Write the general solution corresponding to possibilities (a), (b), and (c) of step 2 using
  (a) $y = C_1 e^{r_1 x} + C_2 e^{r_2 x}.$
  (b) $y = (C_1 + C_2 x) e^{r_1 x}.$
  (c) $y = C_1 e^{(\alpha + i\beta)x} + C_2 e^{(\alpha - i\beta)x}$ or $y = C_1 e^{\alpha x} \cos \beta x + C_2 e^{\alpha x} \sin \beta x.$
[Note that (a) and (c) are really the same case: two distinct roots. It just so happens you have an alternative way to express the functions in case (c).]

## EXERCISES

Find the general solution of the following differential equations.

**1.** $(D^2 - D - 6)y = 0.$
**2.** $(D^2 - D - 2)y = 0.$
**3.** $(D^2 - 2D + 1)y = 0.$
**4.** $(D^2 - 2D - 8)y = 0.$
**5.** $(D^2 + 2D - 8)y = 0.$
**6.** $(D^2 - 6D + 9)y = 0.$
**7.** $y'' - y' - 12y = 0.$
**8.** $y'' - 2y' + 10y = 0.$
**9.** $y'' - 5y' + 6y = 0.$
**10.** $y'' - 6y' + 10y = 0.$
**11.** $y'' + 16y = 0.$
**12.** $y'' - 9y = 0.$
**13.** $6y'' + y' - y = 0.$
**14.** $9y'' - 6y' + y = 0.$
**15.** $y'' + 2y' + 4y = 0.$
**16.** $y'' + 4y' + 2y = 0.$

**17.** Solve the following initial value problems.
  (a) $(D^2 + D - 2)y = 0; \quad y(0) = 0, \quad y'(0) = 3.$
  (b) $(D^2 + 2D - 10)y = 0; \quad y(0) = 0, \quad y'(0) = 4.$
  (c) $y'' + 6y' + 9y = 0; \quad y(0) = 2, \quad y'(0) = 0.$
  (d) $(12D^2 + D - 1)y = 0; \quad y(0) = 4, \quad y'(0) = 0.$
  (e) $y'' + 3y' = 0; \quad y(0) = 4, \quad y'(0) = 3.$
  (f) $y'' - 2\pi y' + 2\pi^2 y = 0; \quad y(0) = 0, \quad y'(0) = -2\pi.$
  (g) $y'' + 10y' + 100y = 0; \quad y(0) = 15, \quad y'(0) = 4.$

**18.** Show that $y = A \cos(\omega x + \phi)$ satisfies

$$y'' + \omega^2 y = 0$$

for any choice of $A$ and $\phi$.

**19.** Give the solution of

$$y'' + 16y = 0, \qquad y(0) = 2\sqrt{2}, \qquad y'(0) = 8\sqrt{2},$$

in the form of exercise 18.

**20.** Find a second order differential equation with real coefficients that has the given general solution
  **(a)** $y = C_1 e^{-x} + C_2 e^{3x}$.
  **(b)** $y = C_1 e^{-2x} \cos 3x + C_2 e^{-2x} \sin 3x$.
  **(c)** $y = C_1 e^{7x} + C_2 x e^{7x}$.
  **(d)** $y = A \sin(3x + \phi)$.
  **(e)** $y = C_1 e^{ax} \cos bx + C_2 e^{ax} \sin bx$.

**21.** For what values of $\omega$ does

$$y'' + \omega^2 y = 0, \qquad y(0) = 0, \qquad y(\pi) = 0,$$

have a nonzero solution? Give these solutions.

**22.** Show that if $r_1$ is a double root of the characteristic equation, the reduction-of-order technique gives a solution of

$$a_2 y'' + a_1 y' + a_0 y = 0,$$

which together with $e^{r_1 x}$ forms a linearly independent set.

**23.** Show that the functions $e^{ax} \cos \beta x$ and $e^{ax} \sin \beta x$ are linearly independent functions.

**24.** **(a)** Show that if $y_1$ and $y_2$ both satisfy the differential equation in exercise 22, then

$$a_2(y_1 y_2'' - y_2 y_1'') + a_1(y_1 y_2' - y_1' y_2) = 0.$$

  **(b)** Since the Wronskian $W[y_1, y_2] = y_1 y_2' - y_1' y_2$, show that $W$ satisfies the differential equation

$$a_2 W' + a_1 W = 0,$$

  with solution $W = Ce^{-a_1 x/a_2}$.

  **(c)** Show that the Wronskian of two solutions of $a_2 y'' + a_1 y' + a_0 y = 0$ is either always zero or never zero.

  **(d)** Show that two solutions of the differential equations in part (c) are linearly independent if and only if their Wronskian is nonzero.

### 4.2.2 Equations of Order Greater Than 2

Flushed with success with equations of orders 1 and 2, we leap to the conclusion that

$$y(x) = Ce^{rx} \tag{4.11}$$

is a possible solution to the $n$th order equation (4.7). Since

$$D^n y = D^n(Ce^{rx}) = r^n Ce^{rx}$$

we substitute (4.11) into (4.7) and find that

$$(a_n r^n + a_{n-1} r^{n-1} + \cdots + a_0)Ce^{rx} = 0.$$

Thus (4.11) is a solution of (4.7) provided that $r$ satisfies

$$a_n r^n + a_{n-1} r^{n-1} + \cdots + a_0 = 0. \tag{4.24}$$

[This is called the *characteristic equation* associated with (4.7).] Equation (4.24) will in general have $n$ roots. If these roots are distinct, we obtain $n$ linearly independent solutions of (4.7)

$$y_j(x) = e^{r_j x}, \qquad j = 1, 2, 3, \ldots, n.$$

(See also exercise 21 at the end of this section.)

If a root $r$ has multiplicity $k$, the $k$ linearly independent solutions of (4.7) associated with $r$ may be taken as

$$e^{rx}, \, xe^{rx}, \, x^2 e^{rx}, \, \ldots, \, x^{k-1}e^{rx}.$$

(This fact will not be proved in general, but there are exercises to verify this for particular differential equations.)

We now summarize the results obtained thus far in this chapter. The general solution of the $n$th order ordinary differential equation with constant coefficients,

$$Ty = (a_n D^n + a_{n-1}D^{n-1} + \cdots + a_1 D + a_0)y = \left( \sum_{j=0}^{n} a_j D^j \right) y = 0 \quad (4.25)$$

is

$$y = \sum_{j=1}^{n} C_j y_j(x), \qquad\qquad (4.26)$$

where $y_j(x)$, $j = 1, 2, 3, \ldots, n$, are $n$ linearly independent solutions of (4.25). A three-step procedure for finding the $y_j(x)$ is given below.

**Step 1.**  Write the characteristic equation associated with (4.25),

$$a_n r^n + a_{n-1}r^{n-1} + \cdots + a_1 r + a_0 = \sum_{j=0}^{n} a_j r^j = 0. \qquad (4.27)$$

**Step 2.**  Solve the characteristic equation and identify:
  (a) All simple roots (real or complex).
  (b) All repeated roots (real or complex).

**Step 3.**  Write the general solution. Here we have two cases:
  (a) The characteristic equation has only simple roots. In this case to each root $r_j$ we assign the function

$$y_j = e^{r_j x}, \qquad j = 1, 2, 3, \ldots, n \qquad (4.28)$$

  and the general solution is given by (4.26).
  (b) The characteristic equation has one or more roots of multiplicity greater than 1. To each root $r_j$ of multiplicity $k$ we assign the function

$$y_j(x) = (C_0 + C_1 x + \cdots + C_{k-1}x^{k-1})e^{r_j x}. \qquad (4.29)$$

  We treat simple roots as in case (a), and the sum of the $y_j$ thus formed by (4.26) is the general solution. Note that the arbitrary constants are included in expression (4.29) and you will not need to multiply by an additional constant when you form the sum in (4.26).

[Note that in either case (a) or (b), for any pair of complex roots $r_j = \alpha_j + i\beta_j$ and $r_{j+1} = \alpha_j - i\beta_j$, you could replace $e^{r_j x}$ and $e^{r_{j+1} x}$ by $e^{\alpha_j x} \cos \beta_j x$ and $e^{\alpha_j x} \sin \beta_j x$.]

We conclude this section with two examples to illustrate this procedure.

**EXAMPLE 4.8**    Consider the differential equation $(D^3 - D^2 - 5D - 3)y = 0$ and use the three-step process to obtain the general solution.

> **Step 1.**    The characteristic equation associated with this equation is
> $$r^3 - r^2 - 5r - 3 = 0.$$

> **Step 2.**    This equation may be factored as
> $$(r + 1)^2(r - 3) = 0,$$

with roots given by

$$r_1 = -1 \ (r_1 \text{ is a double root}), \qquad r_2 = 3.$$

> **Step 3.**    Corresponding to $r_2$, we have the solution
> $$y_2(x) = e^{3x}$$

and corresponding to $r_1$, the solution

$$y_1(x) = (C_0 + C_1 x)e^{-x}.$$

Thus the general solution is

$$y = (C_0 + C_1 x)e^{-x} + C_2 e^{3x}.$$

**EXAMPLE 4.9**    We now solve the fourth order differential equation

$$(D^4 + 8D^2 + 16)y = (D^2 + 4)^2 y = 0$$

using this three-step process.

> **Step 1.**    The characteristic equation is
> $$(r^2 + 4)^2 = 0.$$

> **Step 2.**    This equation factors as $[(r + 2i)(r - 2i)]^2 = 0$ and has complex roots $r_1 = 2i$ and $r_2 = -2i$, each of multiplicity 2.

> **Step 3.**    Thus we have the general solution
> $$y(x) = C_1' e^{2ix} + C_2' x e^{2ix} + C_3' e^{-2ix} + C_4' x e^{-2ix}$$

or

$$y(x) = (C_0 + C_1 x) \cos 2x + (C_2 + C_3 x) \sin 2x.$$

## EXERCISES

Find the general solution of the following differential equations.

1. $(D^3 + D)y = 0$.

2. $(D^3 + 1)y = 0$.

3. $(D^4 + 4D^2 + 4)y = 0$.

4. $(D^4 + 2D^2 - 15)y = 0$.

5. $(D^4 + 2D^2 - 8)y = 0$.

6. $(D^3 - 7D^2 + 19D - 13)y = 0$.

7. $y''' - 2y'' - y' + 2y = 0$.

8. $\dfrac{d^4y}{dx^4} + 8\dfrac{d^2y}{dx^2} - 9y = 0$.

9. $(D^3 + D^2 + 3D - 5)y = 0$.

10. $(D^4 - 5D^2 + 4)y = 0$.

Solve the following initial value problems.

11. $(D^4 + 3D^3 + 2D^2)y = 0$,  $y(0) = y'(0) = y''(0) = 0$,  $y'''(0) = 8$.

12. $(D^4 + 6D^2 + 9)y = 0$,  $y(0) = y'(0) = y''(0) = 0$,  $y'''(0) = 6$.

13. $(D^3 + 6D^2 + 5D - 12)y = 0$,  $y(0) = 0$,  $y'(0) = 4$,  $y''(0) = -8$.

14. $(D^4 - 16)y = 0$,  $y(0) = 0$,  $y'(0) = 0$,  $y''(0) = 0$,  $y'''(0) = 4$.

   (*Hint:* Use hyperbolic and trigonometric functions instead of exponentials when evaluating the arbitrary constants.)

The theorem one uses to show that a set of solutions of a differential equation is linearly independent or dependent is

   **Theorem 4.3.**   Let $\{f_i(x), i = 1, 2, 3, \ldots, n\}$ be a set of solutions of the differential equation

$$\left[\sum_{j=0}^{n} a_j(x)D^j\right]y = 0, \qquad a < x < b.$$

Then this set of solutions is linearly independent if and only if the Wronskian $W[f_1, f_2, f_3, \ldots, f_n] \neq 0$ for some $x$ in $a < x < b$.

Determine if the following solutions of a linear differential equation form a linearly independent set.

15. $\{e^{r_1 x}, e^{r_2 x}\}$,  $r_1 \neq r_2$.

16. $\{e^{rx}, xe^{rx}\}$.

17. $\{e^x, e^{2x}, e^x - e^{2x}\}$.

18. $\{1, x, x^2, x^3, \ldots, x^n\}$.

19. $\{\sin x, \cos x, x \sin x, x \cos x\}$.

20. $\{e^x \sin 2x, e^x \cos 2x\}$.

21. $\{e^{r_1 x}, e^{r_2 x}, e^{r_3 x}, \ldots, e^{r_n x}\}$.  (*Hint:* recall the Vandermonde determinant in Section 4.1.)

22. $\{x^r, x^r \ln x\}$,  $x > 0$.

23. $\{\cos(3 \ln x), \sin(3 \ln x)\}$,  $x > 0$.

24. $\{x^{r_1}, x^{r_2}\}$,  $r_1 \neq r_2$,  $x > 0$.

25. $\{x^2 \cos(\ln x), x^2 \sin(\ln x)\}$,  $x > 0$.

26. $\{1, \ln x, (\ln x)^2\}$,  $x > 0$.

27. Prove Theorem 4.3 for $n = 3$ using the following steps.

(a) Show that $\dfrac{d}{dx} W = \begin{vmatrix} f_1 & f_2 & f_3 \\ f_1' & f_2' & f_3' \\ f_1''' & f_2''' & f_3''' \end{vmatrix}.$

(Recall that the derivative of a determinant is the sum of the determinants obtained by differentiating each row in turn.)

(b) Use the fact that the $f_i$, $i = 1, 2, 3$, satisfy (4.7) with $n = 3$ to show that $dW/dx = -[a_2(x)/a_3(x)]W$.

(c) Integrate the differential equation in part (b) and derive Abel's formula,

$$W = C \exp\left[-\int \frac{a_2(x)}{a_3(x)} dx\right].$$

(d) Use the expression for $W$ in part (c) to complete the proof.

**28.** Prove Theorem 4.3 for any integer $n$.

**29.** Prove Theorem 4.2

(a) For $n = 3$.

(b) For any integer $n$.

## 4.3 THE NONHOMOGENEOUS EQUATION WITH CONSTANT COEFFICIENTS

We now return to (4.1) and lift the restriction that $Q(x) \equiv 0$ while keeping the coefficients $a_i(x)$, $i = 0, \ldots, n$, constant. As in our treatment of the homogeneous equation we will begin by considering some special cases. Recall that we may write the solution of a first order linear differential equation with constant coefficients in two parts, one depending on the forcing function and the other containing an arbitrary constant [note (3.31), (3.34), (3.42)]. This pattern will continue for higher orders as well.

We know that the homogeneous equation has solutions in terms of exponentials, either real or complex. Thus it is natural to ask what the solutions to the non-homogeneous equation might be if

$$Q(x) = e^{sx}.$$

In this case we would have the equation

$$Ty = \left(\sum_{j=0}^{n} a_j D^j\right) y = (a_n D^n + \cdots + a_1 D + a_0)y = e^{sx}.$$

Suppose that we choose $n = 1$, $a_1 = 1$, $a_0 = 1$, $s \neq -1$ and consider the equation

$$(D + 1)y = e^{sx}. \tag{4.30}$$

The homogeneous equation associated with (4.30) is

$$(D + 1)y = 0,$$

with solution

$$y(x) = Ce^{-x}.$$

Recall the property of a linear operator $L$: If $Ly_1 = 0$ and $Ly_p = f$, $L(Cy_1 + y_p) = f$ for all values of $C$. Since $D + 1$ is a linear operator, we look for a function, $y_p$, such that $(D + 1)y_p = e^{sx}$. Such a function is called a *particular solution* to (4.30) and we define the general solution to (4.30) by

$$y(x) = Ce^{-x} + y_p(x).$$

Since we know how to solve homogeneous equations, we seek a method of converting (4.30) to a homogeneous equation. Note that the function on the right-hand side of (4.30), $e^{sx}$, satisfies the differential equation $(D - s)y = 0$.

  Now if $y_p(x)$ satisfies (4.30) it must also satisfy the equation that results from operating on both sides of (4.30) with the operator $(D - s)$. Doing so yields

$$(D - s)(D + 1)y_p = (D - s)e^{sx} = 0,$$

or

$$(D - s)(D + 1)y_p = 0. \tag{4.31}$$

Equation (4.31) is homogeneous with constant coefficients and has the associated characteristic equation

$$(r - s)(r + 1) = 0.$$

Having solutions of the characteristic equation $r = s$ and $r = -1$ gives

$$y_1(x) = e^{-x} \qquad y_2(x) = e^{sx}$$

as solutions of (4.31). Since we want $y_p$ to give the right-hand side of (4.30), and $y_1 = e^{-x}$ satisfies $(D + 1)y_1 = 0$, we try

$$y_p(x) = Ae^{sx}. \tag{4.32}$$

Substituting (4.32) into (4.30) gives

$$(s + 1)Ae^{sx} = e^{sx},$$

which is true if $(s + 1)A = 1$ or $A = 1/(s + 1)$. Thus

$$y_p(x) = \frac{1}{s + 1}e^{sx}$$

does satisfy (4.30) and the general solution to (4.30) may be written as

$$y(x) = Ce^{-x} + \frac{1}{s + 1}e^{sx}.$$

This approach yielded a particular solution because $Q(x) = e^{sx}$ satisfied the equation

$$(D - s)Q(x) = 0$$

so that applying $(D - s)$ to both sides of (4.30) gave us a homogeneous equation that we could solve. Any function which satisfies (4.30) must satisfy (4.31), although there are functions which satisfy (4.31) but not (4.30). [Note that $e^{-x}$ satisfies (4.31) but not (4.30).]

As a second example of how we may use operators to reduce a nonhomogeneous differential equation to a homogeneous differential equation, consider

$$Ly = (D^2 + 3D + 2)y = 3e^{-x} + 40e^{3x}. \qquad (4.33)$$

Note that $e^{-x}$ satisfies $(D + 1)y = 0$, so if we operate on both sides of (4.33) by $D + 1$, we obtain

$$(D + 1)(D^2 + 3D + 2)y = 3(D + 1)e^{-x} + 40(D + 1)e^{3x}$$

$$= 0 + 40(4e^{3x}) = 160e^{3x}.$$

Now since $e^{3x}$ satisfies $(D - 3)y = 0$, we operate on both sides of our last equation by $D - 3$ to obtain

$$(D - 3)(D + 1)(D^2 + 3D + 2)y = (D - 3)160e^{3x} = 0. \qquad (4.34)$$

The characteristic equation associated with this homogeneous differential equation is

$$(r - 3)(r + 1)(r^2 + 3r + 2) = (r - 3)(r + 1)(r + 2)(r + 1) = 0.$$

Thus the general solution of (4.34) is

$$C_1e^{-2x} + C_2e^{-x} + C_3xe^{-x} + C_4e^{3x}.$$

If we use this expression as a possible solution of (4.33), we substitute it into (4.33) and obtain

$$L(C_1e^{-2x} + C_2e^{-x} + C_3xe^{-x} + C_4e^{3x}) = 3e^{-x} + 40e^{3x}.$$

Since $e^{-2x}$ and $e^{-x}$ satisfy $Ly = 0$, this reduces to

$$L(C_3xe^{-x} + C_4e^{3x}) = 3e^{-x} + 40e^{3x},$$

which by performing the indicated differentiation and operations in $L$ becomes

$$C_3e^{-x} + 20C_4e^{3x} = 3e^{-x} + 40e^{3x}.$$

If we write this last equation as

$$(C_3 - 3)e^{-x} + (20C_4 - 40)e^{3x} = 0,$$

we have a solution of (4.33) by choosing

$$C_3 = 3 \quad \text{and} \quad C_4 = 2.$$

(Since $e^{-x}$ and $e^{3x}$ are linearly independent, the only way that a linear combination of the two functions may equal 0 is for both coefficients to equal zero.) Thus the general solution of (4.33) is

$$y = C_1e^{-2x} + C_2e^{-x} + 3xe^x + 2e^{3x}.$$

Notice that we again have the general solution of a nonhomogeneous equation as the sum of the general solution of the homogeneous differential equation plus a particular solution ($3xe^x + 2e^{3x}$). Since we know how to find general solutions of the homogeneous differential equation, we now concentrate on finding a particular solution of the nonhomogeneous differential equation.

For the general case,

$$Ly = (a_n D^n + a_{n-1} D^{n-1} + \cdots + a_1 D + a_0)y = Q(x) \qquad (4.35)$$

we would look for a linear operator such that $TQ(x) = 0$. Then (4.35) would be reduced to the homogeneous differential equation $T(Ly) = TQ = 0$. This method will be successful when $Q(x)$ contains functions which satisfy differential equations with constant coefficients. This will be the case when the terms in $Q(x)$ have the form of *a polynomial, an exponential function, a polynomial times an exponential, or is a combination of exponentials* and/or *polynomials times sines or cosines* [see (4.23), (4.28), and (4.29)]. However, rather than looking for a new operator for each differential equation we solve, we use the fact that we know what types of functions satisfy such equations to develop a straightforward method of solving (4.35). This is called the *method of undetermined coefficients* and is given by the following theorem.

**Theorem 4.4.**     Suppose that $Q(x)$ in (4.35) has the form

$$Q(x) = (p_m x^m + \cdots + p_0)e^{\alpha x} \cos \beta x + (q_m x^m + \cdots + q_0)e^{\alpha x} \sin \beta x. \qquad (4.36)$$

If $\alpha \pm \beta i$ are *not* roots of the characteristic equation, we assume a particular solution of the form

$$y_p(x) = (k_m x^m + k_{m-1} x^{m-1} + \cdots + k_0)e^{\alpha x} \cos \beta x$$
$$+ (\ell_m x^m + \ell_{m-1} x^{m-1} + \cdots + \ell_0)e^{\alpha x} \sin \beta x. \qquad (4.37)$$

The coefficients $k_j$, $\ell_j$, $j = 0, 1, \ldots, m$, are determined by substituting (4.37) into (4.35) and requiring the result to be an identity. If $\alpha \pm \beta i$ are roots of the characteristic equation of multiplicity $h$, we multiply the expression on the right side of (4.37) by $x^h$ and proceed as before.

Note *some* of the special cases included in Theorem 4.4:

| *Conditions* | $Q(x)$ | $y_p(x)$ |
|---|---|---|
| $\alpha = \beta = 0$ | $\sum_{j=0}^{m} p_j x^j$ | $\sum_{j=0}^{m} k_j x^j$ |
| $m = \beta = 0$ | $p_0 e^{\alpha x}$ | $k_0 e^{\alpha x}$ |
| $m = \alpha = 0$ | $p_0 \cos \beta x + q_0 \sin \beta x$ | $k_0 \cos \beta x + \ell_0 \sin \beta x$ |

The other theorem we need is

**Theorem 4.5.**     The general solution of the linear differential equation

$$Ly = \left[ \sum_{j=0}^{n} a_j(x) D^j \right] y = Q(x)$$

has the form

$$y = \sum_{j=1}^{n} C_j y_j(x) + y_p(x),$$

where $y_j(x)$, $j = 1, 2, 3, \ldots, n$, are linearly independent solutions of $Ly = 0$; $C_j, = 1, 2, 3, \ldots, n$, are arbitrary constants; and $y_p(x)$ is any function satisfying $Ly_p = Q(x)$.

(Note the similarity between Theorem 4.5 and the result of exercise 11, page 21.)

**EXAMPLE 4.10**

$$y'' + y' - 2y = 4 \sin 2x. \qquad (4.38)$$

First we look at the homogeneous equation

$$y'' + y' - 2y = 0$$

with the associated characteristic equation

$$r^2 + r - 2 = 0.$$

Factoring this characteristic equation as $(r + 2)(r - 1) = 0$ gives its roots as

$$r_1 = -2, \qquad r_2 = 1$$

and $e^{-2x}$ and $e^x$ satisfy the homogeneous equation.

If we compare $4 \sin 2x$ with (4.36) we see that

$$4 \sin 2x = 4e^{0x} \sin 2x + (0)e^{0x} \cos 2x$$

and

$$p_0 = 0, \qquad \alpha = 0, \qquad \beta = 2, \qquad q_0 = 4, \qquad m = 0.$$

Since $\sin 2x$ is *not* a solution of the homogeneous equation, we try

$$y_p(x) = (\ell_0 x^0)e^{0x} \sin 2x + (k_0 x^0)e^{0x} \cos 2x$$

$$= k \cos 2x + \ell \sin 2x.$$

[The subscripts on $k$ and $\ell$ are not essential here and are omitted in the development of $y_p(x)$.] We calculate

$$y_p'(x) = -2k \sin 2x + 2\ell \cos 2x,$$

$$y_p''(x) = -4k \cos 2x - 4\ell \sin 2x,$$

and substitute these expressions into (4.38) to obtain

$$y_p'' + y_p' - 2y_p = -4k \cos 2x - 4\ell \sin 2x - 2k \sin 2x$$

$$+ 2\ell \cos 2x - 2k \cos 2x - 2\ell \sin 2x$$

$$= (-4\ell - 2k - 2\ell)\sin 2x + (-4k - 2k + 2\ell)\cos 2x$$

$$= 4 \sin 2x.$$

Since $\sin 2x$ and $\cos 2x$ are linearly independent functions, we may equate coefficients to obtain

$$-6\ell - 2k = 4,$$

$$2\ell - 6k = 0.$$

The solution of this system of equations is

$$k = \frac{-1}{5}, \qquad \ell = \frac{-3}{5},$$

giving the particular solution as

$$y_p(x) = -\frac{1}{5}\cos 2x - \frac{3}{5}\sin 2x$$

and the general solution of (4.38) as

$$y = C_1 e^{-2x} + C_2 e^x - \frac{1}{5}\cos 2x - \frac{3}{5}\sin 2x.$$

**EXAMPLE 4.11**

$$(D^2 - 4D + 4)y = 12xe^{2x}. \tag{4.39}$$

First we look at the homogeneous equation

$$(D^2 - 4D + 4)y = 0,$$

with the associated characteristic equation

$$r^2 - 4r + 4 = 0.$$

Since the roots are obtained from $(r - 2)^2 = 0$ as $r_1 = r_2 = 2$, the homogeneous equation has two linearly independent solutions, $e^{2x}$ and $xe^{2x}$. Now compare the right-hand side of the differential equation with $Q(x)$ of (4.36) and obtain

$$12xe^{2x} = (p_m x^m + \cdots + p_0)e^{\alpha x}\cos \beta x$$
$$+ (q_m x^m + \cdots + q_0)e^{\alpha x}\sin \beta x.$$

Thus $\alpha = 2$, $\beta = 0$, $p_1 = 12$, $p_0 = 0$, and $m = 1$, and from (4.37) the trial solution for the nonhomogeneous equation is

$$y_p(x) = (k_1 x + k_0)e^{2x}.$$

However in this case, 2 (the value of $\alpha$) is a root of multiplicity 2 of the characteristic equation associated with the homogeneous differential equation. We must therefore modify our trial solution by multiplying $(k_1 x + k_0)e^{2x}$ by $x^2$ and use

$$y_p(x) = x^2(k_1 x + k_0)e^{2x} = (k_1 x^3 + k_0 x^2)e^{2x}.$$

Substituting this expression into the differential equation (4.39) and equating coefficients of $xe^{2x}$ and $e^{2x}$ gives $k_1 = 2$, $k_0 = 0$. Thus the general solution of (4.39) is

$$y(x) = C_1 e^{2x} + C_2 xe^{2x} + 2x^3 e^{2x}.$$

The procedure used in Examples 4.10 and 4.11 will now be formalized as a three-step procedure as demonstrated with Examples 4.12 and 4.13.

**EXAMPLE 4.12**    Find the general solution of

$$Ly = (D^4 + 2D^3 - 3D^2)y = -36x + 9e^{3x} + 12e^x. \qquad (4.40)$$

**Step 1.**    *Solve the homogeneous differential equation associated with the given equation.*

In this case we have

$$(D^4 + 2D^3 - 3D^2)y = 0,$$

with characteristic equation

$$r^4 + 2r^3 - 3r^2 = r^2(r + 3)(r - 1) = 0.$$

Thus $r = 1$ and $r = -3$ are simple roots and $r = 0$ is a root of multiplicity 2. Thus the four linearly independent solutions of the homogeneous equation are

$$e^x, \quad e^{-3x}, \quad 1, \quad x.$$

(Note that $e^{0x} = 1$.)

**Step 2.**    *Find the proper form for a particular solution.*

In this step we compare the form of $Q(x)$, the nonhomogeneous part of the differential equation, with (4.36).

Comparing each term on the right-hand side of (4.40) with (4.36) gives the following three situations. (Recall the principle of superposition.)

(a)   $-36x = (-36x + 0)e^{0x} \cos 0x$ has a corresponding trial solution, from (4.37), given by

$$k_1 x + k_0.$$

However, since 0 is a root of multiplicity 2, we must multiply $k_1 x + k_0$ by $x^2$ to obtain

$$x^2(k_1 x + k_0) = k_1 x^3 + k_0 x^2$$

as the correct form of the particular solution of $Ly = -36x$.

(b)   $9e^{3x} = 9e^{3x} \cos 0x$. This term requires a particular solution of $Ly = 9e^{3x}$ of the form

$$k_0^* e^{3x}.$$

(c)   $12e^x = 12e^x \cos 0x$ has a corresponding trial solution from (4.37) as

$$k_0^{**} e^x.$$

However, 1 is a simple root of the characteristic equation, so we must multiply by $x$ to obtain

$$k_0^{**} x e^x$$

as the correct form of the trial solution of $Ly = 12e^x$.

**Step 3.** *Determine the constants in each $y_p(x)$, by substitution.*

Since $Ly$ is a linear operator, we have the option of working with the following three equations:

$$L(k_1 x^3 + k_0 x^2) = -36x, \tag{4.41a}$$

$$L(k_0^* e^{3x}) = 9e^{3x}, \tag{4.41b}$$

$$L(k_0^{**} xe^x) = 2e^x, \tag{4.41c}$$

or one equation,

$$L(k_1 x^3 + k_0 x^2 + k_0^* e^{3x} + k_0^{**} xe^x) = -36x + 9e^{3x} + 12e^x.$$

If we start with the first of the three separate equations and calculate

$$D(k_1 x^3 + k_0 x^2) = 3k_1 x^2 + 2k_0 x,$$

$$D^2(k_1 x^3 + k_0 x^2) = 6k_1 x + 2k_0,$$

$$D^3(k_1 x^3 + k_0 x^2) = 6k_1,$$

$$D^4(k_1 x^3 + k_0 x^2) = 0,$$

we find that (4.41a) becomes

$$2(6k_1) - 3(6k_1 x + 2k_0) = -18k_1 x + 12k_1 - 6k_0 = -36x.$$

Thus $k_1 = 2$ and $k_0 = 4$ and the particular solution associated with $-36x$ is

$$2x^3 + 4x^2.$$

Similar calculations in (4.41b) and (4.41c) give particular solutions as

$$\left(\frac{1}{12}\right)e^{3x} \qquad \text{and} \qquad 3xe^x.$$

We now have a particular solution of (4.39) as

$$y_p(x) = 2x^3 + 4x^2 + \left(\frac{1}{12}\right)e^{3x} + 3xe^x$$

and the general solution as

$$y(x) = C_1 e^x + C_2 e^{-3x} + C_3 + C_4 x + 2x^3 + 4x^2 + \left(\frac{1}{12}\right)e^{3x} + 3xe^x.$$

**EXAMPLE 4.13** Find a particular solution of

$$y'' + 2y' + 5y = 4e^{-x} \cos 2x. \tag{4.42}$$

**Step 1.** *Solve the associated homogeneous differential equation.*
Here the equation

$$y'' + 2y' + 5y = 0$$

has a characteristic equation of the form

$$r^2 + 2r + 5 = 0,$$

with solutions

$$r = \frac{-2 \pm \sqrt{4 - 20}}{2} = -1 \pm 2i.$$

The functions $e^{-x} \cos 2x$ and $e^{-x} \sin 2x$ are two linearly independent solutions of the homogeneous equation.

**Step 2.**    *Find the proper form for a particular solution.*
If we write the right-hand side of (4.42) as

$$Q(x) = p_0 e^{\alpha x} \cos \beta x + q_0 e^{\alpha x} \sin \beta x = 4e^{-x} \cos 2x,$$

we see that $\alpha = -1$, $\beta = 2$, $p_0 = 4$, $q_0 = 0$, $m = 0$ and

$$\alpha \pm \beta i = -1 \pm 2i$$

are roots of multiplicity 1 of the characteristic equation associated with the homogeneous differential equation. Thus we must take our trial solution as

$$y_p(x) = x(ke^{-x} \cos 2x + \ell e^{-x} \sin 2x), \tag{4.43}$$

where we have dropped the subscripts on the $k$ and $\ell$.

**Step 3.**    *Determine the constants in $y_p(x)$.*
The derivative of $y_p(x)$ is

$$y_p'(x) = ke^{-x} \cos 2x + \ell e^{-x} \sin 2x - kxe^{-x} \cos 2x - \ell xe^{-x} \sin 2x$$
$$- 2kxe^{-x} \sin 2x + 2\ell xe^{-x} \cos 2x$$
$$= e^{-x}(k - kx + 2\ell x) \cos 2x$$
$$+ e^{-x}(\ell - \ell x - 2kx) \sin 2x$$

and the second derivative is

$$y_p''(x) = -e^{-x}[k + (2\ell - k)x] \cos 2x + e^{-x}(2\ell - k) \cos 2x$$
$$- 2e^{-x}[k + (2\ell - k)x] \sin 2x$$
$$- e^{-x}[\ell - (\ell + 2k)x] \sin 2x - e^{-x}(\ell + 2k) \sin 2x$$
$$+ 2e^{-x}[\ell - (\ell + 2k)x] \cos 2x.$$

Substituting these forms of $y_p(x)$, $y_p'(x)$, and $y_p''(x)$ into (4.42) gives
$$e^{-x}\{ [4\ell - 2k - (3k + 4\ell)x] \cos 2x + [-4k - 2\ell + (4k - 3\ell)x] \sin 2x$$
$$+ 2[k + (2\ell - k)x] \cos 2x + 2[\ell - (\ell + 2k)x] \sin 2x$$
$$+ 5kx \cos 2x + 5\ell x \sin 2x\} = 4e^{-x} \cos 2x.$$

Collecting coefficients results in

$$e^{-x}(4\ell \cos 2x - 4k \sin 2x) = 4e^{-x} \cos 2x.$$

Choosing $\ell = 1$, $k = 0$ makes this equation true and gives the particular solution the form

$$y_p(x) = xe^{-x} \sin 2x.$$

Finally, we have the general solution of (4.42) as

$$y(x) = C_1 e^{-x} \cos 2x + C_2 e^{-x} \sin 2x + xe^{-x} \sin 2x.$$

## EXERCISES

1. Show that if $L$ is a linear differential operator, $f$ is any function satisfying $Ly = 0$ and

$$Lg_1 = q_1(x), \qquad Lg_2 = q_2(x),$$

then

$$y = Cf(x) + Ag_1(x) + Bg_2(x)$$

satisfies $Ly = Aq_1(x) + Bq_2(x)$ for any choice of the constant $C$. (See also exercise 12 on page 21.)

Find the general solution of the following differential equations.

2. $y'' + y' - 6y = x$.
3. $y'' + y' - 6y = e^{3x}$.
4. $y'' + y' - 6y = 2x + \pi e^{3x}$.
5. $y'' + 3y' - 4y = x^2 + 1$.
6. $y'' + y' = 6e^{-x}$.
7. $(D^2 + 4)y = \cos 3x$.
8. $(D^2 + 4)y = 10 \sin 3x$.
9. $(D^2 + 4)y = 6 \cos 3x + 20 \sin 3x$.
10. $(D^2 + 4)y = \sin 2x$.
11. $(D^2 + 4D + 4)y = \cos 2x$.
12. $(4D^2 - 12D + 9)y = 24xe^{3x/2}$.
13. $y''' + 2y'' - y' - 2y = 20 \cos x$.
14. $y^{(4)} - 8y'' - 9y = 6x + 12 \sin x$.
15. $y''' - 7y'' + 19y' - 13y = 2e^{2x} \cos x + 6e^{2x} \sin x$.

Solve the following initial value problems.

16. $y'' + y' - 12y = 8e^{3x}$,   $y(0) = 0$,   $y'(0) = 1$.
17. $y'' + 6y' + 9y = e^{3x}$,   $y(0) = 0$,   $y'(0) = 6$.
18. $(D^2 - 5D + 6)y = 12xe^{-x} - 7e^{-x}$,   $y(0) = y'(0) = 0$.
19. $(D^2 + 4)y = 8 \sin 2x + 8 \cos 2x$,   $y(\pi) = y'(\pi) = 2\pi$.
20. Show that if $\omega \neq \alpha$, a particular solution of

$$(D^2 + \omega^2)y = A \cos \alpha t + B \sin \alpha t$$

is

$$y_p(t) = \frac{A}{\omega^2 - \alpha^2} \cos \alpha t + \frac{B}{\omega^2 - \alpha^2} \sin \alpha t.$$

**21.** Show that the general solution of

$$(D^2 + \omega^2)y = A \cos \omega t$$

is

$$y = C_1 \cos \omega t + C_2 \sin \omega t + \frac{A}{2\omega} t \sin \omega t.$$

**22.** Solve the initial value problem

$$(D^4 - 1)y = \cosh t, \qquad y(0) = y'(0) = y''(0) = y'''(0) = 0.$$

## 4.4 THE CAUCHY–EULER EQUATION

An equation of the form

$$\sum_{j=0}^{n} a_j x^j \frac{d^j y}{dx^j} = a_n x^n \frac{d^n y}{dx^n} + a_{n-1} x^{n-1} \frac{d^{n-1} y}{dx^{n-1}} + \cdots + a_1 x \frac{dy}{dx} + a_0 y = 0, \qquad (4.44)$$

where $a_j$, $j = 0, 1, 2, \ldots, n$, are constants, is often called the *Cauchy–Euler equation*. It is also known as the *Euler equation* and as the *equidimensional equation* (see exercise 14).

If we solve (4.44) for the highest derivative, we have

$$\frac{d^n y}{dx^n} = \frac{-\sum_{j=0}^{n-1} a_j x^j (d^j y / dx^j)}{a_n x^n} = F(x, y, y', y'', \ldots, y^{(n-1)}). \qquad (4.45)$$

It should be clear that $F$ and $\partial F / \partial y^{(j)}$, $j = 0, 1, 2, 3, \ldots, n - 1$, are continuous for any interval that does not contain $x = 0$. Thus in light of Theorem 1.1 we must consider domains of $F$ which do not include $x = 0$.

If $n = 1$, we have the differential equation

$$a_1 x \frac{dy}{dx} + a_0 y = 0$$

with solution

$$y = Cx^{-a_0/a_1}.$$

(Since the equation is both linear and separable, we may use techniques from either Section 2.1 or 3.1 to obtain this solution.) The form of this solution tempts us to try a solution of the second order equation

$$a_2 x^2 \frac{d^2 y}{dx^2} + a_1 x \frac{dy}{dx} + a_0 y = 0 \qquad (4.46)$$

as

$$y = x^r, \qquad (4.47)$$

with $r$ a constant to be determined. If we substitute (4.47) into (4.46), we obtain

$$a_2 x^2 [r(r - 1)x^{r-2}] + a_1 x(rx^{r-1}) + a_0 x^r = [a_2 r^2 + (a_1 - a_2)r + a_0]x^r = 0.$$

This last equation is true for all $x > 0$ if

$$a_2 r^2 + (a_1 - a_2)r + a_0 = 0. \qquad (4.48)$$

Since a quadratic equation either has two distinct roots or one repeated root, we end up with a situation similar to that in which we found ourselves when we solved second order differential equations with constant coefficients (Section 4.1). Rather than continue with this general case, we will make use of the fact that a solution to a second order Cauchy-Euler equation (4.46) has the form of a power of the independent variable, equation (4.47), and work some examples.

**EXAMPLE 4.14**    To solve

$$x^2 y'' + 4xy' - 4y = 0 \qquad (4.49)$$

we assume a solution of the form $y = x^r$ and find by substitution that

$$[r(r - 1) + 4r - 4]x^r = 0.$$

This means that $r$ must satisfy

$$r^2 + 3r - 4 = (r + 4)(r - 1) = 0,$$

and we have two linearly independent solutions, $x^{-4}$ and $x$, of (4.49).

**EXAMPLE 4.15**    Consider solutions to

$$x^2 y'' + 3xy' + 5y = 0 \qquad (4.50)$$

in the form $y = x^r$. This results in

$$r(r - 1) + 3r + 5 = r^2 + 2r + 5 = 0,$$

with solutions

$$r = -1 \pm 2i.$$

Thus $x^{-1+2i}$ and $x^{-1-2i}$ are linearly independent solutions of (4.50). We can obtain two real-valued functions from these solutions by using the two facts

$$(1) \quad x = e^{\ln x}, \quad x > 0,$$

$$(2) \; e^{i\beta} = \cos \beta + i \sin \beta.$$

Thus

$$x^{-1+2i} = x^{-1}x^{2i} = x^{-1}(e^{\ln x})^{2i} = x^{-1}e^{2i \ln x}$$

$$= x^{-1}[\cos(2 \ln x) + i \sin(2 \ln x)]$$

gives

$$x^{-1} \cos(2 \ln x) \quad \text{and} \quad x^{-1} \sin(2 \ln x)$$

as linearly independent solutions of (4.50). Note that since this proce    es
two linearly independent solutions of a second order equation, w        ot
consider $x^{-1-2i}$.

**EXAMPLE 4.16**    If we try a solution of the form $y = x^r$ in

$$x^2y'' - 3xy' + 4y = 0 \tag{4.51}$$

we get the quadratic equation

$$r^2 - 4r + 4 = 0.$$

This equation has $r = 2$ as a root of multiplicity 2 and results in only one solution, $x^2$. Here is where the reduction of order technique from Section 2.5 is useful. We make the change of variable

$$y(x) = x^2v(x) \tag{4.52}$$

and calculate derivatives as

$$y' = 2xv + x^2v',$$

$$y'' = 2v + 4xv' + x^2v''.$$

Substituting the last three equations into (4.51) results in

$$x^2(2v + 4xv' + x^2v'') - 3x(2xv + x^2v') + 4x^2v = 0.$$

Combining terms gives

$$x^4v'' + x^3v' = 0.$$

If we let $v' = w$, we obtain the first order equation in $w$,

$$xw' + w = 0,$$

which is an exact equation $[(xw)' = xw' + w]$. We solve for $w$ as

$$w = C_1x^{-1}$$

and $v$ as

$$v = \int w\, dx = C_1 \ln x + C_2.$$

Since $y = x^2v$, the general solution of (4.51) is

$$y(x) = C_1x^2 \ln x + C_2x^2.$$

Notice that the one solution of (4.51), $x^2 \ln x$, is obtained from our first solution, $x^2$, by multiplication by $\ln x$. The reason for this can be seen from the following theorem.

**Theorem 4.6.**    The change of variables $|x| = e^t$ reduces the Cauchy-Euler equation (4.44) to a differential equation with constant coefficients.

The proof is left for the exercises (see exercises 15 and 17). Note that since the solutions of linear differential equations of order 2 with constant coefficients have the form

$$e^{rt}, \quad e^{\alpha t} \cos \beta t, \quad e^{\alpha t} \sin \beta t, \quad \text{or} \quad te^{rt},$$

solutions for the Cauchy-Euler equation have the form

$$(e^t)^r = |x|^r, \quad |x|^\alpha \cos(\beta \ln |x|), \quad |x|^\alpha \sin(\beta \ln |x|), \quad [\ln |x|] |x|^r.$$

Solutions for Cauchy-Euler equations with $n \geq 3$ can be obtained by assuming a solution of the form $x^r$ or by using the change of variables $x = e^t$ to obtain a differential equation with constant coefficients.

## EXERCISES

Find the general solution of the following differential equations, assuming $x > 0$ for exercises 1 to 10.

1. $x^2 y'' + 2xy' - 6y = 0.$
2. $x^2 y'' + xy' = 0.$
3. $x^2 y'' + 9xy' + 2y = 0.$
4. $x^2 y'' + xy' + 9y = 0.$
5. $x^2 y'' - 5xy' + 25y = 0.$
6. $x^2 y'' + 5xy' + 4y = 0.$
7. $2x^2 y'' - xy' + y = 0.$
8. $x^2 y'' - 5xy' + 5y = 0.$
9. $2x^2 y'' + xy' - y = 0.$
10. $x^2 y'' - xy' + y = 0.$
11. $(x - 1)^2 y'' + 3(x - 1)y' + y = 0.$ [*Hint*: Let $y = (x - 1)^r$.]
12. $(x + 4)^3 y''' - 2(x + 4)y' + 2y = 0.$
13. $(x - 3)^3 y''' + 2(x - 3)^2 y'' - (x - 3)y' + y = 0.$
14. Make the change of variable $x = \beta t$ in (4.44) and show that (4.44) becomes $\sum_{j=0}^{n} a_j t^j (d^j y / dt^j) = 0.$ Notice that the parameter $\beta$ does not enter into the new differential equation. Since we may think of $x = \beta t$ as a change in scale (or dimension), (4.44) is sometimes called the equidimensional equation.
15. Proof of Theorem 4.6 for $n = 2$:
    (a) If $x > 0$ and $x = e^t$, show that
    $$\frac{d}{dx} = \frac{1}{x}\frac{d}{dt} \quad \text{and} \quad \frac{d^2}{dx^2} = \frac{1}{x^2}\left(\frac{d^2}{dt^2} - \frac{d}{dt}\right).$$
    (b) Substitute the expression for the operators $d/dx$ and $d^2/dx^2$ into (4.46) and show that the result is
    $$a_2 \frac{d^2 y}{dt^2} + (a_1 - a_2)\frac{dy}{dt} + a_0 y = 0.$$
    (c) Show that for $x < 0$, letting $x = -e^t$ gives the same differential equation found in part (b).
16. Show that the general solution of (4.46) is
    (a) If $(a_1 - a_2)^2 - 4a_2 a_0 > 0$,
    $$y = C_1 |x|^{r_1} + C_2 |x|^{r_2},$$
    where $r_1$ and $r_2$ are roots of (4.48).

**(b)** If $(a_1 - a_2)^2 - 4a_2a_0 = 0$,

$$y = (C_1 + C_2 \ln |x|) \, |x|^{r_1},$$

with $r_1 = (a_2 - a_1)/(2a_2)$.

**(c)** If $(a_1 - a_2)^2 - 4a_2a_0 < 0$,

$$y = [C_1 \cos (\beta \ln |x|) + C_2 \sin(\beta \ln |x|)] \, |x|^\alpha,$$

with $r = \alpha \pm i\beta$ roots of (4.48).

17. Show that the Wronskian of the functions in exercise 16 is not zero (i.e., each set of functions has two linearly independent members).

(a) $W[x^{r_1}, x^{r_2}]$ for $x > 0$.

(b) $W[x^r, x^r \ln x]$ for $x > 0$.

(c) $W[e^{\alpha x} \cos(\beta \ln x), e^{\alpha x} \sin(\beta \ln x)]$ for $x > 0$.

18. Note that if $D$ denotes differentiation with respect to $t$, the results of part (a) of exercise 15 may be listed as

$$x \frac{d}{dx} = D, \qquad x^2 \frac{d^2}{dx^2} = D^2 - D$$

and the result in part (b) as

$$a_2x^2 \frac{d^2y}{dx^2} + a_1x \frac{dy}{dx} + a_0y = 0$$

becomes

$$(a_2D(D - 1) + a_1D + a_0)y = 0.$$

**(a)** Prove that the transformation $x = e^t$ allows the association

$$x^3 \frac{d^3}{dx^3} = D(D - 1)(D - 2)$$

and use mathematical induction to show that

$$x^n \frac{d^n}{dx^n} = D(D - 1)(D - 2) \cdots (D - n + 1).$$

**(b)** Use the associations in part (a) to show that the Cauchy-Euler equation of order $n$ [equation (4.44)] becomes the $n$th order differential equation with constant coefficients

$$[a_nD(D - 1) \cdots (D - n + 1) + a_{n-1}D(D - 1) \cdots (D - n + 2)$$

$$+ \cdots + a_2D(D - 1) + a_1 D + a_0]y = 0.$$

**(c)** Since solutions of this homogeneous differential equation have the form

$$e^{rt}, \quad t^m e^{rt}, \quad e^{\alpha t} \cos \beta t, \quad e^{\alpha t} \sin \beta t, \quad t^m e^{\alpha t} \cos \beta t, \quad t^m e^{\alpha t} \sin \beta t,$$

for $x > 0$, the solutions of a Cauchy-Euler equation may have the form

$$x^r, \quad (\ln x)^m x^r, \quad x^\alpha \cos(\beta \ln x), \quad x^\alpha \sin(\beta \ln x),$$

$$(\ln x)^m x^\alpha \cos(\beta \ln x), \quad (\ln x)^m x^\alpha \sin(\beta \ln x).$$

**19.** Use the result of exercise 18(b) to solve the following differential equations for $x > 0$.

**(a)** $x^2 \dfrac{d^2y}{dx^2} + 2x \dfrac{dy}{dx} - 6y = 0.$

**(b)** $x^2 \dfrac{d^2y}{dx^2} - 5x \dfrac{dy}{dx} + 25y = 0.$

**(c)** $x^2 y'' + 5xy' + 4y = 0.$

**(d)** $x^3 y''' - x^2 y'' + xy' = 0.$

**(e)** $x^3 y''' - 2x^2 y'' + 8xy' - 8y = 0.$

**(f)** $x^3 y''' - 3x^2 y'' + 6xy' - 6y = 0.$

**(g)** $x^4 y^{(4)} + 6x^3 y''' + 7x^2 y'' + xy' - y = 0.$

**(h)** $x^3 y''' + xy' - y = 0.$

## 4.5 THE METHOD OF VARIATION OF PARAMETERS

To motivate the method of variation of parameters, we take a look at the first order linear differential equation in standard form

$$\frac{dy}{dx} + p(x)y = q(x). \tag{4.53}$$

The general solution of the homogeneous equation

$$\frac{dy}{dx} + p(x)y = 0 \tag{4.54}$$

is

$$y = \frac{C_1}{\rho(x)} = \frac{C_1}{\exp[\int p(x)\, dx]},$$

and the solution of (4.53) is

$$y(x) = \frac{1}{\rho(x)} \int \rho(x)q(x)\, dx + \frac{C_1}{\rho(x)}, \tag{4.55}$$

where $C_1$ is an arbitrary constant [see (3.15)]. What we want to notice is the form of the solution in (4.55). If we let

$$y_1(x) = \frac{1}{\rho(x)}$$

represent a solution of the homogeneous equation, we can write (4.55) as

$$y(x) = y_1(x) \int \frac{q(x)}{y_1(x)}\, dx + C_1 y_1(x). \tag{4.56}$$

Observe that this solution consists of two parts, the solution of the homogeneous equation plus a particular solution of the form $y_1(x)v(x)$, with $v(x)$ depending on both $q(x)$ and $y_1(x)$. That is, if we substitute

$$y_p(x) = y_1(x)v(x) \tag{4.57}$$

into (4.53) we will obtain a particular solution (see exercise 10). We could think of obtaining a trial form for a particular solution by replacing the constant multiplying $y_1(x)$ by an unknown function.

Now we extend this idea to a second order differential equation of the form

$$[a_2(x)D^2 + a_1(x)D + a_0(x)]y = Q(x). \tag{4.58}$$

Here we assume that we know the general solution of the associated homogeneous differential equation is given by

$$C_1 y_1(x) + C_2 y_2(x).$$

In analogy with (4.57), we assume that the particular solution of (4.58) is given by

$$y_p(x) = v_1(x) y_1(x) + v_2(x) y_2(x). \tag{4.59}$$

(The name *variation of constants* or *variation of parameters* is due to this replacement.) We then look for functions $v_1(x)$ and $v_2(x)$ such that (4.59) satisfies (4.58).

Differentiation of (4.59) gives

$$Dy = v_1 y_1' + v_2 y_2' + v_1' y_1 + v_2' y_2,$$

$$D^2 y = v_1 y_1'' + v_2 y_2'' + v_1' y_1' + v_2' y_2' + D(v_1' y_1 + v_2' y_2),$$

and substitution into (4.58) then yields

$$v_1(a_2 y_1'' + a_1 y_1' + a_0 y_1) + v_2(a_2 y_2'' + a_1 y_2' + a_0 y_2) + a_2(v_1' y_1' + v_2' y_2')$$
$$+ D(v_1' y_1 + v_2' y_2)) + a_1(v_1' y_1 + v_2' y_2) = Q(x). \tag{4.60}$$

Because $y_1$ and $y_2$ satisfy the homogeneous equation, we have that the coefficients of $v_1$ and $v_2$ vanish identically. Notice that we have only one equation for the two unknown functions $v_1$ and $v_2$. If we choose a relationship between them such that

$$v_1' y_1 + v_2' y_2 = 0, \tag{4.61}$$

then (4.60) reduces to

$$a_2(v_1' y_1' + v_2' y_2') = Q(x). \tag{4.62}$$

Equations (4.61) and (4.62) form the system of equations

$$\begin{aligned} y_1 v_1' + y_2 v_2' &= 0, \\ y_1' v_1' + y_2' v_2' &= \frac{Q(x)}{a_2}. \end{aligned} \tag{4.63}$$

Equation (4.63) has a unique solution for $v_1'$ and $v_2'$ since

$$\begin{vmatrix} y_1 & y_2 \\ y_1' & y_2' \end{vmatrix} = W[y_1, y_2]$$

is the Wronskian of the functions $y_1$ and $y_2$ and is never zero (recall that $y_1$ and $y_2$ are linearly independent). Thus we obtain the equations

$$v_1' = \frac{-Q(x) y_2(x)}{a_2(x) W[y_1, y_2]}, \qquad v_2' = \frac{Q(x) y_1(x)}{a_2(x) W[y_1, y_2]}, \tag{4.64}$$

giving the derivatives of $v_1(x)$ and $v_2(x)$. All we need do is integrate to get $v_1(x)$ and $v_2(x)$.

What we have done is to obtain (4.64) as the criterion for choosing the functions $v_1(x)$ and $v_2(x)$. Since $y_1(x)$ and $y_2(x)$ are known (because we can solve the homogeneous equation) the entire right-hand side of (4.64) is known and the solution for $v_1$ and $v_2$ may be written down directly as

$$v_1(x) = \int \frac{-Q(x)y_2(x)}{a_2(x)W[y_1,\, y_2]}\, dx, \qquad v_2(x) = \int \frac{Q(x)y_1(x)}{a_2(x)W[y_1,\, y_2]}\, dx. \qquad (4.65)$$

The particular solution then has the form (4.59)

$$y_p(x) = -y_1(x)\int \frac{Q(x)y_2(x)}{a_2(x)W[y_1,\, y_2]}\, dx + y_2(x)\int \frac{Q(x)y_1(x)}{a_2(x)W[y_1,\, y_2]}\, dx,$$

where the Wronskian, $W[y_1,\, y_2]$, can also be a function of $x$. We now work through an example in a series of three steps.

**EXAMPLE 4.17**   Find a particular solution of

$$y'' - 3y' + 2y = -\frac{e^{2x}}{e^x + 1}. \qquad (4.66)$$

**Step 1.**   *Solve the associated homogeneous differential equation.* The homogeneous equation

$$y'' - 3y' + 2y = 0 \qquad (4.67)$$

has the characteristic equation

$$r^2 - 3r + 2 = 0.$$

Since the roots of this characteristic equation are $r = 1$ and $r = 2$, the general solution of (4.67) is

$$C_1 e^x + C_2 e^{2x}.$$

**Step 2.**   *Use the appropriate form for $y_p$ to write down the system of equations for the unknowns $v_1'(x)$ and $v_2'(x)$* (4.63).
   The trial solution

$$y_p(x) = v_1(x)e^x + v_2(x)e^{2x}$$

must satisfy

$$e^x v_1' + e^{2x} v_2' = 0,$$

$$e^x v_1' + 2e^{2x} v_2' = -\frac{e^{2x}}{e^x + 1}.$$

**Step 3.**   *Solve for $v_1'(x)$ and $v_2'(x)$ and integrate to obtain $v_1(x)$ and $v_2(x)$.*
   Here we solve for $v_1'(x)$ in the first equation above as

$$v_1' = \frac{-e^{2x}v_2'}{e^x} = -e^x v_2'$$

and substitute the result into the second equation. This gives

$$e^x v_1' + 2e^{2x} v_2' = -e^x(e^x v_2') + 2e^{2x} v_2' = e^{2x} v_2' = \frac{-e^{2x}}{e^x + 1}$$

or

$$v_2' = \frac{-1}{e^x + 1} = \frac{-e^{-x}}{1 + e^{-x}}.$$

Integrating

$$v_1'(x) = \frac{e^x}{e^x + 1} \quad \text{and} \quad v_2'(x) = -\frac{e^{-x}}{1 + e^{-x}}$$

gives

$$v_1(x) = \int \frac{e^x}{e^x + 1} \, dx = \ln(e^x + 1)$$

and

$$v_2(x) = \int \frac{-e^{-x}}{1 + e^{-x}} \, dx = \ln(1 + e^{-x}).$$

Collecting our results gives the particular solution of (4.66) as

$$y_p(x) = e^x \ln(e^x + 1) + e^{2x} \ln(1 + e^{-x})$$

and the general solution as

$$y(x) = C_1 e^x + C_2 e^{2x} + e^x \ln(e^x + 1) + e^{2x} \ln(1 + e^{-x}).$$

Notice that since the functions $v_1(x)$ and $v_2(x)$ multiply $y_1(x)$ and $y_2(x)$, including arbitrary constants in the integrations for $v_1(x)$ and $v_2(x)$ will not change the form of the general solution.

To illustrate that it is not necessary for the differential equation to have constant coefficients in order for us to apply this method, we include the next example.

**EXAMPLE 4.18**    Find a particular solution and the general solution of

$$x^2 y'' + xy' - y = \frac{1}{x + 1}, \qquad x > 0. \tag{4.68}$$

**Step 1.**    Assuming a solution of the associated homogeneous equation

$$x^2 y'' + xy' - y = 0 \tag{4.69}$$

as $x^r$ gives

$$x^2(r)(r - 1)x^{r-2} + x(r)x^{r-1} - x^r = (r^2 - 1)x^r = 0.$$

Thus the general solution of the associated homogeneous equation is

$$C_1 x + C_2 x^{-1}.$$

**Step 2.**   The trial solution

$$y_p(x) = v_1(x)x + v_2(x)x^{-1}$$

must satisfy

$$xv_1'(x) + x^{-1}v_2'(x) = 0,$$

$$v_1'(x) - x^{-2}v_2'(x) = \frac{1}{x^2(x + 1)}. \qquad (4.70)$$

**Step 3.**   Since the Wronskian $W[x, x^{-1}] = -2x^{-1}$, the solution of (4.70) is

$$v_1'(x) = \frac{1}{2x^2(x + 1)}, \qquad v_2'(x) = \frac{-1}{2(x + 1)}.$$

Integration gives

$$v_1(x) = \frac{1}{2} \int \frac{1}{x^2(x + 1)} \, dx = \frac{1}{2} \int \left( \frac{-1}{x} + \frac{1}{x^2} + \frac{1}{x + 1} \right) dx$$

$$= \frac{1}{2}[-\ln x - x^{-1} + \ln(x + 1)],$$

(Recall partial fractions?)

$$v_2(x) = \frac{1}{2} \int \frac{-1}{x + 1} \, dx = -\frac{1}{2} \ln(x + 1).$$

Collecting results allows us to write the particular solution as

$$y_p(x) = \frac{x}{2}[-\ln x - x^{-1} + \ln(x + 1)] - \frac{x^{-1}}{2}[\ln(x + 1)]$$

$$= \frac{1}{2}[(x - x^{-1}) \ln(x + 1) - x \ln x - 1]$$

and the general solution as

$$y(x) = C_1 x + C_2 x^{-1} + y_p(x).$$

This procedure is formalized in the following theorem for $n$th order linear differential equations.

**Theorem 4.7.**   Let the functions $a_0(x), \ldots, a_n(x)$ and $Q(x)$ be continuous in some interval of the $x$-axis and require that $a_n(x) \neq 0$ in this interval. Let

$$y(x) = \sum_{i=1}^{n} C_i y_i(x) \qquad (4.71)$$

be the general solution to the homogeneous equation

$$(a_n D^n + a_{n-1} D^{n-1} + \cdots + a_1 D + a_0)y = 0. \qquad (4.72)$$

Then a particular solution $y_p(x)$ of the equation

$$(a_n D^n + a_{n-1} D^{n-1} + \cdots + a_1 D + a_0) y = Q(x) \qquad (4.73)$$

is given by

$$y_p(x) = \sum_{i=1}^{n} v_i(x) y_i(x), \qquad (4.74)$$

where the functions $v_i(x)$, $i = 1, 2, 3, \ldots, n$, are chosen to satisfy

$$
\begin{aligned}
y_1 v_1' \quad + \quad y_2 v_2' \quad + y_3 v_3' + \cdots + \quad y_n v_n' &= 0, \\
y_1' v_1' \quad + \quad y_2' v_2' \quad + y_3' v_3' + \cdots + \quad y_n' v_n' &= 0, \\
\vdots \qquad \qquad \vdots \qquad \quad \vdots \qquad \qquad \qquad \vdots \qquad & \quad \vdots \\
y_1^{(n-1)} v_1' + y_2^{(n-1)} v_2' + \quad \cdots \quad + y_n^{(n-1)} v_n' &= Q/a_n.
\end{aligned}
\qquad (4.75)
$$

The general solution of (4.73) is $\sum_{i=1}^{n} C_i y_i(x) + y_p(x)$.

The three-step procedure that follows gives you a guide in finding a particular solution using the method of variation of parameters.

**Step 1.**    Solve the homogeneous differential equation to obtain the $n$ linearly independent functions $y_i(x)$, $i = 1, 2, 3, \ldots, n$.

**Step 2.**    Form the system of equations (4.75).

**Step 3.**    Solve the system of step 2 for the $v_i'(x)$ and integrate to obtain the $v_i(x)$. [Note that since the determinant of the coefficient matrix in (4.75) is the Wronskian, $W[y_1, y_2, \ldots, y_n]$, of the linearly independent solutions of the homogeneous differential equation, it will never be zero and we are always guaranteed a solution of (4.75).]

Once the $v_i(x)$ are determined, the particular solution is given by

$$\sum_{i=1}^{n} v_i(x) y_i(x).$$

The two common situations where we will need to use the method of variation of parameters are:

1. The linear differential operator in (4.73) has constant coefficients but $Q(x)$ is *not* of the form

$$
\begin{aligned}
Q(x) = (p_m x^m + \cdots + p_0) e^{\alpha x} \cos \beta x \\
+ (q_m x^m + \cdots + q_0) e^{\alpha x} \sin \beta x,
\end{aligned}
$$

   so that the method of undetermined coefficients does not apply.
2. The linear differential operator does not have constant coefficients, as for example in the Cauchy-Euler equation.

The following example shows how you can use variation of parameters to find a particular solution after determining a second solution by the reduction-of-order method.

**EXAMPLE 4.19**    Find the general solution of

$$x^3 y'' + xy' - y = e^{1/x}, \qquad x > 0. \tag{4.76}$$

This differential equation does not have constant coefficients, nor is it of the form required for a Cauchy-Euler equation. However, we observe that $y = x$ satisfies the associated homogeneous differential equation

$$x^3 y'' + xy' - y = 0. \tag{4.77}$$

If we use the reduction-of-order method of Section 2.5, we introduce a new dependent variable, $u(x)$, by

$$y(x) = xu(x)$$

and calculate

$$y' = xu' + u,$$

$$y'' = xu'' + 2u'.$$

Substituting these values of $y$, $y'$, and $y''$ into (4.77) gives

$$x^3(xu'' + 2u') + x(xu' + u) - xu = x^4 u'' + (2x^3 + x^2)u' = 0.$$

If we let $w = u'$, we obtain the separable differential equation in $w$ and $x$,

$$x^4 w' = -(2x^3 + x^2)w,$$

or

$$\frac{w'}{w} = \frac{-(2x^3 + x^2)}{x^4} = -\frac{2}{x} - \frac{1}{x^2}.$$

Now we integrate to find

$$\ln|w| = -2\ln|x| + \frac{1}{x} + \ln|C_1|,$$

or $w = C_1 x^{-2} e^{1/x}$. Since $w = u'$, we must integrate once more to find $u(x)$ as

$$u = C_1 \int e^{1/x} x^{-2}\, dx = -C_1 e^{1/x} + C_2$$

and $y(x) = xu = -C_1 x e^{1/x} + C_2 x$. We now have two linearly independent solutions of the homogeneous equation, $x$ and $xe^{1/x}$.

The form of the particular solution to use for variation of parameters is

$$y_p(x) = xv_1(x) + xe^{1/x} v_2(x)$$

where (from 4.75) the functions $v_1'$ and $v_2'$ satisfy

$$xv_1' + xe^{1/x} v_2' = 0,$$

$$v_1' + \left(1 - \frac{1}{x}\right) e^{1/x} v_2' = x^{-3} e^{1/x}. \tag{4.78}$$

Since the Wronskian is

$$W[x, xe^{1/x}] = \begin{vmatrix} x & xe^{1/x} \\ 1 & \left(1 - \dfrac{1}{x}\right)e^{1/x} \end{vmatrix} = -e^{1/x},$$

by Cramer's rule we have

$$v_1' = \frac{1}{-e^{1/x}} \begin{vmatrix} 0 & xe^{1/x} \\ x^{-3}e^{1/x} & \left(1 - \dfrac{1}{x}\right)e^{1/x} \end{vmatrix} = \frac{-x^{-2}e^{2/x}}{-e^{1/x}} = x^{-2}e^{1/x}$$

and from the top equation in (4.78)

$$v_2' = -e^{-1/x}v_1'$$

$$= -x^{-2}.$$

Integration gives $v_1(x) = -e^{1/x}$ and $v_2(x) = x^{-1}$ and allows us to write the particular solution of (4.76) as

$$y_p(x) = x(-e^{1/x}) + xe^{1/x}(x^{-1}) = (1 - x)e^{1/x}$$

and the general solution as

$$y(x) = C_1x + C_2xe^{1/x} + (1 - x)e^{1/x}.$$

## EXERCISES

Find the general solution of the following differential equations.

1. $y'' + y = \tan x,$    $-\pi/2 < x < \pi/2.$
2. $y'' + 6y' + 9y = x^{-1}e^{-3x},$    $x \neq 0.$
3. $y'' - y' = \sec^2 x - \tan x,$    $-\pi/2 < x < \pi/2.$
4. $y'' + y = \sec x,$    $-\pi/2 < x < \pi/2.$
5. $y'' + 2y' + y = 2x^{-2}e^{-x}.$
6. $x^2y'' + 7xy' + 5y = 10 - 4x^{-1},$    $x > 0.$
7. $x^2y'' - xy' + y = x \ln x,$    $x > 0.$
8. $x^2y'' - 2y = \ln x,$    $x > 0.$
9. $x^2y'' + xy' + y = x^3,$    $x > 0.$
10. Show that the substitution $y_p(x) = y_1(x)v(x)$ in (4.53) gives the particular solution found in (4.56).
11. Given $(x^2 + x)y'' + (2 - x^2)y' - (2 + x)y = (x + 1)^2.$
    (a) Show that $e^x$ is a solution of the associated homogeneous differential equation.
    (b) Find a second solution of this homogeneous equation using the reduction-of-order technique.
    (c) Use variation of parameters to find a particular solution and the general solution of the original differential equation.

**12.** Show that the solution of the initial value problem,

$$y'' - y = \frac{1}{x}, \qquad y(1) = A, \qquad y'(1) = B,$$

can be expressed as

$$y = \frac{A+B}{2} e^{x-1} + \frac{A-B}{2} e^{1-x} + \frac{1}{2} e^{x} \int_{1}^{x} e^{-t} t^{-1} \, dt - \frac{1}{2} e^{-x} \int_{1}^{x} e^{t} t^{-1} \, dt.$$

## 4.6 APPLICATIONS

### 4.6.1 Simple Harmonic Motion

We will define simple harmonic motion to be motion in the presence of a restoring force which is proportional to displacement. That is, if $x$ denotes displacement from an equilibrium position, we have

$$F = -kx. \tag{4.79}$$

This is just Hooke's law for an elastic medium. By Newton's second law,

$$F = ma = mx'',$$

where the acceleration is given by $x'' = d^2x/dt^2$. Thus (4.79) may be written as

$$mx'' = -kx$$

or

$$mx'' + kx = 0. \tag{4.80}$$

Equation (4.80) has the characteristic equation

$$r^2 + \frac{k}{m} = 0,$$

with complex roots given by

$$r = \pm i \sqrt{\frac{k}{m}}.$$

Thus the general solution of (4.80) is

$$x(t) = C_1 \cos \sqrt{\frac{k}{m}} t + C_2 \sin \sqrt{\frac{k}{m}} t$$

$$= C_1 \cos \omega t + C_2 \sin \omega t, \tag{4.81}$$

where we have written $\omega = \sqrt{k/m}$. We suppose the initial displacement and the initial velocity are given as

$$x(0) = x_0, \qquad v(0) = x'(0) = v_0. \tag{4.82}$$

Finding the derivative of $x(t)$,

$$x'(t) = -\omega C_1 \sin \omega t + \omega C_2 \cos \omega t,$$

we can solve for the two arbitrary constants $C_1$ and $C_2$ as follows.
    At $t = 0$,

$$v_0 = \omega C_2.$$

Thus

$$C_2 = \frac{v_0}{\omega}$$

and

$$x(0) = x_0 = C_1.$$

This gives us the solution of (4.80), which satisfies the initial conditions (4.82) as

$$x(t) = x_0 \cos \omega t + \frac{v_0}{\omega} \sin \omega t. \qquad (4.83)$$

We could write the solution to (4.80) in a more compact form as

$$x(t) = A \cos(\omega t + \phi) \qquad (4.84)$$

and choose $A$ and $\phi$ so that the initial conditions of (4.82) are satisfied. We would have

$$A \cos \phi = x_0,$$

$$-A\omega \sin \phi = v_0.$$

Thus

$$A = \left(x_0^2 + \frac{v_0^2}{\omega^2}\right)^{1/2}$$

and

$$\phi = \tan^{-1}\frac{-v_0}{x_0 \omega}$$

give the amplitude, $A$, and phase angle, $\phi$, of the periodic function of (4.84). This function has a period of $2\pi/\omega$ and a frequency of $\omega/2\pi$. Notice that (4.83) may be put in the form (4.84) directly by multiplying (4.83) by $A/\sqrt{x_0^2 + v_0^2/\omega^2}$, defining $\phi$ by

$$\cos \phi = \frac{x_0}{\sqrt{x_0^2 + v_0^2/\omega^2}}, \qquad \sin \phi = \frac{-v_0/\omega}{\sqrt{x_0^2 + v_0^2/\omega^2}},$$

and using the trigonometric identity

$$\cos(\omega t + \phi) = \cos \omega t \cos \phi - \sin \omega t \sin \phi.$$

The choice of the form of solution (4.81) or (4.84) depends on the preference of the person solving the differential equation and the type of information desired from the solution.

## EXERCISES

Solve the following initial value problems.

**1.** $\dfrac{d^2x}{dt^2} + 16x = 0, \quad x(0) = 3, \quad x'(0) = 12.$

**2.** $9\dfrac{d^2x}{dt^2} + x = 0, \quad x(0) = -5, \quad x'(0) = \frac{1}{3}.$

**3.** $4\dfrac{dx^2}{dt^2} + 9x = 0, \quad x(0) = \pi, \quad x'(0) = 3.$

**4.** $\dfrac{d^2x}{dt^2} + 100x = 0, \quad x(0) = 0, \quad x'(0) = 0.20.$

The vibration of an object with mass $m$ at the end of a coil spring with spring constant $k$ is governed by (4.80). Here $x$ denotes the displacement from the equilibrium position and is a positive quantity when the spring is elongated. The units of $m$, $k$, $x$, and $t$ must be consistent.

**5.** Consider a spring-mass system that is governed by

$$16\frac{d^2x}{dt^2} + 4x = 0,$$

where motion is started by stretching the spring 2 units and releasing it from rest. Then the initial conditions are $x(0) = 2$, $x'(0) = 0$. What is the solution of this initial value problem?

**6.** Consider a spring-mass system governed by

$$\frac{d^2x}{dt^2} + 8x = 0,$$

where motion is started from the equilibrium position by giving the mass an initial velocity $v_0$. Thus the initial conditions are $x(0) = 0$, $x'(0) = v_0$. Find the function giving the resulting motion.

**7.** Solve (4.80) with $m = 64$, $k = 4$ if the spring-mass system is subject to an initial displacement of $-2$ and initial velocity of 3. What are the amplitude and period of the resulting motion?

**8.** Find the amplitude, period, and phase angle for motion described by

$$x'' + 9x = 0, \qquad x(0) = -2, \qquad x'(0) = -6.$$

**9.** Find the amplitude, period, and phase angle for motion described by

$$x'' + \pi^2 x = 0, \qquad x(0) = 1, \qquad x'(0) = \pi\sqrt{3}.$$

An object floating in a liquid is kept afloat when the upward force due to buoyancy exceeds the downward pull of gravity. If such an object is displaced slightly from its equilibrium position and released, the resulting motion is described by (4.80) when the ratio $k/m$ is replaced by the ratio of the gravitational constant to the displacement of the body at equilibrium. $x$ is positive in the downward direction.

**10.** An oil drum in the form of a right circular cylinder lies upright and half-submerged in a lake. If the cylinder is 4 feet long, and the gravitational constant is taken as 32 feet per second per second, the governing differential equation is

$$\frac{d^2x}{dt^2} + 16x = 0,$$

where $x$ is measured in feet and $t$ in seconds. Find the resulting motion if the cylinder is submersed so that 1 foot is above the water, and then released from rest.

**11.** If the initial conditions of exercise 10 are changed to give the cylinder an initial velocity of 2 ft per second in its equilibrium position, find the resulting solution.

The differential equation for a simple pendulum of length $\ell$ (see Figure 4.1) is

$$\frac{d^2y}{dt^2} + \frac{g}{\ell} \sin y = 0$$

where $g$ is the gravitational constant and the units of $y$ are radians. Recall that the Taylor series for $\sin y$ is

$$\sin y = y - \frac{y^3}{3!} + \frac{y^5}{5!} - \cdots.$$

For oscillations where $y$ is small, we approximate $\sin y$ by $y$ and use the differential equation

$$\frac{d^2y}{dt^2} + \frac{g}{\ell} y = 0$$

for the following exercises.

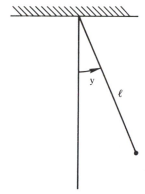

**Figure 4.1**

**12.** Find the solution of the linearized equation for the pendulum if the pendulum is released from rest at an angle of $\frac{1}{10}$ radian.

**13.** The difference between $y$ and $\sin y$ satisfies

$$\frac{y - \sin y}{y} < \frac{y^2}{3!}.$$

What is the maximum relative difference between $y$ and $\sin y$ for the motion in exercise 12?

**14.** Find the solution of the linearized equation if the pendulum is started from its equilibrium position with an initial velocity of 2 radians per second.

**15.** What is the maximum relative difference between $y$ and $\sin y$ for the motion in exercise 14?

If we multiply the differential equation (4.80) by $dx/dt$, we should observe that each term is an exact differential

$$m\left(\frac{dx}{dt}\right)\left(\frac{d^2x}{dt^2}\right) + kx \frac{dx}{dt} = 0.$$

Integration gives

$$\frac{1}{2} m \left(\frac{dx}{dt}\right)^2 + \frac{1}{2} kx^2 = E.$$

The first term on the left-hand side of this equation is the kinetic energy of the object while the second term represents the potential energy. For example, with the spring-mass system $\frac{1}{2}kx^2$ represents the energy stored in the spring. This equation is a statement of conservation of energy, $E$ = total energy.

**16.** Write the conservation of energy for the system in exercise 7 and evaluate the constant $E$. Show that the kinetic energy of the system is a maximum when the potential energy is a minimum.

### 4.6.2 Damped Motion

As a slightly more complicated example we add a frictional force, tending to damp the motion, which is proportional to the velocity. Newton's second law now gives

$$mx'' = -kx - cx'$$

or

$$mx'' + cx' + kx = 0, \tag{4.85}$$

whose characteristic equation is

$$r^2 + \frac{c}{m}r + \frac{k}{m} = 0. \tag{4.86}$$

Equation (4.86) may be solved easily to yield

$$r = \frac{1}{2}\left(-\frac{c}{m} \pm \sqrt{\frac{c^2}{m^2} - 4\frac{k}{m}}\right)$$

or

$$r = \frac{1}{2m}\left(-c \pm \sqrt{c^2 - 4mk}\right). \tag{4.87}$$

The nature of the roots of (4.86) is dependent on $c^2 - 4mk$ and there are only three possibilities to consider:

1. $c^2 - 4mk < 0$: two complex roots (conjugates of each other).
2. $c^2 - 4mk = 0$: one real root, of multiplicity 2.
3. $c^2 - 4mk > 0$: two real and distinct roots.

Consider case 1. If $c^2 - 4mk < 0$, we may write the roots as $r = \alpha \pm \beta i$ where

$$\alpha = -\frac{c}{2m} \quad \text{and} \quad \beta = \frac{1}{2m}\sqrt{4mk - c^2}.$$

Thus, as usual for homogeneous equations with constant coefficients,

$$x(t) = C_1 e^{\alpha t} \cos \beta t + C_2 e^{\alpha t} \sin \beta t. \tag{4.88}$$

The general initial conditions

$$x(0) = x_0, \qquad x'(0) = v_0,$$

yield

$$C_1 = x_0,$$

and since

$$x'(t) = C_1(\alpha e^{\alpha t} \cos \beta t - \beta e^{\alpha t} \sin \beta t) + C_2(\alpha e^{\alpha t} \sin \beta t + \beta e^{\alpha t} \cos \beta t),$$

we have that

$$v_0 = \alpha C_1 + \beta C_2.$$

Hence

$$v_0 = \alpha x_0 + \beta C_2,$$

and thus

$$C_2 = \frac{1}{\beta}(v_0 - \alpha x_0).$$

This allows the general solution to be written as

$$x(t) = e^{\alpha t}[x_0 \cos \beta t + \beta^{-1}(v_0 - \alpha x_0) \sin \beta t].$$

In the special case $v_0 = 0$,

$$C_2 = -\frac{\alpha}{\beta} x_0$$

and

$$x(t) = x_0 e^{\alpha t}\left( \cos \beta t - \frac{\alpha}{\beta} \sin \beta t \right).$$

For reasons mentioned earlier, we could take the form of solution of (4.85) as

$$x(t) = A e^{\alpha t} \cos(\beta t + \phi), \tag{4.89}$$

where $A$ and $\phi$ are the arbitrary constants. Since $|\cos(\beta t + \phi)| \leq 1$, the graph of $x(t)$, given by (4.89), is bounded by those of $A e^{\alpha t}$ and $-A e^{\alpha t}$. This situation is shown in Figure 4.2 for $A > 0$ and $\phi = 0$.

It is clear from the form of solution that for $\alpha < 0$ ($c/m > 0$),

$$\lim_{t \to \infty} x(t) = 0,$$

and we have damped oscillatory motion. For this reason this case with $c^2 - 4mk < 0$ is often called the *underdamped case*.

Since times between two successive maxima, or two successive minima, are all equal and given by $2\pi/\beta$, $2\pi/\beta$ is often called the *quasi-period* of the motion. Notice that if the expression for $\beta$,

$$\beta = \frac{1}{2m}\sqrt{4mk - c^2},$$

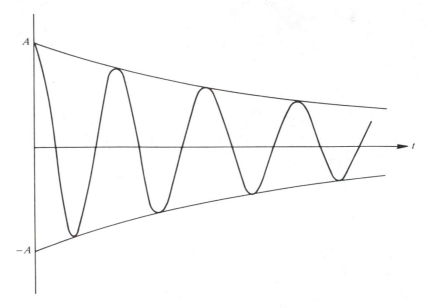

**Figure 4.2**   Graph of $x = Ae^{\alpha t} \cos \beta t$, $\alpha < 0$.

is expanded in a Taylor series,

$$\beta = \frac{1}{2m}\left[4mk\left(1 - \frac{c^2}{4mk}\right)\right]^{1/2}$$

$$= \sqrt{\frac{k}{m}}\left(1 - \frac{c^2}{4mk}\right)^{1/2} = \sqrt{\frac{k}{m}}\left(1 - \frac{c^2}{8mk} + \cdots\right),$$

the difference between $\beta$ and $\sqrt{k/m}$ depends on the comparison of 1 and $c^2/(8mk)$. If the damping factor $c$ is such that $c^2/(8mk) \ll 1$, the quasi-period, $2\pi/\beta$, is very close to the undamped period, $2\pi/\sqrt{k/m}$.

This shows why researchers often ignore friction (let $c = 0$) when they need to find the frequency or period of systems containing small frictional damping.

For case 3, $c^2 - 4mk > 0$ and the general solution of (4.85) can be written as

$$x(t) = Ae^{r_1 t} + Be^{r_2 t}, \tag{4.90}$$

where $r_1$ and $r_2$ are given by

$$\frac{-c \pm \sqrt{c^2 - 4mk}}{2m}.$$

For physical problems where $m$, $c$, and $k$ are all positive, $\sqrt{c^2 - 4mk} < c$ and both $r_1$ and $r_2$ are negative numbers. Since both $e^{r_1 t}$ and $e^{r_2 t}$ will be decreasing monotonic functions, no oscillations can occur and this case is called the *overdamped case*. Typical graphs of the function defined by (4.90) are shown in Figure 4.3. With these curves, the slope for $t = 0$ can change, and any specific curve can have at most one horizontal tangent (see exercise 4).

Case 2, $c^2 - 4mk = 0$, is called the *critically damped case*.

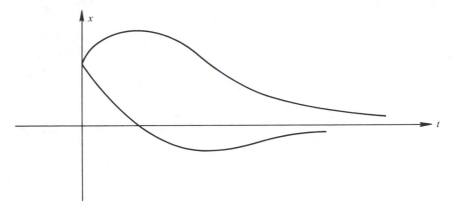

**Figure 4.3**   Overdamped solutions.

Here solutions of (4.85) are given by

$$x(t) = (At + B)e^{rt}, \qquad r = \frac{-c}{2m}. \tag{4.91}$$

Since $r < 0$, $\lim_{t \to \infty} x(t) = 0$ shows that all solutions again decay to zero as time increases.

One way to compare the overdamped, underdamped, and critically damped cases is to consider $m$ and $k$ as fixed and plot solutions of (4.85) subject to initial conditions

$$x(0) = x_0 > 0,$$
$$x'(0) = v_0 > 0,$$

for various values of $c$, as shown in Figure 4.4.

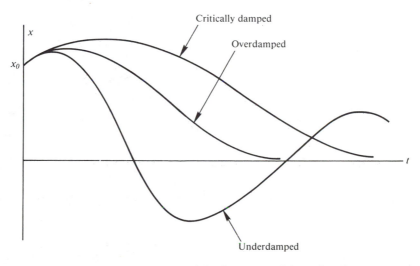

**Figure 4.4**   Comparison of the three cases of damped motion.

## EXERCISES

1. For what values of $c$ are the motions governed by

$$4x'' + cx' + 9x = 0$$

   (a) Overdamped?
   (b) Underdamped?
   (c) Critically damped?

2. (a) Show that for the underdamped case

$$f(t) = \frac{e^{r_2 t} - e^{r_1 t}}{r_2 - r_1}$$

   satisfies (4.85) if $r_1$ and $r_2$ are given by (4.87).

   (b) Set $r_2 = r_1 + \epsilon$ in the formula for $f(t)$ in part (a). Show that in the limit as $\epsilon \to 0$,

$$\lim_{\epsilon \to 0} f(t) = te^{r_1 t}.$$

   (Note that you can use L'Hôpital's rule.)

3. Show that if $m > 0$ and $c > 0$, the general solution of (4.85) approaches 0 as $t \to \infty$ for all nonnegative values of $k$.

4. Show that the solution of (4.85) for the overdamped case with $x(0) = x_0$, $x'(0) = v_0$ may have at *most* one horizontal tangent.

5. Repeat exercise 4 for the critically damped case.

The differential equation (4.85) may be used to model the vibration of an object at the end of a vertical spring, where the term $cx'$ is added to represent the damping effects of air resistance to motion and other retarding forces to the moving object. Solve $mx'' + cx' + kx = 0$ for the following situations.

6. $m = 1$, $c = \frac{1}{8}$, $k = 1$, $x(0) = 0$, $x'(0) = \frac{1}{2}$.
7. $m = 1$, $c = 8$, $k = 16$, $x(0) = 0$, $x'(0) = -3$.
8. $m = 64$, $c = 16$, $k = 17$, $x(0) = 1$, $x'(0) = 0$.
9. $m = 9$, $c = 6$, $k = 37$, $x(0) = 1$, $x'(0) = 1$.
10. If the total energy of the spring-mass system of the exercises in Section 4.6.1 is defined as

$$E = \frac{1}{2}m\left(\frac{dx}{dt}\right)^2 + \frac{1}{2}kx^2,$$

   show that $E$ satisfies

$$\frac{dE}{dt} = -c\left(\frac{dx}{dt}\right)^2$$

   and is therefore a decreasing function of $t$.

The differential equation describing oscillations of a simple pendulum including a damping term is

$$\frac{d^2 y}{dt^2} + 2\alpha\frac{dy}{dt} + \frac{g}{\ell}y = 0,$$

where $g$ is the gravitational constant, $\ell$ the length of the pendulum, and $\alpha$ a constant.

11. Consider a simple pendulum with $g/\ell = 4$.
    (a) For what values of $\alpha$ will the resulting motion of a pendulum with $y(0) = 0.5$, $y'(0) = 0$ be underdamped?
    (b) For what values of $\alpha$ will the pendulum of part (a) always stay on the same side of the vertical line on which it was released?

12. If $\alpha = \frac{1}{10}$, $g/\ell = 4.01$, and the pendulum is set in motion with initial conditions $y(0) = 0$, $y'(0) = 1$, find the maximum value of $|y|$ for the resulting motion.

13. If the initial conditions of exercise 12 are changed to $y(0) = 0$, $y'(0) = \beta$, for what value of $\beta$ will the maximum value of $|y|$ be 0.2 radian?

### 4.6.3 Forced Motion

We now suppose some external force, $F(t)$, acts on the mass to "force" the motion. $F(t)$ is called a *forcing function*. We then obtain an equation of motion of the form

$$mx'' + cx' + kx = F(t). \tag{4.92}$$

Equation (4.92) is a second order, linear, nonhomogeneous ordinary differential equation with constant coefficients. The complete solution to (4.92) will, as we know, consist of the sum of the general solution to the homogeneous equation and any particular solution to (4.92). We already have the solution to the homogeneous equation. To find a particular solution we must make a good guess or use one of our standard techniques, undetermined coefficients or variation of parameters, depending on the form of $F(t)$. In the case where $c = 0$, there is no damping and we have

$$mx'' + kx = F(t). \tag{4.93}$$

Suppose we have a periodic forcing function, say

$$F(t) = A \sin \gamma t.$$

To use our methods for the resulting differential equation,

$$mx'' + kx = A \sin \gamma t, \tag{4.94}$$

we must consider first the homogeneous equation

$$mx'' + kx = 0.$$

This equation has a general solution given by

$$x(t) = C_1 \cos \omega t + C_2 \sin \omega t$$

with $\omega = \sqrt{k/m}$. If $\gamma \neq \omega$, the methods of Section 4.3 tell us to try a particular solution of the form

$$x_p(t) = k_0 \cos \gamma t + \ell_0 \sin \gamma t.$$

Substituting this form of $x_p(t)$ into (4.94) requires that

$$k_0 = 0, \qquad \ell_0 = \frac{A}{m(-\gamma^2 + k/m)} = \frac{A}{m(-\gamma^2 + \omega^2)},$$

so the general solution of (4.49) is (for $\omega^2 \neq \gamma^2$)

$$x(t) = C_1 \cos \omega t + C_2 \sin \omega t + \frac{A}{m(\omega^2 - \gamma^2)} \sin \gamma t. \qquad (4.95)$$

Notice that the particular solution of the differential equation is a periodic function with fundamental period $2\pi/\gamma$; hence it repeats itself at multiples of this period, $2\pi n/\gamma$, $n = 1, 2, 3, \ldots$. The rest of the solution is periodic with fundamental period $2\pi/\omega$ and repeats itself at multiples of this period, $2\pi m/\omega$, $m = 1, 2, 3, \ldots$. Thus the resulting motion will be a combination of these periodic motions, and if $\gamma/\omega$ is a rational number will in turn be a periodic function with period equal to $2\pi n/\gamma$ with $n$ the smallest integer for which

$$\frac{2\pi n}{\gamma} = \frac{2\pi m}{\omega} \qquad (m \text{ also an integer})$$

(see exercises 1 and 8).

If $\omega = \gamma$, this solution is not valid. The particular solution in this case must have the form

$$x_p(t) = t(k_0 \cos \omega t + \ell_0 \sin \omega t).$$

Differentiation gives

$$x_p'(t) = (k_0 + \omega \ell_0 t)\cos \omega t + (-\omega k_0 t + \ell_0)\sin \omega t,$$

$$x_p''(t) = (\omega \ell_0 - \omega^2 k_0 t + \omega \ell_0)\cos \omega t$$
$$+ (-\omega k_0 - \omega^2 \ell_0 t - \omega k_0) \sin \omega t,$$

while substitution into (4.94) gives

$$[m(2\omega \ell_0 - \omega^2 k_0 t) + k k_0 t]\cos \omega t + [m(-2\omega k_0 - \omega^2 \ell_0 t) + k \ell_0 t]\sin \omega t$$

$$= A \sin \omega t.$$

Since $\cos \omega t$, $\sin \omega t$, $t \cos \omega t$, and $t \sin \omega t$ are linearly independent, we can equate coefficients in like terms above to obtain

$$2m\omega \ell_0 = 0,$$

$$(-m\omega^2 + k)k_0 = 0,$$

$$-2m\omega k_0 = A,$$

$$(-m\omega^2 + k)\ell_0 = 0.$$

Since $-m\omega^2 + k = 0$, these linear equations have the solution

$$\ell_0 = 0,$$

$$k_0 = \frac{-A}{2m\omega},$$

so the general solution of (4.94) is

$$x(t) = C_1 \cos \omega t + C_2 \sin \omega t - \frac{A}{2m\omega}t \cos \omega t. \qquad (4.96)$$

This situation, where the forcing function has the same frequency as the natural frequency of the system, is called *resonance*. Notice that the solution given by (4.96) is unbounded as $t \to \infty$. In a physical system this situation corresponds to having a displacement which grows until part of the system breaks.

## EXERCISES

1. Consider the spring-mass system governed by

$$\frac{d^2x}{dt^2} + 16x = 15 \sin t, \qquad x(0) = 0, \quad x'(0) = \sqrt{3}.$$

   (a) Find $x$ as a function of time.
   (b) What is the period of the solution in part (a)?
   (c) Check to see if the period from part (b) agrees with that from page 132.

2. Consider the spring-mass system governed by

$$\frac{d^2x}{dt^2} + 2 \frac{dx}{dt} + 17x = 2 \sin \gamma t.$$

   (a) Find $x$ as a function of time.
   (b) Identify the steady-state and transient parts of the solution in part (a).

3. Consider the initial value problem

$$\frac{d^2x}{dt^2} + 2\alpha \frac{dx}{dt} + (\omega^2 + \alpha^2)x = \sin \gamma t, \qquad x(0) = 0, \quad x'(0) = 0.$$

   (a) Find $x$ as a function of time.
   (b) Identify the steady-state and transient parts of the solution in part (a).
   (c) Determine the amplitude of the steady-state portion of the solution and considering $\alpha$ and $\omega$ fixed, find the value of $\gamma$ which maximizes this amplitude. (*Hint:* Use calculus.)
   (d) Find the value of $\gamma$ in exercise 2 that maximizes the amplitude of the steady-state solution. (When the frequency of the periodic forcing function is chosen to maximize the amplitude of the steady-state response, this damped system is often said to be in resonance.)

4. Find the value of $\gamma$ so that the system governed by

$$\frac{d^2x}{dt^2} + 6 \frac{dx}{dt} + 22x = \cos \gamma t$$

   is in resonance.

Kirchhoff's law may be used to show that the electrical charge on a capacitor, $x(t)$, is governed by the differential equation

$$L \frac{d^2x}{dt^2} + R \frac{dx}{dt} + \frac{1}{C}x = E(t).$$

The constants $L$, $R$, and $C$ are inductance, resistance, and capacitance, respectively, and $E(t)$ is the applied voltage (see Figure 4.5).

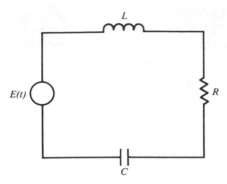

**Figure 4.5**

5. Find $x(t)$ for the electrical circuit having $L = 16$, $R = 8$, $C = \frac{1}{10}$, $E(t) = E_0 \sin \gamma t$, $x(0) = 0$, $x'(0) = 0$.

6. What value of $\gamma$ will maximize the amplitude of the steady-state response in exercise 5? (In practice, a circuit is "tuned" by varying $C$ until the steady-state amplitude is maximized for a fixed value of $\gamma$.)

7. Find $x(t)$ for the electrical circuit described by $L = 1$, $R = 2$, $C = \frac{1}{10}$, $E(t) = 20$ if $x(0) = 2$, $x'(0) = 6$. What is the maximum value of the charge on the capacitor?

8. Consider a system described by

$$\frac{d^2x}{dt^2} + 100x = 99 \cos 9t, \qquad x(0) = 0, \quad x'(0) = 0.$$

(a) Find $x(t)$.

(b) What is the period of the solution in part (a)?

(c) Use the trigonometric identity

$$\cos 2\alpha - \cos 2\beta = -2 \sin(\alpha - \beta) \sin(\alpha + \beta)$$

to write the solution in part (a) as the product of two sine functions. Note that this form of the solution suggests a variable amplitude $[A \sin(19t/2)]$ for $\sin(t/2)$. A graph of such behavior is shown in Figure 4.6. This situation, when the forcing frequency is close to the natural frequency of the system, describes the phenomenon of beats. A typical example of this is when two tuning forks of nearly equal frequencies are struck simultaneously.

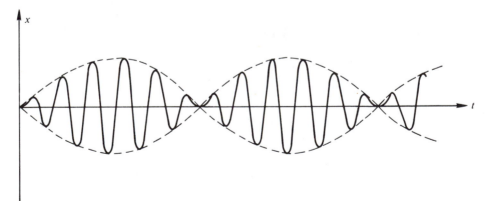

**Figure 4.6**

### 4.6.4 Boundary Value Problems

When we find the solution of an initial value problem, we first find the general solution and then determine the arbitrary constants from the initial conditions. We use this same approach in solving boundary value problems, where the dependent variable, or one or more of its derivatives, is specified at two values of the independent variable. For example, to find a solution of

$$y'' + 3y' - 4y = 0,$$

$$y(0) = 0,$$

$$y'(1) = 2,$$

we would first find the general solution as

$$y(x) = C_1 e^{-4x} + C_2 e^x.$$

The boundary conditions will be satisfied if we take

$$C_1 + C_2 = 0, \qquad -4C_1 e^{-4} + C_2 e = 2,$$

that is,

$$C_1 = -C_2 = \frac{-2}{e + 4e^{-4}}$$

and the desired solution is

$$y(x) = \frac{-2}{e + 4e^{-4}} (e^{-4x} - e^x).$$

We now consider a similar problem:

$$y'' + 4y = 0,$$

$$y(0) = 0,$$

$$y(b) = 2.$$

The general solution in this case is

$$y(x) = C_1 \cos 2x + C_2 \sin 2x.$$

If we choose $C_1 = 0$, we can satisfy $y(0) = 0$, but then we must determine $C_2$ from the condition

$$y(b) = C_2 \sin 2b = 2.$$

If $\sin 2b \neq 0$, we may determine a value for $C_2$ such that the second boundary condition is fulfilled. However, if $\sin 2b = 0$, we cannot solve for $C_2$ and this boundary value problem has no solution.

Notice that if we changed the second boundary condition to

$$y\left(\frac{\pi}{2}\right) = 0,$$

we see that

$$y(x) = C_2 \sin 2x$$

is a solution for *any* choice of the constant $C_2$. Thus boundary value problems may have a unique solution, no solutions, or an infinite number of solutions. Lest we worry that we are doing something incorrect, we should remind ourselves that the theorem that guarantees a solution applies only to initial value problems, not to boundary value problems. (We have included this section to emphasize that boundary value problems for ordinary differential equations are very different from initial value problems.)

*Eigenvalues and eigenfunctions.*    One place where boundary value problems occur is in mathematical models of vibrating systems. Such a typical boundary value problem is

$$y'' + \lambda y = 0,$$
$$y(0) = 0, \tag{4.97}$$
$$y(4) = 0.$$

Here $\lambda$ is an unknown constant and the question to be answered is: For what values of $\lambda$ may we solve this boundary value problem, and what are the corresponding solutions for $y$? (Note that $y = 0$ is a solution to this problem. What we are seeking, then, are nonzero solutions.) We use the word "eigenvalue" for values of $\lambda$ which allow solutions of our boundary value problem and call the associated nonzero solutions for $y$ "eigenfunctions."

To determine the eigenvalues in (4.97), we consider the three possibilities for the constant $\lambda$: positive, negative, or zero.

*1. $\lambda > 0$.*    Here the general solution of the differential equation is

$$y = C_1 \sin \sqrt{\lambda}\, x + C_2 \cos\sqrt{\lambda}\, x$$

and we find that having $C_2 = 0$ satisfies the boundary condition at $x = 0$.
The second boundary condition becomes

$$C_1 \sin 4\sqrt{\lambda} = 0.$$

This means that $C_1 = 0$ or $\sin 4\sqrt{\lambda} = 0$ and $C_1$ is arbitrary. Choosing $C_1 = 0$ gives the trivial solution $y = 0$, so since we are seeking nonzero solutions, we choose $\lambda$ such that

$$\sin 4\sqrt{\lambda} = 0.$$

Since $\sin n\pi = 0$ for $n$ an integer, our eigenvalues are determined from the condition

$$4\sqrt{\lambda} = n\pi \quad \text{or} \quad \lambda = \frac{n^2\pi^2}{16}, \quad n = 1, 2, 3, \ldots$$

and our resulting eigenfunctions are

$$y_n = \sin\frac{n\pi x}{4}, \quad n = 1, 2, 3, \ldots.$$

Note that a constant times an eigenfunction is still an eigenfunction. Note also that we have an infinite number of eigenvalues and an infinite number of eigenfunctions.

**2. $\lambda < 0$.**   Here it is convenient to let $\lambda = -\mu^2$ in (4.97) and write its general solution as

$$y = C_1 \sinh \mu x + C_2 \cosh \mu x.$$

Setting $C_2 = 0$ satisfies the boundary condition at $x = 0$ and the second boundary condition requires

$$C_1 \sinh \mu 4 = 0.$$

Since $\sinh x$ is zero only for $x = 0$ (and $\mu^2 \neq 0$), this is satisfied only for $C_1 = 0$ and we cannot obtain any eigenvalues for this case.

**3. $\lambda = 0$.**   In this case the general solution is

$$y = C_1 + C_2 x$$

and the two boundary conditions require

$$y(0) = C_1 = 0,$$

$$y(4) = C_1 + 4C_2 = 0.$$

Thus $C_1 = C_2 = 0$ is the only solution and we see that all our eigenvalues come from the case $\lambda > 0$.

As an example of a boundary value problem, consider the buckling of a long shaft, or column, under an axial load. The governing equation (F. Hildebrand, *Advanced Calculus for Applications*, 2nd ed., Prentice-Hall, Englewood Cliffs, N.J., 1976) is

$$\frac{d^2}{dx^2}\left(EI\frac{d^2y}{dx^2}\right) + P\frac{d^2y}{dx^2} = 0,$$

where $P$ is the axial compressive force and is assumed to be constant and $EI$ is the flexural rigidity and is also constant. Dividing the equation above by $EI$, we have

$$\frac{d^4y}{dx^4} + \frac{P}{EI}\frac{d^2y}{dx^2} = 0$$

or, with $\lambda = P/(EI) > 0$,

$$\frac{d^4y}{dx^4} + \lambda\frac{d^2y}{dx^2} = 0. \qquad (4.98)$$

Now, (4.98) is a fourth order, linear, homogeneous differential equation with constant coefficients. It has the characteristic equation

$$r^4 + \lambda r^2 = 0,$$

which has 0 as a double root and $\pm i\sqrt{\lambda}$ as complex roots. Thus the general solution of (4.98) is

$$y(x) = C_1 + C_2 x + C_3 \cos(\sqrt{\lambda}x) + C_4 \sin(\sqrt{\lambda}x). \qquad (4.99)$$

***Boundary conditions.***   If the column is hinged on both ends, both the deflection of the beam and its moment must vanish at either end of the beam. This gives boundary conditions

$$y(0) = y''(0) = 0,$$
$$y(\ell) = y''(\ell) = 0. \tag{4.100}$$

If we apply the first of the conditions in (4.100) to our solution in (4.99), we find that

$$C_1 + C_3 = 0,$$
$$C_3 = 0.$$

This means that $C_1 = C_3 = 0$ and our solution at this point is

$$y(x) = C_2 x + C_4 \sin(\sqrt{\lambda} x).$$

Applying the boundary conditions at the other end gives

$$0 = C_2 \ell + C_4 \sin(\sqrt{\lambda} \ell),$$
$$0 = C_4 \lambda \sin(\sqrt{\lambda} \ell). \tag{4.101}$$

Since $\lambda \neq 0$, the second equation in (4.101) implies that either $C_4 = 0$ or

$$\sin(\sqrt{\lambda} \ell) = 0. \tag{4.102}$$

If $C_4 = 0$, the first equation in (4.101) requires that $C_2 = 0$. This results in having $y(x) = 0$, which is the trivial solution. However, we can obtain a nontrivial solution since $\sin \theta = 0$ for $\theta = n\pi$, where $n$ is an integer. This will hold when the value of $\lambda$ is such that

$$\ell\sqrt{\lambda} = n\pi, \qquad n = 1, 2, \ldots .$$

Thus the eigenvalues for this problem are given by

$$\lambda_n = \frac{n^2 \pi^2}{\ell^2}, \qquad n = 1, 2, \ldots \tag{4.103}$$

Since the length of the column, $\ell$, and its characteristics, given by $E$ and $I$, are fixed, and $\lambda = P/(EI)$, this implies that there are an infinite number of loads, $P_n$, which satisfy all the conditions of the problem. These discrete values,

$$P_n = \frac{n^2 \pi^2}{\ell^2} EI, \qquad n = 1, 2, \ldots ,$$

are called the *critical buckling loads*. The smallest of these is $P_1$, given by

$$P_1 = \frac{\pi^2}{\ell^2} EI. \tag{4.104}$$

[For axial loads smaller than $P_1$ the column is stable only in the unbent position; that is, the column will remain straight after the load is applied. The critical value $P_1$, given by (4.104), is also called the *Euler load* and provides an upper limit of stability for the unbent column.]

Thus a nontrivial solution to this problem exists only if $C_2 = 0$ and the $\lambda_n$ are given by (4.103). In this case the eigenfunctions, or *deflection modes*, of the column are given by

$$y_n(x) = \sin \frac{n\pi x}{\ell}, \quad n = 1, 2, \ldots,$$

and the solutions have the form

$$y(x) = C_4 \sin \frac{n\pi x}{\ell}. \tag{4.105}$$

For $n = 1$, (4.105) gives

$$y(x) = C \sin \frac{\pi x}{\ell},$$

which is known as the *fundamental buckling mode*.

Additional boundary value problems appear in Chapter 6.

*Orthogonal functions.*    The eigenfunctions associated with the solution of (4.97) were

$$y_n(x) = \sin \frac{n\pi x}{4}, \qquad n = 1, 2, 3, \ldots . \tag{4.106}$$

Note that these functions have the property

$$\int_0^4 y_n(x) y_m(x)\, dx = \int_0^4 \sin \frac{n\pi x}{4} \sin \frac{m\pi x}{4}\, dx$$

$$= \frac{1}{2} \int_0^4 \left[ \cos \frac{(n-m)\pi x}{4} - \cos \frac{(n+m)\pi x}{4} \right] dx$$

$$= \frac{1}{2} \left[ \frac{\sin(n-m)\pi x/4}{(n-m)\pi/4} - \frac{\sin(n+m)\pi x/4}{(n+m)\pi/4} \right] \Bigg|_0^4 = 0, \qquad n \neq m.$$

This property will be of tremendous importance in later sections and we now give some definitions needed at that time.

A set of functions $\{y_i(x),\ i = 1, 2, 3, \ldots\}$ is said to be *orthogonal with respect to the weight function* $p(x)$ *on the interval* $[a, b]$ if

$$\int_a^b y_n(x) y_m(x) p(x)\, dx = 0, \qquad n \neq m. \tag{4.107}$$

The *norm* of such a set of orthogonal functions is defined by

$$\| y_i(x) \| = \left( \int_a^b [y_i(x)]^2 p(x)\, dx \right)^{1/2}, \qquad i = 1, 2, 3, \ldots . \tag{4.108}$$

If $p(x) = 1$ in (4.107) and (4.108), we simply ignore the phrase about the weight function. Thus for the set of functions from (4.106), we say that $\{\sin(i\pi x/4),\ i = 1, 2, 3, \ldots\}$ is orthogonal on the interval $[0, 4]$ with a norm given by

$$\left\| \sin \frac{i\pi x}{4} \right\| = \sqrt{2}, \qquad i = 1, 2, 3, \ldots .$$

[Note the integral $\int_0^4 \sin^2(i\pi x/4)\, dx = 2$.] The second set of orthogonal functions in the section is found in the eigenfunctions given by (4.105). (Use of the same trig-onometric identity for the product of two sine functions needed above plus a straight-forward integration will show this.) The norm of this set of functions is

$$\left\| \sin \frac{n\pi x}{\ell} \right\| = \sqrt{\frac{\ell}{2}}, \qquad n = 1, 2, 3, \ldots .$$

Orthogonal functions will be seen again in Chapters 6, 10, and 11.

## EXERCISES

Solve the following problems.

**1.** $y'' + 100y = 0$, $y(0) = 4$, $y'(0) = 500$.

**2.** $y'' + 10y' + 26y = 0$, $y(0) = 1$, $y'(0) = 10$.

**3.** $y'' + 3y' + 14y = 0$, $y(0) = 0$, $y'(0) = 5$.

**4.** For what value of $b$ will the boundary value problem

$$y'' + 16y = 0, \qquad y(0) = 0, \qquad y'(b) = 1$$

   **(a)** Have no solution?
   **(b)** Have exactly one solution?
   **(c)** Have an infinite number of solutions?

**5.** For what value of $b$ will the boundary value problem

$$y'' + 16y = 0, \qquad y(0) = 0, \qquad y'(b) = 0$$

   **(a)** Have no solution?
   **(b)** Have exactly one solution?
   **(c)** Have an infinite number of solutions?

**6.** Show that if $f_1(x)$ and $f_2(x)$ satisfy $Ly = 0$, where $L$ is a linear operator, and $f_1(0) = A_1$, $f_1(6) = B_1$, $f_2(0) = A_2$, and $f_2(6) = B_2$, then $\alpha f_1 + \beta f_2$ satisfies the boundary value problem $Ly = 0$,

$$y(0) = \alpha A_1 + \beta A_2, \qquad y(6) = \alpha B_1 + \beta B_2.$$

Find the eigenvalues and corresponding eigenfunctions for the following boundary value problems.

**7.** $y'' + \lambda^2 y = 0$, $y(0) = y(b) = 0$.

**8.** $y'' + \lambda^2 y = 0$, $y'(0) = y(b) = 0$.

**9.** $y'' + \lambda^2 y = 0$, $y(0) = y'(b) = 0$.

**10.** $y'' + \lambda^2 y = 0$, $y'(0) = y'(b) = 0$.

**11.** Ignoring damping, a spring-mass system is described by

$$\frac{d^2 x}{dt^2} + \lambda^2 x = 0, \qquad \lambda^2 = \frac{k}{m}.$$

Find values of the ratio $k/m$ such that any motion that starts at its equilibrium position is at the same position at $t = 3$. [That is, if $x(0) = 0$, $x(3) = 0$ also.]

**12.** Show that the eigenfunctions from exercise 8 form an orthogonal set on $[0, b]$, and find the norm of this set of functions.

13. Show that the eigenfunctions from exercise 9 form an orthogonal set on $[0, b]$, and find the norm of this set of functions.

14. (a) Find an equation that $\lambda$ must satisfy in order that the boundary value problem

$$y'' + \lambda^2 y = 0, \qquad y(0) = 0, \qquad y'(3) + 2y(3) = 0$$

   will have a solution.
   (b) Show that there are an infinite number of eigenvalues. If these eigenvalues are labeled $\lambda_1, \lambda_2, \lambda_3, \ldots,$ what are the associated eigenfunctions?
   (c) Show that this set of eigenfunctions is orthogonal and compute their norm.

## REVIEW EXERCISES

1. For what values of the constant $r$ will the following sets of functions be linearly independent?
   (a) $\{\sin x, \sin rx\}$
   (b) $\{x^2, x^r, x^4\}$
   (c) $\{e^x, e^{2x}, e^{rx}\}$
   (d) $\{1, x, x + r(\sin x)\}$

Find the general solution of the following differential equations.

2. $y'' - 7y' + 6y = 0.$
3. $y'' + 81y = 0.$
4. $x^2 y'' - 2xy' - 4y = 0.$
5. $y'' + 81y = 2e^x + \cos 3x.$
6. $x^2 y'' - 3xy' + 3y = 4x^4.$
7. $(D^2 - 22D + 121)y = 0.$
8. $(D^2 + 4D + 125)y = 0.$
9. $(D^2 - 7D + 6)y = 3e^x.$
10. $(D^4 - 2D^2 - 15)y = 0.$
11. $y'' + 4y' + 13y = \sin x.$
12. $(x^2 D^2 - xD + 1)y = 3 \ln x.$
13. $(D^3 - 64)y = 0.$
14. $(D^4 - k^4)y = 0, \quad k$ is a constant.
15. $(D^3 + 64)y = e^{-4x} + 128 + 64x^2.$
16. $y'' + 9y = 18 \csc 3x.$

Solve the following initial value problems.

17. $y'' + 4y = 0, \quad y(0) = -7, \quad y'(0) = 4.$
18. $(D^2 - D - 5)y = 0, \quad y(0) = 0, \quad y'(0) = 3.$
19. $y'' + 4y = 4 \sin^2 x, \quad y(0) = \frac{1}{2}, \quad y'(0) = 12.$
20. $(D^3 + 3D^2 + 3D + 1)y = x, \quad y(0) = -3, \quad y'(0) = 0, \quad y''(0) = 6.$
21. $(D^4 + 8D^2 + 16)y = 18 \sin x, \quad y(0) = 0, \quad y'(0) = 2, \quad y''(0) = 0, \quad y'''(0) = 14.$
22. Consider the spring–mass system that is governed by

$$\frac{d^2 x}{dt^2} + 4x = 3 \sin t, \qquad x(0) = 0, \quad x'(0) = 7.$$

(a) Find $x$ as a function of $t$.

(b) What is the period of the function in part (a)?

(c) Check to see if the period from part (b) agrees with that from page 132.

23. Consider the system governed by the differential equation

$$\frac{d^2x}{dt^2} + 2\frac{dx}{dt} + 10x = \sin \gamma t.$$

(a) Find $x$ as a function of $t$.

(b) Identify the transient and steady-state parts of the solution in part (a).

(c) Find the value of $\gamma$ so that the above system is in resonance (see exercise 3, page 133).

24. Solve for the charge, $x(t)$, in the circuit of Figure 4.5 if $L = 1$, $R = 2$, $C = \frac{1}{10}$, $E(t) = 5 \sin t$, $x(0) = 0$, and $x'(0) = 0$.

25. For what values of the damping constant, $c$, will the motion described by the differential equation

$$4x'' + cx' + 16x = 0$$

be

(a) Overdamped?

(b) Underdamped?

(c) Critically damped?

# CHAPTER 5

# Systems of Linear Differential Equations

In this chapter we study coupled sets of ordinary differential equations. We show how systems of ordinary differential equations may arise, how a single equation may be written as a system of equations, and give some techniques for solving systems of equations.

## 5.1 INTRODUCTION AND GENERAL REMARKS

We have already seen that a single ordinary differential equation may, at times, be reduced to a pair of equations. For example, the equation

$$D^2y + 6Dy = e^{3x}$$

becomes the pair

$$Dy = v,$$

$$Dv + 6v = e^{3x},$$

upon making the substitution

$$Dy = v.$$

Systems of equations in which each individual member is of first order are called *first order systems* and have the general form

$$\begin{bmatrix} \dot{x}_1 = a_{11}x_1 + a_{12}x_2 + \cdots + a_{1n}x_n + b_1, \\ \dot{x}_2 = a_{21}x_1 + a_{22}x_2 + \cdots + a_{2n}x_n + b_2, \\ \vdots \qquad\qquad \vdots \\ \dot{x}_n = a_{n1}x_1 + a_{n2}x_2 + \cdots + a_{nn}x_n + b_n, \end{bmatrix}$$

where the $a_{ij}$ and the $b_k$ may be functions of $t$ and the dots denote derivatives with respect to $t$. By a *solution* to this system of equations we mean a set of $n$ functions $x_i(t)$, $i = 1, \ldots, n$, defined and differentiable on some interval $I$ of the real line and satisfying each of the equations identically.

Systems of ordinary differential equations arise naturally in biology, mechanics, and electrical engineering, as we shall see later.

Since many operations in this chapter are conveniently expressed (and sometimes necessarily expressed) in terms of matrix notation, we start by considering topics concerning matrices.

## 5.2 MATRICES

Let $R$ denote the set of real numbers and $C$ the set of complex numbers. By an $m \times n$ matrix over $R$ (or $C$) we mean $m \times n$ array of numbers

$$
\begin{bmatrix}
a_{11} & a_{12} & \cdots & a_{1n} \\
a_{21} & a_{22} & \cdots & a_{2n} \\
\vdots & & & \\
a_{m1} & a_{m2} & \cdots & a_{mn}
\end{bmatrix}
\tag{5.1}
$$

having $m$ rows and $n$ columns each of whose *entries* $a_{ij}$ is a real (or complex) number. Synonyms for *entry* are *element* and *component*.

To write $m \times n$ matrices in the form (5.1) for all but the simple cases would soon prove tedious, so we introduce the simpler notation for an $m \times n$ matrix, namely,

$$\mathbf{A} = \{a_{ij}\}, \qquad i = 1, \ldots, m; j = 1, \ldots, n.$$

The index $i$ is the row index and $j$ is the column index.

We may now proceed to define arithmetic operations on matrices in the following way. First we will say what we mean by equality of two matrices.

**Definition 5.1.**   Two $m \times n$ matrices $\{a_{ij}\}$, $\{b_{ij}\}$ are said to be *equal* if and only if

$$a_{ij} = b_{ij}, \qquad \text{for each } i, j.$$

Note that equality is defined for two matrices each of which has the same number of rows and the same number of columns. Also, Definition 5.1 is tantamount to saying that two $m \times n$ matrices are equal if and only if they have the same entries in the same places.

**EXAMPLE 5.1**
   (a) Let

$$\mathbf{A} = \begin{bmatrix} 1 & 2 \\ 2 & 1 \end{bmatrix}, \qquad \mathbf{B} = \begin{bmatrix} 1 & 2 \\ 2 & 1 \end{bmatrix}, \qquad \mathbf{F} = \begin{bmatrix} -1 & 3 \\ -3 & -2 \end{bmatrix},$$

$$\mathbf{C} = \begin{bmatrix} 1 & 2 & 3 \\ 3 & 1 & 2 \end{bmatrix}, \qquad \mathbf{D} = \begin{bmatrix} 1 & 2 & 3 \\ 3 & 2 & 1 \end{bmatrix}.$$

Here $\mathbf{A} = \mathbf{B}$ but $\mathbf{C} \neq \mathbf{D}$. (Why?)

(b) Let $\mathbf{C}$ be as above and

$$\mathbf{E} = \begin{bmatrix} 1 & 3 \\ 2 & 1 \\ 3 & 2 \end{bmatrix}.$$

Thus, even though $\mathbf{C}$ and $\mathbf{E}$ contain the same numbers,

$$\mathbf{C} \neq \mathbf{E}.$$

($\mathbf{C}$ is $2 \times 3$, while $\mathbf{E}$ is $3 \times 2$.)

**Definition 5.2.**    Let $\mathbf{A} = \{a_{ij}\}$ and $\mathbf{B} = \{b_{ij}\}$ be $m \times n$ matrices. We define the sum $\mathbf{A} + \mathbf{B}$ to be the matrix $\mathbf{C} = \{c_{ij}\}$, where

$$c_{ij} = a_{ij} + b_{ij}, \qquad \text{for each } i, j.$$

Thus, using the matrices $\mathbf{A}$ and $\mathbf{F}$ of (a) in Example 5.1,

$$\mathbf{A} + \mathbf{F} = \begin{bmatrix} 1 & 2 \\ 2 & 1 \end{bmatrix} + \begin{bmatrix} -1 & 3 \\ -3 & -2 \end{bmatrix} = \begin{bmatrix} 1-1 & 2+3 \\ 2-3 & 1-2 \end{bmatrix} = \begin{bmatrix} 0 & 5 \\ -1 & -1 \end{bmatrix}.$$

This illustrates that addition of matrices is defined in terms of addition of corresponding components.

**Definition 5.3.**    Let $\mathbf{A} = \{a_{ij}\}$ be an $m \times n$ matrix. Let $r$ be a scalar. We define the scalar multiple of $\mathbf{A}$ by $r$ as that matrix $\mathbf{C}$, $\mathbf{C} = \{c_{ij}\}$, such that

$$c_{ij} = ra_{ij}, \qquad \text{for each } i, j.$$

**EXAMPLE 5.2**    Let

$$\mathbf{A} = \begin{bmatrix} 1 & 3 \\ 4 & 2 \end{bmatrix}, r = 2.$$

Then

$$2\mathbf{A} = 2 \begin{bmatrix} 1 & 3 \\ 4 & 2 \end{bmatrix} = \begin{bmatrix} 2 & 6 \\ 8 & 4 \end{bmatrix}.$$

We may say that scalar multiplication is defined componentwise.

Since the addition of two matrices and the scalar multiple of a matrix are defined in terms of their components, it is not difficult to show, using the corresponding properties of the real or complex numbers, that

(1) $\mathbf{A} + \mathbf{B} = \mathbf{B} + \mathbf{A}$            commutative law

(2) $(\mathbf{A} + \mathbf{B}) + \mathbf{C} = \mathbf{A} + (\mathbf{B} + \mathbf{C})$    associative law

(3) $c(\mathbf{A} + \mathbf{B}) = c\mathbf{A} + c\mathbf{B}$                  distributive laws

(4) $(c + d)\mathbf{A} = c\mathbf{A} + d\mathbf{A}$

are valid whenever the sums are defined.

We may also define a product of matrices in such a way that multiplication of one matrix by another yields a matrix. Here we must exercise some care.

**Definition 5.4.**    Let $\mathbf{A} = \{a_{ij}\}$ be an $m \times n$ matrix and $\mathbf{B} = \{b_{ij}\}$ be an $n \times p$ matrix. The *product matrix* $\mathbf{C} = \mathbf{AB}$ is the $m \times p$ matrix whose elements are given by

$$c_{ij} = \sum_{k=1}^{n} a_{ik}b_{kj}, \qquad \text{for each } i, j.$$

The first thing we must note about Definition 5.4 is that the product is not defined for every two matrices. The number of columns of $\mathbf{A}$ must equal the number of rows of $\mathbf{B}$ in order that $\mathbf{AB}$ be defined. As an example, let

$$\mathbf{A} = \begin{bmatrix} 1 & 2 \\ 3 & 4 \end{bmatrix}, \qquad \mathbf{B} = \begin{bmatrix} 1 & 2 & 3 \\ 4 & 5 & 6 \end{bmatrix}.$$

Then $\mathbf{A}$ is $2 \times 2$, $\mathbf{B}$ is $2 \times 3$, and if we write the dimensions of $\mathbf{A}$ and $\mathbf{B}$ side by side in the order in which we wish to form the product

$$\mathbf{AB}: (2 \times 2)(2 \times 3)$$

we see that the middle two numbers match. Thus the number of columns of $\mathbf{A}$ is equal to the number of rows of $\mathbf{B}$ and the product is defined. In this case

$$\mathbf{AB} = \begin{bmatrix} 1 & 2 \\ 3 & 4 \end{bmatrix}\begin{bmatrix} 1 & 2 & 3 \\ 4 & 5 & 6 \end{bmatrix} = \begin{bmatrix} 1+8 & 2+10 & 3+12 \\ 3+16 & 6+20 & 9+24 \end{bmatrix} = \begin{bmatrix} 9 & 12 & 15 \\ 19 & 26 & 33 \end{bmatrix}.$$

The product $\mathbf{BA}$ is not defined, since

$$\mathbf{BA}: (2 \times 3)(2 \times 2)$$

and the middle two numbers do not match.

In general, if $\mathbf{A}$ is $m \times n$ and $\mathbf{B}$ is $q \times p$, the product

$$\mathbf{AB}: (m \times n)(q \times p)$$

is not defined unless $n = q$. If the product is defined, it will result in a matrix that is $(m \times p)$.

There are some special matrices which we also need, among which are:

1. A *zero matrix*. This is a matrix all of whose entries are zero, $\mathbf{Z} = \{z_{ij}\}$, where

$$z_{ij} = 0, \qquad \text{all } i, j.$$

2. An *identity matrix*. This is an $m \times m$ matrix,

$$\mathbf{I} = \begin{bmatrix} 1 & 0 & 0 & \cdots & 0 \\ 0 & 1 & 0 & \cdots & 0 \\ \vdots & & & & \vdots \\ 0 & 0 & 0 & \cdots & 1 \end{bmatrix}.$$

Here we will write

$$\mathbf{I} = \{\delta_{ij}\}$$

where

$$\delta_{ij} = \begin{cases} 1, & i = j, \\ 0, & i \neq j, \end{cases}$$

is the *Kronecker delta*.

These matrices have the following properties

(1) $\mathbf{A} + \mathbf{Z} = \mathbf{A}$      for all **A** for which the sum is defined (i.e., both **A** and **Z** must be the same size).

(2) $\mathbf{AI} = \mathbf{IA}$      for all square matrices **A**. (Both **A** and **I** are $m \times m$ matrices.)

An identity matrix commutes with any **A** for which the products in Property 2 are defined. In general, as we have seen, matrix multiplication is not commutative.

**Definition 5.5.**    Let $\mathbf{A} = \{a_{ij}\}$ be an $m \times n$ matrix, $i = 1, 2, \ldots, m$; $j = 1, 2, \ldots, n$. $\mathbf{A}^T = \{a_{ji}\}$ is called the *transpose* of **A**.

The transpose of a matrix **A** is easily seen, from the definition, to result from the interchange of the rows and columns of **A**. If

$$\mathbf{A} = \begin{bmatrix} 1 & 2 & 3 \\ 4 & 5 & 6 \end{bmatrix},$$

then

$$\mathbf{A}^T = \begin{bmatrix} 1 & 4 \\ 2 & 5 \\ 3 & 6 \end{bmatrix}.$$

**Definition 5.6.**    Let **A** be an $m \times m$ matrix. If there exists an $m \times m$ matrix **C** such that

$$\mathbf{AC} = \mathbf{CA} = \mathbf{I},$$

we say that **C** is the *inverse* of **A** and write $\mathbf{C} = \mathbf{A}^{-1}$.

Note that inverses are defined only for $m \times m$ (square) matrices. If

$$\mathbf{A} = \begin{bmatrix} 1 & 2 \\ 3 & 4 \end{bmatrix},$$

then

$$\mathbf{A}^{-1} = \begin{bmatrix} -2 & 1 \\ \dfrac{3}{2} & -\dfrac{1}{2} \end{bmatrix}.$$

We may easily verify this by direct calculation. Thus

$$
\mathbf{A}^{-1}\mathbf{A} = \begin{bmatrix} -2 & 1 \\ \dfrac{3}{2} & -\dfrac{1}{2} \end{bmatrix} \begin{bmatrix} 1 & 2 \\ 3 & 4 \end{bmatrix}
$$

$$
= \begin{bmatrix} (-2)(1) + (1)(3) & (-2)(2) + (1)(4) \\ \left(\dfrac{3}{2}\right)(1) + \left(-\dfrac{1}{2}\right)(3) & \left(\dfrac{3}{2}\right)(2) + \left(-\dfrac{1}{2}\right)(4) \end{bmatrix}
$$

$$
= \begin{bmatrix} 1 & 0 \\ 0 & 1 \end{bmatrix} = \mathbf{I}.
$$

The calculation of $\mathbf{A}\mathbf{A}^{-1}$ is similar.

For completeness, we now list other properties of matrix multiplication.

(1) $\mathbf{A}(\mathbf{BC}) = (\mathbf{AB})\mathbf{C}$            associative law

(2) $\mathbf{A}(\mathbf{B} + \mathbf{C}) = \mathbf{AB} + \mathbf{AC}$    ⎫

(3) $(\mathbf{A} + \mathbf{B})\mathbf{C} = \mathbf{AC} + \mathbf{BC}$    ⎬    distributive laws

(4) $r(\mathbf{AB}) = (r\mathbf{A})\mathbf{B} = \mathbf{A}(r\mathbf{B})$     associative law

Before developing techniques for finding the inverse of a matrix, we present a theorem that tells when this inverse exists. To that end we need the following definition.

**Definition 5.7.** An $m \times m$ matrix $\mathbf{A}$, for which det $\mathbf{A} = 0$, is said to be *singular*. Matrices for which det $\mathbf{A} \neq 0$ are nonsingular. (det $\mathbf{A}$ is the determinant of $\mathbf{A}$, see Appendix for a discussion of determinants.)

**Theorem 5.1.** Let $\mathbf{A}$ be an $m \times m$ matrix. Then $\mathbf{A}^{-1}$ exists if and only if $\mathbf{A}$ is nonsingular.

Since nonsingular means det $\mathbf{A} \neq 0$, we need only verify that this is the case in order to be assured of the existence of $\mathbf{A}^{-1}$.

Once we know that a given matrix has an inverse, either of the following two methods will enable us to find it.

*Method 1.* We write out the $m \times m$ matrix to be inverted side by side with the identity matrix of the same dimension, a vertical bar separating them:

$$
\begin{bmatrix} 1 & 2 & 1 & 0 \\ 3 & 4 & 0 & 1 \end{bmatrix}. \tag{5.2}
$$

Using elementary row operations we reduce the system (5.2) until the array to the left of the vertical bar becomes the identity. The matrix to the right of the vertical bar is then the desired inverse. If R2 → R2 − 3R1 represents the operation of replacing row 2 by the result of subtracting 3 times row 1 from it, and so on, we have the following results.

R2 → R2 − 3R1 yields

$$\left[\begin{array}{cc|cc} 1 & 2 & 1 & 0 \\ 0 & -2 & -3 & 1 \end{array}\right].$$

The operation R1 → R1 + R2 then gives

$$\left[\begin{array}{cc|cc} 1 & 0 & -2 & 1 \\ 0 & -2 & -3 & 1 \end{array}\right].$$

Division of row 2 by −2 gives

$$\left[\begin{array}{cc|cc} 1 & 0 & -2 & 1 \\ 0 & 1 & \dfrac{3}{2} & -\dfrac{1}{2} \end{array}\right]. \qquad (5.3)$$

The final step, (5.3), has the 2 × 2 identity matrix to the left of the vertical bar and thus the 2 × 2 matrix to the right of the vertical bar is the desired inverse. Note that if it is impossible to obtain the identity matrix to the left of the vertical bar, the original matrix is singular.

*Method 2*. To use the second method we must first define some terms. Let $\mathbf{A} = \{a_{ij}\}$ be an $m \times m$ matrix which is nonsingular. Let $\alpha_{ij}$ denote the *determinant* of the elements of $\mathbf{A}$ remaining after the $i$th row and $j$th column have been deleted. (The determinant of a 1 × 1 matrix is just the element itself.)

**Definition 5.8.**    $\alpha_{ij}$ is said to be the *minor* of $a_{ij}$ in the matrix $\mathbf{A}$.

In the matrix

$$\begin{bmatrix} 1 & 3 & 6 \\ 4 & 0 & 9 \\ 2 & 1 & 1 \end{bmatrix}$$

we have $a_{11} = 1$, and the minor of $a_{11}$, obtained by deleting the first row and first column, is

$$\alpha_{11} = \begin{vmatrix} 0 & 9 \\ 1 & 1 \end{vmatrix} = -9.$$

Similarly,

$$\alpha_{23} = \begin{vmatrix} 1 & 3 \\ 2 & 1 \end{vmatrix} = 1 - 6 = -5.$$

**Definition 5.9.**    The *cofactor* of $a_{ij}$ is $(-1)^{i+j}\alpha_{ij}$.

In the previous matrix, for $a_{11}$,

$$i = j = 1, \qquad \text{and} \qquad (-1)^{1+1}\alpha_{11} = \alpha_{11}$$

so the cofactor of $a_{11}$ is −9. For $a_{23}$, the cofactor is $(-1)^{2+3}\alpha_{23} = -\alpha_{23} = 5$.

The second method of inverting a matrix is embodied in the following three steps.

**Step 1.**   Replace each element in the matrix by its cofactor.

**Step 2.**   Take the transpose of this matrix of cofactors.

**Step 3.**   Divide this transposed matrix by det $\mathbf{A}$.

For a $2 \times 2$ matrix the process is particularly simple. Again look at

$$\mathbf{A} = \begin{bmatrix} 1 & 2 \\ 3 & 4 \end{bmatrix}.$$

The elements with their associated cofactors are:

| Element | Cofactor |
|:---:|:---:|
| 1 | 4 |
| 2 | −3 |
| 3 | −2 |
| 4 | 1 |

The result of applying step 1 yields the array

$$\begin{bmatrix} 4 & -3 \\ -2 & 1 \end{bmatrix},$$

whose transpose is

$$\begin{bmatrix} 4 & -2 \\ -3 & 1 \end{bmatrix}.$$

Now det $\mathbf{A} = 4 - 6 = -2$, so step 3 gives us

$$\mathbf{A}^{-1} = \frac{1}{-2} \begin{bmatrix} 4 & -2 \\ -3 & 1 \end{bmatrix} = \begin{bmatrix} -2 & 1 \\ \frac{3}{2} & -\frac{1}{2} \end{bmatrix}.$$

We may now use the inverse to solve the system of equations

$$x + 2y = 1,$$
$$3x + 4y = 2.$$

As may easily be seen, this system of equations may be written in the form

$$\begin{bmatrix} 1 & 2 \\ 3 & 4 \end{bmatrix} \begin{bmatrix} x \\ y \end{bmatrix} = \begin{bmatrix} 1 \\ 2 \end{bmatrix}. \tag{5.4}$$

Since the coefficient matrix $\mathbf{A}$ is nonsingular it has an inverse and

$$\mathbf{A}^{-1} = \begin{bmatrix} -2 & 1 \\ \frac{3}{2} & -\frac{1}{2} \end{bmatrix}.$$

If we multiply both sides of (5.4) by $\mathbf{A}^{-1}$, we get

$$\begin{bmatrix} x \\ y \end{bmatrix} = \begin{bmatrix} -2 & 1 \\ \dfrac{3}{2} & -\dfrac{1}{2} \end{bmatrix} \begin{bmatrix} 1 \\ 2 \end{bmatrix} = \begin{bmatrix} -2+2 \\ \dfrac{3}{2}-1 \end{bmatrix} = \begin{bmatrix} 0 \\ \dfrac{1}{2} \end{bmatrix}.$$

Thus $x = 0$, $y = \frac{1}{2}$ is the unique solution to the system of equations (5.4).

We finish this section by computing the inverse of a $3 \times 3$ matrix by these two different methods. Let

$$\mathbf{A} = \begin{bmatrix} 1 & -1 & 0 \\ 0 & 1 & 1 \\ 2 & 2 & 0 \end{bmatrix}.$$

To compute $\mathbf{A}^{-1}$ using method 1 [see (5.2)], we proceed as follows. First write $\mathbf{A}$ side by side with the $3 \times 3$ identity matrix.

$$\begin{bmatrix} 1 & -1 & 0 & | & 1 & 0 & 0 \\ 0 & 1 & 1 & | & 0 & 1 & 0 \\ 2 & 2 & 0 & | & 0 & 0 & 1 \end{bmatrix}.$$

Since we want to reduce the left side of this matrix to the identity matrix, we perform the operation R3 → R3 − 2R1, obtaining

$$\begin{bmatrix} 1 & -1 & 0 & | & 1 & 0 & 0 \\ 0 & 1 & 1 & | & 0 & 1 & 0 \\ 0 & 4 & 0 & | & -2 & 0 & 1 \end{bmatrix}.$$

The operation R3 → R3 − 4R2 applied to the array above gives

$$\begin{bmatrix} 1 & -1 & 0 & | & 1 & 0 & 0 \\ 0 & 1 & 1 & | & 0 & 1 & 0 \\ 0 & 0 & -4 & | & -2 & -4 & 1 \end{bmatrix}.$$

R1 → R1 + R2 gives

$$\begin{bmatrix} 1 & 0 & 1 & | & 1 & 1 & 0 \\ 0 & 1 & 1 & | & 0 & 1 & 0 \\ 0 & 0 & -4 & | & -2 & -4 & 1 \end{bmatrix}.$$

R3 → $-\frac{1}{4}$R3 gives

$$\begin{bmatrix} 1 & 0 & 1 & | & 1 & 1 & 0 \\ 0 & 1 & 1 & | & 0 & 1 & 0 \\ 0 & 0 & 1 & | & \dfrac{1}{2} & 1 & -\dfrac{1}{4} \end{bmatrix}.$$

The operations $R1 \to R1 - R3$ and $R2 \to R2 - R3$ will reduce the matrix to the left of the vertical bar to the identity; thus

$$\left[\begin{array}{ccc|ccc} 1 & 0 & 0 & \dfrac{1}{2} & 0 & \dfrac{1}{4} \\[2mm] 0 & 1 & 0 & -\dfrac{1}{2} & 0 & \dfrac{1}{4} \\[2mm] 0 & 0 & 1 & \dfrac{1}{2} & 1 & -\dfrac{1}{4} \end{array}\right]$$

and

$$\mathbf{A}^{-1} = \left[\begin{array}{ccc} \dfrac{1}{2} & 0 & \dfrac{1}{4} \\[2mm] -\dfrac{1}{2} & 0 & \dfrac{1}{4} \\[2mm] \dfrac{1}{2} & 1 & -\dfrac{1}{4} \end{array}\right].$$

The reader should verify that $\mathbf{A}^{-1}\mathbf{A} = \mathbf{I}$ and $\mathbf{A}\mathbf{A}^{-1} = \mathbf{I}$.

Using method 2, we would first calculate the determinant of $\mathbf{A}$ as

$$\det \mathbf{A} = \begin{vmatrix} 1 & -1 & 0 \\ 0 & 1 & 1 \\ 2 & 2 & 0 \end{vmatrix} = (1)\begin{vmatrix} 1 & 1 \\ 2 & 0 \end{vmatrix} - (-1)\begin{vmatrix} 0 & 1 \\ 2 & 0 \end{vmatrix} + (0)\begin{vmatrix} 0 & 1 \\ 2 & 2 \end{vmatrix}$$

$$= -4.$$

The cofactor matrix of $\mathbf{A}$ is

$$\begin{bmatrix} -2 & 2 & -2 \\ 0 & 0 & -4 \\ -1 & -1 & 1 \end{bmatrix}$$

while the transpose of this matrix is

$$\begin{bmatrix} -2 & 0 & -1 \\ 2 & 0 & -1 \\ -2 & -4 & 1 \end{bmatrix}.$$

Dividing each entry in this last matrix by the determinant of $\mathbf{A}$ results in

$$\mathbf{A}^{-1} = \left[\begin{array}{ccc} \dfrac{1}{2} & 0 & \dfrac{1}{4} \\[2mm] -\dfrac{1}{2} & 0 & \dfrac{1}{4} \\[2mm] \dfrac{1}{2} & 1 & -\dfrac{1}{4} \end{array}\right].$$

## EXERCISES

**1.** If

$$\mathbf{A} = \begin{bmatrix} 2 & 1 \\ -1 & 2 \end{bmatrix}, \quad \mathbf{B} = \begin{bmatrix} 1 & 2 \\ 3 & 4 \end{bmatrix}, \quad \mathbf{C} = \begin{bmatrix} 4 & -2 \\ -3 & 1 \end{bmatrix},$$

$$\mathbf{R} = \begin{bmatrix} -2 & -1 \\ 1 & -2 \end{bmatrix}, \quad \mathbf{D} = \begin{bmatrix} 1 & -1 \\ 2 & 3 \end{bmatrix}, \quad \mathbf{I} = \begin{bmatrix} 1 & 0 \\ 0 & 1 \end{bmatrix}, \quad \mathbf{Z} = \begin{bmatrix} 0 & 0 \\ 0 & 0 \end{bmatrix},$$

calculate the following.

(a) **AB**       (b) **AC**       (c) **BC**       (d) **CB**       (e) **BA**
(f) **A + B**    (g) **A + R**    (h) **R + A**    (i) **B + A**    (j) **D + Z**
(k) **2A + C**   (l) **3C − 2B**  (m) **DZ**       (n) **IC**       (o) **CI**

**2.** If

$$\mathbf{A} = \begin{bmatrix} 2 & 0 & 1 \\ -1 & 1 & 3 \\ 0 & -2 & 1 \end{bmatrix}, \quad \mathbf{B} = \begin{bmatrix} 0 & 2 & 1 \\ -1 & 0 & -2 \\ 0 & 3 & 0 \end{bmatrix}, \quad \mathbf{C} = \begin{bmatrix} 1 & 1 & 1 \\ 1 & 1 & 1 \\ 1 & 1 & 1 \end{bmatrix},$$

$$\mathbf{D} = \begin{bmatrix} 6 & 3 & -4 \\ 0 & 0 & -1 \\ -3 & 0 & 2 \end{bmatrix}, \quad \mathbf{E} = \begin{bmatrix} -4 & 0 & -2 \\ 2 & -2 & -6 \\ 0 & 4 & -2 \end{bmatrix}, \quad \mathbf{Z} = \begin{bmatrix} 0 & 0 & 0 \\ 0 & 0 & 0 \\ 0 & 0 & 0 \end{bmatrix},$$

$$\mathbf{I} = \begin{bmatrix} 1 & 0 & 0 \\ 0 & 1 & 0 \\ 0 & 0 & 1 \end{bmatrix},$$

calculate the following.

(a) **2A + B**   (b) **E + 2C**   (c) **2A + E**   (d) **D + Z**
(e) **A − I**    (f) **AB**       (g) **BC**       (h) **DI**       (i) **BD**
(j) **BA**       (k) **CB**       (l) **ID**       (m) **DB**       (n) **EZ**

**3.** All questions here refer to matrices given in exercise 2.
   (a) What is the cofactor of 3 in matrix **A**?
   (b) What is the cofactor of −1 in matrix **B**?
   (c) What is the cofactor of 2 in matrix **B**?
   (d) What is the cofactor of −4 in matrix **D**?
   (e) What is the cofactor of −3 in matrix **D**?
   (f) What is the cofactor of −4 in matrix **E**?
   (g) What is the cofactor of 2 in matrix **E**?

**4.** Find the following (if they exist) for the 2 × 2 matrices in exercise 1.
   (a) $\mathbf{A}^T$           (b) $\mathbf{I}^T$           (c) $(\mathbf{A} + \mathbf{R})^T$       (d) $\mathbf{A}^T + \mathbf{R}$
   (e) $\mathbf{A}^T + \mathbf{R}^T$   (f) $\mathbf{A}^{-1}$        (g) $\mathbf{B}^{-1}$        (h) $\mathbf{C}^{-1}$
   (i) $\mathbf{R}^{-1}$        (j) $\mathbf{D}^{-1}$        (k) $\mathbf{I}^{-1}$        (l) $(\mathbf{B} + \mathbf{C})^{-1}$
   (m) $\mathbf{B}^{-1} + \mathbf{C}^{-1} - (\mathbf{B} + \mathbf{C})^{-1}$

**5.** Find the following (if they exist) for the 3 × 3 matrices in exercise 2.
   (a) $\mathbf{A}^T$           (b) $\mathbf{E}^T$           (c) $\mathbf{C}^T$           (d) $\mathbf{I}^T$           (e) $\mathbf{A}^{-1}$
   (f) $\mathbf{B}^{-1}$        (g) $\mathbf{C}^{-1}$        (h) $\mathbf{D}^{-1}$        (i) $\mathbf{E}^{-1}$        (j) $\mathbf{A}^{-1} + 2\mathbf{E}^{-1}$
   (k) $(\mathbf{A} - \mathbf{C})^{-1}$       (l) $(\mathbf{A} - \mathbf{C})^{-1} - \mathbf{A}^{-1} + \mathbf{C}^{-1}$

**6.** (a) Prove matrix properties (1) to (4) on page 145 for the general 2 × 2 matrix.
   (b) Prove matrix properties (1) to (4) on page 148 for the general 2 × 2 matrix.

**7.** (a) Prove matrix properties (1) to (4) on page 145 for the general $m \times n$ matrix.
   (b) Prove matrix properties (1) to (4) on page 148 for the general $m \times n$ matrix.

## 5.3 ADDITIONAL CONSIDERATIONS

### 5.3.1 Linear Independence of a Set of Vectors

Consider the set of vectors

$$\{\mathbf{u}_i \,|\, i = 1, 2, 3, \ldots, n\}$$

in $\mathbf{R}^m$.

**Definition 5.10.** The set $\{\mathbf{u}_i \,|\, i = 1, 2, 3, \ldots, n\}$ is said to be linearly independent if the only way

$$\sum_{i=1}^{n} c_i \mathbf{u}_i = \mathbf{0}$$

is for $c_i$ to equal zero for every $i = 1, 2, 3, \ldots, n$.

The set of vectors which is not linearly independent is said to be *linearly dependent*.

An infinite set of vectors is said to be *linearly independent* if every finite subset is independent.

**EXAMPLE 5.3**    Determine whether the vectors

$$\mathbf{u}_1 = (1, 0, 1), \qquad \mathbf{u}_2 = (2, 1, 0), \text{ and } \qquad \mathbf{u}_3 = (4, 3, -2),$$

are linearly independent or dependent.

Consider

$$\sum_{i=1}^{3} c_i \mathbf{u}_i = (c_1, 0, c_1) + (2c_2, c_2, 0) + (4c_3, 3c_3, -2c_3)$$

$$= (c_1 + 2c_2 + 4c_3, \, c_2 + 3c_3, \, c_1 - 2c_3).$$

Now $\sum_{i=1}^{3} c_i \mathbf{u}_i = \mathbf{0}$ if each component vanishes, that is,

$$c_1 + 2c_2 + 4c_3 = 0,$$

$$c_2 + 3c_3 = 0,$$

and

$$c_1 - 2c_3 = 0.$$

In solving this system of equations using Gaussian elimination, we first write the associated augmented matrix

$$\begin{bmatrix} 1 & 2 & 4 & 0 \\ 0 & 1 & 3 & 0 \\ 1 & 0 & -2 & 0 \end{bmatrix}.$$

Replacing the last row by the result of subtracting the first row from it (R3 → R3 − R1) yields

$$\begin{bmatrix} 1 & 2 & 4 & 0 \\ 0 & 1 & 3 & 0 \\ 0 & -2 & -6 & 0 \end{bmatrix}.$$

Now replace the last row by the result of adding it to twice the second row (R3 → R3 + 2R2). This gives

$$\begin{bmatrix} 1 & 2 & 4 & 0 \\ 0 & 1 & 3 & 0 \\ 0 & 0 & 0 & 0 \end{bmatrix}.$$

This is equivalent to having

$$c_2 = -3c_3,$$

$$c_1 = -2c_2 - 4c_3 = 6c_3 - 4c_3 = 2c_3.$$

In other words, $c_3$ can be chosen arbitrarily and then $c_1 = 2c_3$ and $c_2 = -3c_3$. Since we satisfy the condition

$$\sum_{i=1}^{3} c_i \mathbf{u}_i = \mathbf{0},$$

with at least one $c_i$ not zero, this set of vectors is linearly dependent.

**EXAMPLE 5.4**    Determine whether the vectors

$$\mathbf{u}_1 = (1, 2, 3), \qquad \mathbf{u}_2 = (0, 2, 2), \text{ and } \qquad \mathbf{u}_3 = (0, -1, 1)$$

are linearly independent or dependent.

Consider

$$\sum_{i=1}^{3} c_i \mathbf{u}_i = (c_1, 2c_1, 3c_1) + (0, 2c_2, 2c_2) + (0, -c_3, c_3).$$

Thus $\sum_{i=1}^{3} c_i \mathbf{u}_i = \mathbf{0}$ if

$$c_1 = 0,$$

$$2c_1 + 2c_2 - c_3 = 0,$$

$$3c_1 + 2c_2 + c_3 = 0.$$

Since the determinant of the coefficient matrix equals 4,

$$\begin{vmatrix} 1 & 0 & 0 \\ 2 & 2 & -1 \\ 3 & 2 & 1 \end{vmatrix} = 1 \begin{vmatrix} 2 & -1 \\ 2 & 1 \end{vmatrix} = 4,$$

Cramer's rule gives the unique solution of this system as

$$c_1 = c_2 = c_3 = 0.$$

Thus $(1, 2, 3)$ $(0, 2, 2)$, and $(0, -1, 1)$ are linearly independent.

Example 5.4 illustrates the following theorem, which can be used to determine whether a set of $n$ vectors in $R^n$ is linearly independent or dependent. In this theorem the notation $\det[v_1 v_2 v_3 \cdots v_n]$ denotes the determinant whose first column contains the components of $v_1$, second column those of $v_2$, and so on.

**Theorem 5.2.**  Let $v_1$, $v_2$, $v_3$, . . . , $v_n$ be elements of $R^n$.
(a) If $\det[v_1 v_2 v_3 \cdots v_n] \neq 0$, this set of elements is linearly independent.
(b) If $\det[v_1 v_2 v_3 \cdots v_n] = 0$, this set of elements is linearly dependent.

*Proof.*  The vector equation

$$\sum_{i=1}^{n} c_i v_i = \sum_{i=1}^{n} v_i c_i = 0$$

is equivalent to the matrix equation

$$\mathbf{AC} = \mathbf{0},$$

where $\mathbf{A}$ is the matrix whose $i$th column contains the components of $v_i$, that is,

$$\mathbf{A} = [v_1 v_2 v_3 \cdots v_n].$$

We now solve this system of equations using Gaussian elimination. The final form of the solution will either be

$$\begin{vmatrix} 1 & - & - & \cdots & - & & 0 \\ 0 & 1 & - & \cdots & - & & 0 \\ 0 & 0 & 1 & \cdots & - & & 0 \\ \vdots & & & & & & \vdots \\ & & & & 1 & - & 0 \\ 0 & 0 & 0 & \cdots & 0 & 0 & 1 & 0 \end{vmatrix}$$

or a similar-looking augmented matrix except that the last row will contain all zeros. If $\det \mathbf{A} \neq 0$, we end up with the last row of the augmented matrix above containing all zeros except for a 1 in the next-to-last column. (This is because elementary row operations may only change the magnitude of a determinant by a nonzero multiple. Changing the sign corresponds to multiplying by $-1$. These operations cannot change a singular determinant to a nonsingular one, or vice versa.) This is the situation for part (a) of the theorem, and back substitution gives the unique solution

$$c_1 = c_2 = c_3 = \cdots = c_n = 0.$$

If $\det \mathbf{A} = 0$, the last row of the augmented matrix must contain all zeros. Thus we can choose $c_n \neq 0$ and choose $c_1$, $c_2$, . . . , $c_{n-1}$ so that the first $n - 1$ equations represented by the rows of the augmented matrix are satisfied. Thus we have a nontrivial solution (at least $c_n \neq 0$) of the system of equations, which means the set of vectors is linearly dependent.

# EXERCISES

1. Are the following sets of vectors independent or dependent?
   (a) $\{(1, 0, 0), (1, -1, 1), (1, 2, -1)\}$
   (b) $\{(1, 1, 1), (1, -1, 1), (-1, 1, -1)\}$
   (c) $\{(1, 1, 1), (1, -1, -1), (3, 1, 1)\}$
   (d) $\{(1, 0, 0), (0, 1, -1), (1, 0, 1)\}$
   (e) $\{(1, 0, 0, 0), (0, 1, 0, 0), (1, 1, 0, 0), (0, 1, 1, 0)\}$
   (f) $\{(0, 1, 0, 0), (1, 1, 0, 0), (0, 1, 1, 0), (0, 0, 0, 1)\}$

2. Determine whether $(1, 0, 2)$ may be written as a linear combination of the vectors in exercise 1(a) to (d).

3. Show that any set of three vectors in $R^2$ must be a dependent set.

4. Show that if any set in $R^n$ contains the zero vector, the set is linearly dependent.

5. Let $v_1, v_2, v_3$ be vectors in $R^3$. If the sets $\{v_1, v_2\}$ and $\{v_2, v_3\}$ are both linearly independent sets, find examples to show that $\{v_1, v_2, v_3\}$ may be an independent or a dependent set.

6. Find $k$ such that the set of vectors will be dependent.
   (a) $(k, 0, 1), (1, 1, 1), (0, -1, 0)$
   (b) $(k, k + 1, 1), (1, 1, 1), (1, 0, 0)$
   (c) $(1, 1, 1), (1, k, -1), (k + 4, 1, 1)$

7. Find $k$ such that the set of vectors from exercise 6 will be independent.

8. For what value of $k$ is $\{(0, 1, 1), (2k, k, 0), (k^2, k, 0)\}$ a linearly independent set.

9. Are the following sets of vectors with complex components linearly independent or dependent? Note here that multiplication by a scalar corresponds to multiplication by a complex number.
   (a) $\{(1, i), (i, 1)\}$
   (b) $\{(1, i), (-i, 1)\}$
   (c) $\{(1 + i, 3), (1 - i, 3)\}$
   (d) $\{(2 - i, 3 + i), (-i, i)\}$
   (e) $\{(2 + i, i), (i - 3, 1 - 2i)\}$
   (f) $\{(i, i^2), (i^3, 1)\}$

10. Prove that $v_1 = (a, b)$ and $v_2 = (c, d)$ are linearly dependent if and only if $ad = bc$.

11. Show that each of the following sets is linearly independent and add a third element to each set so that it forms a linearly independent set in $R^3$.
    (a) $\{(1, 1, 0), (1, 0, 1)\}$
    (b) $\{(1, 1, 1), (1, -1, 1)\}$

12. Show that each of the following sets is linearly independent and add a fourth element to each set so that it forms a linearly independent set in $R^4$.
    (a) $\{(1, 0, 1, 0), (0, 0, 1, 1), (1, 0, -1, 0)\}$
    (b) $\{(0, 0, 0, 1), (0, 0, 1, -2), (0, 1, 1, 0)\}$
    (c) $\{(0, 0, 1, 1), (0, 1, 1, 0), (0, 1, -1, 0)\}$
    (d) $\{(0, 0, 1, 1), (0, 1, 0, 1), (1, 0, 1, -1)\}$

13. Prove that any set of $n + 1$ vectors in $R^n$ is linearly dependent.

## 5.3.2 The Eigenvalue Problem

Let $\mathbf{A}$ be a square matrix and $\mathbf{X}$ a column vector for which the product $\mathbf{AX}$ is defined. Then $\mathbf{AX}$ is again a column vector and we may ask the following question: What are the nonzero vectors $\mathbf{X}$ with the property that $\mathbf{AX}$ is a multiple of $\mathbf{X}$? In mathematical notation this question becomes: Find all nonzero vectors $\mathbf{X}$ and scalars $\lambda$ such that

$$\mathbf{AX} = \lambda\mathbf{X}. \tag{5.5}$$

A nonzero vector $\mathbf{X}$ which satisfies (5.5) for some $\lambda$ is said to be an *eigenvector* of $\mathbf{A}$ belonging to the *eigenvalue* $\lambda$. $\lambda$ may be a real or a complex number.

Using $\mathbf{I}$ to denote the $n \times n$ identity matrix, we may write (5.5) as

$$\mathbf{AX} = \lambda\mathbf{IX} \tag{5.6}$$

which is equivalent to

$$(\mathbf{A} - \lambda\mathbf{I})\mathbf{X} = \mathbf{0}. \tag{5.7}$$

Now (5.7) is just an $n \times n$ homogeneous system of equations and thus we know that if its coefficient matrix, $\mathbf{A} - \lambda\mathbf{I}$, is nonsingular, the zero vector will be the unique solution. We are interested, then, in the case in which (5.7) has nontrivial solutions, and thus we will require that

$$|\mathbf{A} - \lambda\mathbf{I}| = 0. \tag{5.8}$$

The eigenvalues $\lambda$ will be solutions of (5.8). [Equation (5.8) is a polynomial in $\lambda$ of degree $n$.]

**EXAMPLE 5.5**    Let

$$\mathbf{A} = \begin{bmatrix} -1 & 1 \\ 4 & 2 \end{bmatrix}.$$

Then (5.5) becomes

$$\begin{bmatrix} -1 & 1 \\ 4 & 2 \end{bmatrix}\begin{bmatrix} x_1 \\ x_2 \end{bmatrix} = \begin{bmatrix} \lambda x_1 \\ \lambda x_2 \end{bmatrix},$$

(5.6) becomes

$$\begin{bmatrix} -1 & 1 \\ 4 & 2 \end{bmatrix}\begin{bmatrix} x_1 \\ x_2 \end{bmatrix} = \begin{bmatrix} \lambda & 0 \\ 0 & \lambda \end{bmatrix}\begin{bmatrix} x_1 \\ x_2 \end{bmatrix}$$

and (5.8) is

$$\begin{vmatrix} -1 - \lambda & 1 \\ 4 & 2 - \lambda \end{vmatrix} = 0. \tag{5.9}$$

Expanding (5.9), we get

$$(-1 - \lambda)(2 - \lambda) - 4 = 0,$$

$$-2 - \lambda + \lambda^2 - 4 = 0,$$

$$\lambda^2 - \lambda - 6 = 0.$$

The eigenvalues $\lambda_1 = 3$ and $\lambda_2 = -2$ are solutions of this quadratic equation. To find the eigenvectors, we write out (5.7) using each of $\lambda_1$ and $\lambda_2$ in turn. For $\lambda_1 = 3$ we get

$$\begin{bmatrix} -4 & 1 \\ 4 & -1 \end{bmatrix}\begin{bmatrix} x_1 \\ x_2 \end{bmatrix} = \begin{bmatrix} 0 \\ 0 \end{bmatrix}$$

or

$$-4x_1 + x_2 = 0,$$

$$4x_1 - x_2 = 0.$$

This system is easily seen to be a dependent system equivalent to the single equation

$$x_2 = 4x_1.$$

Thus the eigenvectors corresponding to the eigenvalue $\lambda_1 = 3$ have the form

$$\begin{bmatrix} a \\ 4a \end{bmatrix},$$

where $a$ is any nonzero constant.

For $\lambda_2 = -2$ we get

$$\begin{bmatrix} 1 & 1 \\ 4 & 4 \end{bmatrix}\begin{bmatrix} x_1 \\ x_2 \end{bmatrix} = \begin{bmatrix} 0 \\ 0 \end{bmatrix}$$

and it is easy to see that $x_1 = -x_2$. Thus the eigenvectors corresponding to the eigenvalue $\lambda_2 = -2$ are of the form

$$\begin{bmatrix} b \\ -b \end{bmatrix}, \qquad b \neq 0.$$

**EXAMPLE 5.6**    Let

$$\mathbf{A} = \begin{bmatrix} 1 & 2 & -1 \\ 0 & -2 & 0 \\ 0 & -5 & 2 \end{bmatrix}.$$

The eigenvalue equation becomes

$$\begin{vmatrix} 1 - \lambda & 2 & -1 \\ 0 & -2 - \lambda & 0 \\ 0 & -5 & 2 - \lambda \end{vmatrix} = 0$$

or, by evaluating the determinant,

$$(1 - \lambda)(-2 - \lambda)(2 - \lambda) = 0.$$

Thus $A$ has the three eigenvalues

$$\lambda_1 = 1, \qquad \lambda_2 = -2, \qquad \lambda_3 = 2.$$

For $\lambda_1 = 1$ we have

$$2x_2 - x_3 = 0,$$

$$-3x_2 \qquad = 0,$$

$$-5x_2 + x_3 = 0.$$

Thus $x_1$ is arbitrary and $x_2 = x_3 = 0$. The eigenvectors belonging to the eigenvalue $\lambda_1 = 1$ have the form

$$\begin{bmatrix} b \\ 0 \\ 0 \end{bmatrix}, \quad b \neq 0.$$

For $\lambda_2 = -2$ we have

$$3x_1 + 2x_2 - x_3 = 0,$$
$$0x_2 = 0,$$
$$-5x_2 + 4x_3 = 0.$$

Choose $x_1 = a \neq 0$ so that the above equations reduce to

$$2x_2 - x_3 = -3a,$$
$$-5x_2 + 4x_3 = 0.$$

Now solving for $x_2$ gives

$$x_2 = -4a,$$

while the last equation determines $x_3$ as

$$x_3 = -5a.$$

Thus the eigenvectors corresponding to the eigenvalue $\lambda_2 = -2$ have the form

$$\begin{bmatrix} a \\ -4a \\ -5a \end{bmatrix}, \quad a \neq 0.$$

For $\lambda_3 = 2$ we get

$$-x_1 + 2x_2 - x_3 = 0,$$
$$-4x_2 = 0,$$
$$-5x_2 + 0x_3 = 0.$$

From this we see immediately that

$$x_3 = -x_1, \quad x_2 = 0$$

so that the eigenvectors corresponding to the eigenvalue $\lambda_3 = 2$ have the form

$$\begin{bmatrix} c \\ 0 \\ -c \end{bmatrix}, \quad c \neq 0.$$

**EXAMPLE 5.7**     Let

$$A = \begin{bmatrix} 1 & -2 \\ 2 & 1 \end{bmatrix}.$$

The eigenvalues of $\mathbf{A}$ are determined from

$$\det[\mathbf{A} - \lambda\mathbf{I}] = \begin{vmatrix} 1 - \lambda & -2 \\ 2 & 1 - \lambda \end{vmatrix} = (1 - \lambda)^2 + 4 = 0.$$

Then $1 - \lambda = \pm 2i$ or $\lambda = 1 \mp 2i$ are the eigenvalues.

If we label $\lambda_1 = 1 - 2i$, the system of equations giving the corresponding eigenvector is

$$\begin{bmatrix} 1 - (1 - 2i) & -2 \\ 2 & 1 - (1 - 2i) \end{bmatrix}\begin{bmatrix} x_1 \\ x_2 \end{bmatrix} = \begin{bmatrix} 0 \\ 0 \end{bmatrix}.$$

This dependent system of equations, namely,

$$2ix_1 - 2x_2 = 0,$$
$$2x_1 + 2ix_2 = 0,$$

has a solution of the form

$$a\begin{bmatrix} -i \\ 1 \end{bmatrix}, \qquad \text{with } a \neq 0.$$

For $\lambda_2 = 1 + 2i$, we have the system of equations

$$\begin{bmatrix} 1 - (1 + 2i) & -2 \\ 2 & 1 - (1 + 2i) \end{bmatrix}\begin{bmatrix} x_1 \\ x_2 \end{bmatrix} = \begin{bmatrix} 0 \\ 0 \end{bmatrix}$$

or

$$-2ix_1 - 2x_2 = 0,$$
$$2x_1 - 2ix_2 = 0.$$

Thus the eigenvector associated with $1 + 2i$ is

$$b\begin{bmatrix} i \\ 1 \end{bmatrix}, \qquad \text{where } b \neq 0.$$

Notice that in Examples 5.5 to 5.7 the eigenvectors corresponding to different eigenvalues are linearly independent. This illustrates the general result given in the following theorem.

**Theorem 5.3.**   Let $\lambda_1, \lambda_2, \lambda_3, \ldots, \lambda_k$ be distinct eigenvalues of a matrix $\mathbf{A}$ and let $\mathbf{X}_1, \mathbf{X}_2, \mathbf{X}_3, \ldots, \mathbf{X}_k$ be nonzero eigenvectors associated with these eigenvalues. Then $\mathbf{X}_1, \mathbf{X}_2, \mathbf{X}_3, \ldots, \mathbf{X}_k$ are linearly independent.

*Proof.* We need to show that the only solution of

$$\sum_{i=1}^{k} c_i\mathbf{X}_i = c_1\mathbf{X}_1 + c_2\mathbf{X}_2 + c_3\mathbf{X}_3 + \cdots + c_k\mathbf{X}_k = \mathbf{0} \qquad (5.10)$$

is $c_i = 0$ for all $i$. Multiplying (5.10) by the matrix $\mathbf{A}$ and using the fact that $\mathbf{A}\mathbf{X}_i = \lambda_i\mathbf{X}_i$ gives

$$c_1\lambda_1\mathbf{X}_1 + c_2\lambda_2\mathbf{X}_2 + c_3\lambda_3\mathbf{X}_3 + \cdots + c_k\lambda_k\mathbf{X}_k = \mathbf{0}.$$

Repeating this process gives the additional equations

$$c_1\lambda_1^2\mathbf{X}_1 + c_2\lambda_2^2\mathbf{X}_2 + c_3\lambda_3^2\mathbf{X}_3 + \cdots + c_k\lambda_k^2\mathbf{X}_k = \mathbf{0},$$

$$\vdots \qquad \vdots \qquad \vdots \qquad\qquad \vdots$$

$$c_1\lambda_1^{k-1}\mathbf{X}_1 + c_2\lambda_2^{k-1}\mathbf{X}_2 + c_3\lambda_3^{k-1}\mathbf{X}_3 + \cdots + c_k\lambda_k^{k-1}\mathbf{X}_k = \mathbf{0}.$$

If we write this as the matrix equation

$$
\begin{bmatrix}
1 & 1 & 1 & \cdots & 1 \\
\lambda_1 & \lambda_2 & \lambda_3 & \cdots & \lambda_k \\
\lambda_1^2 & \lambda_2^2 & \lambda_3^2 & \cdots & \lambda_k^2 \\
\vdots & \vdots & & & \vdots \\
\lambda_1^{k-1} & \lambda_2^{k-1} & \lambda_3^{k-1} & \cdots & \lambda_k^{k-1}
\end{bmatrix}
\begin{bmatrix}
c_1\mathbf{X}_1 \\
c_2\mathbf{X}_2 \\
c_3\mathbf{X}_3 \\
\vdots \\
c_k\mathbf{X}_k
\end{bmatrix}
=
\begin{bmatrix}
\mathbf{0} \\
\mathbf{0} \\
\mathbf{0} \\
\vdots \\
\mathbf{0}
\end{bmatrix}
$$

we recognize the coefficient matrix as the Vandermonde matrix. Since the $\lambda_i$ are all distinct, its determinant is not zero (see the exercises on page 88) and we may multiply this matrix equation by its inverse to obtain

$$
\begin{bmatrix}
c_1\mathbf{X}_1 \\
c_2\mathbf{X}_2 \\
c_3\mathbf{X}_3 \\
\vdots \\
c_k\mathbf{X}_k
\end{bmatrix}
=
\begin{bmatrix}
\mathbf{0} \\
\mathbf{0} \\
\mathbf{0} \\
\vdots \\
\mathbf{0}
\end{bmatrix}.
$$

Since none of the $\mathbf{X}_i$ are zero, this gives $c_i = 0$ for all $i$ and the set of vectors is linearly independent.

## EXERCISES

**1.** Find all eigenvalues and corresponding eigenvectors for the following matrices.

(a) $\begin{bmatrix} 0 & 2 \\ 1 & 1 \end{bmatrix}$
   
(b) $\begin{bmatrix} 0 & 2 \\ 2 & 3 \end{bmatrix}$
   
(c) $\begin{bmatrix} -1 & -9 \\ 1 & 5 \end{bmatrix}$
   
(d) $\begin{bmatrix} 1 & 0 \\ 0 & 1 \end{bmatrix}$

(e) $\begin{bmatrix} 0 & 0 \\ 0 & 0 \end{bmatrix}$
   
(f) $\begin{bmatrix} 1 & 4 \\ 1 & 1 \end{bmatrix}$
   
(g) $\begin{bmatrix} 1 & 1 \\ 2 & 2 \end{bmatrix}$
   
(h) $\begin{bmatrix} 1 & 3 \\ -2 & 1 \end{bmatrix}$

(i) $\begin{bmatrix} 3 & -1 \\ 1 & 1 \end{bmatrix}$
   
(j) $\begin{bmatrix} 0 & i \\ -i & 0 \end{bmatrix}$

**2.** Find all the eigenvalues and eigenvectors for the following matrices.

(a) $\begin{bmatrix} 1 & 0 & 1 \\ 0 & 1 & 0 \\ 1 & 0 & 1 \end{bmatrix}$
   
(b) $\begin{bmatrix} 1 & 0 & 0 \\ 0 & 1 & 0 \\ 0 & 0 & 1 \end{bmatrix}$
   
(c) $\begin{bmatrix} 0 & 0 & 0 \\ 0 & 0 & 0 \\ 0 & 0 & 0 \end{bmatrix}$

(d) $\begin{bmatrix} -5 & 6 & 6 \\ 1 & -4 & 0 \\ -3 & 6 & 4 \end{bmatrix}$
   
(e) $\begin{bmatrix} 1 & 2 & 3 \\ 0 & 2 & 3 \\ 0 & 0 & 2 \end{bmatrix}$
   
(f) $\begin{bmatrix} -5 & -1 & -1 \\ 3 & -1 & 3 \\ 2 & 2 & 2 \end{bmatrix}$

$$
\textbf{(g)} \begin{bmatrix} 3 & 1 & 0 \\ 1 & 3 & 0 \\ 0 & 0 & 2 \end{bmatrix}
\qquad
\textbf{(h)} \begin{bmatrix} 1 & 0 & 1 \\ 0 & 1 & 0 \\ 1 & 0 & -1 \end{bmatrix}
\qquad
\textbf{(i)} \begin{bmatrix} 1 & 1 & 1 \\ 1 & 0 & -2 \\ 1 & -1 & 1 \end{bmatrix}
$$

$$
\textbf{(j)} \begin{bmatrix} 1 & 0 & 0 \\ -1 & 0 & 3 \\ 1 & -3 & 0 \end{bmatrix}
$$

3. Show that if $A$ is a real matrix with a complex eigenvalue $\lambda_1$ and eigenvector $X_1$, then the complex conjugates $\bar{\lambda}_1$ and $\bar{X}_1$ are also an eigenvalue and eigenvector of $A$.

4. Show that the eigenvalues of

$$
\begin{bmatrix} a & b & c \\ 0 & d & e \\ 0 & 0 & f \end{bmatrix}
$$

are $a$, $d$, and $f$.

5. For what values of $r$ will the matrix

$$
\begin{bmatrix} 1 & 2 \\ r & 3 \end{bmatrix}
$$

have

(a) Two distinct real eigenvalues?

(b) Two distinct complex eigenvalues?

(c) One repeated real eigenvalue?

6. Show that if the eigenvalues of $A$ are $\lambda_1, \lambda_2, \lambda_3, \ldots, \lambda_n$, the eigenvalues of $\alpha A$ are $\alpha\lambda_1$, $\alpha\lambda_2, \alpha\lambda_3, \ldots, \alpha\lambda_n$.

7. Show that if the eigenvalues of $A$ are $\lambda_1, \lambda_2, \lambda_3, \ldots, \lambda_n$ and $\det A \neq 0$, the eigenvalues of $A^{-1}$ are $1/\lambda_1, 1/\lambda_2, 1/\lambda_3, \ldots, 1/\lambda_n$. (*Hint:* Multiply $AX = \lambda X$ by $A^{-1}$.)

8. Prove Theorem 5.3 as follows:

(a) Show that if $X_i$ is the eigenvector corresponding to the eigenvalue $\lambda_i$ of the matrix $A$, then

$$
(A - \lambda_j I)X_i = (\lambda_i - \lambda_j)X_i
$$

and

$$
(A - \lambda_i I)X_i = 0.
$$

(b) Apply the product with $k - 1$ factors,

$$
L = (A - \lambda_1 I)(A - \lambda_2 I) \cdots (A - \lambda_{i-1} I)(A - \lambda_{i+1} I) \cdots (A - \lambda_{k-1} I)(A - \lambda_k I)
$$

to $X_j$ and show that

$$
LX_j = 0, \qquad j \neq i,
$$

$$
LX_i = (\lambda_i - \lambda_1)(\lambda_i - \lambda_2) \cdots (\lambda_i - \lambda_{i-1})(\lambda_i - \lambda_{i+1}) \cdots (\lambda_i - \lambda_{k-1})(\lambda_i - \lambda_k)X_i
$$

(c) Apply $L$ to (5.10) to show that

$$
0 = c_i(\lambda_i - \lambda_1)(\lambda_i - \lambda_2) \cdots (\lambda_i - \lambda_{i-1})(\lambda_i - \lambda_{i+1}) \cdots (\lambda_i - \lambda_{k-1})(\lambda_i - \lambda_k)X_i.
$$

(d) Since all the eigenvalues are distinct, $\lambda_i \neq \lambda_j$, $i \neq j$, and $X_i \neq 0$, conclude that $c_i = 0$. Thus letting $i = 1, 2, 3, \ldots, k$ shows that $c_i = 0$, $i = 1, 2, 3, \ldots, k$, and the eigenvectors are linearly independent.

## 5.4 FIRST ORDER SYSTEMS (THE ELIMINATION METHOD)

We will consider three different ways of solving systems of linear differential equations. Two of these ways, the elimination method of this section and the Laplace transform method of Chapter 7, may be used on systems of any order. The third way uses fundamental matrices of Section 5.5 and is limited to first order systems of the form

$$\frac{dx_1}{dt} = a_{11}(t)x_1(t) + a_{12}(t)x_2(t) + \cdots + a_{1n}(t)x_n(t) + b_1(t),$$

$$\frac{dx_2}{dt} = a_{21}(t)x_1(t) + a_{22}(t)x_2(t) + \cdots + a_{2n}(t)x_n(t) + b_2(t),$$

$$\vdots \qquad \vdots \qquad \vdots \qquad \vdots \qquad (5.11)$$

$$\frac{dx_n}{dt} = a_{n1}(t)x_1(t) + a_{n2}(t)x_2(t) + \cdots + a_{nn}(t)x_n(t) + b_n(t).$$

The theorem for systems of linear differential equations which is analogous to Theorem 1.1 for a single $n$th order equation is as follows.

**Theorem 5.4.**    If the functions $a_{ij}(t)$, $i, j = 1, 2, 3, \ldots, n$, are continuous on the open interval $a < t < b$ containing the point $t = t_0$, then there exists a unique solution of the system (5.11), valid throughout this interval, which also satisfies the initial conditions $x_1(t_0) = \alpha_1$, $x_2(t_0) = \alpha_2$, $x_3(t_0) = \alpha_3, \ldots, x_n(t_0) = \alpha_n$.

A system of first order linear differential equations has the form

$$\frac{dx_1}{dt} = a_{11}(t)x_1(t) + a_{12}(t)x_2(t) + \cdots + a_{1n}(t)x_n(t) + b_1(t),$$

$$\frac{dx_2}{dt} = a_{21}(t)x_1(t) + a_{22}(t)x_2(t) + \cdots + a_{2n}(t)x_n(t) + b_2(t),$$

$$\vdots \qquad \vdots \qquad \vdots \qquad \vdots \qquad (5.11)$$

$$\frac{dx_n}{dt} = a_{n1}(t)x_1(t) + a_{n2}(t)x_2(t) + \cdots + a_{nn}(t)x_n(t) + b_n(t).$$

Comparison of (5.11) to a system of algebraic equations suggests the notation

$$\dot{\mathbf{X}} = \mathbf{A}(t)\mathbf{X}(t) + \mathbf{b}(t) \qquad (5.12)$$

where

$$\dot{\mathbf{X}} = D\begin{bmatrix} x_1 \\ \vdots \\ x_n \end{bmatrix}, \qquad \mathbf{b} = \begin{bmatrix} b_1(t) \\ \vdots \\ \vdots \\ b_n(t) \end{bmatrix}, \qquad \mathbf{X} = \begin{bmatrix} x_1(t) \\ \vdots \\ x_n(t) \end{bmatrix},$$

and

$$\mathbf{A}(t) = \{a_{ij}\} = \begin{bmatrix} a_{11}(t) & \cdots & a_{1n}(t) \\ \vdots & & \vdots \\ a_{n1}(t) & \cdots & a_{nn}(t) \end{bmatrix}.$$

As with algebraic systems the matrix **A** is called the *coefficient matrix* of the system. We give the following definitions:

1. The system (5.12) is said to be *homogeneous* if the column vector **b** = **0**.
2. The system (5.12) is said to be *nonhomogeneous* if **b** ≠ **0**.

We note immediately, however, that the situation is somewhat more complicated here in that the entries in the coefficient matrix, the vector **X** and the vector **b**, may all be functions of the independent variable, $t$, with respect to which the differentiation takes place. (This differentiation is denoted by a dot, $D$, or $d/dt$.) Initially, we shall be concerned with the case in which **b** = **0**, that is, the homogeneous system, and the case in which the matrix **A** has entries which are all constants. An example of such a system is the following:

$$Dx_1 = 3x_1 + x_2 - 4x_3,$$
$$Dx_2 = x_1 + 2x_2 - x_3, \qquad (5.13)$$
$$Dx_3 = 2x_1 + x_2 - 3x_3,$$

or

$$\dot{\mathbf{X}} = \mathbf{AX}, \qquad \mathbf{X} = \begin{bmatrix} x_1 \\ x_2 \\ x_3 \end{bmatrix}, \qquad \mathbf{A} = \begin{bmatrix} 3 & 1 & -4 \\ 1 & 2 & -1 \\ 2 & 1 & -3 \end{bmatrix}.$$

Rearranging gives

$$(D - 3)x_1 - x_2 + 4x_3 = 0,$$
$$-x_1 + (D - 2)x_2 + x_3 = 0, \qquad (5.14)$$
$$-2x_1 - x_2 + (D + 3)x_3 = 0,$$

as a form more closely resembling a system of algebraic equations in $x_1(t)$, $x_2(t)$, and $x_3(t)$. The first method we shall use to solve this system is the *method of elimination*. The idea here is to eliminate all but one of the variables from an equation, solve it, and use that solution to obtain solutions for the remaining equations. To do so we may perform the following operations, just as with algebraic systems:

1. Interchange any two rows.
2. Replace any row by the result of adding to that row any constant multiple of another row.

In addition, we may operate upon any row with a differential operator. For example, multiplying the second of equations (5.14) by $-2$ and adding it to the third equation eliminates the function $x_1(t)$ from the third equation, reducing the system to

$$(D - 3)x_1 - x_2 + 4x_3 = 0,$$
$$-x_1 + (D - 2)x_2 + x_3 = 0, \qquad (5.15)$$
$$(-2D + 3)x_2 + (D + 1)x_3 = 0.$$

We now apply the operator $(D - 3)$ to the second of equations (5.15) and add it to the first, obtaining the system

$$[-1 + (D - 3)(D - 2)]x_2 + [4 + (D - 3)]x_3 = 0,$$
$$-x_1 + (D - 2)x_2 + x_3 = 0,$$
$$(-2D + 3)x_2 + (D + 1)x_3 = 0,$$

or

$$(D^2 - 5D + 5)x_2 + (D + 1)x_3 = 0,$$
$$-x_1 + (D - 2)x_2 + x_3 = 0,$$
$$(-2D + 3)x_2 + (D + 1)x_3 = 0.$$

We now replace the first equation by the difference of the first and third equations.

$$(D^2 - 3D + 2)x_2 = 0,$$
$$-x_1 + (D - 2)x_2 + x_3 = 0, \qquad (5.16)$$
$$(-2D + 3)x_2 + (D + 1)x_3 = 0.$$

The first of equations (5.16) is second order and linear with constant coefficients. It has the characteristic equation

$$r^2 - 3r + 2 = 0,$$

with roots $r_1 = 1$ and $r_2 = 2$, so the solution is

$$x_2(t) = C_1 e^t + C_2 e^{2t}. \qquad (5.17)$$

Substitution of $x_2(t)$ in the third of equations (5.16) gives

$$(D + 1)x_3 = (2D - 3)x_2 = -C_1 e^t + C_2 e^{2t}. \qquad (5.18)$$

Now $e^{-t}$ satisfies $(D + 1)x = 0$, so the method of undetermined coefficients gives the proper form of the particular solution of (5.18) as

$$\ell e^t + k e^{2t}.$$

Substitution into (5.18) gives $\ell = -C_1/2$, $k = C_2/3$ and the general solution to (5.18) as

$$x_3(t) = C_3 e^{-t} - \frac{1}{2}C_1 e^t + \frac{1}{3}C_2 e^{2t}. \qquad (5.19)$$

From the second of equations (5.16) we have

$$x_1(t) = x_3(t) + (D - 2)x_2(t),$$

and since $x_2(t)$ and $x_3(t)$ are given by (5.17) and (5.19), substitution gives

$$x_1(t) = C_3 e^{-t} - \frac{3}{2}C_1 e^t + \frac{1}{3}C_2 e^{2t}. \qquad (5.20)$$

Notice that the solution may be written as the sum of arbitrary constants times three column vectors:

$$
\begin{bmatrix} x_1(t) \\ x_2(t) \\ x_3(t) \end{bmatrix} = C_1 \begin{bmatrix} -\dfrac{3}{2}e^t \\ e^t \\ -\dfrac{1}{2}e^t \end{bmatrix} + C_2 \begin{bmatrix} \dfrac{1}{3}e^{2t} \\ e^{2t} \\ \dfrac{1}{3}e^{2t} \end{bmatrix} + C_3 \begin{bmatrix} e^{-t} \\ 0 \\ e^{-t} \end{bmatrix}
$$

$$
= \begin{bmatrix} -\dfrac{3}{2}e^t & \dfrac{1}{3}e^{2t} & e^{-t} \\ e^t & e^{2t} & 0 \\ -\dfrac{1}{2}e^t & \dfrac{1}{3}e^{2t} & e^{-t} \end{bmatrix} \begin{bmatrix} C_1 \\ C_2 \\ C_3 \end{bmatrix}. \tag{5.21}
$$

Each of the column vectors of the $3 \times 3$ matrix in (5.21) is a solution of the original system, (5.13), so we see the similarity of the solution to that of a single differential equation.

As a second example we consider the nonhomogeneous system

$$
\begin{aligned} Dx_1 &= -4x_1 - 6x_2 + 6e^{2t}, \\ Dx_2 &= x_1 + x_2 + 3e^{2t}, \end{aligned} \tag{5.22}
$$

subject to the initial conditions

$$
x_1(0) = -2, \qquad x_2(0) = -4. \tag{5.23}
$$

As in the previous example, we write (5.22) in operator form.

$$
\begin{aligned} (D + 4)x_1 + 6x_2 &= 6e^{2t}, \\ -x_1 + (D - 1)x_2 &= 3e^{2t}. \end{aligned} \tag{5.24}
$$

Operating on the second of equations (5.24) with $D + 4$ yields the system

$$
\begin{aligned} (D + 4)x_1 + 6x_2 &= 6e^{2t}, \\ -(D + 4)x_1 + (D + 4)(D - 1)x_2 &= (D + 4)3e^{2t} = 18e^{2t}. \end{aligned} \tag{5.25}
$$

If we now simply add the equations in (5.25), we eliminate $x_1$ and obtain

$$
[(D + 4)(D - 1) + 6]x_2 = 24e^{2t},
$$

or

$$
(D^2 + 3D + 2)x_2 = 24e^{2t}. \tag{5.26}
$$

The general solution of (5.26) is

$$
x_2(t) = C_1 e^{-2t} + C_2 e^{-t} + 2e^{2t}. \tag{5.27}
$$

We now solve the second of equations (5.24) for $x_1(t)$, obtaining

$$
x_1(t) = (D - 1)x_2(t) - 3e^{2t}. \tag{5.28}
$$

Since $x_2(t)$ is given by (5.27), we substitute directly into (5.28) to obtain

$$x_1(t) = -3C_1e^{-2t} - 2C_2e^{-t} - e^{2t}. \tag{5.29}$$

We may write the solution in vector form as

$$\begin{bmatrix} x_1(t) \\ x_2(t) \end{bmatrix} = \begin{bmatrix} -3e^{-2t} & -2e^{-t} \\ e^{-2t} & e^{-t} \end{bmatrix} \begin{bmatrix} C_1 \\ C_2 \end{bmatrix} + \begin{bmatrix} -e^{2t} \\ 2e^{2t} \end{bmatrix}. \tag{5.30}$$

The initial conditions (5.23) are now applied and (5.30) becomes

$$\begin{bmatrix} -2 \\ -4 \end{bmatrix} = \begin{bmatrix} -3 & -2 \\ 1 & 1 \end{bmatrix} \begin{bmatrix} C_1 \\ C_2 \end{bmatrix} + \begin{bmatrix} -1 \\ 2 \end{bmatrix},$$

or

$$\begin{bmatrix} -3 & -2 \\ 1 & 1 \end{bmatrix} \begin{bmatrix} C_1 \\ C_2 \end{bmatrix} = \begin{bmatrix} -1 \\ -6 \end{bmatrix}. \tag{5.31}$$

Now (5.31) is just a system of linear algebraic equations for the constants $C_1$ and $C_2$, which are easily found to be

$$C_1 = 13, \qquad C_2 = -19.$$

The method just described is straightforward but tedious and for large numbers of equations can be a chore to use. Fortunately for us, there are other methods available, although for simple systems the method of elimination can be quite useful.

As another example of a nonhomogeneous system, consider

$$(D + 2)x_1 + Dx_2 = 16e^{-2t},$$
$$2Dx_1 + (3D + 5)x_2 = 15. \tag{5.32}$$

If we operate on the first equation by $3D + 5$ and the second by $D$, and subtract, $x_2$ is eliminated and we obtain

$$(3D + 5)(D + 2)x_1 - 2D^2x_1 = (3D + 5)(16e^{-2t}) - (D)15,$$

or

$$(D^2 + 11D + 10)x_1 = -16e^{-2t}.$$

The general solution of this equation is

$$x_1(t) = C_1e^{-10t} + C_2e^{-t} + 2e^{-2t}. \tag{5.33}$$

If this form of $x_1(t)$ is substituted into the first of equations (5.32), we obtain

$$Dx_2 = 16e^{-2t} - (-8C_1e^{-10t} + C_2e^{-t}) = 16e^{-2t} + 8C_1e^{-10t} - C_2e^{-t}$$

with a general solution

$$x_2(t) = -8e^{-2t} - \frac{4}{5}C_1e^{-10t} + C_2e^{-t} + C_3. \tag{5.34}$$

The constant $C_3$ is not arbitrary, but can be determined by substituting (5.34) and (5.33) into the second of equations (5.32) as

$$C_3 = 3.$$

If the solution for $x_1(t)$ from (5.33) were substituted into the second of equations (5.32), the same solution would be obtained.

Also note that the solution may be written as

$$\begin{bmatrix} x_1(t) \\ x_2(t) \end{bmatrix} = \begin{bmatrix} 1 \\ -4/5 \end{bmatrix} e^{-10t}C_1 + \begin{bmatrix} 1 \\ 1 \end{bmatrix} e^{-t}C_2 + \begin{bmatrix} 2e^{-2t} \\ -8e^{-2t} + 3 \end{bmatrix}. \tag{5.35}$$

We should point out that the elimination method may also be applied to linear differential equations of order 2 or higher. For the system of equations

$$L_1 x_1 + L_2 x_2 = b_1(t), \tag{5.36a}$$

$$L_3 x_1 + L_4 x_2 = b_2(t), \tag{5.36b}$$

where $L_1$, $L_2$, $L_3$, and $L_4$ are differential operators with constant coefficients, operating on the first equation by $L_4$ and the second equation by $L_2$ and subtracting the results yields

$$(L_4 L_1 - L_2 L_3)x_1 = L_4 b_1(t) - L_2 b_2(t).$$

[Notice that if $L_1$, $L_2$, $L_3$, and $L_4$ were constants, this equation would give $x_1$ as the solution of (5.36a) and (5.36b) using Cramer's rule.] Similarly operating on the first equation by $L_3$ and the second equation by $L_1$ and subtracting the results yields

$$(L_1 L_4 - L_3 L_2)x_2 = L_1 b_2(t) - L_3 b_1(t).$$

Recall from Chapter 1 that since the differential operators above have constant coefficients, they commute, that is,

$$L_4 L_1 = L_1 L_4 \qquad \text{and} \qquad L_3 L_2 = L_2 L_3.$$

Thus the operators $L_4 L_1 - L_2 L_3$ and $L_1 L_4 - L_3 L_2$ in the above equations are equal. From Section 4.2.2 we know that the general solution of an $N$th order linear differential equation with constant coefficients has $N$ arbitrary constants. The number of arbitrary constants in the general solution of the system of linear differential equations with constant coefficients (5.36a) and (5.36b) will equal the order of the operator $L_1 L_4 - L_3 L_2$. This result is particularly easy to remember since $L_1 L_4 - L_3 L_2$ is the "determinant" of the operators on the left-hand side of (5.36a) and (5.36b). If $x_1(t)$ and $x_2(t)$ are each determined using the above method, the $N$ arbitrary constants in each function are related to each other. These relationships may be obtained by substituting the functions back into one or both of the original equations. To illustrate this procedure, consider

$$L_1 x_1 + L_2 x_2 = (D - 3)x_1 + Dx_2 = -4e^{2t},$$

$$L_3 x_1 + L_4 x_2 = Dx_1 + D^2 x_2 = 10e^{-t}.$$

Thus $(L_4 L_1 - L_2 L_3)x_1 = L_4 b_1(t) - L_2 b_2(t)$ becomes

$$[(D - 3)D^2 - D^2]x_1 = D^2(-4e^{2t}) - D(10e^{-t})$$

or

$$(D^3 - 4D^2)x_1 = -16e^{2t} + 10e^{-t}.$$

The general solution of this differential equation is

$$x_1(t) = C_1 + C_2 t + C_3 e^{4t} + 2e^{2t} - 2e^{-t}.$$

On the other hand, if we solve for $x_2(t)$, we have

$$(L_1 L_4 - L_3 L_2)x_2 = L_1 b_2(t) - L_3 b_1(t),$$

$$(D^3 - 4D^2)x_2 = (D - 3)(10e^{-t}) - D(-4e^{2t})$$

$$= -40e^{-t} + 8e^{2t}.$$

The general solution of this equation is

$$x_2(t) = C_4 + C_5 t + C_6 e^{4t} + 8e^{-t} - e^{2t}.$$

To determine the relationship between the two sets of arbitrary constants, we substitute the expressions for $x_1(t)$ and $x_2(t)$ into the first of the differential equations in our system and obtain

$$-3C_1 + (C_2 - 3C_2 t) + C_3 e^{4t} - 2e^{2t} + 8e^{-t} + C_5 + 4C_6 e^{4t} - 8e^{-t} - 2e^{2t}$$

$$= -4e^{2t}.$$

This equation simplifies to

$$(-3C_1 + C_2 + C_5) - 3C_2 t + (C_3 + 4C_6)e^{4t} = 0.$$

Since $1$, $t$, and $e^{4t}$ are linearly independent, we obtain

$$C_2 = 0, \qquad C_3 = -4C_6, \qquad C_5 = 3C_1,$$

with $C_1$, $C_4$, and $C_6$ chosen arbitrarily. Note that the number of arbitrary constants, 3, is the same as the order of the operator $L_1 L_4 - L_3 L_2 = D^3 - 4D^2$. Note also that if the forms of $x_1(t)$ and $x_2(t)$ are substituted into the second of the original differential equations, we obtain

$$C_2 = 0 \qquad \text{and} \qquad C_3 = -4C_6$$

but no information about the rest of the constants. The other condition, $C_5 = 3C_1$, may be obtained only when the first equation is used. Thus the fact that we know the number of arbitrary constants required in this solution may save us some work.

## EXERCISES

Find the general solution of the following systems of equations.

**1.** $Dx_1 - 2x_1 - x_2 = 0,$
$\quad x_1 + Dx_2 - 2x_2 = 0.$

**2.** $Dx_1 + 2x_1 - x_2 = 0,$
$\quad 3x_1 + Dx_2 - 2x_2 = \cos t.$

**3.** $Dx_1 - x_1 + x_2 = 0,$
$\quad 2x_1 - Dx_2 - x_2 = t.$

**4.** $\quad Dx_1 - Dx_2 = -e^t,$
$\quad 2Dx_1 - 2Dx_2 - x_2 = 8.$

**5.** $Dx_1 - 3x_1 + 2x_2 = 0,$
$\quad 2x_1 - Dx_2 - 2x_2 = 0.$

**6.** $Dx_1 - 5x_1 + 2x_2 = 3,$
$\quad 6x_1 - Dx_2 - 2x_2 = 0.$

**7.** $(D - 1)x_1 - \qquad x_2 - \qquad 2x_3 = 0,$
$\qquad (D - 2)x_2 - \qquad 2x_3 = 0,$
$\qquad x_1 - \qquad x_2 + (D - 3)x_3 = 0.$

**8.** $(D + 2)x_1 - \qquad 2x_2 - \qquad 2x_3 = 0,$
$\qquad x_1 - (D + 1)x_2 \qquad\qquad = 0,$
$\qquad x_1 + \qquad\qquad (D + 1)x_3 = 0.$

**9.** $(D - 1)x_1 - \qquad\qquad x_3 = 0,$
$\qquad (D - 1)x_2 - \qquad 2x_3 = 0,$
$\qquad x_1 + \qquad 2x_2 - (D - 5)x_3 = 0.$

**10.** $(D + 7)x_1 + \qquad\qquad 6x_3 = 0,$
$\qquad (D - 5)x_2 \qquad\qquad = 0,$
$\qquad 3x_1 - \qquad\qquad (D + 1)x_3 = 0.$

**11.** $(D + 1)x_1 + \qquad 2x_2 - \qquad 2x_3 = 0,$
$\qquad 2x_1 + (D + 1)x_2 - \qquad 2x_3 = 0,$
$\qquad 2x_1 + \qquad 2x_2 + (D - 3)x_3 = 0.$

**12.** $(D - 1)x_1 + \qquad x_2 + \qquad x_3 = 0,$
$\qquad (D - 1)x_2 - \qquad 3x_3 = 0,$
$\qquad 3x_2 - (D - 1)x_3 = 0.$

**13.** $(D + 2)x_1 + \qquad Dx_2 = 0,$
$\qquad 2Dx_1 + (3D + 5)x_2 = 0.$

**14.** $Dx_1 + (D + 2)x_2 = 0,$
$\qquad x_1 + \qquad Dx_2 = 1.$

**15.** $tDx_1 + \qquad Dx_2 = 4t,$
$\qquad Dx_1 - (tD + 1)x_2 = 0.$

**16.** $(3D - 5)x_1 + (D + 2)x_2 = 0,$
$\qquad (D + 3)x_1 + 5(D + 2)x_2 = 0.$

**17.** $(D - 3)x_1 + \qquad x_2 = e^{2t},$
$\qquad x_1 - (D - 4)x_2 = 4.$

**18.** $\qquad Dx_1 - (D + 6)x_2 = 0,$
$\qquad (D - 3)x_1 + \qquad 2Dx_2 = 0.$

**19.** $(5D^2 - 8)x_1 - \qquad 3Dx_2 = 0,$
$\qquad (5D - 4)x_1 - (4D - 5)x_2 = 0.$

**20.** $(D^2 + 2)x_1 - \qquad 4Dx_2 = 0,$
$\qquad Dx_1 + (D^2 - 4)x_2 = 0.$

**21.** $\qquad (D - 3)x_1 + \qquad\qquad Dx_2 = t^2 + 1,$
$\qquad (D^2 - 4D - 12)x_1 + 2(D^2 + 2D)x_2 = 4t + 6.$

**22.** $(2D - 1)x_1 + \qquad (D - 1)x_3 = 0,$
$\qquad (D + 3)x_1 + (D - 4)x_2 + 3x_3 = 0,$
$\qquad D^2x_1 + (D - 1)x_2 \qquad\qquad = 0.$

## 5.5 HOMOGENEOUS SYSTEMS WITH CONSTANT COEFFICIENTS (THE FUNDAMENTAL MATRIX)

Let matrices $\mathbf{A}(t) = \{a_{ij}(t)\}$ and $\mathbf{B}(t) = \{b_{ij}(t)\}$ and a scalar $f(t)$ be given and recall the following from linear algebra.

1. If $f(t)$ and $\mathbf{A}(t)$ are defined and differentiable in the same domain (see exercises 25 and 26):

   (a) $f(t)\mathbf{A}(t) = \mathbf{A}(t)f(t)$.

   (b) $\dfrac{d}{dt}[f(t)\mathbf{A}(t)] = \dfrac{df}{dt}\mathbf{A}(t) + f(t)\dfrac{d\mathbf{A}}{dt}$

   or

   $D[f(t)\mathbf{A}(t)] = [Df]\mathbf{A}(t) + f(t)D\mathbf{A}$
   where $D\mathbf{A} = d\mathbf{A}/dt = \{da_{ij}/dt\}$.

2. Because $D$ is a linear differential operator

$$D[\mathbf{A} + \mathbf{B}] = D\mathbf{A} + D\mathbf{B}$$

and we may show (see exercises 27 and 28) that

$$D[\mathbf{AB}] = [D\mathbf{A}]\mathbf{B} + \mathbf{A}[D\mathbf{B}].$$

We will now apply some of the techniques of linear algebra to systems of equations. Consider the system of first order, linear ODEs with constant coefficients

$$\frac{d\mathbf{X}}{dt} = \frac{d}{dt}\begin{bmatrix} x_1 \\ x_2 \end{bmatrix} = \begin{bmatrix} 1 & 3 \\ 2 & 2 \end{bmatrix}\begin{bmatrix} x_1 \\ x_2 \end{bmatrix} = \mathbf{AX}. \tag{5.37}$$

Since the arbitrary constants in the three examples of Section 5.4 [equations (5.21), (5.30), and (5.35)] all multiplied terms of the form $\mathbf{k}e^{\lambda t}$, where $\mathbf{k}$ is a constant vector, we assume $\mathbf{X} = \mathbf{k}e^{\lambda t}$ in (5.37), where $\mathbf{k}$ is a nonzero constant vector to be determined. Thus if $\mathbf{X} = \mathbf{k}e^{\lambda t}$ is a solution of (5.37), we have

$$\lambda\mathbf{k}e^{\lambda t} = \mathbf{A}\mathbf{k}e^{\lambda t},$$

or

$$(\mathbf{A} - \lambda\mathbf{I})\mathbf{k}e^{\lambda t} = \mathbf{0}.$$

Since $e^{\lambda t} \neq 0$, $(\mathbf{A} - \lambda\mathbf{I})\mathbf{k} = \mathbf{0}$ and the system has nontrivial solutions for $\mathbf{k}$ only if

$$\det(\mathbf{A} - \lambda\mathbf{I}) = 0,$$

so we see that $\lambda$ must be an eigenvalue of the matrix $\mathbf{A}$. Now $(\mathbf{A} - \lambda\mathbf{I})\mathbf{k} = \mathbf{0}$ becomes

$$\begin{bmatrix} 1 - \lambda & 3 \\ 2 & 2 - \lambda \end{bmatrix}\begin{bmatrix} k_1 \\ k_2 \end{bmatrix} = \begin{bmatrix} 0 \\ 0 \end{bmatrix} \tag{5.38}$$

so that

$$\det(\mathbf{A} - \lambda\mathbf{I}) = \lambda^2 - 3\lambda + 2 - 6$$

$$= \lambda^2 - 3\lambda - 4 = (\lambda - 4)(\lambda + 1)$$

and we have eigenvalues $\lambda_1 = 4$, $\lambda_2 = -1$. The corresponding eigenvectors are then found by substituting these eigenvalues into (5.38). For $\lambda_1 = 4$, we get the system of equations

$$\begin{bmatrix} -3 & 3 \\ 2 & -2 \end{bmatrix}\begin{bmatrix} k_1 \\ k_2 \end{bmatrix} = \begin{bmatrix} 0 \\ 0 \end{bmatrix}$$

or

$$-3k_1 = -3k_2,$$

$$2k_1 = 2k_2.$$

Thus $k_2 = k_1$ and $k_1$ is arbitrary. We choose $k_1 = 1$ so that $k_2 = 1$.

For $\lambda = -1$ we have the system of equations

$$\begin{bmatrix} 2 & 3 \\ 2 & 3 \end{bmatrix}\begin{bmatrix} k_1 \\ k_2 \end{bmatrix} = \begin{bmatrix} 0 \\ 0 \end{bmatrix}.$$

Thus $2k_1 = -3k_2$, and the eigenvectors may be taken to have the form $(3a, -2a)$ where $a \neq 0$ is arbitrary. For the specific choice $a = -1$ we have the eigenvector

$$\begin{bmatrix} -3 \\ 2 \end{bmatrix}.$$

The eigenvectors are

$$\text{for } \lambda = 4, \mathbf{k} = \begin{bmatrix} 1 \\ 1 \end{bmatrix}, \quad \text{for } \lambda = -1, \mathbf{k} = \begin{bmatrix} -3 \\ 2 \end{bmatrix}.$$

The corresponding solutions of (5.37) are

$$\begin{bmatrix} 1 \\ 1 \end{bmatrix} e^{4t} \quad \text{and} \quad \begin{bmatrix} -3 \\ 2 \end{bmatrix} e^{-t}$$

and the general solution is given by

$$\mathbf{X}(t) = \begin{bmatrix} e^{4t} \\ e^{4t} \end{bmatrix} C_1 + \begin{bmatrix} -3e^{-t} \\ 2e^{-t} \end{bmatrix} C_2. \tag{5.39}$$

Defining

$$\mathbf{U}(t) = [\mathbf{X}_1(t), \mathbf{X}_2(t)] = \begin{bmatrix} e^{4t} & -3e^{-t} \\ e^{4t} & 2e^{-t} \end{bmatrix}$$

allows us to write the general solution as

$$\mathbf{X}(t) = \mathbf{U}(t)\mathbf{C} \quad \text{with } \mathbf{C} = \begin{bmatrix} C_1 \\ C_2 \end{bmatrix}.$$

In general, the system $D\mathbf{X} = \mathbf{AX}$ will have a solution of the form $\mathbf{X} = e^{\lambda t}\mathbf{k}$ if $\mathbf{k}$ is an eigenvector of $\mathbf{A}$ corresponding to the eigenvalue $\lambda$.

If $\mathbf{A}$ is an $n \times n$ matrix, the general solution of

$$D\mathbf{X} = \mathbf{AX} \tag{5.40}$$

will have the form

$$\mathbf{X}(t) = [\mathbf{X}_1(t), \mathbf{X}_2(t), \ldots, \mathbf{X}_n(t)] \begin{bmatrix} C_1 \\ C_2 \\ \vdots \\ C_n \end{bmatrix},$$

where the $n$ vectors, $\mathbf{X}_i(t)$, $i = 1, 2, 3, \ldots, n$, all satisfy (5.40) and are linearly independent. The matrix

$$\mathbf{U}(t) = [\mathbf{X}_1(t), \mathbf{X}_2(t), \ldots, \mathbf{X}_n(t)] \tag{5.41}$$

is called *a fundamental matrix* associated with (5.40). If $\det \mathbf{U}(t_0) \neq 0$ for all $t_0$, the set of functions $\{\mathbf{X}_i(t) \mid i = 1, 2, 3, \ldots, n\}$ will be linearly independent. However, since all the $\mathbf{X}_i(t)$ satisfy (5.40), we have the stronger result that either $\det \mathbf{U}(t_0) \neq 0$ for all $t_0$ or $\det \mathbf{U}(t_0) = 0$ for all $t_0$. Thus to see if $\mathbf{U}(t)$ forms a fundamental matrix, compute $\mathbf{U}(t_0)$ for a convenient choice of $t_0$. If $\det \mathbf{U}(t_0) \neq 0$, the answer is yes, but if $\det \mathbf{U}(t_0) = 0$, the set $\{\mathbf{X}_i(t) \,|\, i = 1, 2, 3, \ldots, n\}$ is a dependent set and the corresponding $\mathbf{U}(t)$ is not a fundamental matrix.

The role of a fundamental matrix in solving systems of linear differential equations is summarized in the following theorem.

**Theorem 5.5.**    The general solution of the system of equations (5.11) may be expressed as $\mathbf{X}(t) = \mathbf{U}(t)\mathbf{C}$, where $\mathbf{U}(t)$ is a fundamental matrix and $\mathbf{C}$ is a column vector of arbitrary constants.

While all the examples in this section are concerned with coefficient matrices that contain only constants, Theorem 5.5 is also valid for nonconstant-coefficient matrices as long as $\mathbf{U}(t)$ is a fundamental matrix [i.e., is composed of linearly independent solutions of (5.11)].

As another example, consider the homogeneous system

$$DX = AX \tag{5.42}$$

with

$$\mathbf{A} = \begin{bmatrix} -2 & 4 \\ 1 & 1 \end{bmatrix}, \qquad \mathbf{X} = \begin{bmatrix} x_1 \\ x_2 \end{bmatrix}$$

and an initial condition

$$\mathbf{X}(0) = \begin{bmatrix} -2 \\ 3 \end{bmatrix}. \tag{5.43}$$

We substitute a solution of the form

$$\mathbf{X} = \mathbf{k}e^{\lambda t}$$

into (5.42) and get the system of equations

$$\begin{bmatrix} -2 - \lambda & 4 \\ 1 & 1 - \lambda \end{bmatrix} \begin{bmatrix} k_1 \\ k_2 \end{bmatrix} = \begin{bmatrix} 0 \\ 0 \end{bmatrix}. \tag{5.44}$$

We require

$$\det \begin{bmatrix} -2 - \lambda & 4 \\ 1 & 1 - \lambda \end{bmatrix} = 0,$$

which gives

$$(\lambda + 2)(\lambda - 1) - 4 = \lambda^2 + \lambda - 2 - 4 = 0,$$

or

$$(\lambda + 3)(\lambda - 2) = 0.$$

Thus the eigenvalues of $\mathbf{A}$ are

$$\lambda_1 = -3, \qquad \lambda_2 = 2.$$

Substituting into (5.44) we obtain for $\lambda_1 = -3$,

$$\begin{bmatrix} 1 & 4 \\ 1 & 4 \end{bmatrix} \begin{bmatrix} k_1 \\ k_2 \end{bmatrix} = \begin{bmatrix} 0 \\ 0 \end{bmatrix},$$

or

$$-k_1 = 4k_2,$$
$$k_1 = -4k_2,$$

so we have solutions of the form

$$\begin{bmatrix} -4k_2 \\ k_2 \end{bmatrix},$$

where $k_2$ is arbitrary. If we choose $k_2 = 1$, we get the eigenvector

$$\begin{bmatrix} -4 \\ 1 \end{bmatrix}$$

and the corresponding solution

$$\mathbf{X}_1(t) = \begin{bmatrix} -4e^{-3t} \\ e^{-3t} \end{bmatrix}.$$

For $\lambda_2 = 2$,

$$\begin{bmatrix} -4 & 4 \\ 1 & -1 \end{bmatrix} \begin{bmatrix} k_1 \\ k_2 \end{bmatrix} = \begin{bmatrix} 0 \\ 0 \end{bmatrix}$$

gives

$$k_1 = k_2$$

so we get eigenvectors of the form

$$\begin{bmatrix} k_2 \\ k_2 \end{bmatrix},$$

where $k_2$ is arbitrary. Choosing $k_2 = 1$, we get the eigenvector

$$\begin{bmatrix} 1 \\ 1 \end{bmatrix}$$

and the solution

$$\mathbf{X}_2(t) = \begin{bmatrix} e^{2t} \\ e^{2t} \end{bmatrix}.$$

Thus we have a fundamental matrix

$$\mathbf{U}(t) = \begin{bmatrix} -4e^{-3t} & e^{2t} \\ e^{-3t} & e^{2t} \end{bmatrix}.$$

Since $\det \mathbf{U}(0) = -5 \neq 0$, $\mathbf{X}_1(t)$ and $\mathbf{X}_2(t)$ are linearly independent, and the general solution of (5.42) is given by

$$\mathbf{X}(t) = \mathbf{U}(t)\mathbf{C}, \qquad \mathbf{C} = \begin{bmatrix} C_1 \\ C_2 \end{bmatrix}.$$

The initial condition $\mathbf{X}(0) = \begin{bmatrix} -2 \\ 3 \end{bmatrix}$ enables us to determine $C_1$ and $C_2$. Since

$$\begin{bmatrix} -2 \\ 3 \end{bmatrix} = \begin{bmatrix} -4 & 1 \\ 1 & 1 \end{bmatrix}\begin{bmatrix} C_1 \\ C_2 \end{bmatrix}$$

or

$$-4C_1 + C_2 = -2,$$
$$C_1 + C_2 = 3,$$

we have

$$-5C_1 = -5,$$

so

$$C_1 = 1$$

and

$$C_2 = 2.$$

Hence

$$\mathbf{X}(t) = \begin{bmatrix} -4e^{-3t} & e^{2t} \\ e^{-3t} & e^{2t} \end{bmatrix}\begin{bmatrix} 1 \\ 2 \end{bmatrix} = \begin{bmatrix} -4e^{-3t} + 2e^{2t} \\ e^{-3t} + 2e^{2t} \end{bmatrix}$$

is the solution of (5.42) and (5.43).

As our next example, consider

$$D\mathbf{X} = D\begin{bmatrix} x_1 \\ x_2 \end{bmatrix} = \begin{bmatrix} 2 & 1 \\ -1 & 2 \end{bmatrix}\begin{bmatrix} x_1 \\ x_2 \end{bmatrix}, \qquad \mathbf{X}(0) = \begin{bmatrix} 2 \\ 3 \end{bmatrix}.$$

Assuming $\mathbf{X} = \mathbf{k}e^{\lambda t}$ as a solution gives the system of algebraic equations

$$\begin{bmatrix} 2 - \lambda & 1 \\ -1 & 2 - \lambda \end{bmatrix}\begin{bmatrix} k_1 \\ k_2 \end{bmatrix} = \begin{bmatrix} 0 \\ 0 \end{bmatrix}.$$

Thus the eigenvalues are given by the solution of

$$(2 - \lambda)(2 - \lambda) + 1 = 5 - 4\lambda + \lambda^2 = 0$$

as

$$\lambda_1 = 2 + i \qquad \text{and} \qquad \lambda_2 = 2 - i.$$

If we use the eigenvalue $\lambda_1 = 2 + i$ in the system of algebraic equations above, we obtain

$$\begin{bmatrix} -i & 1 \\ -1 & -i \end{bmatrix}\begin{bmatrix} k_1 \\ k_2 \end{bmatrix} = \begin{bmatrix} 0 \\ 0 \end{bmatrix}$$

or

$$ik_1 = k_2,$$
$$-k_1 = ik_2.$$

If we choose $k_1 = 1$, then $k_2 = i$, the eigenvector is

$$\begin{bmatrix} 1 \\ i \end{bmatrix}$$

and the corresponding solution is

$$\begin{bmatrix} 1 \\ i \end{bmatrix} e^{(2+i)t} = \begin{bmatrix} 1 \\ i \end{bmatrix} e^{2t} e^{it} = \begin{bmatrix} 1 \\ i \end{bmatrix} e^{2t}(\cos t + i \sin t).$$

If we write the above solution in terms of real and imaginary parts, we obtain

$$\begin{bmatrix} 1 \\ i \end{bmatrix} e^{(2+i)t} = \begin{bmatrix} e^{2t} \cos t \\ -e^{2t} \sin t \end{bmatrix} + i \begin{bmatrix} e^{2t} \sin t \\ e^{2t} \cos t \end{bmatrix}.$$

Since the original system of differential equations contained only real numbers, both the real and imaginary parts of this solution must satisfy the system (see exercise 11). Thus we may form a fundamental matrix as

$$\mathbf{U}(t) = \begin{bmatrix} e^{2t} \cos t & e^{2t} \sin t \\ -e^{2t} \sin t & e^{2t} \cos t \end{bmatrix}$$

(note that det $\mathbf{U}(t) = e^{4t} \neq 0$ for any finite value of $t$) and write our solution as

$$\mathbf{X}(t) = \mathbf{U}(t)\mathbf{C} = \begin{bmatrix} e^{2t} \cos t & e^{2t} \sin t \\ -e^{2t} \sin t & e^{2t} \cos t \end{bmatrix} \begin{bmatrix} C_1 \\ C_2 \end{bmatrix}.$$

To satisfy the initial condition, we set

$$\mathbf{X}(0) = \begin{bmatrix} 2 \\ 3 \end{bmatrix}$$

and obtain

$$\mathbf{U}(0)\mathbf{C} = \begin{bmatrix} 1 & 0 \\ 0 & 1 \end{bmatrix} \begin{bmatrix} C_1 \\ C_2 \end{bmatrix} = \begin{bmatrix} 2 \\ 3 \end{bmatrix}.$$

Thus $C_1 = 2$, $C_2 = 3$, and the solution of the initial value problem is given by

$$\begin{bmatrix} x_1(t) \\ x_2(t) \end{bmatrix} = \begin{bmatrix} e^{2t} \cos t & e^{2t} \sin t \\ -e^{2t} \sin t & e^{2t} \cos t \end{bmatrix} \begin{bmatrix} 2 \\ 3 \end{bmatrix} = \begin{bmatrix} 2e^{2t} \cos t + 3e^{2t} \sin t \\ -2e^{2t} \sin t + 3e^{2t} \cos t \end{bmatrix}.$$

Notice that for this complex eigenvalue, we obtained two linearly independent solutions by considering only one of the eigenvalues. If we had considered the other eigenvalue, $2 - i$, we would also have obtained two linearly independent solutions. Solving the initial value problem using this eigenvalue will give the same solution as above (see exercise 12). [Recall that in using the characteristic equation to solve second order differential equations with constant coefficients, we found two linearly independent solutions from one complex root. For example, $y''(t) + 4y'(t) + 13y(t) = 0$ has a characteristic equation of $r^2 + 4r + 13 = 0$ and roots $r = 2 \pm i3$. From $r = 2 + i3$ we have $e^{(2+i3)t} = e^{2t}(\cos 3t + i \sin 3t)$, so $e^{2t} \cos 3t$ and $e^{2t} \sin 3t$ are linearly independent solutions.]

For some systems the details of solution are very simple. The system

$$D\begin{bmatrix} x_1 \\ x_2 \end{bmatrix} = \begin{bmatrix} -2 & 0 \\ 0 & -2 \end{bmatrix}\begin{bmatrix} x_1 \\ x_2 \end{bmatrix},$$

for example, has a characteristic equation

$$(-2 - \lambda)^2 = 0$$

and one eigenvalue, $\lambda = -2$, of multiplicity 2. Substitution into the equation for the eigenvectors,

$$\begin{bmatrix} -2 - \lambda & 0 \\ 0 & -2 - \lambda \end{bmatrix}\begin{bmatrix} k_1 \\ k_2 \end{bmatrix} = \begin{bmatrix} 0 \\ 0 \end{bmatrix},$$

yields the system

$$0k_1 + 0k_2 = 0,$$
$$0k_1 + 0k_2 = 0.$$

Since $k_1$ and $k_2$ are both arbitrary, they may be chosen independently. To get a set of two linearly independent vectors we take vectors of the form

$$\begin{bmatrix} 1 \\ 0 \end{bmatrix} \quad \text{and} \quad \begin{bmatrix} 0 \\ 1 \end{bmatrix}$$

as the simplest possible choice; hence a fundamental matrix is

$$\mathbf{U}(t) = \begin{bmatrix} e^{-2t} & 0 \\ 0 & e^{-2t} \end{bmatrix}.$$

Not all systems with eigenvalues of multiplicity 2 may be so easy. Consider the simple system

$$DX = D\begin{bmatrix} x_1 \\ x_2 \end{bmatrix} = \begin{bmatrix} 1 & 1 \\ -1 & 3 \end{bmatrix} = \mathbf{A}\mathbf{X}. \tag{5.45}$$

The coefficient matrix of (5.45) has the single eigenvalue $\lambda = 2$ of multiplicity 2 and the usual procedure yields a single family of eigenvectors of the form

$$a\begin{bmatrix} 1 \\ 1 \end{bmatrix}$$

and a single solution ($a = 1$)

$$\mathbf{X}_1(t) = \begin{bmatrix} e^{2t} \\ e^{2t} \end{bmatrix}. \tag{5.46}$$

To construct the general solution to (5.45), we must find a second solution which is independent of (5.46). By analogy with what we did for single equations whose characteristic equations had multiple roots, we try solutions of the form

$$\mathbf{X}_2(t) = (t\mathbf{k}_1 + \mathbf{k}_2)e^{2t}$$
$$= \left( t\begin{bmatrix} \alpha_1 \\ \beta_1 \end{bmatrix} + \begin{bmatrix} \alpha_2 \\ \beta_2 \end{bmatrix} \right)e^{2t}, \tag{5.47}$$

where $\alpha_1$, $\beta_1$, $\alpha_2$, $\beta_2$ are the components of $\mathbf{k}_1$ and $\mathbf{k}_2$ subject to the requirement that $\mathbf{X}_1(t)$ and $\mathbf{X}_2(t)$ be linearly independent. Substitution of (5.47) into (5.45) yields

$$D[(t\mathbf{k}_1 + \mathbf{k}_2)e^{2t}] = \mathbf{k}_1 e^{2t} + (t\mathbf{k}_1 + \mathbf{k}_2)2e^{2t} = \mathbf{A}(t\mathbf{k}_1 + \mathbf{k}_2)e^{2t}.$$

Since $e^{2t}$ and $te^{2t}$ are linearly independent, we may equate coefficients of like terms to obtain

$$\mathbf{k}_1 + 2\mathbf{k}_2 = \mathbf{A}\mathbf{k}_2,$$
$$2\mathbf{k}_1 = \mathbf{A}\mathbf{k}_1.$$

In terms of the components of $\mathbf{k}_1$ and $\mathbf{k}_2$, we have

$$\alpha_1 + 2\alpha_2 = \alpha_2 + \beta_2, \qquad 2\alpha_1 = \alpha_1 + \beta_1,$$

and                                                                                                    (5.48)

$$\beta_1 + 2\beta_2 = -\alpha_2 + 3\beta_2, \qquad 2\beta_1 = -\alpha_1 + 3\beta_1,$$

which is equivalent to

$$\alpha_2 - \beta_2 = -\alpha_1, \qquad \alpha_1 - \beta_1 = 0,$$
$$\alpha_2 - \beta_2 = -\beta_1, \qquad \alpha_1 - \beta_1 = 0.$$

Thus $\alpha_1 = \beta_1$ and solving for $\alpha_2$ gives $\alpha_2 = \beta_2 - \beta_1$, and the components of $\mathbf{k}_1$ and $\mathbf{k}_2$ are related by

$$\alpha_1 = \beta_1,$$
                                                                                                       (5.49)
$$\alpha_2 = \beta_2 - \beta_1.$$

If in (5.49) we choose $\beta_1 = 1$, $\beta_2 = 0$, we get

$$\alpha_1 = 1, \qquad \beta_1 = 1, \qquad \alpha_2 = -1, \qquad \beta_2 = 0$$

and a second solution is

$$\mathbf{X}_2(t) = te^{2t}\begin{bmatrix} 1 \\ 1 \end{bmatrix} + e^{2t}\begin{bmatrix} -1 \\ 0 \end{bmatrix}$$

$$= \begin{bmatrix} (t - 1)e^{2t} \\ te^{2t} \end{bmatrix}.$$                                             (5.50)

Now from (5.46) and (5.50) we form the matrix

$$\mathbf{U}(t) = [\mathbf{X}_1(t), \mathbf{X}_2(t)] = \begin{bmatrix} e^{2t} & (t - 1)e^{2t} \\ e^{2t} & te^{2t} \end{bmatrix}.$$   (5.51)

Since

$$\det \mathbf{U}(0) = \begin{vmatrix} 1 & -1 \\ 1 & 0 \end{vmatrix} = 1$$

$\mathbf{X}_1(t)$ and $\mathbf{X}_2(t)$ are linearly independent and (5.51) is a fundamental matrix for (5.45). Thus the general solution of (5.45) is

$$\mathbf{X}(t) = \begin{bmatrix} e^{2t} & (t - 1)e^{2t} \\ e^{2t} & te^{2t} \end{bmatrix}\begin{bmatrix} C_1 \\ C_2 \end{bmatrix}.$$

This procedure is generally valid. If we have an eigenvector of multiplicity $h$, we try solutions of the form

$$\mathbf{X}(t) = (t^{h-1}\mathbf{k}_h + t^{h-2}\mathbf{k}_{h-1} + \cdots + \mathbf{k}_1)e^{\lambda t}. \tag{5.52}$$

Substitution of (5.52) into the system to be solved permits the $h$ vectors $\mathbf{k}_i$, $i = 1, 2, 3, \ldots, h$, to be determined.

To conclude this section, consider the task of finding the fundamental matrix for the system

$$D\begin{bmatrix} x_1 \\ x_2 \\ x_3 \\ x_4 \end{bmatrix} = \begin{bmatrix} 1 & 1 & 0 & 0 \\ 0 & 1 & 2 & 0 \\ 0 & 0 & 1 & 0 \\ 0 & -2 & 0 & 1 \end{bmatrix}\begin{bmatrix} x_1 \\ x_2 \\ x_3 \\ x_4 \end{bmatrix}, \tag{5.53}$$

or $D\mathbf{X} = \mathbf{AX}$. Assuming a solution of the form $\mathbf{k}e^{\lambda t}$ gives the matrix equation

$$\begin{bmatrix} 1-\lambda & 1 & 0 & 0 \\ 0 & 1-\lambda & 2 & 0 \\ 0 & 0 & 1-\lambda & 0 \\ 0 & -2 & 0 & 1-\lambda \end{bmatrix}\begin{bmatrix} k_1 \\ k_2 \\ k_3 \\ k_4 \end{bmatrix} = \begin{bmatrix} 0 \\ 0 \\ 0 \\ 0 \end{bmatrix}. \tag{5.54}$$

Now $\det[\mathbf{A} - \lambda\mathbf{I}] = (1 - \lambda)^4 = 0$ shows that $\lambda = 1$ is a root of multiplicity 4. Finding eigenvectors from (5.54) gives

$$\begin{bmatrix} 0 & 1 & 0 & 0 \\ 0 & 0 & 2 & 0 \\ 0 & 0 & 0 & 0 \\ 0 & -2 & 0 & 0 \end{bmatrix}\begin{bmatrix} k_1 \\ k_2 \\ k_3 \\ k_4 \end{bmatrix} = \begin{bmatrix} 0 \\ 0 \\ 0 \\ 0 \end{bmatrix}.$$

Since $k_2 = k_3 = 0$ and $k_1$ and $k_4$ are arbitrary, the set of all eigenvectors of (5.54) is given by a linear combination of

$$\begin{bmatrix} 1 \\ 0 \\ 0 \\ 0 \end{bmatrix} \quad \text{and} \quad \begin{bmatrix} 0 \\ 0 \\ 0 \\ 1 \end{bmatrix},$$

and

$$\mathbf{X}_1(t) = \begin{bmatrix} e^t \\ 0 \\ 0 \\ 0 \end{bmatrix} \quad \text{and} \quad \mathbf{X}_2(t) = \begin{bmatrix} 0 \\ 0 \\ 0 \\ e^t \end{bmatrix}$$

are two linearly independent solutions.

While the general form of (5.52) suggests a solution of the form

$$(t^3\mathbf{k}_4 + t^2\mathbf{k}_3 + t\mathbf{k}_2 + \mathbf{k}_1)e^t,$$

we need only two more linearly independent solutions, so we try to get by with

$$\mathbf{X} = (t^2\mathbf{k}_3 + t\mathbf{k}_2 + \mathbf{k}_1)e^t. \tag{5.55}$$

If this does not give these two solutions, we try the more complicated form.

Substituting (5.55) into $D\mathbf{X} = \mathbf{A}\mathbf{X}$ gives

$$(2t\mathbf{k}_3 + t^2\mathbf{k}_3 + \mathbf{k}_2 + t\mathbf{k}_2 + \mathbf{k}_1) = \mathbf{A}(t^2\mathbf{k}_3 + t\mathbf{k}_2 + \mathbf{k}_1) \tag{5.56}$$

and since $1$, $t$, and $t^2$ are linearly independent, this is equivalent to

$$\mathbf{k}_3 = \mathbf{A}\mathbf{k}_3, \qquad 2\mathbf{k}_3 + \mathbf{k}_2 = \mathbf{A}\mathbf{k}_2, \qquad \mathbf{k}_2 + \mathbf{k}_1 = \mathbf{A}\mathbf{k}_1. \tag{5.57}$$

If $\mathbf{k}_i$ has components

$$\begin{bmatrix} \alpha_i \\ \beta_i \\ \gamma_i \\ \delta_i \end{bmatrix}, \; i = 1, 2, 3,$$

(5.57) is equivalent to

$$\alpha_3 = \alpha_3 + \beta_3,$$
$$\beta_3 = \beta_3 + 2\gamma_3,$$
$$\gamma_3 = \gamma_3,$$
$$\delta_3 = \delta_3 - 2\beta_3,$$
$$2\alpha_3 + \alpha_2 = \alpha_2 + \beta_2,$$
$$2\beta_3 + \beta_2 = \beta_2 + 2\gamma_2,$$
$$2\gamma_3 + \gamma_2 = \gamma_2,$$
$$2\delta_3 + \delta_2 = \delta_2 - 2\beta_2,$$
$$\alpha_2 + \alpha_1 = \alpha_1 + \beta_1,$$
$$\beta_2 + \beta_1 = \beta_1 + 2\gamma_1,$$
$$\gamma_2 + \gamma_1 = \gamma_1,$$
$$\delta_2 + \delta_1 = \delta_1 - 2\beta_1,$$

with solution given by

$$\beta_3 = \gamma_3 = \gamma_2 = 0,$$
$$2\alpha_3 = \beta_2,$$
$$\alpha_2 = \beta_1,$$
$$\beta_2 = 2\gamma_1,$$
$$\delta_3 = -\beta_2,$$
$$\delta_2 = -2\beta_1.$$

Thus $\beta_1$, $\alpha_1$, $\gamma_1$, and $\delta_1$ may be chosen arbitrarily.

If we let $\beta_1 = 1$, $\delta_1 = \gamma_1 = \alpha_1 = 0$, we obtain the solution

$$\mathbf{X}_3(t) = \begin{bmatrix} te^t \\ e^t \\ 0 \\ -2te^t \end{bmatrix},$$

while the choice $\gamma_1 = 1$, $\beta_1 = 2$, $\alpha_1 = \delta_1 = 0$ gives

$$\mathbf{X}_4(t) = \begin{bmatrix} (t^2 + 2t)e^t \\ (2t + 2)e^t \\ e^t \\ (-2t^2 - 4t)e^t \end{bmatrix}.$$

Thus a fundamental matrix is

$$\mathbf{U}(t) = \begin{bmatrix} e^t & 0 & te^t & (t^2 + 2t)e^t \\ 0 & 0 & e^t & (2t + 2)e^t \\ 0 & 0 & 0 & e^t \\ 0 & e^t & -2te^t & (-2t^2 - 4t)e^t \end{bmatrix}.$$

Since $\det \mathbf{U}(0) = 1$, $\mathbf{U}(t)$ is composed of linearly independent columns. Notice that the choice $\beta_1 = \gamma_1 = 0$ and $\alpha_1 = \delta_1 = 1$ gives rise to a vector that is the sum of $\mathbf{X}_1(t)$ and $\mathbf{X}_2(t)$ and thus is not a new linearly independent vector.

## EXERCISES

Solve the following initial value problems by first finding a fundamental matrix.

1. $D\mathbf{X} = \mathbf{AX}$,  with $\mathbf{A} = \begin{bmatrix} 2 & 5 \\ 1 & 6 \end{bmatrix}$,  $\mathbf{X}(0) = \begin{bmatrix} 3 \\ -2 \end{bmatrix}$.

2. $D\mathbf{X} = \mathbf{AX}$,  with $\mathbf{A} = \begin{bmatrix} 2 & -1 \\ -6 & 1 \end{bmatrix}$,  $\mathbf{X}(0) = \begin{bmatrix} 2 \\ 0 \end{bmatrix}$.

3. $D\mathbf{X} = \mathbf{AX}$,  with $\mathbf{A} = \begin{bmatrix} 5 & 6 \\ 1 & 4 \end{bmatrix}$,  $\mathbf{X}(0) = \begin{bmatrix} 5 \\ 10 \end{bmatrix}$.

4. $D\mathbf{X} = \mathbf{AX}$,  with $\mathbf{A} = \begin{bmatrix} 5 & -3 \\ -1 & -1 \end{bmatrix}$,  $\mathbf{X}(0) = \begin{bmatrix} 0 \\ 1 \end{bmatrix}$.

5. $D\mathbf{X} = \mathbf{AX}$,  with $\mathbf{A} = \begin{bmatrix} 2 & 9 \\ -1 & -4 \end{bmatrix}$,  $\mathbf{X}(0) = \begin{bmatrix} 3 \\ 7 \end{bmatrix}$.

6. $D\mathbf{X} = \mathbf{AX}$,  with $\mathbf{A} = \begin{bmatrix} 2 & 1 \\ 1 & -4 \end{bmatrix}$,  $\mathbf{X}(0) = \begin{bmatrix} 1 \\ 0 \end{bmatrix}$.

7. $D\mathbf{X} = \mathbf{AX}$,  with $\mathbf{A} = \begin{bmatrix} 4 & -2 \\ 5 & -2 \end{bmatrix}$,  $\mathbf{X}(0) = \begin{bmatrix} 1 \\ 2 \end{bmatrix}$.

8. $D\mathbf{X} = \mathbf{AX}$,  with $\mathbf{A} = \begin{bmatrix} 2 & 1 \\ -4 & 2 \end{bmatrix}$,  $\mathbf{X}(0) = \begin{bmatrix} 3 \\ 4 \end{bmatrix}$.

9. $D\mathbf{X} = \mathbf{AX}$,  with $\mathbf{A} = \begin{bmatrix} 3 & 5 \\ -5 & 3 \end{bmatrix}$,  $\mathbf{X}(0) = \begin{bmatrix} 0 \\ 1 \end{bmatrix}$.

**10.** $DX = AX$, with $A = \begin{bmatrix} 1 & 1 \\ 1 & 3 \end{bmatrix}$, $X(0) = \begin{bmatrix} 0 \\ 3 \end{bmatrix}$.

**11.** Show that if $X = U + iV$ is a solution of $DX = AX$, where $A$ is a real matrix, then both $U$ and $V$ are also solutions.

**12.** Solve

$$DX = \begin{bmatrix} 2 & 1 \\ -1 & 2 \end{bmatrix} \begin{bmatrix} x_1 \\ x_2 \end{bmatrix}, \qquad X(0) = \begin{bmatrix} 2 \\ 3 \end{bmatrix}$$

using the eigenvalue $2 - i$.

Find the general solution of the following problems in terms of a fundamental matrix.

**13.** $DX = AX$, with $A = \begin{bmatrix} 1 & 2 & 1 \\ 0 & -1 & 6 \\ 1 & 2 & 1 \end{bmatrix}$.

**14.** $DX = AX$, with $A = \begin{bmatrix} 12 & -3 & -3 \\ -3 & 9 & 0 \\ -3 & 0 & 9 \end{bmatrix}$.

**15.** $DX = AX$, with $A = \begin{bmatrix} 3 & -1 & -1 \\ -1 & 3 & -1 \\ -1 & -1 & 3 \end{bmatrix}$.

**16.** $DX = AX$, with $A = \begin{bmatrix} 2 & 0 & 9 \\ 0 & 3 & 0 \\ 1 & 0 & 2 \end{bmatrix}$.

**17.** $DX = AX$, with $A = \begin{bmatrix} 2 & -2 & 0 \\ 1 & -2 & -1 \\ -2 & 1 & -2 \end{bmatrix}$.

**18.** $DX = AX$, with $A = \begin{bmatrix} -1 & 0 & 0 \\ 2 & -1 & 0 \\ 3 & 5 & -1 \end{bmatrix}$.

**19.** $DX = AX$, with $A = \begin{bmatrix} 2 & 0 & 5 \\ 0 & 1 & 2 \\ -4 & 5 & 0 \end{bmatrix}$.

**20.** $DX = AX$, with $A = \begin{bmatrix} 0 & 3 & 3 \\ 2 & -1 & -3 \\ 0 & -1 & -1 \end{bmatrix}$.

Solve the following initial value problems.

**21.** Exercise 15 with $X(0) = \begin{bmatrix} 1 \\ 0 \\ 0 \end{bmatrix}$.

**22.** Exercise 16 with $X(0) = \begin{bmatrix} 1 \\ 0 \\ 1 \end{bmatrix}$.

**23.** Exercise 19 with $X(0) = \begin{bmatrix} 0 \\ 1 \\ 1 \end{bmatrix}$.

**24.** Exercise 20 with $X(0) = \begin{bmatrix} 3 \\ 0 \\ 7 \end{bmatrix}$.

**25.** Given

$$\mathbf{A}(t) = \begin{bmatrix} t & e^t \\ 3 & 3t^2 \end{bmatrix}, \qquad \mathbf{B}(t) = \begin{bmatrix} 3^{3t} & e^{2t} \\ e^{-t} & e^{4t} \end{bmatrix}.$$

(a) Compute $t^2 \mathbf{A}(t)$.

(b) Compute $\dfrac{d\mathbf{A}}{dt} = D\mathbf{A}$.

(c) Verify that $D[t^2 \mathbf{A}(t)] = 2t\mathbf{A}(t) + t^2 D\mathbf{A}$.

(d) Compute $e^{-2t}\mathbf{B}(t)$.

(e) Compute $\dfrac{d\mathbf{B}}{dt} = D\mathbf{B}$.

(f) Verify that $D[e^{-2t}\mathbf{B}(t)] = -2e^{-2t}\mathbf{B}(t) + e^{-2t}D\mathbf{B}$.

**26.** If $\mathbf{A}$ is an $m \times n$ matrix whose entries are functions of $t$, prove that

$$\frac{d}{dt}[f(t)\mathbf{A}] = f'(t)\mathbf{A} + f(t)\frac{d\mathbf{A}}{dt}.$$

**27.** For $\mathbf{A}(t)$ and $\mathbf{B}(t)$ as in exercise 25, show that

$$D[\mathbf{AB}] = [D\mathbf{A}]\mathbf{B} + \mathbf{A}[D\mathbf{B}].$$

**28.** If $\mathbf{A}$, $\mathbf{B}$, and $\mathbf{C}$ are functions of $t$ defined as follows:

$$\mathbf{A} = \{a_{ij}\}, \qquad i = 1, 2, 3, \ldots, n, \quad j = 1, 2, 3, \ldots, p$$

$$\mathbf{B} = \{b_{jk}\}, \qquad j = 1, 2, 3, \ldots, p, \quad k = 1, 2, 3, \ldots, m$$

and $\mathbf{AB} = \mathbf{C}$, where $\mathbf{C} = \{c_{ik}\}$, $c_{ik} = \sum_{j=1}^{p} a_{ij} b_{jk}$, and $D = d/dt$, prove that $D[\mathbf{AB}] = [D\mathbf{A}]\mathbf{B} + \mathbf{A}[D\mathbf{B}]$.

## 5.6 NONHOMOGENEOUS SYSTEMS: VARIATION OF PARAMETERS

The general solution of

$$D\mathbf{X} = \mathbf{AX} + \mathbf{B} \tag{5.58}$$

is given by

$$\mathbf{X} = \mathbf{UC} + \mathbf{X}_p,$$

where $\mathbf{U}$ is a fundamental matrix, $\mathbf{C}$ a vector of arbitrary constants, and $\mathbf{X}_p$ a particular solution of (5.58). The technique of undetermined coefficients also applies to systems of equations, as witnessed by the following theorem.

**Theorem 5.3.**   In (5.58), if $\mathbf{B} = \mathbf{b}e^{\omega t}$, with $\mathbf{b}$ a constant vector and $\omega$ not an eigenvalue of $\mathbf{A}$, then a particular solution of (5.58) has the form $\mathbf{k}e^{\omega t}$.

**EXAMPLE 5.8**   Find a particular solution of

$$D\mathbf{X} = \mathbf{AX} + \mathbf{B} \tag{5.59}$$

if

$$A = \begin{bmatrix} -2 & 4 \\ 1 & 1 \end{bmatrix} \quad \text{and} \quad B = \begin{bmatrix} 8e^{-2t} \\ 3e^{-2t} \end{bmatrix}.$$

In an earlier example [see (5.44)], we determined the eigenvalues of $A$ as $\lambda = 2$ and $-3$, so the theorem above applies and we substitute $X_p = ke^{-2t}$ into (5.59). This gives

$$-2 \begin{bmatrix} k_1 \\ k_2 \end{bmatrix} = \begin{bmatrix} -2 & 4 \\ 1 & 1 \end{bmatrix} \begin{bmatrix} k_1 \\ k_2 \end{bmatrix} + \begin{bmatrix} 8 \\ 3 \end{bmatrix}$$

or

$$-2k_1 = -2k_1 + 4k_2 + 8,$$

$$-2k_2 = k_1 + k_2 + 3.$$

The solution of this system of equations is $k_2 = -2$, $k_1 = 3$ and a particular solution of (5.59) is

$$X_p = \begin{bmatrix} 3e^{-2t} \\ -2e^{-2t} \end{bmatrix}.$$

Because

$$\frac{d}{dt} e^{i\omega t} = i\omega e^{i\omega t},$$

it is often easier to use exponential functions instead of $\sin \omega t$ and $\cos \omega t$ to find particular solutions when the $B$ in (5.58) contains trigonometric functions. This method works if $A$ and $b$ in

$$DX = AX + be^{i\omega t} \tag{5.60}$$

contain only real numbers. The reason is that if $X$ satisfies (5.60), then by taking the real and imaginary parts, we obtain

$$D(\text{Re } X) = A[\text{Re } X] + b \cos \omega t, \tag{5.61a}$$

$$D(\text{Im } X) = A[\text{Im } X] + b \sin \omega t. \tag{5.61b}$$

That is, if $X$ satisfies (5.60), the real part of $X$ satisfies (5.61a) and the imaginary part of $X$ satisfies (5.61b). To illustrate this technique, consider

$$DX = AX + B, \tag{5.62}$$

where

$$A = \begin{bmatrix} -2 & 4 \\ 1 & 1 \end{bmatrix} \quad \text{and} \quad B = \begin{bmatrix} 2 \cos t \\ 0 \end{bmatrix}.$$

The solution proceeds in the following steps.

**Step 1.**   Write the associated differential equation.

**Step 2.**    Solve the associated differential equation.

**Step 3.**    Take the Re[$X_p$] or Im[$X_p$] as the appropriate solution.

For this example the steps are performed as follows.

**Step 1.**

$$DX = AX + \begin{bmatrix} 2 \\ 0 \end{bmatrix} e^{it}. \tag{5.63}$$

**Step 2.**    Since the eigenvalues of $A$ are 2 and $-3$, try $X_p = ke^{it}$ and solve

$$i\begin{bmatrix} k_1 \\ k_2 \end{bmatrix} = \begin{bmatrix} -2 & 4 \\ 1 & 1 \end{bmatrix}\begin{bmatrix} k_1 \\ k_2 \end{bmatrix} + \begin{bmatrix} 2 \\ 0 \end{bmatrix}.$$

Thus

$$(-2 - i)k_1 + \qquad 4k_2 = -2,$$
$$k_1 + (1 - i)k_2 = \quad 0,$$

or equivalently,

$$k_1 = -(1 - i)k_2,$$
$$[(2 + i)(1 - i) + 4]k_2 = -2.$$

This gives

$$k_2 = \frac{-2}{7 - i} = \frac{-2}{50}(7 + i) = \frac{-7 - i}{25}$$

and

$$k_1 = (1 - i)\frac{7 + i}{25} = \frac{-6i + 8}{25},$$

so the particular solution of (5.63) is

$$X_p = \begin{bmatrix} \dfrac{8 - 6i}{25} \\ \dfrac{-7 - i}{25} \end{bmatrix} e^{it} = \begin{bmatrix} \dfrac{8 - 6i}{25} \\ \dfrac{-(7 + i)}{25} \end{bmatrix} (\cos t + i \sin t). \tag{5.64}$$

**Step 3.**    Since

$$\begin{bmatrix} 2 \cos t \\ 0 \end{bmatrix} = \text{Re}\left[\begin{bmatrix} 2 \\ 0 \end{bmatrix} e^{it}\right],$$

here we need the real part of (5.64) for our particular solution of (5.62), namely,

$$
\mathbf{X}_p = \begin{bmatrix} \dfrac{8}{25}\cos t + \dfrac{6}{25}\sin t \\[2mm] \dfrac{-7}{25}\cos t + \dfrac{1}{25}\sin t \end{bmatrix}.
$$

It should be apparent that you could also try a particular solution of (5.62) in the form $\mathbf{k}_1 \cos t + \mathbf{k}_2 \sin t$.

Consider now the case where $\mathbf{B}$ is an arbitrary function of $t$ in

$$
\dot{\mathbf{X}} = D\mathbf{X} = \mathbf{AX} + \mathbf{B}. \tag{5.65}
$$

As with the scalar equations, the general solution to this system will consist of two parts: a fundamental solution to the homogeneous equation and any particular solution to (5.65). If the $n$ functions $\mathbf{X}_i$, $i = 1, \ldots, n$, are linearly independent solutions of $D\mathbf{X} = \mathbf{AX}$, a fundamental matrix has the form

$$
\mathbf{U} = [\mathbf{X}_1, \mathbf{X}_2, \ldots, \mathbf{X}_n].
$$

To find a particular solution $\mathbf{X}_p$, we will use a technique based on variation of parameters. To do so, we need to introduce the following *definition:* Let

$$
\mathbf{v}(t) = \begin{bmatrix} v_1(t) \\ \vdots \\ v_n(t) \end{bmatrix}
$$

then

$$
\int_a^b \mathbf{v}(t)\, dt = \begin{bmatrix} \displaystyle\int_a^b v_1(t)\, dt \\ \vdots \\ \displaystyle\int_a^b v_n(t)\, dt \end{bmatrix}.
$$

We now assume that we have solved the homogeneous equation and denote by $\mathbf{X}_i$, $i = 1, \ldots, n$, the linearly independent column vectors of a fundamental matrix $\mathbf{U}$. We assume a solution of the nonhomogeneous equation of the form

$$
\mathbf{X}_p(t) = \sum_{i=1}^n \mathbf{X}_i(t)\mathbf{C}_i(t) = \mathbf{U}(t)\begin{bmatrix} C_1(t) \\ \vdots \\ C_n(t) \end{bmatrix} = \mathbf{U}(t)\mathbf{C}(t). \tag{5.66}
$$

That is, we multiply each vector, $\mathbf{X}_i$, by an unknown function of $t$, $C_i(t)$, and sum from $i = 1$ to $n$. Substituting (5.66) into (5.65), we get

$$\dot{\mathbf{U}}(t)\mathbf{C}(t) + \mathbf{U}(t)\dot{\mathbf{C}}(t) = \mathbf{A}[\mathbf{U}(t)\mathbf{C}(t)] + \mathbf{B}(t)$$
$$= [\mathbf{A}\mathbf{U}(t)]\mathbf{C}(t) + \mathbf{B}(t). \qquad (5.67)$$

But $\mathbf{U}$ is a fundamental matrix of the homogeneous equation, that is,

$$\dot{\mathbf{U}}(t) = \mathbf{A}\mathbf{U}(t).$$

Thus

$$\dot{\mathbf{U}}(t)\mathbf{C}(t) = \mathbf{A}\mathbf{U}(t)\mathbf{C}(t) \qquad (5.68)$$

and applying (5.68) to (5.67), we obtain

$$\mathbf{U}(t)\dot{\mathbf{C}}(t) = \mathbf{B}(t). \qquad (5.69)$$

But $\mathbf{U}$ is a fundamental matrix and so is nonsingular. Hence

$$\dot{\mathbf{C}}(t) = \mathbf{U}^{-1}(t)\mathbf{B}(t)$$

and

$$\mathbf{C}(t) = \int_{t_0}^{t} \mathbf{U}^{-1}(s)\mathbf{B}(s)\ ds.$$

These steps constitute the method of variation of parameters for systems of equations and are summarized in the following procedure.

*Procedure for Variation of Parameters for Systems of ODEs*

**Step 1.**   Solve the homogeneous system.

**Step 2.**   Form $\mathbf{X}_p = \sum_{i=1}^{n} \mathbf{X}_i(t)C_i(t) = \mathbf{U}(t)\mathbf{C}(t)$, where $\mathbf{U}(t)$ is a fundamental matrix for $\dot{\mathbf{X}} = \mathbf{A}\mathbf{X}$.

**Step 3.**   Calculate $\mathbf{C}(t) = \int_{t_0}^{t} \mathbf{U}^{-1}(s)\mathbf{B}(s)\ ds$.

**Step 4.**   $\mathbf{X}_p = \mathbf{U}(t)\mathbf{C}(t) = \mathbf{U}(t) \int_{t_0}^{t} \mathbf{U}^{-1}(s)\mathbf{B}(s)\ ds$.

**EXAMPLE 5.9**

$$\mathbf{A} = \begin{bmatrix} 2 & 1 \\ -3 & -2 \end{bmatrix}, \qquad \mathbf{B} = \begin{bmatrix} 2 \\ 4 \end{bmatrix} e^t, \qquad \text{in} \qquad \dot{\mathbf{X}} = \mathbf{A}\mathbf{X} + \mathbf{B}.$$

**Step 1.**   We first find the eigenvalues of $A$ by assuming $\mathbf{X} = k e^{\lambda t} = \begin{bmatrix} k_1 \\ k_2 \end{bmatrix} e^{\lambda t}$ and obtaining

$$\begin{bmatrix} 2 - \lambda & 1 \\ -3 & -2 - \lambda \end{bmatrix} \begin{bmatrix} k_1 \\ k_2 \end{bmatrix} = \begin{bmatrix} 0 \\ 0 \end{bmatrix}.$$

Thus

$$-(2 - \lambda)(2 + \lambda) + 3 = \lambda^2 - 1 = 0$$

gives the eigenvalues

$$\lambda_1 = 1, \qquad \lambda_2 = -1.$$

For $\lambda_1 = 1$, we have

$$k_1 + k_2 = 0,$$
$$-3k_1 - 3k_2 = 0,$$

so

$$\begin{bmatrix} 1 \\ -1 \end{bmatrix}$$

is an eigenvector and

$$\mathbf{X}_1(t) = \begin{bmatrix} 1 \\ -1 \end{bmatrix} e^t.$$

For $\lambda_2 = -1$,

$$3k_1 + k_2 = 0,$$
$$-3k_1 - k_2 = 0,$$

gives an eigenvector of the form

$$\begin{bmatrix} 1 \\ -3 \end{bmatrix},$$

so

$$\mathbf{X}_2(t) = \begin{bmatrix} 1 \\ -3 \end{bmatrix} e^{-t}.$$

**Step 2.**

$$\mathbf{U}(t) = \begin{bmatrix} e^t & e^{-t} \\ -e^t & -3e^{-t} \end{bmatrix}, \qquad \mathbf{U}^{-1}(t) = -\frac{1}{2}\begin{bmatrix} -3e^{-t} & -e^{-t} \\ e^t & e^t \end{bmatrix}.$$

**Step 3.**

$$\mathbf{C}(t) = -\frac{1}{2}\int_{t_0}^t \begin{bmatrix} -3e^{-s} & -e^{-s} \\ e^s & e^s \end{bmatrix}\begin{bmatrix} 2e^s \\ 4e^s \end{bmatrix} ds$$

$$= -\frac{1}{2}\int_{t_0}^t \begin{bmatrix} -6 - 4 \\ (2 + 4)e^{2s} \end{bmatrix} ds$$

$$= -\frac{1}{2}\int_{t_0}^t \begin{bmatrix} -10 \\ 6e^{2s} \end{bmatrix} ds.$$

Thus

$$C_1(t) = 5(t - t_0)$$

and

$$C_2(t) = -\frac{3}{2}(e^{2t} - e^{2t_0}).$$

If we choose $t_0 = 0$, $C_1(t) = 5t$, $C_2(t) = -\frac{3}{2}(e^{2t} - 1)$.

**Step 4.**

$$\mathbf{X}_p(t) = \begin{bmatrix} e^t & e^{-t} \\ -e^t & -3e^{-t} \end{bmatrix} \begin{bmatrix} 5t \\ -\frac{3}{2}(e^{2t} - 1) \end{bmatrix}$$

$$= \begin{bmatrix} 5te^t - \frac{3}{2}(e^t - e^{-t}) \\ -5te^t + \frac{9}{2}(e^t - e^{-t}) \end{bmatrix}.$$

Thus

$$\mathbf{X}_p(t) = 5te^t \begin{bmatrix} 1 \\ -1 \end{bmatrix} + \frac{1}{2}(e^t - e^{-t}) \begin{bmatrix} -3 \\ 9 \end{bmatrix}$$

is a particular solution.

## EXERCISES

Find a particular solution to $D\mathbf{X} = \mathbf{AX} + \mathbf{B}$ for $\mathbf{A}$ and $\mathbf{B}$ as specified.

1. $\mathbf{A} = \begin{bmatrix} -2 & 4 \\ 1 & 1 \end{bmatrix}$,     $\mathbf{B} = \begin{bmatrix} 7e^{3t} \\ -2e^{3t} \end{bmatrix}$.

2. $\mathbf{A} = \begin{bmatrix} -2 & 4 \\ 1 & 1 \end{bmatrix}$,     $\mathbf{B} = \begin{bmatrix} 0 \\ -\sin 2t \end{bmatrix}$.

3. $\mathbf{A} = \begin{bmatrix} -2 & 4 \\ 1 & 1 \end{bmatrix}$,     $\mathbf{B} = \begin{bmatrix} e^{2t} \\ -2e^{2t} \end{bmatrix}$.

4. $\mathbf{A} = \begin{bmatrix} -2 & 4 \\ 1 & 1 \end{bmatrix}$,     $\mathbf{B} = 2\begin{bmatrix} 7e^{3t} \\ -2e^{3t} \end{bmatrix} + \pi\begin{bmatrix} 0 \\ -\sin 2t \end{bmatrix}$.

5. $\mathbf{A} = \begin{bmatrix} 2 & 1 \\ -3 & -2 \end{bmatrix}$,     $\mathbf{B} = \begin{bmatrix} e^{\pi t} \\ 3e^{\pi t} \end{bmatrix}$.

6. $\mathbf{A} = \begin{bmatrix} 2 & 1 \\ -3 & -2 \end{bmatrix}$,     $\mathbf{B} = \begin{bmatrix} 3e^t \\ -e^t \end{bmatrix}$.

7. $A = \begin{bmatrix} 2 & 1 \\ -3 & -2 \end{bmatrix}$, $B = \begin{bmatrix} \cos t \\ 0 \end{bmatrix}$.

8. $A = \begin{bmatrix} 2 & 1 \\ -3 & -2 \end{bmatrix}$, $B = \begin{bmatrix} 3e^t \\ -e^t \end{bmatrix} + 17 \begin{bmatrix} e^{\pi t} \\ 3e^{\pi t} \end{bmatrix}$.

## 5.7 APPLICATIONS

Systems of differential equations may occur when mathematical models are used to describe physical, biological, chemical, and other phenomena. We conclude this chapter with a few such examples.

### 5.7.1 Coupled Spring-Mass Systems

Consider a mass $m_1$ suspended vertically from a rigid support by a weightless spring with a second mass $m_2$ suspended from the first mass by means of a second weightless spring (see Figure 5.1). We treat the two masses, $m_1$, $m_2$, as point masses and assume the two springs obey Hooke's law with respective spring constants $k_1$, $k_2$. We consider the vertical oscillations of this system and let $x_1(t)$ be the displacement (at time $t$) of mass $m_1$ from its equilibrium position and $x_2(t)$ be the displacement (at time $t$) of mass $m_2$ from its equilibrium position. The net force acting on $m_1$ is $-k_1x_1(t) + k_2[x_2(t) - x_1(t)]$ and the net force acting on $m_2$ is $-k_2[x_2(t) - x_1(t)]$, so by Newton's second law of motion we have

$$m_1\ddot{x}_1(t) = -k_1x_1(t) + k_2[x_2(t) - x_1(t)],$$
$$m_2\ddot{x}_2(t) = -k_2[x_2(t) - x_1(t)]. \qquad (5.70)$$

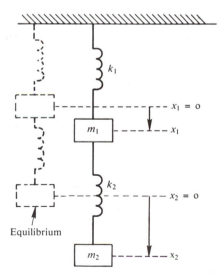

Figure 5.1

In operator form (5.70) becomes

$$[m_1 D^2 + (k_1 + k_2)]x_1(t) \qquad\quad - k_2 x_2(t) = 0,$$
$$-k_2 x_1(t) + (m_2 D^2 + k_2)x_2(t) = 0. \tag{5.71}$$

By the results of the preceding section we know that the general solution of (5.71) will have four arbitrary constants. If we specify initial values of displacement, $x_1(0)$, $x_2(0)$, and velocity, $\dot{x}_1(0)$, $\dot{x}_2(0)$, of the two masses, the values of the four arbitrary constants may be determined.

We now solve (5.71) for the case in which $m_1 = m_2 = 1$, $k_1 = 5$, $k_2 = 6$ and the initial conditions are $x_1(0) = 2$, $\dot{x}_1(0) = 5$, $x_2(0) = -10$, and $\dot{x}_2(0) = 1$. The resulting differential equations are

$$(D^2 + 11)x_1 \qquad\quad - 6x_2 = 0,$$
$$-6x_1 + (D^2 + 6)x_2 = 0. \tag{5.72}$$

If we eliminate $x_2$ by operating on the first equation in (5.72) by $D^2 + 6$, on the second equation by 6, and adding the results, we obtain

$$[(D^2 + 6)(D^2 + 11) - 36]x_1 = 0,$$

or

$$(D^4 + 17D^2 + 30)x_1 = 0. \tag{5.73}$$

The characteristic equation associated with (5.73) is

$$r^4 + 17r^2 + 30 = (r^2 + 15)(r^2 + 2) = 0,$$

with solutions

$$r = \pm(\sqrt{15})i, \qquad \pm(\sqrt{2})i.$$

Thus

$$x_1(t) = C_1 \cos \sqrt{2}t + C_2 \sin \sqrt{2}t + C_3 \cos \sqrt{15}t + C_4 \sin \sqrt{15}t, \tag{5.74}$$

and from the first of the equations (5.72),

$$x_2(t) = \tfrac{1}{6}(9C_1 \cos \sqrt{2}t + 9C_2 \sin \sqrt{2}t - 4C_3 \cos \sqrt{15}t - 4C_4 \sin \sqrt{15}t). \tag{5.75}$$

If we now choose $C_1$, $C_2$, $C_3$, and $C_4$ so that the initial conditions are satisfied, we obtain

$$C_1 + C_3 = 2,$$
$$\sqrt{2}C_2 + \sqrt{15}C_4 = 5,$$
$$\frac{3}{2}C_1 - \frac{2}{3}C_3 = -10, \tag{5.76}$$
$$\frac{3}{2}\sqrt{2}C_2 - \frac{2}{3}\sqrt{15}C_4 = 1.$$

Solving the first and third of equations (5.76) gives

$$C_1 = -4, \qquad C_3 = 6,$$

while solving the second and fourth equations gives

$$C_2 = \sqrt{2}, \qquad C_4 = \frac{3}{\sqrt{15}}.$$

Thus the solution of the initial value problem is

$$x_1(t) = -4 \cos \sqrt{2}t + \sqrt{2} \sin \sqrt{2}t + 6 \cos \sqrt{15}t + \frac{3}{\sqrt{15}} \sin \sqrt{15}t,$$

$$x_2(t) = -6 \cos \sqrt{2}t + \frac{3\sqrt{2}}{2} \sin \sqrt{2}t - 4 \cos \sqrt{15}t - \frac{2}{\sqrt{15}} \sin \sqrt{15}t. \tag{5.77}$$

### 5.7.2 Electrical Networks

In earlier chapters we solved problems involving the flow of charge or current in a single-loop electric circuit involving resistors, inductors, and capacitors. The three circuit laws used were

$$V = RI, \qquad R \text{ is the resistance (ohms),}$$

$$V = L\frac{dI}{dt}, \qquad L \text{ is the inductance (henrys),}$$

$$q = CV \quad \text{or} \quad I = C\frac{dV}{dt}, \qquad C \text{ is the capacitance (farads),}$$

where $V$ is the voltage drop across the circuit element, $I$ is the resulting current, and $q$ is the charge. (Note $I = dq/dt$.) In addition, we use the fact that the algebraic sum of the voltage drops around a closed electric circuit is zero. This fact is also called Kirchhoff's law.

These laws are also used in deriving differential equations for electric circuits consisting of more than one loop. Consider, for example, the circuit of Figure 5.2. There are three closed loops in this figure, ABGHA, ABDFGHA, and BDFGB. Places where the current may branch are called *junctions, branch points,* or *nodes.* We denote

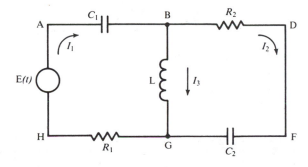

Figure 5.2

the current in the three branches of Figure 5.2 by $I_1$, $I_2$, and $I_3$, with the positive direction of the current as indicated. If $E(t)$ denotes the electromotive force, we may use Kirchhoff's law in loop ABGHA to obtain

$$\frac{1}{C_1}q_1 + L\frac{dI_3}{dt} + R_1I_1 - E(t) = 0, \tag{5.78}$$

where $q_1$ is the charge in the first branch. In loop BDFGB we have the differential equation

$$R_2I_2 + \frac{1}{C_2}q_2 - L\frac{dI_3}{dt} = 0, \tag{5.79}$$

while in loop ABDFGHA we have

$$\frac{1}{C_1}q_1 + R_2I_2 + \frac{1}{C_2}q_2 + R_1I_1 - E(t) = 0. \tag{5.80}$$

[Notice that these three equations are not independent, as if we add (5.78) and (5.79), we obtain (5.80).] To eliminate the variables $q_1$ and $q_2$, we differentiate (5.78) and (5.79) and use the fact that $dq_1/dt = I_1$ and $dq_2/dt = I_2$ to obtain

$$\frac{1}{C_1}I_1 + L\frac{d^2I_3}{dt^2} + R_1\frac{dI_1}{dt} - \frac{dE}{dt} = 0, \tag{5.81}$$

$$R_2\frac{dI_2}{dt} + \frac{1}{C_2}I_2 - L\frac{d^2I_3}{dt^2} = 0. \tag{5.82}$$

To obtain a third equation in our three dependent variables, we use another of Kirchhoff's laws, which says that the algebraic sum of currents at any branch point is zero. Then at branch point B we have

$$I_1 = I_2 + I_3. \tag{5.83}$$

(Notice that we would also obtain this equation by considering the currents at branch point $G$.) If we use (5.83) in (5.81) and rearrange, we obtain the following differential equations for $I_2$ and $I_3$:

$$R_1\frac{dI_2}{dt} + \frac{1}{C_1}I_2 + L\frac{d^2I_3}{dt^2} + R_1\frac{dI_3}{dt} + \frac{1}{C_1}I_3 = \frac{dE}{dt}, \tag{5.84}$$

$$R_2\frac{dI_2}{dt} + \frac{1}{C_2}I_2 - L\frac{d^2I_3}{dt^2} = 0. \tag{5.85}$$

Solutions of this system of differential equations will be explored in the exercises for specific values of $R_1$, $R_2$, $C_1$, $C_2$, $L$, and $E(t)$.

### 5.7.3 Mixture Problems

In Section 2.6.1 we discovered how to determine the concentration of a solute in a single container when the concentration and rate of input and rate of output were known. Now we consider the case where there are two containers, as shown in Figure 5.3. Let $x_1$ = amount of substance in container A at time $t$ and $x_2$ = amount of substance in container B at time $t$. We assume a solution with concentration of 3 pounds per gallon is entering container A at a rate of 4 gallons per minute and the well-stirred mixture leaves container B at the same rate. If there are 100 gallons in each container and the mixture flows from container A to container B at a rate of 5 gallons per minute and leaks back from container B to container A at a rate of 1 gallon per minute, the differential equations governing this system may be derived using a continuity equation. [See (2.111), which says the rate of change = input − output.]

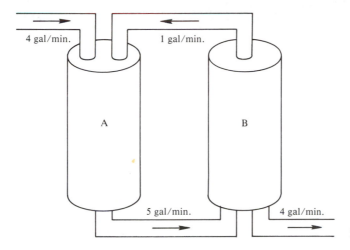

4 gal/min.          1 gal/min.

A          B

5 gal/min.          4 gal/min.

**Figure 5.3**

Thus for container A we have

$$\frac{dx_1}{dt} = (3)(4) - \left(\frac{x_1}{100}\right)(5) + \left(\frac{x_2}{100}\right)(1) \tag{5.86}$$

and for container B

$$\frac{dx_2}{dt} = \left(\frac{x_1}{100}\right)(5) - \left(\frac{x_2}{100}\right)(1) - \left(\frac{x_2}{100}\right)(4). \tag{5.87}$$

We now solve the system of equations with the added stipulation that the initial amount of substance in container A is 300 pounds and that in container B is 0 pounds. If $D = d/dt$, we may write our system of differential equations as

$$\mathbf{DX} = \mathbf{AX} + \mathbf{B}, \quad \text{where } \mathbf{X} = \begin{bmatrix} x_1 \\ x_2 \end{bmatrix}, \quad \mathbf{A} = \begin{bmatrix} -\dfrac{1}{20} & \dfrac{1}{100} \\ \dfrac{1}{20} & -\dfrac{1}{20} \end{bmatrix}, \quad \mathbf{B} = \begin{bmatrix} 12 \\ 0 \end{bmatrix}.$$

To solve the system of equations we find a fundamental matrix for the associated homogeneous system of assuming $\mathbf{X} = \mathbf{k}e^{\lambda t}$. This gives the system of algebraic equations

$$
\begin{bmatrix} -\dfrac{1}{20} - \lambda & \dfrac{1}{100} \\[2ex] \dfrac{1}{20} & -\dfrac{1}{20} - \lambda \end{bmatrix} \begin{bmatrix} k_1 \\[2ex] k_2 \end{bmatrix} = \begin{bmatrix} 0 \\[2ex] 0 \end{bmatrix}. \tag{5.88}
$$

Solving (5.88) gives the eigenvalues $\lambda_1 = -\frac{1}{20}(1 + 1/\sqrt{5})$, $\lambda_2 = -\frac{1}{20}(1 - 1/\sqrt{5})$ with the associated eigenvectors

$$
\begin{bmatrix} \sqrt{5} \\ -5 \end{bmatrix} \quad \text{and} \quad \begin{bmatrix} \sqrt{5} \\ 5 \end{bmatrix}
$$

respectively. This gives a fundamental matrix

$$
\mathbf{U}(t) = \begin{bmatrix} \sqrt{5}e^{\lambda_1 t} & \sqrt{5}e^{\lambda_2 t} \\ -5e^{\lambda_1 t} & 5e^{\lambda_2 t} \end{bmatrix}.
$$

A particular solution of the system (5.86) and (5.87) may be obtained by inspection as

$$
\mathbf{X}_p(t) = \begin{bmatrix} 300 \\ 300 \end{bmatrix}
$$

and the general solution is

$$
\mathbf{X}(t) = \mathbf{U}(t)\mathbf{C} + \mathbf{X}_p(t) \tag{5.89}
$$

where

$$
\mathbf{C} = \begin{bmatrix} C_1 \\ C_2 \end{bmatrix}.
$$

Since at time zero, container A has 300 pounds of the substance while container B has none, we have initial conditions $x_1(0) = 300$, $x_2(0) = 0$. Thus from (5.89), $C_1$ and $C_2$ are given by the solution of

$$
\begin{bmatrix} 300 \\ 0 \end{bmatrix} = \begin{bmatrix} \sqrt{5} & \sqrt{5} \\ -5 & 5 \end{bmatrix} \begin{bmatrix} C_1 \\ C_2 \end{bmatrix} + \begin{bmatrix} 300 \\ 300 \end{bmatrix}
$$

as $C_1 = 30$, $C_2 = -30$ and the solution is

$$
\mathbf{X}(t) = \begin{bmatrix} \sqrt{5}e^{\lambda_1 t} & \sqrt{5}e^{\lambda_2 t} \\ -5e^{\lambda_1 t} & 5e^{\lambda_2 t} \end{bmatrix} \begin{bmatrix} 30 \\ -30 \end{bmatrix} + \begin{bmatrix} 300 \\ 300 \end{bmatrix}. \tag{5.90}
$$

Notice that since $\lambda_1$ and $\lambda_2$ in (5.90) are both negative, $\lambda_1, \lambda_2 = -\frac{1}{20}(1 \pm 1/\sqrt{5})$,

$$
\lim_{t \to \infty} \mathbf{X}(t) = \begin{bmatrix} 300 \\ 300 \end{bmatrix}.
$$

That is, after a long time, both containers have the same concentration, $300/100$, as that of the input solution.

### 5.7.4 Population Dynamics

In Sections 2.6.3 and 2.6.4 we introduced models of growth and decay of the population of a single species. However, in many situations of interest in population dynamics we need to study the interaction of two species. For example, consider the predator-prey problem, where one species is the primary food source for the other. You could think of rabbits as the prey and fox as the predator, a small fish as the prey for a larger fish, and so on.

To derive a system of the differential equations for such a situation, let

$$x_1 = \text{the prey population at time } t,$$

$$x_2 = \text{the predator population at time } t.$$

We assume that the rate of change of the prey, $dx_1/dt$, is given by

$$\frac{dx_1}{dt} = a_{11}x_1 - a_{12}x_2, \tag{5.91}$$

where $a_{11}$ takes into account the difference between the birth and death rates of the prey. We assume $a_{12}$ is greater than zero since an increase in the predator population should cause the rate of change of prey to diminish.

Similarly, we have

$$\frac{dx_2}{dt} = a_{21}x_1 - a_{22}x_2, \tag{5.92}$$

where $a_{21}$ is positive since an increase in the prey population should increase the population of the predator. Also, $a_{22}$ represents the difference between birth and death rates of the predator if the prey were absent. Taking $a_{22}$ positive would account for the fact that the predators would die out altogether if there were no prey.

If the population of prey and predator were known at some time as $B_1$ and $B_2$, respectively, the population of subsequent times would be modeled by the solution of (5.91) and (5.92) with initial conditions $x_1(0) = B_1$, $x_2(0) = B_2$.

To solve a specific problem, assume that $a_{11} = 2$, $a_{12} = 5$, $a_{21} = 2$, and $a_{22} = 4$, so the system of differential equations becomes

$$\frac{d\mathbf{X}}{dt} = \begin{bmatrix} 2 & -5 \\ 2 & -4 \end{bmatrix} \mathbf{X} = \mathbf{AX}. \tag{5.93}$$

The eigenvalues of the matrix $\mathbf{A}$ in (5.93) are determined by

$$(2 - \lambda)(-4 - \lambda) + 10 = \lambda^2 + 2\lambda + 2 = (\lambda + 1)^2 + 1 = 0.$$

An eigenvector associated with

$$\lambda_1 = -1 + i \quad \text{is} \quad \begin{bmatrix} 5 \\ 3 - i \end{bmatrix},$$

so

$$\begin{bmatrix} 5 \\ 3 - i \end{bmatrix} e^{(-1+i)t} = e^{-t} \begin{bmatrix} 5 \\ 3 - i \end{bmatrix} (\cos t + i \sin t)$$

$$= e^{-t} \begin{bmatrix} 5 \cos t \\ 3 \cos t + \sin t \end{bmatrix} + i \begin{bmatrix} 5 \sin t \\ 3 \sin t - \cos t \end{bmatrix} e^{-t}$$

and a fundamental matrix is

$$\mathbf{U}(t) = \begin{bmatrix} 5e^{-t} \cos t & 5e^{-t} \sin t \\ e^{-t}(3 \cos t + \sin t) & e^{-t}(3 \sin t - \cos t) \end{bmatrix}.$$

Thus the solution of (5.93) is $\mathbf{X}(t) = \mathbf{U}(t)\mathbf{C}$, where by choosing

$$\mathbf{C} = \begin{bmatrix} 0.2B_1 \\ 0.6B_1 - B_2 \end{bmatrix}$$

we satisfy the initial conditions $x_1(0) = B_1$, $x_2(0) = B_2$. Computing $\mathbf{UC}$ gives

$$x_1(t) = B_1 e^{-t}(\cos t + 3 \sin t) - 5B_2 e^{-t} \sin t,$$

$$x_2(t) = 2B_1 e^{-t} \sin t - B_2 e^{-t}(3 \sin t - \cos t).$$

The graph of $x_1(t)$ and $x_2(t)$ as functions of $t$ for $B_1 = 2B_2$ as shown in Figure 5.4.

Notice that $x_1(\tan^{-1} 2) = 0$, so the species represented by $x_1$ has vanished when $t = \tan^{-1} 2$. Obviously, the model is meaningless for larger times, and in fact may only be used for small values of time.

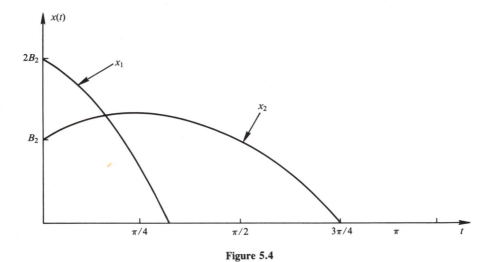

**Figure 5.4**

# EXERCISES

**1.** If the springs and masses shown in Figure 5.1 are given by $m_1 = m_2 = 2$, $k_1 = 6$, and $k_2 = 4$, find solutions of (5.71) which satisfy the initial conditions $x_1(0) = 0$, $\dot{x}_1(0) = 4$, $x_2(0) = 0$, $\dot{x}_2(0) = -7$.

2. Show that for positive values of $m_1$, $m_2$, $k_1$, and $k_2$ the solutions of (5.71) will always be oscillatory (i.e., consist only of sines and cosines of real arguments).

3. The differential system which describes the motion of the spring-mass system shown in Figure 5.5 is

$$m_1\ddot{x}_1 = -k_1x_1 + k_2(x_2 - x_1),$$
$$m_2\ddot{x}_2 = -k_2(x_2 - x_1) - k_3x_2. \tag{5.94}$$

   (a) If $m_1 = m_2 = 1$, $k_1 = k_3 = 9$, and $k_2 = 8$, solve (5.94) subject to the initial values $x_1(0) = 2$, $\dot{x}_1(0) = -2$, $x_2(0) = 6$, and $\dot{x}_2(0) = 8$.
   (b) If $m_1 = 1$, $m_2 = 2$, $k_1 = 1$, $k_2 = k_3 = 2$, solve (5.94) subject to the initial values $x_1(0) = 3$, $\dot{x}_1(0) = 3$, $x_2(0) = 0$, and $\dot{x}_2(0) = -3$.
   (c) Show that for positive values of $m_1$, $m_2$, $k_1$, $k_2$, and $k_3$, the solutions of (5.94) will always be oscillatory.

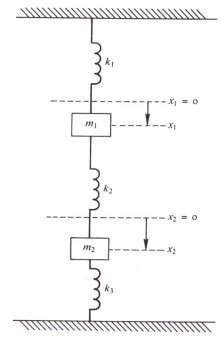

**Figure 5.5**

4. Consider the two identical pendulums of mass $m$ and length $\ell$ (Figure 5.6) that are connected by means of a spring with spring constant $k$. The small oscillations of this system are governed by

$$m\ell\ddot{x}_1 + mgx_1 = -k(\ell x_1 - \ell x_2),$$
$$m\ell\ddot{x}_2 + mgx_2 = k(\ell x_1 - \ell x_2). \tag{5.95}$$

   (a) Find the solution of (5.95) if $g/\ell = 4$, $k/m = 5/2$ and $x_1(0) = 0$, $\dot{x}_1(0) = \dot{x}_2(0) = 0$, $x_2(0) = 0.1$.
   (b) Find the solution of (5.95) if $g/\ell = 4$, $k/m = 5/2$ and $x_1(0) = x_2(0) = 0$, $\dot{x}_1(0) = -\dot{x}_2(0) = 0.01$.
   (c) Find the solution of (5.95) if $g/\ell = 9$, $k = 7/2$ and $x_1(0) = -x_2(0) = 0.1$, $\dot{x}_1(0) = \dot{x}_2(0) = 0$.

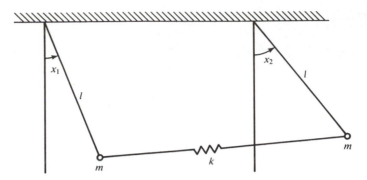

**Figure 5.6**

(d) Find the solution of (5.95) if $g/\ell = 9$, $k = 7/2$ and $x_1(0) = x_2(0) = 0.1$, $\dot{x}_1(0) = \dot{x}_2(0) = -0.01$.

5. If a damping force proportional to the velocity is included in the motion of the pendulums of exercise 4, we obtain

$$m\ell\ddot{x}_1 + c\ell\dot{x}_1 + mgx_1 = -k(\ell x_1 - \ell x_2),$$
$$m\ell\ddot{x}_2 + c\ell\dot{x}_2 + mgx_2 = k(\ell x_1 - \ell x_2). \tag{5.96}$$

Show that if $c/m$, $g/\ell$, and $k/m$ are all positive, all solutions of (5.96) approach zero in the limit as $t$ approaches infinity.

6. A projectile of mass $m$ is fired into the air with an initial velocity $v_0$ and at an angle $\alpha$ from the horizontal. If an $x_1 - x_2$ coordinate system is placed as shown in Figure 5.7 and if air resists the motion with a force per unit mass of $c$ times the velocity, the resulting equations of motion are

$$m\ddot{X} = -mc X - \begin{bmatrix} 0 \\ mg \end{bmatrix}, \quad \text{where } X = \begin{bmatrix} x_1 \\ x_2 \end{bmatrix} \text{ and } c > 0.$$

Solve this system of equations subject to initial conditions

$$X(0) = 0, \quad \dot{X}(0) = \begin{bmatrix} v_0 \cos \alpha \\ v_0 \sin \alpha \end{bmatrix}.$$

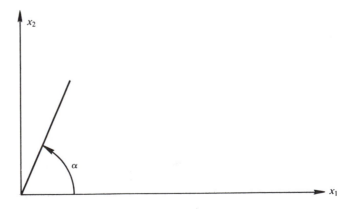

**Figure 5.7**

**7. (a)** Show that if $I_1$, $I_2$, and $I_3$ are the currents in the three branches of the circuit shown in Figure 5.8, they obey

$$RI_1 + LDI_2 = E(t),$$

$$RDI_1 + \frac{1}{C}I_3 = DE(t),$$

$$I_1 = I_2 + I_3.$$

**(b)** Solve the system of equations in part (a) if $R = 50$ ohms, $C = 2 \times 10^{-4}$ farad, $L = 4 \times 10^{-2}$ henry, $E = 100$ volts, and $I_1(0) = I_2(0) = 0$. Explain what happens for large values of time.

**(c)** Solve the system of equations if $R, L$, and $C$ are as in part (b) but $E(t) = 100 \sin(100t)$.

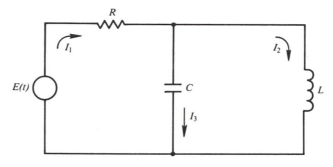

**Figure 5.8**

**8. (a)** Show that if $I_1$, $I_2$, and $I_3$ arc the currents in the three branches of the circuit as shown in Figure 5.9, they obey

$$R_1 I_1 + L_1 DI_3 = E(t),$$

$$R_2 I_2 + L_2 DI_2 - L_1 DI_3 = 0,$$

$$I_1 = I_2 + I_3.$$

**(b)** Solve the system in part (a) if $R_1 = 10$ ohms, $R_2 = \frac{40}{9}$ ohms, $L_1 = \frac{5}{4}$ henrys, $L_2 = \frac{10}{9}$ henrys, $E = 12$ volts, and $I_1(0) = I_2(0) = 0$.

**(c)** Solve the system of part (b) if $E = E_0 \sin t$.

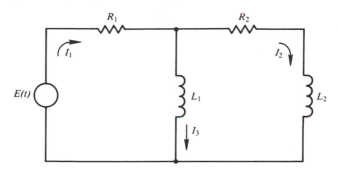

**Figure 5.9**

9. Find the currents in the circuit of Figure 5.10 if $R_1 = R_2 = 1000$ ohms, $C_1 = 2 \times 10^{-6}$ farad, $C_2 = 3 \times 10^{-6}$ farad, $E(t) = E_0 \sin t$, and $I_1(0) = I_2(0) = 0$.

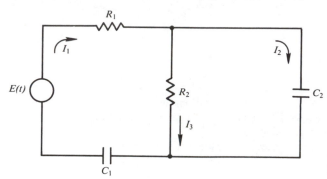

**Figure 5.10**

10. Solve the example in Section 5.7.3 if the solution entering container A is pure water and all other aspects of that example remain the same.

11. Consider the two containers of Figure 5.11 each of which hold 100 gallons of liquid.
    (a) Initially, one container has 10 pounds of salt and the other has none. How long after the valves at A and B are opened will the number of pounds in one container be reduced from 10 to 8? (The flow rate through valves A and B is 4 gallons per minute.)
    (b) Determine the amount of salt in each container as time approaches infinity.

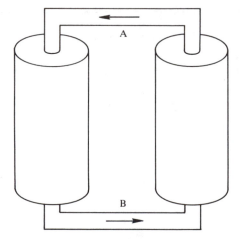

**Figure 5.11**

12. Work exercise 11 if a salt solution 0.02 pound per gallon is added at a rate of 3 gallons per minute to the container with 10 pounds of salt and the well-stirred mixture of the other container is drained at 3 gallons per minute.

13. Let $x_1$ denote the number of fish in a lake and $x_2$ the number of fishermen at the lake. If the rate of stocking the lake is denoted by $S(t)$, the differential equations describing $x_1$ and $x_2$ are

$$\frac{dx_1}{dt} = a_{11}x_1 - a_{12}x_2 + S(t),$$

$$\frac{dx_2}{dt} = a_{21}x_1 - a_{22}x_2. \qquad (5.97)$$

Here $a_{11}$ is the difference between the natural birth and death rates of fish and $a_{12}$ is the catch rate of the fishermen. $a_{21}$ and $a_{22}$ are both positive since the rate of change of fishermen should increase with the number of fish present and decrease with the number of fishermen.

(a) If $a_{11} = a_{21} = 2$, $a_{12} = 8$, $a_{22} = 6$ and the stocking rate is approximated by $S(t) = S_0[1 - \cos(t/10)]$, solve (5.97) with initial conditions $x_1(0) = A$, $x_2(0) = 0$.

(b) Note that the solution for part (a) has terms which decrease rapidly with time (called the transient solution) as well as those that do not (called the steady-state solution). What values of $S_0$ will guarantee that the number of fishermen, using the steady-state solution of this model, will be always less than 300?

(c) Solve (5.97) if $a_{11} = 4$, $a_{21} = 5$, $a_{12} = a_{22} = 2$, $S(t) = S_0(1 - \cos t)$, $x_1(0) = A$, $x_2(0) = 0$.

14. (a) Solve equations (5.91) and (5.92) if $a_{11} = 5$, $a_{12} = 3$, $a_{21} = 10$, $a_{22} = 8$, $x_1(0) = 10,000$, $x_2(0) = 100$.

(b) What is the minimum value of $t$ for which either the predator or the prey vanishes according to the model?

15. (a) Solve exercise 14(a) if the initial conditions are changed to $x_1(0) = A$, $x_2(0) = B$.

(b) Are there choices of $A$ and $B$ that give $x_1(t) > 0$ and $x_2(t) > 0$ for all $t$?

16. The system of differential equations

$$\frac{dx_1}{dt} = (a_1 - a_2 x_2)x_1,$$

$$\frac{dx_2}{dt} = (-a_3 + a_4 x_1)x_2, \tag{5.98}$$

is also used to model the predator-prey situation. Here the reasoning is that the "growth rate" of the prey, $a_1 - a_2 x_2$, is diminished by the presence of a predator $(-a_2 x_2)$, while the "growth rate" of the predator, $-a_3 + a_4 x_1$, is increased by the presence of the prey $(a_4 x_1)$. Here $a_1$, $a_2$, $a_3$, and $a_4$ are positive constants. Notice that

$$x_2 = \frac{a_1}{a_2}, \qquad x_1 = \frac{a_3}{a_4},$$

gives a solution of (5.98). This solution is called an *equilibrium solution,* is independent of time, and may be found by solving (5.98) after setting $dx_1/dt = dx_2/dt = 0$.

(a) Show that $x_1 = x_2 = 0$ is also an equilibrium solution of (5.98).

(b) Show that letting

$$x_1 = \frac{a_1}{a_2} + y_1, \qquad x_2 = \frac{a_3}{a_4} + y_2,$$

in (5.98) gives a pair of linear equations

$$\frac{dy_1}{dt} = -\frac{a_2 a_3}{a_4}y_2, \qquad \frac{dy_2}{dt} = \frac{a_4 a_1}{a_2}y_1. \tag{5.99}$$

(c) Solve (5.99) subject to the initial conditions $x_1(0) = A$, $x_2(0) = B$.

(d) Use the solution from part (c) to sketch $x_1(t)$ and $x_2(t)$ as functions of $t$.

## REVIEW EXERCISES

Solve the following initial value problems.

1.  $(D - 1)x_1 - 3x_2 = 0,$
    $-5x_1 + (D - 3)x_2 = 0,$    $x_1(0) = 1,$    $x_2(0) = -1.$

2.          $Dx_1 + (D - 5)x_2 = 0,$    $x_1(0) = 3,$    $x_2(0) = 2,$
    $(D + 1)x_1 + (2D - 8)x_2 = 0.$

3.  $(D - 3)x_1 + 4x_2 = 0,$
    $4x_1 - (D + 7)x_2 = 0,$    $x_1(0) = 1,$    $x_2(0) = 2.$

4.  $(D - 1)x_1 + x_2 \qquad\quad = 0,$
    $\qquad (D + 1)x_2 - 3x_3 = 0,$    $x_1(0) = 4,$    $x_2(0) = 6,$    $x_3(0) = 0.$
    $\quad x_1 - x_2 + Dx_3 = 0,$

5.  $(D + 2)x_1 + \quad (D + 3)x_2 = -4,$
    $(2D - 6)x_1 + (3D - 4)x_2 = \quad 2,$    $x_1(0) = 3,$    $x_2(0) = -3.$

6.  $(D - 2)x_1 - 6x_2 = 0,$
    $2x_1 + (D + 5)x_2 = 0,$    $x_1(0) = 0,$    $x_2(0) = 1.$

7.  $(D - 4)x_1 + 3x_2 = 0,$
    $5x_1 + (D + 4)x_2 = 0,$    $x_1(0) = -1,$    $x_2(0) = -3.$

8.  $(D - 7)x_1 - 6x_2 = -10e^{3t},$
    $2x_1 - (D - 6)x_2 = \quad 5e^{3t},$    $x_1(0) = -1,$    $x_2(0) = 0.$

9.  $(D - 2)x_1 - x_2 = 0,$
    $\qquad (D - 2)x_2 = e^{2t},$    $x_1(0) = 1,$    $x_2(0) = 0.$

10.         $Dx_1 + tDx_2 = 2t,$
    $(tD + 1)x_1 - Dx_2 = \quad 0,$    $x_1(0) = 0,$    $x_2(0) = 7.$

11. $(D - 1)x_1 - x_2 + x_3 = 0,$
    $\qquad\qquad Dx_2 - x_3 = 0,$    $x_1(0) = 1,$    $x_2(0) = 0,$    $x_3(0) = 2.$
    $\qquad 2x_2 + (D + 3)x_3 = 0,$

# CHAPTER 6

# Series Solutions

At this point we have techniques for solving special types of first order differential equations—exact, separable, linear, and so on—as well as two types of linear equations of higher order: Cauchy-Euler type and those with constant coefficients. However, if we need to find the solution of a differential equation that is not one of these forms, we have only the method of isoclines (to give a graphical solution) or Picard's iteration method at our disposal. In this chapter we discuss the development of a Taylor series solution for linear or nonlinear differential equations as well as a power series solution for specific types of second order differential equations. Since many series solutions have coefficients that are defined recursively, a computer may easily compute as many terms as desired.

## 6.1 DIRECT COMPUTATION OF A TAYLOR SERIES SOLUTION

Recall in Chapter 1 that we used Picard's iteration method to obtain an approximate solution of the initial value problem

$$y' = 2xy, \qquad y(0) = 3$$

and ended up showing that the approximate solution was the $n$th partial sum of the Taylor series expansion of $3e^{x^2}$. Also, in Section 1.3, we have a theorem that gives conditions under which the initial value problem

$$y' = F(x, y)$$
$$y(x_0) = y_0$$

has a unique solution. If $F(x, y)$ is such that the hypotheses to Theorem 1.1 are satisfied and is analytic near $(x_0, y_0)$, but none of our previous techniques for solving specific types of first order differential equations apply, we will develop the Taylor series expansion of $y$ as powers of $x - x_0$. [Recall that $F(x, y)$ is analytic near $(x_0, y_0)$ if it may be expanded in a convergent series in powers of $x - x_0$, $y - y_0$, that is, $\sum_{m=0}^{\infty} \sum_{n=0}^{\infty} a_{mn}(x - x_0)^m (y - y_0)^n$ converges for $|x - x_0| < h, |y - y_0| < h$.] Since the Taylor series for $f(x)$ about $x = x_0$ is

$$\sum_{k=0}^{\infty} \frac{f^{(k)}(x_0)}{k!}(x - x_0)^k,$$

where $f^{(k)}(x_0)$ is the $k$th derivative of $f(x)$ evaluated at $x_0$, we use the initial condition to find $y(x_0)$ and the differential equation to obtain $y'(x_0)$ as $F(x_0, y_0)$. The derivatives of order greater than 1 are obtained by differentiating the differential equation implicitly with respect to $x$. If these derivatives may be evaluated for $x = x_0$, we may obtain as many terms in the Taylor series expansion of the solution as desired.

**EXAMPLE 6.1**    Find the Taylor series expansion for the solution of the initial value problem

$$y' = 1 + y^2,$$

$$y(0) = 1. \tag{6.1}$$

If we write the Taylor series for the solution as

$$\sum_{k=0}^{\infty} \frac{y^{(k)}(0)}{k!}(x - 0)^k,$$

we have $y(0) = 1$ and from the differential equation,

$$y'(0) = 1 + y^2(0) = 1 + 1 = 2.$$

To find $y''(0)$, differentiate (6.1) with respect to $x$ to obtain

$$y'' = 2yy',$$

so

$$y''(0) = 2y(0)y'(0) = 2(1)(2) = 4.$$

Continuing this process gives

$$y''' = 2(y')^2 + 2yy'',$$

with

$$y'''(0) = 2(y'(0))^2 + 2y(0)y''(0) = 2(2)^2 + 2(1)(4) = 16.$$

$$y^{(4)} = 4y'y'' + 2y'y'' + 2yy''',$$

with

$$y^{(4)}(0) = 6(2)(4) + 2(1)(16) = 80,$$

$$y^{(5)} = 6(y'')^2 + 6y'y''' + 2y'y''' + 2yy^{(4)},$$

with

$$y^{(5)}(0) = 6(4)^2 + 8(2)(16) + 2(1)(80) = 512.$$

Thus the Taylor series for the solution to order 5 is given by

$$1 + 2x + 2x^2 + \frac{8x^3}{3} + \frac{10x^4}{3} + \frac{64x^5}{15}. \tag{6.2}$$

In Example 6.1 the pattern for the value of the $n$th derivative is not apparent, so we cannot write a general formula for the $n$th partial sum and write an expression for the entire Taylor series. However, in Example 6.1, we may compare our Taylor series in (6.2) with that of the exact solution of (6.1), namely

$$y = \tan\left(x + \frac{\pi}{4}\right).$$

The series in (6.2) contains the first six nonzero terms in the Taylor series of $\tan(x + \pi/4)$. We also note that the polynomial of (6.2) agrees with that found by Picard iteration in exercise 7 of Section 1.6.

Taylor series solutions may also be obtained for differential equations of order 2 and higher. Consider the initial value problem

$$y'' = F(x, y, y'), \qquad y(x_0) = y_0, \qquad y'(x_0) = y_1, \tag{6.3}$$

where $F$, $\partial F/\partial y$, and $\partial F/\partial y'$, are all continuous for

$$|x - x_0| < h, \qquad |y - y_0| < h, \qquad |y - y_1| < h,$$

and $F$ is analytic near $(x_0, y_0, y_1)$. Theorem 1.1 guarantees a unique solution and we assume that this solution has a Taylor series expansion. The procedure to obtain this Taylor series expansion will be similar to that used on a first order equation, except now we have both $y(x_0)$ and $y'(x_0)$ given by the initial conditions, and $y''(x_0)$ given by the differential equation. Derivatives of orders greater than 2 may be obtained by differentiating the differential equation implicity with respect to $x$. To illustrate this technique we obtain a Taylor series solution to the initial value problem describing the motion of a pendulum given by

$$y'' = -\sin y,$$

$$y(0) = 0.2, \tag{6.4}$$

$$y'(0) = -0.15.$$

These initial conditions describe releasing the pendulum at an angle of 0.2 radian with an initial angular velocity of $-0.15$ radian per second.

From the differential equation we can evaluate $y''(0) = -\sin 0.2 = -0.1987$. To find $y'''$ we differentiate (6.4) implicitly with respect to $x$ to obtain

$$y''' = (-\cos y)y'$$

and evaluate $y'''(0)$ as

$$y'''(0) = (-\cos 0.2)(-0.15) = 0.1470.$$

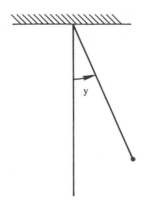

**Figure 6.1**

If we desire five terms in our Taylor series, we need one more differentiation to obtain

$$y^{(4)} = (-\cos y)y'' + (\sin y)(y')^2$$

and

$$y^{(4)}(0) = (-\cos 0.2)(-0.1987) + (\sin 0.2)(-0.15)^2 = 0.1992.$$

This gives the solution as

$$y = 0.2 - 0.15x - \frac{0.1987}{2}x^2 + \frac{0.1470}{6}x^3 + \frac{0.1992}{24}x^4$$

$$= 0.2 - 0.15x - 0.0993x^2 + 0.0245x^3 + 0.0083x^4. \tag{6.5}$$

If we consider situations where $y$ is small and approximate $\sin y$ by $y$ in the differential equation (6.4), we may write the solution of this resulting equation which satisfies $y(0) = 0.2$ and $y'(0) = -0.15$ as

$$y = 0.2 \cos x - 0.15 \sin x. \tag{6.6}$$

Substitution of the Taylor series for $\cos x$ and $\sin x$ in (6.6) gives the first five terms in this solution as

$$y = 0.2 - 0.15x - 0.1x^2 + 0.025x^3 + 0.0083x^4. \tag{6.7}$$

A comparison of the coefficients of the powers of $x$ in (6.7) and (6.5) shows that for the small oscillations resulting from this initial value problem, there is little difference in using $y$ for $\sin y$ in the differential equation.

The biggest drawback in finding Taylor series solutions of initial value problems is that we often cannot determine a general formula for $y^{(k)}(x_0)$ or the radius of convergence of the series. Fortunately, for first order differential equations, we have the following theorem, which gives a lower bound for the radius of convergence.

**Theorem 6.1.**   Let $F(x, y)$ in the initial value problem

$$\frac{dy}{dx} = F(x, y), \qquad y(x_0) = y_0, \tag{6.8}$$

be an analytic function of $x$ and $y$ for $|x - x_0| < h_1$ and $|y - y_0| < h_2$. If $|F(x, y)| \leq M$ for $x$ and $y$ in these intervals, the Taylor series for the solution of (6.8) converges at least for

$$|x - x_0| < h_1\left[1 - \exp\left(\frac{-h_2}{2Mh_1}\right)\right].$$

Applying Theorem 6.1 to Example 6.1 shows that if we take $h_1 = h_2 = 1$ and $1 + y^2 \leq 5$, the radius of convergence of the Taylor series solution is at least

$$1 - \exp\left(\frac{-1}{10}\right) \approx 0.095.$$

## EXERCISES

1. Find the first four terms in the Taylor series expansion of the solution to

$$y' = x^2 + y^2, \qquad y(-1) = -1,$$

   and compare this expression to the approximate solution by Picard's iteration method for exercise 3 in Section 1.6.

2. Find the Taylor series solution to

$$y' = 2xy, \qquad y(0) = 1,$$

   and show that it may be expressed as $e^{x^2}$.

3. (a) Find the first four terms in the Taylor series expansion of the solution to

$$y' = \sqrt{x^2 + y^2}, \qquad y(0) = 1.$$

   (b) Find a minimum for the radius of convergence for this series using Theorem 6.1 for the region $|x| \leq 1, |y - 1| \leq 1$.

4. (a) Find the first four terms in the Taylor series expansion of the solution to

$$y' = e^x y^2 + 3 \sin x, \qquad y(0) = 1.$$

   (b) Find a minimum for the radius of convergence for this series using Theorem 6.1 for the region $|x| \leq 1, |y - 1| \leq 1$.

5. (a) Find the first five terms in the Taylor series expansion of the solution to

$$y'' = -\sin y, \qquad y(0) = 1, \qquad y'(0) = 0.$$

   (b) Compare your solution in part (a) with the solution of $y'' = -y, y(0) = 1, y'(0) = 0$.

6. Find the Taylor series solution to

$$y'' = -4x^2 y - 2 \sin x^2, \qquad y(0) = 1, \quad y'(0) = 0,$$

   and show it may be represented as $\cos x^2$.

7. Find the first five terms in the Taylor series expansion of the solution to

$$y'' = y' \ln y + x, \qquad y(0) = 1, \quad y'(0) = 2.$$

## 6.2 THE METHOD OF UNDETERMINED COEFFICIENTS

In this section we develop an alternative method to finding the coefficients in the Taylor series expansion of a solution to a differential equation. This method is based on the fact that the radius of a convergence of power series is not affected by term-by-term differentiation. Thus if

$$f(x) = \sum_{k=0}^{\infty} c_k(x - x_0)^k$$

converges for $|x - x_0| < h$, then

$$f'(x) = \sum_{k=1}^{\infty} kc_k(x - x_0)^{k-1}$$

converges for $|x - x_0| < h$ also. [Note that the lower limit in the index of summation for $f'(x)$ begins at $k = 1$. This is because the term for $k = 0$ in the expression for $f(x)$ is a constant.] This process may be repeated and use of the ratio test shows that the power series for $f^{(n)}(x)$ will have $h$ as its radius of convergence. To use this fact in the method of undetermined coefficients to solve

$$y' = F(x, y),$$

$$y(x_0) = y_0,$$

we assume the solution has the form

$$y = c_0 + c_1(x - x_0) + c_2(x - x_0)^2 + \cdots = \sum_{k=0}^{\infty} c_k(x - x_0)^k \qquad (6.9a)$$

and compute $y'$ as

$$y' = c_1 + 2c_2(x - x_0) + 3c_3(x - x_0)^2 + \cdots = \sum_{k=1}^{\infty} kc_k(x - x_0)^{k-1}. \qquad (6.9b)$$

Notice that choosing $c_0 = y_0$ satisfies the initial condition and the rest of the $c_k$, $k = 1, 2, 3, \ldots$, are chosen so that the infinite series for $y$ and $y'$ satisfy the differential equation. If $F(x, y)$ is a polynomial in $x$ and $y$, this procedure is straightforward, but works best when $F$ is a linear function of $y$. The reason for this is that when $F(x, y)$ is a linear function of $y$, the algebraic equations for the $c_k$, $k = 1, 2, 3, \ldots$, will also be linear. This is illustrated by the following two examples.

**EXAMPLE 6.2**

$$y' + 2xy = 1,$$

$$y(0) = y_0. \qquad (6.10)$$

If we use the expressions for the series from (6.9a) and (6.9b) (with $x_0 = 0$) in (6.10), we obtain

$$c_1 + 2c_2x + 3c_3x^2 + 4c_4x^3 + \cdots = 1 - 2x(c_0 + c_1x + c_2x^2 + \cdots)$$

$$= 1 - 2c_0x - 2c_1x^2 - 2c_2x^3 - \cdots.$$

Since power series expansions are unique, we may equate the coefficient of $x^k$, $k = 0, 1, 2, \ldots$, on the left-hand side of this equation with the coefficient of $x^k$ on the right-hand side of the equation. This yields

$$c_1 = 1, \qquad 2c_2 = -2c_0,$$
$$3c_3 = -2c_1, \qquad 4c_4 = -2c_2, \cdots,$$

and a resulting power series

$$c_0 + c_1 x + c_2 x^2 + c_3 x^3 + c_4 x^4 + \cdots$$
$$= c_0 + x - c_0 x^2 - \frac{2x^3}{3} + \frac{c_0 x^4}{2} + \cdots$$
$$= x - \frac{2x^3}{3} + \cdots + c_0\left(1 - x^2 + \frac{x^4}{2} + \cdots\right).$$

It should be clear that choosing $c_0 = y_0$ will satisfy the initial condition. However, with these few terms we are unable to recognize this power series as representing any familiar function. For much of our work with series solutions, it is advantageous to keep the summation notation intact when expressions (6.9a) and (6.9b) are substituted into the differential equation. Redoing the substitution in this manner gives

$$\sum_{k=1}^{\infty} k c_k x^{k-1} = 1 - 2x \sum_{k=0}^{\infty} c_k x^k. \tag{6.11}$$

If we bring the $2x$ on the right-hand side of (6.11) under the summation sign and transpose the entire series to the other side of the equation, we obtain

$$\sum_{k=1}^{\infty} k c_k x^{k-1} + \sum_{k=0}^{\infty} 2c_k x^{k+1} = 1.$$

In order to combine the two series into a single sum, we change the index of summation in the second series by replacing $k$ by $k - 2$ and raising the lower limit of that sum by 2 (see exercise 1). This gives

$$\sum_{k=1}^{\infty} k c_k x^{k-1} + \sum_{k=2}^{\infty} 2c_{k-2} x^{k-1} = 1$$

or

$$c_1 + \sum_{k=2}^{\infty} k c_k x^{k-1} + \sum_{k=2}^{\infty} 2c_{k-2} x^{k-1} = 1.$$

Combining these series gives

$$c_1 + \sum_{k=2}^{\infty} (k c_k + 2c_{k-2}) x^{k-1} = 1$$

and the uniqueness of power series allows us to conclude that

$$c_1 = 1,$$
$$kc_k + 2c_{k-2} = 0, \qquad k = 2, 3, 4, \ldots. \tag{6.12}$$

This last equation is called a *recurrence relation* and from it we may obtain the coefficients for even subscripts as

$$c_2 = -2\frac{c_0}{2} = -c_0,$$

$$c_4 = \frac{-2c_2}{4} = \frac{(-1)^2 c_0}{2 \cdot 1},$$

$$c_6 = \frac{-2c_4}{6} = \frac{(-1)^3 c_0}{3 \cdot 2 \cdot 1},$$

$$\vdots$$

$$c_{2n} = \frac{(-1)^n c_0}{n(n-1) \cdots (3)(2)(1)} = \frac{(-1)^n c_0}{n!}, \qquad n = 1, 2, 3, \ldots,$$

and the ones for the odd subscripts as

$$c_3 = \frac{-2c_1}{3} = -\frac{2}{3},$$

$$c_5 = \frac{-2c_3}{5} = \frac{(-2)^2}{5 \cdot 3},$$

$$c_7 = \frac{-2c_5}{7} = \frac{(-2)^3}{7 \cdot 5 \cdot 3},$$

$$\vdots$$

$$c_{2n+1} = \frac{(-2)^n}{(2n+1)(2n-1) \cdots (5)(3)(1)}, \qquad n = 1, 2, 3, \ldots.$$

Thus the solution of (6.10) is

$$y = c_0 \sum_{n=0}^{\infty} \frac{(-1)^n x^{2n}}{n!} + \sum_{n=0}^{\infty} \frac{(-2)^n x^{2n+1}}{(2n+1)(2n-1) \cdots (5)(3)(1)} \qquad (6.13)$$

and $y(0) = y_0$ is satisfied by choosing $c_0 = y_0$. The first series is recognized as the Taylor series expansion for $e^{-x^2}$,

$$\left[ e^t = \sum_{n=0}^{\infty} \frac{t^n}{n!} \quad \text{so} \quad e^{-x^2} = \sum_{n=0}^{\infty} \frac{(-x^2)^n}{n!} = \sum_{n=0}^{\infty} \frac{(-1)^n x^{2n}}{n!} \right],$$

and we note that $e^{-x^2}$ satisfies $y' + 2xy = 0$, that is, is a solution of the homogeneous differential equation associated with (6.10). The second series in (6.13) is the particular solution and converges for all $x$ [see exercise 3(a)]. If we solve (6.10) using an integrating factor, we see that this series is the Taylor series for

$$e^{-x^2} \int_0^x e^{t^2} \, dt.$$

To demonstrate the method of undetermined coefficients for second order differential equations, consider the following example.

**EXAMPLE 6.3**

$$y'' + 2(x - 1)y' + y = 0,$$
$$y(1) = A, \qquad (6.14)$$
$$y'(1) = B.$$

Since the initial conditions are given at $x = 1$, we assume a solution of the form given by (6.9a) with $x_0 = 1$. Then substituting the expressions for $y$ and $y'$ from (6.9a) and (6.9b) together with that for the second derivative,

$$y'' = \sum_{k=2}^{\infty} k(k - 1)c_k(x - 1)^{k-2},$$

into (6.14) gives

$$\sum_{k=2}^{\infty} k(k - 1)c_k(x - 1)^{k-2} + \sum_{k=1}^{\infty} 2kc_k(x - 1)^k + \sum_{k=0}^{\infty} c_k(x - 1)^k = 0.$$

Raising the index in the first series by 2 (and reducing the lower limit of summation by 2) gives

$$\sum_{k=0}^{\infty} (k + 2)(k + 1)c_{k+2}(x - 1)^k + \sum_{k=1}^{\infty} 2kc_k(x - 1)^k + \sum_{k=0}^{\infty} c_k(x - 1)^k = 0.$$

Since the first and third series start with a lower limit of 0 and the middle series starts with 1, we write the two terms corresponding to $k = 0$ separately and combine the rest of the terms as

$$2c_2 + c_0 + \sum_{k=1}^{\infty} [(k + 2)(k + 1)c_{k+2} + (2k + 1)c_k](x - 1)^k = 0.$$

Equating the coefficient of each power of $(x - 1)^k$ to zero gives

$$2c_2 + c_0 = 0,$$
$$(k + 2)(k + 1)c_{k+2} + (2k + 1)c_k = 0,$$

or

$$c_{k+2} = \frac{-(2k + 1)}{(k + 2)(k + 1)} c_k, \qquad k = 1, 2, 3, \ldots.$$

Thus

$$c_2 = \frac{-1}{2 \cdot 1} c_0, \qquad c_3 = \frac{-3}{3 \cdot 2 \cdot 1} c_1,$$

and all of the other coefficients may be written in terms of $c_0$ and $c_1$. It is clear that $c_0$ and $c_1$ are arbitrary constants, with the remaining coefficients determined by the above recurrence relation. The even subscripted terms are given by

$$c_4 = \frac{-5}{4 \cdot 3} c_2 = \frac{(-5)(-1)}{4 \cdot 3 \cdot 2 \cdot 1} c_0,$$

$$c_6 = \frac{-9}{6 \cdot 5} c_4 = \frac{(-1)^3 9 \cdot 5 \cdot 1}{6!} c_0,$$

$$c_8 = \frac{-13}{8 \cdot 7} c_6 = \frac{(-1)^4 13 \cdot 9 \cdot 5 \cdot 1}{8!} c_0,$$

$$\vdots$$

$$c_{2n} = \frac{(-1)^n (4n - 3)(4n - 7) \cdots (9)(5)(1)}{(2n)!} c_0, \qquad n = 1, 2, 3, \ldots ,$$

while the odd coefficients are given by

$$c_5 = \frac{-7}{5 \cdot 4} c_3 = \frac{(-1)^2 7 \cdot 3}{5!} c_1,$$

$$c_7 = \frac{-11}{7 \cdot 6} c_5 = \frac{(-1)^3 11 \cdot 7 \cdot 3}{7!} c_1,$$

$$\vdots$$

$$c_{2n+1} = \frac{(-1)^n (4n - 1)(4n - 5) \cdots (7)(3)}{(2n + 1)!} c_1, \qquad n = 1, 2, 3, \ldots .$$

Thus the solution

$$y = c_0 \left[ 1 + \sum_{n=1}^{\infty} \frac{(-1)^n (4n - 3)(4n - 7) \cdots (9)(5)(1)}{(2n)!} (x - 1)^{2n} \right]$$

$$+ c_1 \left[ x - 1 + \sum_{n=1}^{\infty} \frac{(-1)^n (4n - 1)(4n - 5) \cdots (7)(3)}{(2n + 1)!} (x - 1)^{2n+1} \right]$$

is the general solution to (6.14). Choosing the arbitrary constants as $c_0 = A$, $c_1 = B$ satisfies the initial conditions. The ratio test shows that both of these series converge for all values of $x$ [see exercise 3(b)].

The procedure for solving nonlinear problems follows along the same lines as illustrated by these two examples. We are limited only by our ability to solve the nonlinear algebraic equations for the constants $c_k$. For the nonlinear differential equations given in the exercises, this will not be a problem.

# EXERCISES

1. Make a change in the index of summation to show the validity of the following equations.

(a) $\displaystyle\sum_{k=0}^{\infty} c_k x^{k+2} = \sum_{m=2}^{\infty} c_{m-2} x^m.$

**(b)** $\displaystyle\sum_{k=4}^{\infty} c_{k-1}x^{k-2} = \sum_{m=2}^{\infty} c_{m+1}x^m.$

**(c)** $\displaystyle\sum_{m=n}^{\infty} c_m x^m = \sum_{k=0}^{\infty} c_{k+n}x^{k+n}.$ (That is, if we reduce the lower limit of the summation by $n$, we increase the index to the right of the summation sign by $n$.)

**(d)** $\displaystyle\sum_{m=n}^{\infty} c_m x^{m-p} = \sum_{k=n-p}^{\infty} c_{k+p}x^k.$ (That is, if we increase the exponent by $p$, we increase the corresponding subscript by $p$ and reduce the lower limit by $p$.)

2. Show that

**(a)** $\displaystyle\sum_{k=2}^{\infty} c_k + A \sum_{k=0}^{\infty} a_k = A(a_0 + a_1) + \sum_{k=2}^{\infty} (c_k + Aa_k).$

**(b)** $\displaystyle\sum_{k=0}^{\infty} c_k = \sum_{k=0}^{n} c_k + \sum_{k=n+1}^{\infty} c_k$ for any positive integer $n$.

3. **(a)** Use the ratio test to show the second series in (6.13) converges for all values of $x$.
   **(b)** Use the ratio test to show that both series in the last example of this section converge for all values of $x$.

Use the method of undetermined coefficients in the following exercises.

4. $y' = -2x^2 y, \quad y(0) = -2.$
5. $y' = x^3 y + 3x, \quad y(0) = 4.$
6. $y' = (1 + 2x + 2x^2)y, \quad y(0) = 3.$
7. $y' = 2(x - 1)^2 y, \quad y(1) = 4.$
8. $y' = (x - \pi)y + \cos x, \quad y(\pi) = 2.$
9. $y' = y \tan x, \quad y(\pi/4) = -3.$
10. $y' = (3x + 1)y, \quad y(0) = 4.$
11. $y' = 2x^2 - y^2, \quad y(1) = 2.$
12. $y'' - xy' + 3y = 0, \quad y(0) = 2, \quad y'(0) = 0.$
13. $y'' + 2x^2 y' + xy = 0, \quad y(0) = 0, \quad y'(0) = 3.$
14. $y'' - (\sin x)y' + y = 0, \quad y(0) = 0, \quad y'(0) = -1.$
15. $y'' + (y')^2 = e^x y^2, \quad y(0) = 0, \quad y'(0) = 1.$
16. $y'' - (y')^2 = y^3 + 3, \quad y(0) = 0, \quad y'(0) = 0.$

## 6.3 ORDINARY AND SINGULAR POINTS

In Section 6.2 we used the method of undetermined coefficients to find power series solutions to differential equations. However, if we substitute

$$y = \sum_{k=0}^{\infty} c_k x^k$$

into

$$x^3 y'' + y = 0, \tag{6.15}$$

we find that

$$\sum_{k=2}^{\infty} k(k - 1)c_k x^{k+1} + \sum_{k=0}^{\infty} c_k x^k = 0.$$

Changing the index of summation in the first series and combining terms gives

$$c_0 + c_1 x + c_2 x^2 + \sum_{k=3}^{\infty} [(k-1)(k-2)c_{k-1} + c_k]x^k = 0$$

and the only way this equation is satisfied is for

$$c_i = 0, \qquad i = 0, 1, 2, \ldots .$$

Obviously, it would be useful to know if the method of undetermined coefficients works before we go through all the details trying to find a solution. To that end we focus our attention on homogeneous second order linear differential equations since the nonhomogeneous part plays no role in the classification. Such an equation may be taken as

$$a_2(x)y'' + a_1(x)y' + a_0(x)y = 0 \qquad (6.16)$$

or, since $a_2(x)$ is not identically zero, we will divide through by $a_2(x)$ and use

$$y'' + p(x)y' + q(x)y = 0 \qquad (6.17)$$

as the standard form. [Note that

$$p(x) = \frac{a_1(x)}{a_2(x)} \qquad \text{and} \qquad q(x) = \frac{a_0(x)}{a_2(x)} \Bigg].$$

**Definition 6.1.**    $x = x_0$ is called an *ordinary point* of the second order linear differential equation if $p(x)$ and $q(x)$ are *both* analytic at $x_0$. If $x = x_0$ is not an ordinary point, it is called a *singular point*.

Now a function $f$ is analytic at $x_0$ if its Taylor series about $x_0$ converges to $f(x)$. Thus if $p(x)$ and $q(x)$ are both ratios of polynomials in $x$ with all common factors removed, the singular points will occur at the zeros of the denominators of $p(x)$ and $q(x)$. All other points will be ordinary points.

For example, to find the ordinary and singular points of

$$x^2(x-1)y'' + \frac{x(x-1)}{(x+3)}y' + \frac{(x-1)^2}{(x+2)}y = 0 \qquad (6.18)$$

we first divide by $x^2(x-1)$ to obtain the standard form of the differential equation as

$$y'' + \frac{1}{x(x+3)}y' + \frac{(x-1)}{x^2(x+2)}y = 0.$$

Thus, $0$, $-2$, and $-3$ are singular points of (6.18) and all other points are ordinary points.

With Definition 6.1 the question about when to try the method of undetermined coefficients is answered by the following theorem.

**Theorem 6.2.**    Given the initial value problem

$$y'' + p(x)y' + q(x)y = b(x),$$

$$y(x_0) = y_0, \qquad y'(x_0) = y_1. \qquad (6.19)$$

If $p(x)$, $q(x)$, and $b(x)$ are all analytic at $x = x_0$, the solution $y(x)$ has a Taylor series about $x_0$ with a radius of convergence at least as large as the minimum of the radii of convergence for $p(x)$, $q(x)$, and $b(x)$ about $x_0$.

Thus if the hypothesis of Theorem 6.2 is satisfied, we may confidently proceed with calculation of the Taylor series for $y(x)$, either by the method of undetermined coefficients or by using implicit differentiation on the differential equation itself. Notice that in the differential equation (6.15), where the method of undetermined coefficients failed, the point $x = 0$ is a singular point and the hypothesis of Theorem 6.2 is not satisfied.

**EXAMPLE 6.4**    Find the Taylor series solution to

$$xy'' + (\sin x)y' + 2xy = 3x^2,$$

$$y(0) = A, \qquad y'(0) = B,$$

(6.20)

and determine its radius of convergence. To check the hypothesis of Theorem 6.2, we divide by $x$ to obtain the standard form of the equation as

$$y'' + \frac{\sin x}{x}y' + 2y = 3x.$$

Since the coefficients

$$\frac{\sin x}{x} = \frac{1}{x}\sum_{k=0}^{\infty} \frac{(-1)^k x^{2k+1}}{(2k+1)!} = \sum_{k=0}^{\infty} \frac{(-1)^k x^{2k}}{(2k+1)!},$$

2, and $3x$ are convergent Taylor series about $x = 0$ for all values of $x$, the hypothesis of Theorem 6.2 is fulfilled and we know before we start that the Taylor series for the solution has an infinite radius of convergence. That is nice to know, because for this example we will not be able to obtain a "nice" form for the $n$th term and use the ratio test to show convergence. Assuming a solution of the form

$$y = \sum_{k=0}^{\infty} c_k x^k$$

and substituting it into (6.20) gives

$$\sum_{k=2}^{\infty} k(k-1)c_k x^{k-1} + \sum_{k=0}^{\infty} \frac{(-1)^k x^{2k+1}}{(2k+1)!}\sum_{k=1}^{\infty} kc_k x^{k-1} + \sum_{k=0}^{\infty} 2c_k x^{k+1} = 3x^2.$$

If we desire the Taylor series for the solution to terms up to $x^5$, we truncate the infinite series of the previous equation to

$$2c_2 x + 6c_3 x^2 + 12c_4 x^3 + 20c_5 x^4$$

$$+ \left(x - \frac{x^3}{6} + \frac{x^5}{120}\right)(c_1 + 2c_2 x + 3c_3 x^2 + 4c_4 x^3)$$

$$+ 2c_0 x + 2c_1 x^2 + 2c_2 x^3 + 2c_3 x^4 = 3x^2.$$

Equating coefficients of $x^n$, $n = 1, 2, 3, 4$, on both sides of this equation gives

$$2c_2 + c_1 + 2c_0 = 0, \qquad 6c_3 + 2c_2 + 2c_1 = 3,$$

$$12c_4 + 3c_3 - \frac{c_1}{6} + 2c_2 = 0, \qquad 20c_5 + 4c_4 - \frac{c_2}{3} + 2c_3 = 0.$$

Notice that this system of four equations in six unknowns is a dependent system. If we let $c_0$ and $c_1$ be chosen arbitrarily, then

$$c_2 = -\frac{1}{2}(c_1 + 2c_0), \qquad c_3 = \frac{-(c_1 - 2c_0)}{6} + \frac{1}{2},$$

$$c_4 = \frac{5c_1}{36} + \frac{c_0}{12} - \frac{1}{8}, \qquad c_5 = \frac{-(1/2 + 7c_1/18 + 4c_0/3)}{20},$$

and

$$y(x) = c_0\left(1 - x^2 + \frac{x^3}{3} + \frac{x^4}{12} - \frac{x^5}{15} + \cdots\right)$$

$$+ c_1\left(x - \frac{x^2}{2} - \frac{x^3}{6} + \frac{5x^4}{36} - \frac{7x^5}{360} + \cdots\right)$$

$$+ \frac{x^3}{2} - \frac{x^4}{8} - \frac{x^5}{40} + \cdots.$$

Notice that the Taylor series has the form

$$y(x) = c_0 y_1(x) + c_1 y_2(x) + y_p(x),$$

where $y_1(x)$ and $y_2(x)$ represent Taylor series expansions of the two linearly independent solutions of the homogeneous differential equation, while $y_p(x)$ is a particular solution of (6.20). Choosing $c_0 = A$ and $c_1 = B$ satisfies the given initial conditions.

When the $a_0(x)$, $a_1(x)$, and $a_2(x)$ of (6.16) are polynomials, we have another useful theorem to aid us in determining the radius of convergence of the Taylor series expansion of our solution.

**Theorem 6.3.**   Given

$$a_2(x)y'' + a_1(x)y' + a_0(x)y = 0, \tag{6.21}$$

where $a_2(x)$, $a_1(x)$, and $a_0(x)$ are the polynomials in $x$ remaining after all factors common to all three have been removed. If $a_2(x_0) \neq 0$, the general solution of (6.21) may be obtained as a Taylor series about $x_0$ and may be written as

$$\sum_{k=0}^{\infty} c_k(x - x_0)^k = c_0 y_1(x) + c_1 y_2(x),$$

where $y_1(x)$ and $y_2(x)$ are linearly independent functions and $c_0$ and $c_1$ are arbitrary constants. This series converges at least for $|x - x_0| < h$, where $h$ equals the distance between $x_0$ and the zero of $a_2(x)$, which is closest to $x_0$. In calculating the zeros of $a_2(x)$, we must also include complex zeros.

**EXAMPLE 6.5**    Later in this chapter we will consider the differential equation

$$(1 - x^2)y'' - 2xy' + \lambda y = 0,$$

where $\lambda$ is a constant. Notice that $x = 0$ is an ordinary point of this differential equation and $a_2(x) = 1 - x^2 = (1 - x)(1 + x)$ has zeros at 1 and $-1$. Thus according to Theorem 6.3, the Taylor series about $x = 0$ will converge at least for $|x| < 1$.

**EXAMPLE 6.6**    Consider the differential equation

$$(4 + x^2)y'' + (3x^2 - 2)y' + 16y = 0,$$

with $x = 1$ as an ordinary point. Since $4 + x^2$ has complex roots, $\pm 2i$, the Taylor series solution about 1 will converge for $|x - 1| < \sqrt{1^2 + 2^2} = \sqrt{5}$. Note that $x = 0$ is also an ordinary point of the equation above. Thus the Taylor series solution about 0 will converge for $|x| < 2$.

## EXERCISES

1. List the singular points for the following differential equations.

   (a) $(1 - x)y'' + \dfrac{3x}{x + 2}y' + \dfrac{(1 - x)^2}{x + 3}y = 0.$

   (b) $(x^2 + x)y'' + \dfrac{x^3}{x - 1}y' + \dfrac{x^4 + 3x}{x + 2}y = 0.$

   (c) $\dfrac{1}{x}y'' + \dfrac{3(x - 4)}{x + 6}y' + \dfrac{x^2(x - 2)}{x - 1}y = 0.$

   (d) $(x^2 + 3x + 2)y'' + \dfrac{x + 2}{x - 1}y' + \dfrac{x - 2}{x}y = 0.$

   (e) $(x^2 + 9)y'' + \dfrac{x - 2}{x + 7}y' + \dfrac{x^2 + 3}{2}y = 0.$

   (f) $e^x y'' + \dfrac{3x - 4}{x + 4}y' + \dfrac{x}{x - 4}y = 0.$

   (g) $(\sin x)y'' + \dfrac{x}{x - 2}y' + \dfrac{x - 2}{x}y = 0.$

2. Theorem 6.3 concluded that $\sum_{k=0}^{\infty} c_k(x - x_0)^k$ converged for $|x - x_0| < h$. For each of the differential equations in exercise 1, find the $h$ for the value of $x_0$ given below.

   (a) $x_0 = 1.$                                    (b) $x_0 = 4.$
   (c) $x_0 = -1.$                                   (d) $x_0 = 2.$
   (e) $x_0 = 0.$                                    (f) $x_0 = 0.$
   (g) $x_0 = 3.$

3. Find the first five terms in a Taylor series solution of the following differential equations

and give a value of $h$ such that the series solution converges for $|x - x_0| < h$. (Use the method of undetermined coefficients.)

(a) $y'' + x^2 y' + (\sin x)y = 3$,  $y(0) = 0$,  $y'(0) = 3$.

(b) $y'' + 3xy' + (\cos x)y = x^2$,  $y(0) = 1$,  $y'(0) = 0$.

(c) $y'' - 2xy' + e^x y = e^x$,  $y(0) = 0$,  $y'(0) = -2$.

(d) $y'' + (\sin x)y' + xy = 0$,  $y(0) = 3$,  $y'(0) = 0$.

(e) $y'' - xy' + (\ln x)y = 0$,  $y(1) = 0$,  $y'(1) = 2$.

(f) $y'' - xy' + \dfrac{y}{1 - 2x} = 0$,  $y(0) = 1$,  $y'(0) = 0$.

4. Find the general solution of the following differential equations as Taylor series about the given point and give the radius of convergence. (Use the method of undetermined coefficients.)

(a) $y'' + x^2 y = 0$,  $x_0 = 0$.

(b) $y'' - 3xy' + y = 0$,  $x_0 = 0$.

(c) $(1 - x^2)y'' + 2xy' + 5y = 0$,  $x_0 = 0$.

(d) $xy'' + x^2 y' + y = 0$,  $y(2) = 0$,  $y'(2) = 1$.

(e) $x^2 y'' + y' + 2y = 0$,  $y(1) = 0$,  $y'(1) = 1$.

(f) $(x^2 + 1)y'' + xy' + 3y = 0$,  $y(2) = 0$,  $y'(2) = 3$.

### 6.3.1 Regular Singular Points

If we try to find a power series solution to

$$x^2 y'' - y = 0 \tag{6.22}$$

in the form $\sum_{k=0}^{\infty} c_k x^k$, we discover that

$$\sum_{k=2}^{\infty} (k)(k - 1)c_k x^k - \sum_{k=0}^{\infty} c_k x^k = 0.$$

This requires that $c_0 = 0$, $c_1 = 0$, and $(k(k - 1) - 1)c_k = 0$, $k = 2, 3, 4, \ldots$, and we obtain the trivial solution to (6.22). However, we recognize (6.22) as a Cauchy-Euler equation, so we try a solution of the form $x^r$. Letting $y = x^r$ in (6.22) gives

$$r^2 - r - 1 = 0,$$

or

$$r = \frac{1 \pm \sqrt{5}}{2}.$$

Thus the general solution of (6.22) is

$$y = c_0 x^{(1+\sqrt{5})/2} + c_1 x^{(1-\sqrt{5})/2}.$$

Notice that neither of these functions is analytic at $x = 0$ and that $x = 0$ is a singular point of (6.22).

We want to find conditions for which differential equations with singular points have solutions in the form of a series of powers of $x$, finite or infinite. Since we are dealing with Taylor series about a real number $x_0$, we consider only those singular points that are real and give the following definition. (Note that in using Theorems 6.3 and 6.6 we need to consider singular points which are complex numbers. However, we will not worry whether a complex number is a regular or an irregular singular point.)

**Definition 6.2.**    Given that $x = x_0$ is a singular point of the differential equation

$$y'' + p(x)y' + q(x)y = 0, \qquad (6.17)$$

if the functions

$$(x - x_0)p(x) \qquad \text{and} \qquad (x - x_0)^2 q(x) \qquad (6.23)$$

are *both* analytic at $x_0$, then $x_0$ is a regular singular point of (6.17). If at least one of the products in (6.23) is not analytic at $x_0$, then $x_0$ is called an irregular singular point of (6.17).

**EXAMPLE 6.7**    The differential equation

$$(1 - x^2)y'' - 2xy' + \lambda y = 0$$

has singular points at 1 and $-1$ since

$$p(x) = \frac{-2x}{1 - x^2} \qquad \text{and} \qquad q(x) = \frac{\lambda}{1 - x^2}.$$

These are both regular singular points since

$$(x - 1)p(x) = \frac{-2x}{-(1 + x)} \qquad \text{and} \qquad (x - 1)^2 q(x) = \frac{\lambda(x - 1)}{-(1 + x)}$$

are both analytic at 1, while

$$(x + 1)p(x) = \frac{-2x}{1 - x} \qquad \text{and} \qquad (x + 1)^2 q(x) = \frac{\lambda(x + 1)}{1 - x}$$

are both analytic at $-1$. All other points are ordinary points of this differential equation.

**EXAMPLE 6.8**    The differential equation

$$x(\cos x)y'' + 3(x + 1)y' + \frac{2}{x^2(x - 4)^2}y = 0 \qquad (6.24)$$

has singular points at 0, 4, and $(2n + 1)\pi/2, n = 0, \pm 1, \pm 2, \ldots$, since $p(x)$ and $q(x)$ are given by

$$p(x) = \frac{3(x + 1)}{x \cos x}, \qquad q(x) = \frac{2}{x^3(x - 4)^2 \cos x}.$$

We have an irregular singular point at 0 since

$$x^2 q(x) = \frac{2}{x(x - 4)^2 \cos x}$$

is *not* analytic at 0. $x = 4$ is a regular singular point since

$$(x - 4)p(x) = \frac{3(x + 1)(x - 4)}{x \cos x} \quad \text{and} \quad (x - 4)^2 q(x) = \frac{2}{x^3 \cos x}$$

are both analytic at 4. To determine the nature of the singularity at $(2n + 1)\pi/2$, $n = 0, \pm 1, \pm 2, \ldots$, calculate

$$\frac{(x - (2n + 1)\pi/2)}{\cos x} \frac{(3)(x + 1)}{x} \quad \text{and} \quad \frac{[x - (2n + 1)\pi/2]^2}{\cos x} \frac{2}{x^3(x - 4)^2}.$$

$$(6.25)$$

Since we know that the quotient of two analytic functions is analytic at all points except possibly where the denominator is zero, we need only check to see if both expressions are defined at $x = (2n + 1)\pi/2$. Using L'Hôpital's rule on the term on the left of (6.25) gives

$$\lim_{x \to (2n+1)\pi/2} \frac{(x - (2n + 1)\pi/2)}{\cos x} \frac{3(x + 1)}{x}$$

$$= \frac{1}{-\sin(2n + 1)\pi/2} \frac{3((2n + 1)\pi/2 + 1)}{(2n + 1)\pi/2}.$$

This limit gives a finite number and the limit as $x \to (2n + 1)\pi/2$ of the term on the right in (6.25) is

$$\lim_{x \to (2n+1)\pi/2} [x - (2n + 1)\pi/2] \lim_{x \to (2n+1)\pi/2} \frac{(x - (2n + 1)\pi/2)2}{(\cos x)(x^3)(x - 4)^2}$$

$$= (0)(\text{some finite number}) = 0.$$

Since both limits are finite, $(2n + 1)\pi/2$, $n = 0, \pm 1, \pm 2, \ldots$, are all regular singular points of (6.24).

For the situation where $p(x)$ and $q(x)$ in (6.17) are rational functions, we can restate Definition 6.2 in a simpler manner.

**Definition 6.3.** If $p(x)$ and $q(x)$ are rational functions (i.e., quotients of polynomials) and $x_0$ is a singular point of

$$y'' + p(x)y' + q(x)y = 0,$$

then $x_0$ is a regular singular point if

$$\lim_{x \to x_0} (x - x_0)p(x) \quad \text{and} \quad \lim_{x \to x_0} (x - x_0)^2 q(x) \qquad (6.26)$$

*both* exist.

To use Definition 6.3, note that we first find singular points where at least one of $p(x)$ or $q(x)$ is not defined (i.e., has zeros in the denominator). Then take the limits in (6.26) to see if these are regular singular points.

**EXAMPLE 6.9** The singular points of

$$y'' + \frac{x^4 + 1}{x^4 - 1}y' + \frac{x^4}{(4 - x)^4}y = 0$$

are $\pm 1$, $\pm i$, and 4, since

$$x^4 - 1 = (x^2 + 1)(x^2 - 1) = (x + i)(x - i)(x + 1)(x - 1);$$

4 is an irregular singular point since

$$\lim_{x \to 4} \frac{(x - 4)^2 x^4}{(4 - x)^4}$$

does not exist; 1 is a regular singular point since

$$\lim_{x \to 1} \frac{(x - 1)(x^4 + 1)}{x^4 - 1} = \lim_{x \to 1} \frac{(x^4 + 1)}{(x^2 + 1)(x + 1)} = \frac{2}{4}$$

and

$$\lim_{x \to 1} \frac{(x - 1)^2 (x^4)}{(4 - x)^4} = 0.$$

Similar calculations show that $-1$ is also a regular singular point.

## EXERCISES

Determine all the regular and irregular singular points for the following differential equations.

**1.** $x^2(x^2 - 1)^2 y'' + \dfrac{x(x + 1)}{x - 4} y' + \dfrac{3(x - 1)}{x^2 - 16} y = 0.$

**2.** $x(x^2 - 3x - 10)y'' + \dfrac{x + 4}{x - 2} y' + 16y = 0.$

**3.** $x(\sin x)y'' + \dfrac{3(x - 1)}{x + 1} y' + (\cos x)y = 0.$

**4.** $(\sin x)y'' + \dfrac{x \cos x}{x + 1} y' - \dfrac{x^2}{x - 2} y = 0.$

**5.** $x^2(x^2 - 16)y'' + \dfrac{(x - 2)}{x + 2} y' + 32y = 0.$

**6.** $e^x y'' + 3xy' + \dfrac{1}{1 - e^x} y = 0.$

**7.** $x(x - 1)y'' + \dfrac{x + 1}{(x - 4)^2} y' + \dfrac{1}{x^2} y = 0.$

**8.** $(x^2 + x - 6)y'' + \dfrac{14}{1 - x^2} y' + \dfrac{12}{1 + x^2} y = 0.$

## 6.4 THE METHOD OF FROBENIUS

### 6.4.1 The Method of Frobenius, Part 1

We noted in Section 6.3.1 that trying to find a power series solution of

$$x^2 y'' - y = 0$$

led to the trivial solution. This was due to the fact that $x = 0$ is a singular point for this differential equation. However, $x = 0$ is a regular singular point and the equation has solutions of the form $x^r$, where $r = (1 \pm \sqrt{5})/2$. The idea behind a regular singular point was that even though at least one of $p(x)$ and $q(x)$ was not analytic at 0, multiplying by a power of $x$ resulted in two analytic functions. Combining this idea with the fact that $x^r$ gave solutions to the above differential equation suggests that for differential equations with regular singular points at 0, we try solutions of the form

$$y = x^r \sum_{k=0}^{\infty} c_k x^k. \tag{6.27}$$

In (6.27), $r$ and the $c_k$, $k = 0, 1, 2, \ldots$, are all constants to be determined so this series satisfies the pertinent differential equation. We will try a solution of this type for the following differential equation with a regular singular point at $x = 0$:

$$9x^2 y'' + 9x^2 y' + 2y = 0. \tag{6.28}$$

If we combine the $x^r$ with the $x^k$ under the summation sign in (6.27) and differentiate, we obtain

$$y' = \sum_{k=0}^{\infty} (k + r)c_k x^{k+r-1} \quad \text{and} \quad y'' = \sum_{k=0}^{\infty} (k + r)(k + r - 1)c_k x^{k+r-2}.$$

(Note that since $r$ is unknown, our lower limits do not change upon differentiation as they may for a power series.) Substituting these three series into (6.28) yields

$$\sum_{k=0}^{\infty} 9(k + r)(k + r - 1)c_k x^{k+r} + \sum_{k=0}^{\infty} 9(k + r)c_k x^{k+r+1} + \sum_{k=0}^{\infty} 2c_k x^{k+r} = 0.$$

$$\tag{6.29}$$

To obtain the same exponent on $x$ in all three series, we lower the index in the middle series by 1 (and raise the lower limit of summation by 1) to obtain

$$\sum_{k=0}^{\infty} 9(k + r)c_k x^{k+r+1} = \sum_{k=1}^{\infty} 9(k - 1 + r)c_{k-1} x^{k+r}.$$

Notice that the lower limit of summation on the first and third series in (6.29) starts at 0, while the new form of the middle series starts at 1. If the terms for $k = 0$ are written separately, we may combine the three series and rewrite (6.29) as

$$[9(r)(r - 1) + 2]c_0 x^r + \sum_{k=1}^{\infty} \{[9(k + r)(k + r - 1) + 2]c_k$$

$$+ 9(k - 1 + r)c_{k-1}\}x^{k+r} = 0.$$

Equating the coefficient of each power of $x$ to zero gives

$$[9(r)(r - 1) + 2]c_0 = 0, \tag{6.30}$$

$$[9(k + r)(k + r - 1) + 2]c_k + 9(k - 1 + r)c_{k-1} = 0, \qquad k = 1, 2, 3, \ldots. \tag{6.31}$$

Since $c_0$ is the coefficient of the lowest power of $x$, that is, $x^r$, there is no advantage to setting it equal to zero. Thus we choose $r$ such that

$$9(r)(r - 1) + 2 = 0. \tag{6.32}$$

This equation may be factored as

$$(3r - 2)(3r - 1) = 0$$

with two possible values for $r$, $\frac{2}{3}$ and $\frac{1}{3}$. Since either value of $r$ satisfies (6.30), $c_0$ is an arbitrary constant and the $c_k$, $k = 1, 2, 3, \ldots$, are found from (6.31) separately for each choice of $r$. Thus for $r = \frac{2}{3}$ we have the recurrence relation

$$c_k = \frac{-9(k - 1/3)}{9(k + 2/3)(k - 1/3) + 2} c_{k-1}$$

$$= \frac{-3(3k - 1)}{9k^2 + 3k - 2 + 2} c_{k-1}$$

$$= \frac{-(3k - 1)}{k(3k + 1)} c_{k-1}, \qquad k = 1, 2, 3, \ldots .$$

Solving for a few coefficients gives

$$c_1 = -\frac{(2)}{1(4)} c_0.$$

$$c_2 = \frac{-(5)}{2(7)} c_1 = \frac{(-1)^2 (5)(2)}{(2 \cdot 1)(7 \cdot 4)} c_0,$$

$$c_3 = \frac{-(8)}{3(10)} c_2 = \frac{(-1)^3 (8)(5)(2)}{(3 \cdot 2 \cdot 1)(10 \cdot 7 \cdot 4)} c_0,$$

$$\vdots$$

$$c_n = \frac{(-1)^n (3n - 1)(3n - 4) \cdots (5)(2)}{n! \, (3n + 1)(3n - 2) \cdots (7)(4)} c_0, \qquad n = 1, 2, 3, \ldots .$$

Thus we have a solution

$$c_0 x^{2/3} \left[ 1 + \sum_{n=1}^{\infty} \frac{(-1)^n (3n - 1)(3n - 4) \cdots (5)(2)n}{n! \, (3n + 1)(3n - 2) \cdots (7)(4)} x^n \right]. \tag{6.33}$$

If we now use the other value of $r$, which makes (6.30) zero, $r = \frac{1}{3}$, in (6.31) we obtain the recurrence relation

$$c_k = \frac{-9(k - 2/3)}{9(k + 1/3)(k - 2/3) + 2} c_{k-1}$$

$$= \frac{-3(3k - 2)}{9k^2 - 3k - 2 + 2} c_{k-1}$$

$$= \frac{-(3k - 2)}{k(3k - 1)} c_{k-1}, \qquad k = 1, 2, 3, \ldots .$$

Solving for a few coefficients gives

$$c_1 = \frac{-(1)}{1(2)} c_0,$$

$$c_2 = \frac{-(4)}{2(5)} c_1 = \frac{(-1)^2(4 \cdot 1)}{(2 \cdot 1)(5 \cdot 2)} c_0,$$

$$c_3 = \frac{-(7)}{3(8)} c_2 = \frac{(-1)^3(7 \cdot 4 \cdot 1)}{3 \cdot 2 \cdot 1 (8 \cdot 5 \cdot 2)} c_0,$$

$$\vdots$$

$$c_n = \frac{(-1)^n(3n - 2)(3n - 5) \cdots (4)(1)}{n! \, (3n - 1)(3n - 4) \cdots (5)(2)} c_0, \qquad n = 1, 2, 3, \ldots,$$

and a second solution of the form

$$c_0^* x^{1/3} \left[ 1 + \sum_{n=1}^{\infty} \frac{(-1)^n(3n - 2)(3n - 5) \cdots (4)(1)}{n! \, (3n - 1)(3n - 4) \cdots (5)(2)} x^n \right]. \tag{6.34}$$

Note that we relabeled the lowest coefficient $c_0^*$ since it may be chosen arbitrarily and need not equal the arbitrary constant $c_0$ in (6.33). The general solution of (6.28) is now given by the sum of the series in (6.33) and (6.34).

The procedure we used in solving this differential equation is an example of the *method of Frobenius*. Let us list the steps used in the *method of Frobenius* for developing a series solution about a regular singular point, $x = 0$, of the equation

$$a_2(x)y'' + a_1(x)y' + a_0(x)y = 0, \tag{6.21}$$

where all common factors of $x$ have been removed. (We use the equation in this form since it is easier to multiply a power series by a polynomial than to divide by a polynomial.)

**Step 1.**   Assume that $0 < x < h$ and assume a solution of the form

$$y = x^r \sum_{k=0}^{\infty} c_k x^k. \tag{6.35}$$

(If $r$ turns out to be a complex number, or if we allow $x$ to have negative values, we must write $x^r$ as $|x|^r$.)

**Step 2.**   Determine the value of $r$. The equation giving the values of $r$ is called the *indicial equation*. This equation is a quadratic equation and may be determined by substituting the series in (6.35) into (6.21) and equating to 0 the coefficient of the term containing the smallest power of $x$, or by a method we will develop shortly.

**Step 3.**   Determine the recurrence relation for one value of $r$. This recurrence relation is obtained by substituting the series of (6.35) into (6.21) and equating to 0 the coefficients of each power of $x$.

**Step 4.**   Use the recurrence relation to find as many terms in the series as desired.

**Step 5.**   If needed, repeat steps 3 and 4 for the other value of $r$.

In our first example in this section we determined $r$ from the indicial equation (6.30) which we obtained by substituting the trial series solution into the differential equation. Since the behavior of the solution is greatly influenced by the value of $r$, it would be nice to have a simple means to obtain the indicial equation. (Note the difference in the behavior of the solution near $x = 0$ between an $r$ of $\frac{1}{2}$ and an $r$ of $-\frac{1}{2}$.)

The method is based on the fact that $x = 0$ is a regular singular point of

$$a_2(x)y'' + a_1(x)y' + a_0(x)y = 0 \qquad (6.36)$$

if

$$x\frac{a_1(x)}{a_2(x)} \qquad \text{and} \qquad x^2\frac{a_0(x)}{a_2(x)}$$

are analytic at 0. Thus we multiply (6.36) by an appropriate power of $x$ so that it has the form

$$x^2 b_2(x)y'' + x b_1(x)y' + b_0(x)y = 0 \qquad (6.37)$$

where $b_2(x)$, $b_1(x)$, and $b_0(x)$ are analytic functions written as

$$\begin{aligned}
b_2(x) &= \alpha_0 + \alpha_1 x + \alpha_2 x^2 + \cdots, &&\text{with } \alpha_0 \neq 0, \\
b_1(x) &= \beta_0 + \beta_1 x + \beta_2 x^2 + \cdots, && \\
b_0(x) &= \gamma_0 + \gamma_1 x + \gamma_2 x^2 + \cdots.
\end{aligned} \qquad (6.38)$$

If we try a solution of the form $x^r \sum_{k=0}^{\infty} c_k x^k$ in (6.37), we obtain as the coefficient of the lowest power of $x$, namely $x^r$, the following expression as the indicial equation:

$$\alpha_0 r(r - 1) + \beta_0 r + \gamma_o = 0. \qquad (6.39)$$

We state these results as a theorem.

**Theorem 6.4.**    If $x = 0$ is a regular singular point of a second order differential equation which is written as

$$x^2 b_2(x)y'' + x b_1(x)y' + b_0(x)y = 0, \qquad (6.37)$$

where $b_2(0) = \alpha_0 \neq 0$, then the indicial equation is given by

$$\alpha_0 r(r - 1) + \beta_0 r + \gamma_0 = 0, \qquad (6.39)$$

where $\beta_0 = b_1(0)$ and $\gamma_0 = b_0(0)$.

To use this on our previous example, we note that (6.28) is already in the form required by (6.37) as

$$9x^2 y'' + 9x^2 y' + 2y = 0, \qquad (6.28)$$

with $\alpha_0 = 9$, $\beta_0 = b_1(0) = 0$, and $\gamma_0 = 2$. Thus the indicial equation is

$$9r(r - 1) + 2 = 0$$

and is identical with what we found before with (6.30).

An alternative way to determine the indicial equation is to observe that when the expression

$$x^r \sum_{k=0}^{\infty} c_k x^k = c_0 x^r + c_1 x^{r+1} + \cdots$$

is substituted into the differential equation (6.36), the term containing $x$ to the smallest power will involve operations only on $c_0 x^r$. (This is because for a regular singular point $a_0(x)/a_2(x)$ has at most a factor of $1/x^2$ and $a_1(x)/a_2(x)$ has at most a factor of $1/x$.) Thus we may state the following theorem.

**Theorem 6.5.**   If $x = 0$ is a regular singular point of a second order differential equation $Ly = 0$ where

$$L = a_2(x)D^2 + a_1(x)D + a_0(x),$$

then the indicial equation may be obtained by applying $L$ to $x^r$ and in the resulting expression equating to zero the coefficient of the lowest power of $x$.

**EXAMPLE 6.10**   To find the indicial equation for

$$Ly = (2xD^2 + 6D - 9x)y = 0, \tag{6.40}$$

evaluate $Lx^r$ as

$$2xr(r - 1)x^{r-2} + 6rx^{r-1} - 9x^{r+1} = 2r(r - 1)x^{r-1} + 6rx^{r-1} - 9x^{r+1}.$$

This gives the indicial equation as

$$2r(r - 1) + 6r = 2r^2 + 4r = 0. \tag{6.41}$$

To use the other method, observe that we must multiply (6.40) by $x$ to obtain the form required by (6.37) as

$$2x^2y'' + 6xy' - 9x^2y = 0.$$

Here $\alpha_0 = b_2(0) = 2$, $\beta_0 = b_1(0) = 6$, and $\gamma_0 = b_0(0) = 0$, and the indicial equation is

$$\alpha_0 r(r - 1) + \beta_0 r + \gamma_0 = 2(r)(r - 1) + 6r = 0.$$

To conclude this section we use the form $x^r \sum_{k=0}^{\infty} c_k x^k$ to find the general solution of

$$4xy'' + 2y' - 7y = 0 \tag{6.42}$$

using the method of Frobenius. If we calculate

$$Lx^r = 4xr(r - 1)x^{r-2} + 2rx^{r-1} - 7x^r$$
$$= 4r(r - 1)x^{r-1} + 2rx^{r-1} - 7x^r,$$

we obtain the indicial equation

$$4r(r - 1) + 2r = 4r^2 - 2r = 2r(2r - 1) = 0,$$

with roots $r = 0, \frac{1}{2}$. To find the recurrence relation we calculate derivatives of our solution

$$y = \sum_{k=0}^{\infty} c_k x^{k+r}$$

as

$$y' = \sum_{k=0}^{\infty} (k+r)c_k x^{k+r-1}, \qquad y'' = \sum_{k=0}^{\infty} (k+r)(k+r-1)c_k x^{k+r-2}.$$

Substituting these series into the differential equation gives

$$\sum_{k=0}^{\infty} 4(k+r)(k+r-1)c_k x^{k+r-1} + \sum_{k=0}^{\infty} 2(k+r)c_k x^{k+r-1} - \sum_{k=0}^{\infty} 7c_k x^{k+r} = 0.$$

$$(6.43)$$

If we lower the index of the last series by 1 and increase the lower limit of summation by 1, we obtain

$$\sum_{k=0}^{\infty} 7c_k x^{k+r} = \sum_{k=1}^{\infty} 7c_{k-1} x^{k+r-1}.$$

Substituting this new form of the series into (6.43) and combining all three series under a single expression gives

$$(4(r)(r-1) + 2r)c_0 x^{r-1} + \sum_{k=1}^{\infty} \{[4(k+r)(k+r-1)$$
$$+ 2(k+r)]c_k - 7c_{k-1}\}x^{k+r-1} = 0.$$

Notice that the term on the left equals zero if $r$ is chosen to satisfy the indicial equation, while we obtain the recurrence relation by requiring that the coefficient of $x^{k+r-1}$ be zero. This gives

$$[4(k+r)(k+r-1) + 2(k+r)]c_k = 7c_{k-1}, \qquad k = 1, 2, 3, \ldots . \qquad (6.44)$$

**1. $r = \frac{1}{2}$.**    If we use the largest root of the indicial equation, the recurrence relation becomes

$$c_k = \frac{7}{(2k+1)(2k)}c_{k-1}, \qquad k = 1, 2, 3, \ldots .$$

The values of the first few coefficients are

$$c_1 = \frac{7}{3 \cdot 2 \cdot 1}c_0,$$

$$c_2 = \frac{7}{5 \cdot 4}c_1 = \frac{7^2}{5 \cdot 4 \cdot 3 \cdot 2 \cdot 1}c_0,$$

$$c_3 = \frac{7}{7 \cdot 6}c_2 = \frac{7^3}{7 \cdot 6 \cdot 5 \cdot 4 \cdot 3 \cdot 2 \cdot 1}c_0,$$

and from this pattern we may write

$$c_n = \frac{7^n}{(2n+1)!}c_0, \qquad n = 1, 2, 3, \ldots .$$

Thus

$$c_0 x^{1/2}\left[1 + \sum_{n=1}^{\infty} \frac{7^n x^n}{(2n+1)!}\right] = c_0 x^{1/2} \sum_{n=0}^{\infty} \frac{7^n x^n}{(2n+1)!} \qquad (6.45)$$

is a solution to (6.42).

**2. $r = 0$.** To obtain a second solution, we use the other root of the indicial equation ($r = 0$) in (6.44) to obtain the recurrence relation

$$c_k = \frac{7}{2k(2k-1)} c_{k-1}, \qquad k = 1, 2, 3, \ldots .$$

The values of the first few coefficients are

$$c_1 = \frac{7}{2 \cdot 1} c_0,$$

$$c_2 = \frac{7}{4 \cdot 3} c_1 = \frac{7^2}{4 \cdot 3 \cdot 2 \cdot 1} c_0,$$

$$c_3 = \frac{7}{6 \cdot 5} c_2 = \frac{7^3}{6 \cdot 5 \cdot 4 \cdot 3 \cdot 2 \cdot 1} c_0$$

and from this pattern we have the $n$th term given by

$$c_n = \frac{7^n}{(2n)!} c_0, \qquad n = 1, 2, 3, \ldots \qquad (6.46)$$

Thus if $c_0^*$ is used as the arbitrary constant in (6.46) in place of $c_0$, the general solution to (6.42) is

$$y = c_0 x^{1/2} \sum_{n=0}^{\infty} \frac{7^n x^n}{(2n+1)!} + c_0^*\left[1 + \sum_{n=1}^{\infty} \frac{7^n x^n}{(2n)!}\right].$$

## EXERCISES

Use the method of Frobenius to find general solutions to the following differential equations. For series for which the formula for the $n$th term is not apparent, find terms in the series expansion for the solution up to $x^4$.

**1.** $4xy'' + 2(1 + x)y' + y = 0$.
**2.** $xy'' - (3 + x)y' + 2y = 0$.
**3.** $3xy'' + y' - y = 0$.
**4.** $xy'' + (x^3 - 1)y' + x^2 y = 0$.
**5.** $4xy'' + 2y' + y = 0$.
**6.** $4x^2 y'' - 4xy' + (3 - 4x^2)y = 0$.
**7.** $2x^2 y'' + (2x^2 + x)y' + 2xy = 0$.
**8.** $x^2 y'' + x^3 y' - 2y = 0$.
**9.** $2x^2 y'' - xy' + (x - 5)y = 0$.
**10.** $4x^2 y'' + (4x - 2x^2)y' - (25 + 3x)y = 0$.

Find the general solution of the following differential equations using the method of Frobenius with expansions about the given point.

**11.** $2(x - 1)y'' - y' + e^x y = 0$, $x_0 = 1$.

**12.** $2(x - 1)^2 y'' + (5x - 5)y' + xy = 0$, $x_0 = 1$.

**13.** $2(x + 1)^2 y'' - 3(x^2 + 3x + 2)y' + (3x + 5)y = 0$, $x_0 = -1$.

**14.** $x(x + 1)^2 y'' - (x^2 + 3x + 2)y' + 9y = 0$, $x_0 = -1$.

**15.** $(x - 1)^2 y'' - (x - 1)(x^2 - 2x)y' + (2x - x^2)y = 0$, $x_0 = 1$.

**16. (a)** Show that $x = 0$ is an irregular singular point of

$$x^2 y'' + (4x - 1)y' + 2y = 0.$$

**(b)** Find coefficients of $y = \sum_{k=0}^{\infty} c_k x^k$ by substituting this series into the differential equation.

**(c)** Show that the series from part (b) converges only for $x = 0$.

Since the indicial equation is a real quadratic equation in $r$, if we have one complex root, $r_1$, the other root will be its complex conjugate $\bar{r}_1$. If the coefficients of the derivatives in the differential equation are all real and we define

$$|x|^r = |x|^{\alpha + i\beta} = |x|^\alpha e^{i\beta \ln |x|}$$

$$= |x|^\alpha [\cos(\beta \ln |x|) + i \sin(\beta \ln |x|)],$$

two linearly independent solutions of the differential equation may be obtained from the real and imaginary parts of the solution corresponding to one complex root. (The coefficients in the Frobenius series may also be complex numbers.) Find the general solution of the following differential equations where the roots of the indicial equation will be complex numbers.

**17.** $x^2 y'' + xy' + (1 + x)y = 0$.

**18.** $x^2(1 + x)y'' + xy' + (1 - x)(x^2 + 1)y = 0$.

### 6.4.2 The Method of Frobenius, Part 2

In the examples and exercises of Section 6.4.1, using the two roots of the indicial equation gave series solutions which were linearly independent. However, in the case where the indicial equation has a double root, it is obvious that the method of Frobenius yields only one solution. Such is the case of the next example.

**EXAMPLE 6.11**    Find the general solution of

$$Ly = x^2 y'' + (3x + x^2)y' + (1 + x^2)y = 0 \qquad (6.47)$$

by the method of Frobenius using a series about the regular singular point $x = 0$. Calculating $Lx^r$ gives

$$Lx^r = r(r - 1)x^r + 3rx^r + rx^{r+1} + x^r + x^{r+2},$$

so the indicial equation is

$$r(r - 1) + 3r + 1 = r^2 + 2r + 1 = 0.$$

Thus $r = -1$ is a double root of the indicial equation. To find the recurrence relation for this value of $r$, substitute

$$y = \sum_{k=0}^{\infty} c_k x^{k-1}$$

into (6.47) to obtain

$$\sum_{k=0}^{\infty} (k-1)(k-2)c_k x^{k-1} + \sum_{k=0}^{\infty} 3(k-1)c_k x^{k-1} + \sum_{k=0}^{\infty} (k-1)c_k x^k$$

$$+ \sum_{k=0}^{\infty} c_k x^{k-1} + \sum_{k=0}^{\infty} c_k x^{k+1} = 0.$$

Next we change the index of summation in the third and fifth series and combine the other three series to obtain

$$\sum_{k=0}^{\infty} [k^2 - 3k + 2 + 3k - 3 + 1]c_k x^{k-1} + \sum_{k=1}^{\infty} (k-2)c_{k-1} x^{k-1}$$

$$+ \sum_{k=2}^{\infty} c_{k-2} x^{k-1} = 0.$$

Writing the terms for $k = 0$ and $k = 1$ from the first two series separately and combining the terms remaining in these three series gives

$$c_1 - c_0 + \sum_{k=2}^{\infty} [k^2 c_k + (k-2)c_{k-1} + c_{k-2}]x^{k-1} = 0. \tag{6.48}$$

Thus $c_0$ is arbitrary, $c_1 = c_0$, and

$$c_k = \frac{-(k-2)c_{k-1} - c_{k-2}}{k^2}, \qquad k = 2, 3, 4, \ldots .$$

Calculating the first few terms gives

$$c_2 = \frac{-c_0}{4},$$

$$c_3 = \frac{-c_2 - c_1}{9} = \frac{-3c_0/4}{9} = \frac{-c_0}{12},$$

$$c_4 = \frac{-2c_3 - c_2}{16} = \frac{\frac{1}{6} + \frac{1}{4}}{16} c_0 = \frac{5}{(12)(16)} c_0,$$

and the solution is

$$x^{-1} \sum_{k=0}^{\infty} c_k x^k = c_0 x^{-1} \left( 1 + x - \frac{x^2}{4} - \frac{x^3}{12} - \frac{5x^4}{192} + \cdots \right). \tag{6.49}$$

Recall that in Chapter 2 we used the reduction-of-order technique to find a second solution of a second order differential equation when we knew one solution. The result was [see (2.109) on page 51] that if $y_1(x)$ satisfied

$$a_2(x)y'' + a_1(x)y' + a_0(x)y = 0,$$

so did

$$y_2(x) = y_1(x) \int \frac{\exp[-\int(a_1(x)/a_2(x))\, dx]}{y_1^2(x)}\, dx. \tag{6.50}$$

If we use (6.50) to find our second solution of (6.47), we notice that

$$\int \frac{a_1(x)}{a_2(x)}\, dx = \int \frac{3x + x^2}{x^2}\, dx = \int \left(\frac{3}{x} + 1\right) dx = 3 \ln x + x,$$

so

$$\exp\left[-\int \frac{a_1(x)}{a_2(x)}\, dx\right] = e^{-x} x^{-3}.$$

In order to compute the first few terms in the remaining integrand in (6.50), we use the Taylor series for $e^{-x}$, write the right-hand side of (6.49) as $c_0 y_1(x)$, square the resulting $y_1(x)$, and use long division. This yields

$$\frac{\exp[-\int a_1(x)/a_2(x)\, dx]}{y_1^2(x)} = \frac{x^{-3}(1 - x + x^2/2 - x^3/6 + \cdots)}{x^{-2}(1 + 2x + x^2/2 - 2x^3/3 + \cdots)}$$

$$= x^{-1}(1 - 3x + 6x^2 - 10x^3 + \cdots). \tag{6.51}$$

If we use the expression (6.51) in (6.50) and integrate, we obtain

$$y_2(x) = y_1(x)\left(\ln|x| - 3x + 3x^2 - \frac{10x^3}{3} + \cdots\right)$$

$$= y_1(x) \ln|x| + x^{-1}\left(-3x + \frac{5x^3}{12} + \cdots\right). \tag{6.52}$$

Thus the general solution of 6.47 is

$$y = c_0 y_1(x) + c_1 y_2(x),$$

where $c_0 y_1(x)$ is given by (6.49), $y_2(x)$ by (6.52), and $c_0$ and $c_1$ are arbitrary constants.

Although we could find our second solution this way for double roots of the indicial equation, it is usually easier to use the following theorem. This theorem will not be proved. We use Example 6.11 to motivate the conclusion of the theorem regarding repeated roots.

**Theorem 6.6.**    Given the second order differential equation

$$a_2(x) \frac{d^2 y}{dx^2} + a_1(x) \frac{dy}{dx} + a_0(x)y = 0 \tag{6.53}$$

with $x = 0$ as a regular singular point. Let $h$ denote the smallest radius of convergence of

$$\frac{xa_1(x)}{a_2(x)} \quad \text{and} \quad \frac{x^2 a_0(x)}{a_2(x)}$$

and $r_1$ and $r_2$ be solutions of the indicial equation with $\text{Re}(r_1) \geq \text{Re}(r_2)$. Then (6.53) has two linearly independent solutions, $y_1(x)$ and $y_2(x)$, whose series converge for at least $0 < |x| < h$. For all cases,

$$y_1(x) = |x|^{r_1}\left(1 + \sum_{k=1}^{\infty} c_k x^k\right), \tag{6.54}$$

while the form for $y_2(x)$ depends on the relation between the roots of the indicial equation as follows.

(a) If $r_1 - r_2$ is not equal to an integer, then

$$y_2(x) = |x|^{r_2}\left(1 + \sum_{k=1}^{\infty} c_k^* x^k\right). \tag{6.55}$$

(b) If $r_1 = r_2$, then

$$y_2(x) = y_1(x) \ln|x| + |x|^{r_1} \sum_{k=1}^{\infty} c_k^* x^k. \tag{6.56}$$

(c) If $r_1 - r_2$ equals a positive integer, then

$$y_2(x) = Cy_1(x) \ln|x| + |x|^{r_2}\left(c_0^* + \sum_{k=1}^{\infty} c_k^* x^k\right), \tag{6.57}$$

where, depending on the differential equation, the constant $C$ may or may not equal zero.

In Example 6.11 we found the second solution for an indicial equation with a double root by reduction of order. Let us now work a problem using the assumed form of $y_2(x)$ from Theorem 6.6.

**EXAMPLE 6.12**   Find the general solution of

$$Ly = xy'' + y' + 4xy = 0. \tag{6.58}$$

The indicial equation is obtained by considering

$$Lx^r = r(r-1)x^{r-1} + rx^{r-1} + 4x^{r+1}$$

as

$$r(r-1) + r = r^2 = 0.$$

Since $r = 0$, we assume a solution of the form

$$y = \sum_{k=0}^{\infty} c_k x^k,$$

calculate $y'$ and $y''$, and substitute them into (6.58). This results in

$$\sum_{k=0}^{\infty} k(k-1)c_k x^{k-1} + \sum_{k=0}^{\infty} kc_k x^{k-1} + \sum_{k=0}^{\infty} 4c_k x^{k+1} = 0. \tag{6.59}$$

If we lower the index in the last series by 2 and raise the lower limit of summation by 2, we obtain

$$\sum_{k=0}^{\infty} 4c_k x^{k+1} = \sum_{k=2}^{\infty} 4c_{k-2} x^{k-1}.$$

If we now use this expression in (6.59) and write terms for $k = 0$ and $k = 1$ separately, we obtain

$$c_1 + \sum_{k=2}^{\infty} \{[k(k-1) + k]c_k + 4c_{k-2}\}x^{k-1} = 0.$$

Thus

$$c_1 = 0$$

and

$$c_k = \frac{-4}{k^2}c_{k-2}, \qquad k = 2, 3, 4, \ldots . \tag{6.60}$$

From this recurrence relation we see that all the coefficients with odd subscripts equal zero and the even ones are given by

$$c_2 = \frac{-4}{2^2}c_0,$$

$$c_4 = \frac{-4}{4^2}c_2 = \frac{(-4)^2}{4^2 2^2}c_0,$$

$$c_6 = \frac{-4}{6^2}c_4 = \frac{(-4)^3}{6^2 4^2 2^2}c_0,$$

$$\vdots$$

$$c_{2n} = \frac{(-4)^n}{(2n)^2(2n-2)^2 \cdots (6^2)(4^2)(2^2)}c_0$$

$$= \frac{(-1)^n 4^n}{(2 \cdot n)^2(2(n-1))^2 \cdots (2 \cdot 3)^2(2 \cdot 2)^2(2 \cdot 1)^2}c_0$$

$$= \frac{(-1)^n 4^n}{(2^2)^n(n!)^2}c_0 = \frac{(-1)^n}{(n!)^2}c_0, \qquad n = 1, 2, 3, \ldots .$$

If we let $c_0 = 1$, we obtain our first solution as

$$y_1(x) = 1 + \sum_{n=1}^{\infty} \frac{(-1)^n}{(n!)^2}x^{2n} = \sum_{n=0}^{\infty} \frac{(-1)^n}{(n!)^2}x^{2n}. \tag{6.61}$$

To find a second solution, we use the form of $y_2(x)$ from (6.56), assume $x > 0$, and differentiate to obtain (note that we use $c_k$ rather than $c_k^*$)

$$y_2(x) = y_1(x) \ln x + \sum_{k=1}^{\infty} c_k x^k,$$

$$y_2'(x) = y_1'(x) \ln x + \frac{y_1(x)}{x} + \sum_{k=1}^{\infty} k c_k x^{k-1},$$

$$y_2''(x) = y_1'' \ln x + \frac{2y_1'(x)}{x} - \frac{y_1(x)}{x^2} + \sum_{k=1}^{\infty} k(k-1) c_k x^{k-2}.$$

Substituting these values into the original differential equation (6.58) gives

$$(xy_1'' + y_1' + 4xy_1) \ln x + 2y_1'(x) - \frac{y_1(x)}{x} + \frac{y_1(x)}{x}$$

$$+ \sum_{k=1}^{\infty} k(k-1) c_k x^{k-1} + \sum_{k=1}^{\infty} k c_k x^{k-1} + \sum_{k=1}^{\infty} 4 c_k x^{k+1} = 0. \qquad (6.62)$$

Notice that since $y_1$ satisfies (6.58), the coefficient of $\ln x$ equals zero. Changing the index of the last series in (6.62) and combining terms gives

$$2y_1'(x) + c_1 + 4c_2 x + \sum_{k=3}^{\infty} (k^2 c_k + 4c_{k-2}) x^{k-1} = 0. \qquad (6.63)$$

If we substitute the value of $2y_1'(x)$ from (6.61) as

$$\sum_{k=1}^{\infty} \frac{(-1)^k (4k)}{(k!)^2} x^{2k-1}$$

into (6.63) and equate coefficients, we obtain

$$c_1 = 0,$$

$$4c_2 = 4,$$

$$k^2 c_k + 4c_{k-2} = 0, \qquad k = 3, 5, 7, \ldots,$$

$$(2k)^2 c_{2k} + 4c_{2k-2} = \frac{-(-1)^k (4k)}{(k!)^2}, \qquad k = 2, 3, 4, \ldots.$$

Thus $c_k = 0$, $k = 1, 3, 5, \ldots$, and

$$c_2 = 1,$$

$$c_4 = \frac{-4c_2 - 8/(2!)^2}{16} = \frac{-3}{8},$$

$$c_6 = \frac{-4c_4 + 12/(3!)^2}{36} = \frac{11}{216},$$

$$c_8 = \frac{-4c_6 - 16/(4!)^2}{64} = \frac{-25}{6912},$$

$$\vdots$$

Thus $y_2(x)$ has the form

$$y_2(x) = y_1(x) \ln x + x^2 - \frac{3x^4}{8} + \frac{11x^6}{216} - \frac{25x^8}{6912} + \cdots \qquad (6.64)$$

and the general solution of (6.58) is

$$y = C_1 y_1(x) + C_2 y_2(x),$$

with $y_1(x)$ given by (6.61) and $y_2(x)$ by (6.64). Since $xa_1(x)/a_2(x) = 1$ and $x^2 a_0(x)/a_2(x) = -4x^2$ are both polynomials (and Taylor series about the origin), they have an infinite radius of convergence ($h = \infty$) and the series expressions in $y_1(x)$ and $y_2(x)$ converge for all values of $x$.

When the two roots of the indicial equation differ by an integer, we are guaranteed a solution if we use the larger root. However, sometimes we may obtain two linearly independent solutions by using the smaller root. Such is the case in the next example, where the constant $C$ in the expression for $y_2(x)$ in (6.57) [case (c)] is zero.

**EXAMPLE 6.13**    Use the method of Frobenius to find the general solution of

$$(x - 1)^2 y'' + (3x^2 - 4x + 1)y' - 2y = 0 \qquad (6.65)$$

by expanding about $x = 1$.

Since we assume a solution of the form

$$y = |x - 1|^r \sum_{k=0}^{\infty} c_k (x - 1)^k, \qquad (6.66)$$

we first rewrite the coefficient of $y'$ in terms of powers of $x - 1$. Thus

$$\begin{aligned}
3x^2 - 4x + 1 &= 3x^2 - 3x - x + 1 \\
&= 3x(x - 1) - (x - 1) \\
&= 3(x - 1 + 1)(x - 1) - (x - 1) \\
&= 3(x - 1)^2 + 2(x - 1)
\end{aligned}$$

and (6.65) may be written as

$$Ly = (x - 1)^2 y'' + [3(x - 1)^2 + 2(x - 1)]y' - 2y = 0. \qquad (6.67)$$

To find the indicial equation, we compute

$$L(x - 1)^r = r(r - 1)(x - 1)^r + 3r(x - 1)^{r+1} + 2r(x - 1)^r - 2(x - 1)^r.$$

Thus the indicial equation is

$$r(r - 1) + 2r - 2 = r^2 + r - 2 = (r + 2)(r - 1) = 0.$$

Since the roots, $-2$ and $1$, differ by a positive integer, we use the smaller root and try to obtain two linearly independent solutions with this root. Since

$$(x - 1)p(x) = 3(x - 1) + 2 \qquad \text{and} \qquad (x - 1)^2 q(x) = -2,$$

we are assured that the series appearing in the solution converge for all $x \neq 1$. Assume that $r = -2$, $x > 0$, and substitute (6.66) into (6.67) to obtain

$$\sum_{k=0}^{\infty} (k-2)(k-3)c_k(x-1)^{k-2} + \sum_{k=0}^{\infty} 3(k-2)c_k(x-1)^{k-1}$$

$$+ \sum_{k=0}^{\infty} 2(k-2)c_k(x-1)^{k-2} - \sum_{k=0}^{\infty} 2c_k(x-1)^{k-2} = 0. \qquad (6.68)$$

Changing the index of summation in the second series gives

$$\sum_{k=0}^{\infty} 3(k-2)c_k(x-1)^{k-1} = \sum_{k=1}^{\infty} 3(k-3)c_{k-1}(x-1)^{k-2}$$

and (6.68) may be expressed as

$$(6c_0 - 4c_0 - 2c_0)(x-1)^{-2} + \sum_{k=1}^{\infty} \{[(k-2)(k-3) + 2(k-2) - 2]c_k$$

$$+ 3(k-3)c_{k-1}\}(x-1)^{k-2} = 0.$$

Thus the recurrence relation is

$$(k-3)c_k = -\frac{3(k-3)}{k}c_{k-1}, \qquad k = 1, 2, 3, \ldots.$$

[Note that division by $(k-3)$ is not allowed in general since 3 is one of the possible values of $k$.] Calculating the first few coefficients gives

$$c_1 = -3c_0,$$

$$c_2 = -\frac{3}{2}c_1 = \frac{(-3)^2}{2}c_0,$$

$$(0)c_3 = \frac{-3(0)}{3}c_2,$$

$$c_4 = \frac{-3}{4}c_3,$$

$$c_5 = \frac{-3}{5}c_4 = \frac{(-3)^2}{5\cdot4}c_3,$$

$$c_6 = \frac{-3}{6}c_5 = \frac{(-3)^3}{6\cdot5\cdot4}c_3 = \frac{(-3)^3 3!}{6!}c_3,$$

$$\vdots$$

$$c_n = \frac{-3}{n}c_{n-1} = \frac{(-3)^{n-3}3!}{n!}c_3, \qquad n = 4, 5, \ldots.$$

Note that the third equation in this list is satisfied for *all* values of $c_3$. Therefore, $c_3$ is arbitrary and our general solution to (6.65) is

$$y = c_0(x - 1)^{-2}\left[1 - 3(x - 1) + \frac{9(x - 1)^2}{2}\right]$$

$$+ c_3(x - 1)^{-2}\sum_{n=3}^{\infty}\frac{(-3)^{n-3}3!}{n!}(x - 1)^n$$

$$= c_0(x - 1)^{-2}\left[1 - 3(x - 1) + \frac{9(x - 1)^2}{2}\right]$$

$$+ c_3(x - 1)\sum_{k=0}^{\infty}\frac{(-3)^k 3!}{(k + 3)!}(x - 1)^k, \tag{6.69}$$

where we made an index change by replacing $n - 3$ by $k$ in the infinite series. We now see that (6.69) is of the form given in (6.57) with $C = 0$.

We conclude this section with an example where the roots of the indicial equation differ by a positive integer but the $C$ in (6.57) is not zero.

**EXAMPLE 6.14**    Use the method of Frobenius to find the general solution of

$$Ly = xy'' + (x - 1)y' - 2y = 0 \tag{6.70}$$

by expanding about the origin.

To determine the indicial equation, compute $Lx^r$ as

$$Lx^r = r(r - 1)x^{r-1} + rx^r - rx^{r-1} - 2x^r.$$

Thus the indicial equation is

$$r(r - 1) - r = r(r - 2) = 0,$$

with roots 0 and 2.

If we try the smaller root, 0, and hope to obtain two linearly independent solutions of (6.70), we get only the trivial solution (see exercise 13). Thus we substitute

$$y = \sum_{k=0}^{\infty} c_k x^{k+2}$$

into (6.70) and obtain

$$\sum_{k=0}^{\infty}(k + 2)(k + 1)c_k x^{k+1} + \sum_{k=0}^{\infty}(k + 2)c_k x^{k+2}$$

$$- \sum_{k=0}^{\infty}(k + 2)c_k x^{k+1} - \sum_{k=0}^{\infty}2c_k x^{k+2} = 0$$

or

$$\sum_{k=0}^{\infty}(k + 2)(k)c_k x^{k+1} + \sum_{k=0}^{\infty}kc_k x^{k+2} = 0.$$

If we raise the index of summation in the first series, we have

$$\sum_{k=0}^{\infty} (k+3)(k+1)c_{k+1}x^{k+2} + \sum_{k=0}^{\infty} kc_k x^{k+2} = 0$$

or

$$\sum_{k=0}^{\infty} [(k+3)(k+1)c_{k+1} + kc_k]x^{k+2} = 0.$$

This gives the recurrence relation as

$$(k+3)(k+1)c_{k+1} = -kc_k, \qquad k = 0, 1, 2, \ldots.$$

Writing this out for the first few values of $k$ gives

$$3c_1 = 0,$$
$$8c_2 = -c_1,$$
$$15c_3 = -2c_2,$$
$$c_{k+1} = \frac{-k}{(k+3)(k+1)}c_k, \qquad k = 3, 4, 5, \ldots.$$

Thus $c_0$ is arbitrary and all the other coefficients are zero. Choosing $c_0 = 1$ gives our first solution as

$$y_1(x) = x^2. \tag{6.71}$$

To find our second solution, use (6.57) as

$$y_2(x) = Cx^2 \ln x + |x|^0 \left( c_0 + \sum_{k=1}^{\infty} c_k x^k \right) \tag{6.72}$$

and differentiate to obtain

$$y_2'(x) = C(x + 2x \ln x) + \sum_{k=1}^{\infty} kc_k x^{k-1}$$

and

$$y_2''(x) = C(3 + 2 \ln x) + \sum_{k=2}^{\infty} k(k-1)c_k x^{k-2}.$$

Substituting these expressions in (6.70) yields

$$C(2x + x^2) + \sum_{k=1}^{\infty} k(k-1)c_k x^{k-1} + \sum_{k=1}^{\infty} kc_k x^k$$
$$-\sum_{k=1}^{\infty} kc_k x^{k-1} - 2c_0 - \sum_{k=1}^{\infty} 2c_k x^k = 0$$

or

$$-2c_0 + 2Cx + Cx^2 + \sum_{k=1}^{\infty} k(k-2)c_k x^{k-1} + \sum_{k=1}^{\infty} (k-2)c_k x^k = 0.$$

Raising the index of summation on the first series and combining the series gives

$$-2c_0 + 2Cx + Cx^2 - c_1 + \sum_{k=1}^{\infty} [(k + 1)(k - 1)c_{k+1} + (k - 2)c_k]x^k = 0.$$

Equating to zero the coefficients of like powers of $x$ gives

$$-2c_0 - c_1 = 0,$$
$$2C - c_1 = 0,$$
$$C + 3c_3 + 0c_2 = 0,$$
$$c_{k+1} = \frac{-(k - 2)}{(k + 1)(k - 1)} c_k, \qquad k = 3, 4, 5, \dots .$$

If we chose $c_0 = 1$, then

$$c_1 = -2, \qquad C = -1, \qquad c_3 = \tfrac{1}{3}, \qquad c_2 \text{ is arbitrary},$$

$$c_4 = \frac{-1}{(4)(2)} c_3 = \frac{-1}{(4)(3)(2)},$$

$$c_5 = \frac{-2}{(5)(3)} c_4 = \frac{(-2)(-1)}{(5)(4)(3)(2)(3)},$$

$$c_6 = \frac{-3}{(6)(4)} c_5 = \frac{(-3)(-2)(-1)}{(6)(5)(4)(3)(2)(4)(3)},$$

$$\vdots$$

$$c_n = \frac{(-1)^{n-3}(n - 3)! \,(2)}{n! \,(n - 2)!} = \frac{(-1)^{n-3}(2)}{n! \,(n - 2)}, \qquad n = 4, 5, 6, \dots ,$$

and $y_2(x)$ may be written as

$$y_2(x) = -x^2 \ln x + 1 - 2x + c_2 x^2 + \frac{x^3}{3} + 2 \sum_{n=4}^{\infty} \frac{(-1)^{n-3} x^n}{n! \,(n - 2)}. \qquad (6.73)$$

Notice that the coefficient of $x^2$ in $y_2(x)$, namely $c_2$, was arbitrary. Since $y_1(x) = x^2$, this $c_2$ could be combined with the arbitrary constant multiplying $y_1(x)$, or equivalently set $c_2 = 0$. Thus the general solution to (6.70) is

$$y = c_0 x^2 + c_1 y_2(x),$$

where $y_2(x)$ is given by (6.73). Since

$$xp(x) = x - 1 \qquad \text{and} \qquad x^2 q(x) = -2x$$

are both polynomials, the infinite series in (6.73) converges for all values of $x$. You could also use the ratio test to discover this.

## EXERCISES

Find the general solution to the following differential equations for $x > 0$. For series for which the formula for the $n$th term is not apparent, find terms in the series up to $x^4$. Also find $h$ such that the series converge for $0 < x < h$.

1. $xy'' + y' - xy = 0$.
2. $xy'' + y' - 4xy = 0$.
3. $(x^2 - x)y'' + 3y' - 2y = 0$.
4. $x^2y'' + 2xy' + xy = 0$.
5. $(x^2 + x^4)y'' + (9x + 5x^3)y' + (16 + 4x^2)y = 0$.
6. $xy'' + (x - 1)y' - y = 0$.
7. $x^2y'' + (x^2 - x)y' - (x - 1)y = 0$.
8. $x^2y'' + (x^2 - x)y' + y = 0$.
9. $xy'' + xy' + y = 0$.
10. $(x - x^2)y'' - 3xy' - y = 0$.
11. Show that the answer to exercise 10 may also be written as

$$y = c_0 x(1 - x)^{-2} + c_1[x(1 - x)^{-2}\ln x + (1 - x)^{-1}].$$

12. Show that one solution of exercise 8 is $y_1(x) = xe^{-x}$ and use a formula for the second solution from the reduction-of-order technique to find this second solution.
13. Show that using $r = 0$ in a Frobenius series solution of (6.70) gives the trivial solution.

## 6.5 THE LEGENDRE AND BESSEL EQUATIONS

### 6.5.1 The Legendre Equation

One differential equation which occurs in various applications, such as boundary value problems associated with a sphere, is the Legendre equation

$$(1 - x^2)y'' - 2xy' + \nu(\nu + 1)y = 0. \tag{6.74}$$

(See Example 11.13). The constant in (6.74) is written as $\nu(\nu + 1)$ to simplify some of our calculations. Notice that (6.74) has singular points at 1 and $-1$, so if we find a series solution about the ordinary point $x = 0$, it will converge for $|x| < 1$.

Assuming a series of the form

$$y = \sum_{k=0}^{\infty} c_k x^k \tag{6.75}$$

and substituting it into (6.74) gives

$$\sum_{k=2}^{\infty} k(k - 1)c_k x^{k-2} - \sum_{k=0}^{\infty} k(k - 1)c_k x^k - \sum_{k=0}^{\infty} 2kc_k x^k + \sum_{k=0}^{\infty} \nu(\nu + 1)c_k x^k = 0.$$

If we raise the index of summation in the first series by 2, we may combine all four series as

$$\sum_{k=0}^{\infty} [(k + 2)(k + 1)c_{k+2} - (k^2 - k + 2k - \nu(\nu + 1))c_k]x^k = 0.$$

This gives our recurrence relation as

$$c_{k+2} = \frac{k^2 - \nu^2 + k - \nu}{(k + 2)(k + 1)} c_k = \frac{-(\nu - k)(k + \nu + 1)}{(k + 2)(k + 1)} c_k, \qquad k = 0, 1, 2, \ldots .$$

$$(6.76)$$

Calculating the first few coefficients gives

$$c_2 = \frac{-\nu(\nu + 1)}{2 \cdot 1} c_0,$$

$$c_4 = \frac{-(\nu - 2)(\nu + 3)}{4 \cdot 3} c_2 = \frac{(-1)^2 \nu(\nu - 2)(\nu + 1)(\nu + 3)}{4!} c_0,$$

$$c_6 = \frac{-(\nu - 4)(\nu + 5)}{6 \cdot 5} c_4 = \frac{(-1)^3 \nu(\nu - 2)(\nu - 4)(\nu + 1)(\nu + 3)(\nu + 5)}{6!} c_0,$$

and for general values of $m$,

$$c_{2m} = \frac{(-1)^m \nu(\nu - 2)(\nu - 4) \cdots (\nu - 2m + 2)(\nu + 1)(\nu + 3) \cdots (\nu + 2m - 1)}{(2m)!} c_0,$$

$$(6.77)$$

while

$$c_3 = \frac{-(\nu - 1)(\nu + 2)}{3 \cdot 2} c_1,$$

$$c_5 = \frac{-(\nu - 3)(\nu + 4)}{5 \cdot 4} c_3 = \frac{(-1)^2 (\nu - 1)(\nu - 3)(\nu + 2)(\nu + 4)}{5!} c_1,$$

$$c_7 = \frac{-(\nu - 5)(\nu + 6)}{7 \cdot 6} c_5 = \frac{(-1)^3 (\nu - 1)(\nu - 3)(\nu - 5)(\nu + 2)(\nu + 4)(\nu + 6)}{7!} c_1,$$

with the resulting pattern

$$c_{2m+1} = \frac{(-1)^m (\nu - 1)(\nu - 3) \cdots (\nu - 2m + 1)(\nu + 2)(\nu + 4) \cdots (\nu + 2m)}{(2m + 1)!} c_1.$$

$$(6.78)$$

Thus the general solution of (6.74) is

$$y = c_0 y_1(x) + c_1 y_2(x)$$

where

$$y_1(x) = 1 +$$

$$\sum_{m=1}^{\infty} \frac{(-1)^m \nu(\nu - 2)(\nu - 4) \cdots (\nu - 2m + 2)(\nu + 1)(\nu + 3) \cdots (\nu + 2m - 1)}{(2m)!} x^{2m}$$

$$(6.79)$$

$$y_2(x) = x +$$

$$\sum_{m=1}^{\infty} \frac{(-1)^m(\nu - 1)(\nu - 3) \cdots (\nu - 2m + 1)(\nu + 2)(\nu + 4) \cdots (\nu + 2m)}{(2m + 1)!} x^{2m + 1}.$$

$$(6.80)$$

Theorem 6.6 guarantees that both these series converge for $|x| < 1$, but if $\nu$ is not an integer, neither of them converges on the entire closed interval $-1 \leq x \leq 1$. Since the solution of (6.74) for use in boundary value problems must be bounded on this closed interval, we now turn our attention to the case where this will be true, namely $\nu$ equal to a positive integer.

If $\nu = n$, a positive integer, the recurrence relation (6.76) for $k = n$ gives

$$c_{n+2} = \frac{-(n - n)(n + n + 1)}{(n + 2)(n + 1)} c_n = 0.$$

Thus all the coefficients in our series which are greater than $n$ by a multiple of 2 will be zero, and one of the solutions to the differential equation will be a polynomial of degree $n$. If $\nu = n =$ an even integer, $c_0 y_1(x)$ is a polynomial and the infinite series for $c_1 y_2(x)$ in (6.80) with $c_1 = (-1)^{n/2} 2^n [(n/2)!]^2/n!$ is defined as $Q_n(x)$. If $\nu = n$, an odd integer, $c_1 y_2(x)$ is a polynomial and the infinite series for $c_0 y_1(x)$ in (6.77) with $c_0 = (-1)^{(n-1)/2} 2^{n-1} [(n/2 - \frac{1}{2})!]^2/n!$ is defined as $Q_n(x)$. $Q_n(x)$ is called the *Legendre function of the second kind* and converges for $-1 < x < 1$.

The Legendre polynomials are defined as the polynomial solutions of (6.74) with specific choices of the constants $c_0$ and $c_1$. When $n$ is an even integer, the choice for $c_0$ gives $c_0 y_1(x)$ the value 1 at $x = 1$ and the resulting polynomial,

$$P_n(x) = \frac{(-1)^{n/2} n!}{2^n [(n/2)!]^2}$$

$$\times \left[ 1 + \sum_{m=1}^{n/2} \frac{(-1)^m n(n - 2)(n - 4) \cdots (n - 2m + 2)(n + 1)(n + 3) \cdots (n + 2m - 1)}{(2m)!} x^{2m} \right],$$

$$(6.81)$$

is called the *Legendre polynomial of degree n*. For $n$ odd, $c_1$ is chosen so that $c_1 y_2(1) = 1$, and the Legendre polynomial is defined as

$$P_n(x) = \frac{(-1)^{(n-1)/2} n!}{2^{n-1} [(n/2 - 1/2)!]^2}$$

$$\times \left[ x + \sum_{m=1}^{M} \frac{(-1)^m (n - 1)(n - 3) \cdots (n - 2m + 1)(n + 2)(n + 4) \cdots (n + 2m)}{(2m + 1)!} x^{2m+1} \right],$$

$$(6.82)$$

where $M = (n - 1)/2$. Although these sums may look formidable, it is a fairly easy task to obtain the first few Legendre polynomials from (6.81) and (6.82) as

$$P_0(x) = 1, \qquad P_1(x) = x,$$

$$P_2(x) = \frac{(-1)2!}{2^2(1!)^2}\left[1 + \frac{(-1)(2)(3)}{2!}x^2\right] = \frac{3x^2 - 1}{2},$$

$$P_3(x) = \frac{(-1)3!}{2^2(1!)^2}\left[x + \frac{(-1)(2)(5)}{3!}x^3\right] = \frac{5x^3 - 3x}{2},$$

$$P_4(x) = \frac{4!}{2^4(2!)^2}\left[1 + \frac{(-1)(4)(5)}{2!}x^2 + \frac{(4)(2)(5)(7)}{4!}x^4\right]$$

$$= \frac{35x^4 - 30x^2 + 3}{8},$$

$$P_5(x) = \frac{5!}{2^4(2!)^2}\left[x + \frac{(-1)(4)(7)}{3!}x^3 + \frac{(4)(2)(7)(9)}{5!}x^5\right]$$

$$= \frac{63x^5 - 70x^3 + 15x}{8}.$$

The behavior of the first five Legendre polynomials is shown in Figure 6.2. Other expressions for the Legendre polynomials are developed in the exercises as

$$P_n(x) = \frac{1}{2^n}\sum_{k=0}^{N}\frac{(-1)^k(2n - 2k)!}{k!(n - 2k)!(n - k)!}x^{n-2k}, \qquad n = 0, 1, 2, \ldots, \qquad (6.83)$$

where $N = n/2$ for $n$ even and $N = (n - 1)/2$ for $n$ odd, and

$$P_n(x) = \frac{1}{2^n n!}\frac{d^n}{dx^n}(x^2 - 1)^n, \qquad n = 0, 1, 2, \ldots. \qquad (6.84)$$

(6.84) is called a *Rodrigues' formula* for Legendre polynomials and is very useful for obtaining various identities. If we denote $dP_n(x)/dx$ by $P_n'(x)$, to derive

$$P_{n+1}'(x) = (2n + 1)P_n(x) + P_{n-1}'(x), \qquad n = 1, 2, 3, \ldots, \qquad (6.85)$$

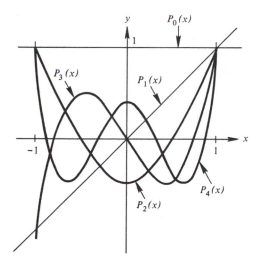

**Figure 6.2**

we find $P'_{n+1}(x)$ from (6.84) as

$$P'_{n+1}(x) = \frac{d}{dx}\left[\frac{1}{2^{n+1}(n+1)!}\frac{d^{n+1}}{dx^{n+1}}(x^2-1)^{n+1}\right]$$

$$= \frac{1}{2^{n+1}(n+1)!}\frac{d^{n+1}}{dx^{n+1}}[(n+1)(x^2-1)^n(2x)]$$

and calculate

$$P'_{n+1}(x) = \frac{1}{2^n n!}\frac{d^{n+1}}{dx^{n+1}}[x(x^2-1)^n]$$

$$= \frac{1}{2^n n!}\frac{d^n}{dx^n}[(x^2-1)^n + 2nx^2(x^2-1)^{n-1}]$$

$$= \frac{1}{2^n n!}\frac{d^n}{dx^n}[(x^2-1)^n + 2n(x^2-1+1)(x^2-1)^{n-1}]$$

$$= \frac{1}{2^n n!}\frac{d^n}{dx^n}[(x^2-1)^n(1+2n) + 2n(x^2-1)^{n-1}]$$

$$= (2n+1)P_n(x) + \frac{1}{2^{n-1}(n-1)!}\frac{d}{dx}\left[\frac{d^{n-1}}{dx^{n-1}}(x^2-1)^{n-1}\right]$$

$$= (2n+1)P_n(x) + P'_{n-1}(x).$$

Another identity uses the fact that

$$\frac{d^{n+1}}{dx^{n+1}}[xf(x)] = x\frac{d^{n+1}}{dx^{n+1}}f(x) + (n+1)\frac{d^n}{dx^n}f(x). \qquad (6.86)$$

In deriving (6.85), at step two we had

$$P'_{n+1}(x) = \frac{1}{2^n n!}\frac{d^{n+1}}{dx^{n+1}}[x(x^2-1)^n].$$

If we use (6.86) to expand the right-hand side, we obtain

$$P'_{n+1}(x) = \frac{1}{2^n n!}\left[x\frac{d^{n+1}}{dx^{n+1}}(x^2-1)^n + (n+1)\frac{d^n}{dx^n}(x^2-1)^n\right]$$

$$= x\frac{d}{dx}\frac{1}{2^n n!}\frac{d^n}{dx^n}(x^2-1)^n + \frac{n+1}{2^n n!}\frac{d^n}{dx^n}(x^2-1)^n$$

$$= xP'_n(x) + (n+1)P_n(x), \qquad n = 0, 1, 2, \ldots . \qquad (6.87)$$

One of the reasons the Legendre polynomials are so useful is that they form an orthogonal set on $(-1, 1)$, that is,

$$\int_{-1}^{1} P_n(x)P_m(x)dx = 0, \qquad n \neq m. \qquad (6.88)$$

To prove this orthogonality, we write down the differential equations satisfied by $P_n(x)$ and $P_m(x)$ as

$$\frac{d}{dx}\left[(1 - x^2)\frac{d}{dx}P_n(x)\right] + n(n + 1)P_n(x) = 0, \tag{6.89}$$

$$\frac{d}{dx}\left[(1 - x^2)\frac{d}{dx}P_m(x)\right] + m(m + 1)P_m(x) = 0, \tag{6.90}$$

where we combined the first and second derivatives. We now multiply (6.89) by $P_m(x)$ and (6.90) by $P_n(x)$, subtract the resulting equations, and integrate from $x = -1$ to $x = 1$. This gives

$$\int_{-1}^{1}\left(P_m(x)\frac{d}{dx}\left[(1 - x^2)\frac{dP_n(x)}{dx}\right] - P_n(x)\frac{d}{dx}\left[(1 - x^2)\frac{dP_m(x)}{dx}\right]\right) dx$$

$$+ [n(n + 1) - m(m + 1)]\int_{-1}^{1} P_n(x)P_m(x)\, dx = 0. \tag{6.91}$$

If we integrate the first integral in (6.91) by parts, we obtain

$$\int_{-1}^{1} P_m(x)\frac{d}{dx}\left[(1 - x^2)\frac{dP_n(x)}{dx}\right] dx$$

$$= P_m(x)(1 - x^2)\frac{dP_n(x)}{dx}\bigg|_{-1}^{1} - \int_{-1}^{1}\frac{dP_m(x)}{dx}(1 - x^2)\frac{dP_n(x)}{dx}\, dx \tag{6.92}$$

and

$$\int_{-1}^{1} P_n(x)\frac{d}{dx}\left[(1 - x^2)\frac{dP_m(x)}{dx}\right] dx$$

$$= P_n(x)(1 - x^2)\frac{d}{dx}P_m(x)\bigg|_{-1}^{1} - \int_{-1}^{1}\frac{dP_n(x)}{dx}(1 - x^2)\frac{dP_m(x)}{dx}\, dx. \tag{6.93}$$

Notice that the terms evaluated at $-1$ and 1 vanish and the integrals remaining after integration by parts are identical. Thus if we use expressions (6.92) and (6.93) in (6.91), we obtain

$$(n(n + 1) - m(m + 1))\int_{-1}^{1} P_n(x)P_m(x)dx = 0.$$

Thus for $n \neq m$, the integral must vanish and we have shown the Legendre polynomials are orthogonal.

In terms of earlier terminology (see Section 4.6), we can say that the $P_n(x)$ are the eigenfunctions corresponding to the eigenvalues $\lambda_n = n(n + 1)$ of the differential equation

$$(1 - x^2)y'' - 2xy' + \lambda y = 0 \tag{6.94}$$

on the interval $-1 \leq x \leq 1$. Notice that since $-1$ and 1 are both regular singular points, we did not need any boundary conditions, only a requirement of a bounded solution on the entire interval. [Note that neither series in the general solution for $\lambda = \nu(\nu + 1)$ for $\nu$ not an integer, as given by (6.79) and (6.80), converges on the entire interval $-1 \leq x \leq 1$, and the resulting functions are not bounded. Thus the eigenvalues, $\lambda_n = n(n + 1)$, are determined by a boundedness condition on the eigenfunctions. For $\nu = n$, the general solution is

$$y = C_1 P_n(x) + C_2 Q_n(x),$$

but since $Q_n(x)$ is not bounded for $-1 \le x \le 1$, we must choose $C_2 = 0$ and that leaves the Legendre polynomials as our eigenfunctions.]

To calculate the norm of these orthogonal functions (see Section 4.6.4, page 139), we use one of our identities (see exercise 2)

$$(n + 1)P_{n+1}(x) + nP_{n-1}(x) = (2n + 1)xP_n(x), \qquad n = 1, 2, 3, \ldots . \tag{6.95}$$

If we multiply (6.95) by $P_{n-1}$, integrate from $x = -1$ to $x = 1$, and use the fact that the $P_n$ are orthogonal [see (6.88)], we obtain

$$n \int_{-1}^{1} P_{n-1}^2(x) \, dx = (2n + 1) \int_{-1}^{1} xP_n(x)P_{n-1}(x) \, dx. \tag{6.96}$$

Now we rewrite (6.95), replacing $n$ by $n - 1$, as

$$nP_n(x) + (n - 1)P_{n-2}(x) = (2n - 1)xP_{n-1}(x),$$

multiply by $P_n(x)$, and integrate from $x = -1$ to $x = 1$. This gives

$$n \int_{-1}^{1} P_n^2(x) \, dx = (2n - 1) \int_{-1}^{1} xP_{n-1}(x)P_n(x) \, dx \tag{6.97}$$

and if we use (6.96) we have the identity

$$(2n + 1) \int_{-1}^{1} P_n^2(x) \, dx = (2n - 1) \int_{-1}^{1} P_{n-1}^2(x) \, dx.$$

Denoting the norm of $P_n(x)$ by $\|P_n\|$, we may rewrite this last equation as

$$(2n + 1)\|P_n\|^2 = (2n - 1)\|P_{n-1}\|^2, \qquad n = 1, 2, 3, \ldots . \tag{6.98}$$

Using this equation shows that

$$3\|P_1\|^2 = \|P_0\|^2,$$
$$5\|P_2\|^2 = 3\|P_1\|^2,$$
$$7\|P_3\|^2 = 5\|P_2\|^2,$$
$$\vdots$$
$$(2n - 1)\|P_{n-1}\|^2 = (2n - 3)\|P_{n-2}\|^2,$$
$$(2n + 1)\|P_n\|^2 = (2n - 1)\|P_{n-1}\|^2.$$

Thus we have

$$\|P_n\|^2 = \frac{1}{2n + 1}\|P_0\|^2$$

and since

$$\| P_0 \|^2 = \int_{-1}^{1} 1 \, dx = 2,$$

$$\| P_n \| = \sqrt{\frac{2}{2n + 1}}.$$

In closing we mention that another use of Legendre polynomials is in approximating functions. This is because Legendre polynomials give the best "least-squares" polynomial approximation to any given continuous function on $[-1, 1]$.

## EXERCISES

1. Show that using (6.83) and (6.84) gives the correct values of $P_0(x)$, $P_1(x)$, $P_2(x)$, $P_3(x)$, $P_4(x)$, and $P_5(x)$.

2. Use (6.85) and (6.87) to obtain the identity

$$nP_n(x) = xP'_n(x) - P'_{n-1}(x), \qquad n = 1, 2, 3, \ldots .$$

3. Use the identities mentioned in exercise 2 to show that

$$(n + 1)P_{n+1}(x) + nP_{n-1}(x) = (2n + 1)xP_n(x), \qquad n = 1, 2, 3, \ldots .$$

4. Use the identity in exercise 3 with $P_0(x) = 1$, $P_1(x) = x$ to calculate $P_2(x)$, $P_4(x)$, and $P_5(x)$.

5. Prove the identity

$$(x^2 - 1)P'_n(x) = nxP_n(x) - nP_{n-1}(x), \qquad n = 1, 2, 3, \ldots .$$

6. A generating function for Legendre polynomials is

$$(1 - 2xt + t^2)^{-1/2} = \sum_{k=0}^{\infty} P_k(x)t^k.$$

Use $y = t^2 - 2xt$ in the binomial expansion

$$(1 + y)^{\nu} = 1 + \nu y + \frac{\nu(\nu - 1)}{2!}y^2 + \frac{\nu(\nu - 1)(\nu - 2)}{3!}y^3 + \cdots$$

to obtain expressions for the first four Legendre polynomials.

7. Use the generating function to show that $P_n(1) = 1$ and $P_n(-1) = (-1)^n$.

8. (a) Write $x^2$ and $x^3$ as linear combinations of Legendre polynomials.
   (b) Use the fact that $x^m$ may be expressed as a linear combination of $P_0(x)$, $P_1(x)$, $\ldots$, $P_m(x)$, that is,

$$x^m = \sum_{k=0}^{m} a_k P_k(x),$$

   to show that

$$\int_{-1}^{1} x^m P_n(x) \, dx = 0, \qquad m < n.$$

**(c)** Show that

$$\int_{-1}^{1} P_n(x)\, dx = 0, \qquad n > 0.$$

**(d)** Use the results above, together with (6.83), to show that

$$x^n = \frac{2^n(n!)^2}{(2n)!} P_n(x) + g(x),$$

where $g(x)$ is a polynomial of degree $n - 2$, and

$$\int_{-1}^{1} x^n P_n(x)\, dx = \frac{2^{n+1}(n!)^2}{(2n + 1)!}.$$

**9.** Use (6.85) and (6.95) to show that

$$(2n + 1)xP_n(x) = \frac{n + 1}{2n + 3} P'_{n+2}(x)$$

$$+ \frac{2n + 1}{(2n + 3)(2n - 1)} P'_n(x) - \frac{n}{2n - 1} P'_{n-2}(x), \qquad n = 2, 3, 4, \ldots.$$

**10.** Use results from (6.85) and exercise 9 to show that

$$(2n + 1)\int P_n(x)\, dx = P_{n+1}(x) - P_{n-1}(x) + C,$$

$$(2n + 1)\int xP_n(x)\, dx = \frac{n + 1}{2n + 3} P_{n+2}(x) + \frac{2n + 1}{(2n + 3)(2n - 1)} P_n(x) - \frac{n}{2n - 1} P_{n-2}(x) + C.$$

**11.** Use the change of variable $x = \cos\theta$ to show that the Legendre differential equation is transformed to

$$\frac{d^2 y}{d\theta^2} + \cot\theta \frac{dy}{d\theta} + n(n + 1)y = 0,$$

with the general solution

$$y = C_1 P_n(\cos\theta) + C_2 Q_n(\cos\theta).$$

**12.** Find an expression for a polynomial solution of (6.74), with $v = n$, an integer, as a series in $(x - 1)$.

**13.** Use reduction of order to obtain a second solution of (6.74) as

$$Q_n(x) = P_n(x) \int \frac{dx}{(1 - x^2)P_n^2(x)}.$$

**14.** Use exercise 13 to show that

**(a)** $Q_0(x) = \dfrac{1}{2} \ln \dfrac{1 + x}{1 - x}.$

**(b)** $Q_1(x) = xQ_0(x) - 1.$

**15.** The Schrödinger equation for a one-dimensional harmonic oscillator of mass $m$ and frequency $\omega$ is

$$-\frac{\hbar^2}{2m} \frac{d^2 u}{dz^2} + \left(\frac{m\omega^2 z^2}{2} - E\right)u = 0,$$

where $\hbar = $ (Planck's constant)$/(2\pi)$ and $E = $ total energy are constants.

(a) Let $x = z\sqrt{m\omega/\hbar}$, $u = e^{-x^2/2}y$, and $2\lambda = 2E/(\hbar\omega) - 1$, to show that the differential equation above becomes

$$\frac{d^2y}{dx^2} - 2x\frac{dy}{dx} + 2\lambda y = 0.$$

(This is called Hermite's differential equation.)

(b) Find values of $\lambda$ that give polynomial solutions of Hermite's differential equation. [Note that these values of $\lambda$ correspond to discrete energy levels for $E (E = (\lambda + \frac{1}{2})\hbar\omega)$].

(c) The solutions in part (b) are called Hermite polynomials. Find the first five Hermite polynomials.

(d) Show that the polynomial solutions may be expressed as

$$H_n(x) = \sum_{k=0}^{N} \frac{(-1)^k}{k!(n - 2k)!}(2x)^{n-2k},$$

where
$$N = \frac{n}{2}, \quad n \text{ even}; \qquad N = \frac{n-1}{2}, \quad n \text{ odd}.$$

(e) Show that the Rodrigues' formula,

$$H_n(x) = (-1)^n e^{x^2} \frac{d^n}{dx^n} e^{-x^2},$$

gives the first five Hermite polynomials.

17. (a) Show that $x = 0$ is a regular singular point of Laguerre's equation

$$xy'' + (1 - x)y' + \lambda y = 0.$$

(b) Find the values of $\lambda$ that give polynomial solutions of this differential equation.

(c) Show that the polynomial solutions are given by

$$L_n(x) = (n!)^2 \sum_{k=0}^{n} \frac{(-1)^k}{k![(n - k)!]^2} x^{n-k}.$$

(These are called *Laguerre polynomials*.)

(d) Find the first four Laguerre polynomials.

(e) Show that these four polynomials may be also obtained by a Rodrigues' formula,

$$L_n(x) = (-1)^n e^x \frac{d^n}{dx^n}(x^n e^{-x}).$$

18. The Tchebysheff equation is

$$(1 - x^2)y'' - xy' + \lambda y = 0.$$

(a) Find values of $\lambda$ which give polynomial solutions.

(b) Find the first four polynomial solutions.

## 6.5.2 The Bessel Equation

A differential equation which occurs in boundary value problems associated with a right circular cylinder (see Example 11.10, page 406) is the *Bessel equation*,

$$Ly = x^2 y'' + xy' + (x^2 - \mu^2)y = 0. \tag{6.99}$$

Note that $x = 0$ is a regular singular point. To find the indicial equation, we compute

$$Lx^r = r(r - 1)x^r + rx^r + x^{r+2} - \mu^2 x^r$$

and determine the indicial equation as

$$r^2 - \mu^2 = 0.$$

Thus $\mu$ and $-\mu$ are roots of the indicial equation.
   If we use the larger root, we substitute

$$y = \sum_{k=0}^{\infty} c_k x^{k+\mu}$$

into (6.99) to obtain

$$\sum_{k=0}^{\infty} (k + \mu)(k + \mu - 1)c_k x^{k+\mu} + \sum_{k=0}^{\infty} (k + \mu)c_k x^{k+\mu}$$

$$+ \sum_{k=0}^{\infty} c_k x^{k+\mu+2} - \sum_{k=0}^{\infty} \mu^2 c_k x^{k+\mu} = 0.$$

If we lower the index of summation by 2 in the third series and write the first two terms in the other three series separately, we have

$$[\mu(\mu - 1) + \mu - \mu^2]c_0 x^\mu + [(1 + \mu)(\mu) + 1 + \mu - \mu^2]c_1 x^{\mu+1}$$

$$+ \sum_{k=2}^{\infty} \{[(k + \mu)^2 - \mu^2]c_k + c_{k-2}\}x^{k+\mu} = 0.$$

Thus

$$(1 + 2\mu)c_1 = 0, \tag{6.100}$$

$$k(k + 2\mu)c_k = -c_{k-2}, \qquad k = 2, 3, 4, \ldots, \tag{6.101}$$

give the recurrence relation. If $\mu \neq -\frac{1}{2}$, $c_1 = 0$, and (6.101) requires that all the odd coefficients are zero. If $2\mu$ is not a negative integer, we can solve for all the even coefficients as

$$c_2 = \frac{-1}{2(2 + 2\mu)} c_0,$$

$$c_4 = \frac{-1}{4(4 + 2\mu)} c_2 = \frac{(-1)^2}{4 \cdot 2(2 + 2\mu)(4 + 2\mu)} c_0,$$

$$c_6 = \frac{-1}{6(6 + 2\mu)} c_4 = \frac{(-1)^3}{6 \cdot 4 \cdot 2(2 + 2\mu)(4 + 2\mu)(6 + 2\mu)} c_0,$$

$$\vdots$$

$$c_{2m} = \frac{(-1)^m}{2^m m! \, 2^m (1 + \mu)(2 + \mu)(3 + \mu) \cdots (m + \mu)} c_0, \qquad m = 1, 2, 3, \ldots.$$

Thus

$$y_1(x) = c_0 \left[ 1 + \sum_{m=1}^{\infty} \frac{(-1)^m}{2^{2m} m! \, (1 + \mu)(2 + \mu)(3 + \mu) \cdots (m + \mu)} \, x^{2m+\mu} \right]$$

(6.102)

is a solution of (6.99). Similar calculations will determine a solution using the root $-\mu$ of the indicial equation. We will concentrate on the case where $\mu = n$, a nonnegative integer, and the constant $c_0$ in (6.102) is chosen as $1/(2^n n!)$. The resulting function, called the *Bessel function of order n of the first kind,* is given by

$$J_n(x) = \sum_{m=0}^{\infty} \frac{(-1)^m}{m! \, (m + n)!} \left( \frac{x}{2} \right)^{2m+n}.$$

(6.103)

Graphs of $J_n(x)$ versus $x$ for $n = 0, 1$, and $2$ appear in Figure 6.3. Note that $J_0(0) = 1$, but $J_n(0) = 0$ for $n \geq 1$.

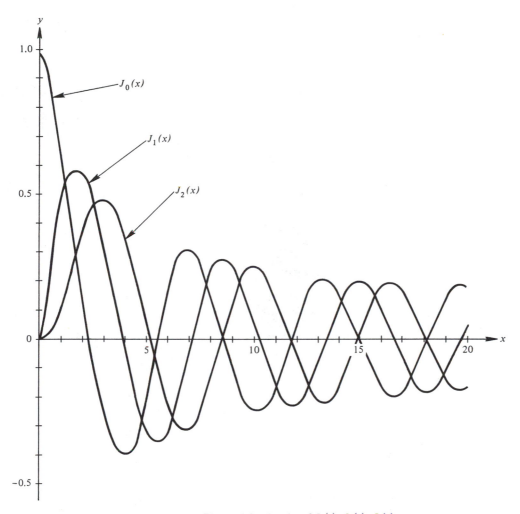

**Figure 6.3**    Graphs of $J_0(x)$, $J_1(x)$, $J_2(x)$.

The Frobenius series associated with the root $-n$ of the indicial equation does not produce a function linearly independent of $J_n(x)$. Thus by cases (b), if $n = 0$, and (c) of Theorem 6.6, we know that the second solution of Bessel's equation will have the form

$$y_2(x) = CJ_n(x) \ln x + x^{-n}\left(c_0^* + \sum_{k=1}^{\infty} c_k^* x^k\right),$$

where $C = 1$ and $c_0^* = 0$ if $n = 0$. [Note that this includes the cases $n = 0$ or $n > 0$ from (6.56) and (6.57), and shows that $y_2(0)$ is not bounded.] The most commonly used second solution to Bessel's differential equation (6.99) when $n$ is an integer is a linear combination of $J_n(x)$ and $y_2(x)$, given above with $C = 1$, $c_0^* = -1$, namely

$$Y_n(x) = \frac{2}{\pi}[y_2(x) + (\gamma - \ln 2)J_n(x)], \qquad n = 0, 1, 2, \ldots .$$

$Y_n(x)$ is called the Bessel function of the second kind of order $n$. The constant $\gamma$ occurring in the expression for $Y_n(x)$ is defined as

$$\gamma = \lim_{n \to \infty} \left(\sum_{k=1}^{n} \frac{1}{k} - \ln n\right) \approx 0.5772156649\ldots$$

and is usually called the Euler constant. Other names in use are the Mascheroni constant or Euler–Mascheroni constant. Thus the general solution of (6.99) for $n$ a positive integer is

$$y = C_1 J_n(x) + C_2 Y_n(x),$$

where $Y_n(x)$ is not bounded at $x = 0$.

The series in (6.103) is easier to work with than that for Legendre polynomials and we will use it to derive an identity for Bessel functions. If we multiply both sides of (6.103) by $x^{-n}$ and differentiate with respect to $x$, we obtain

$$\frac{d}{dx}[x^{-n}J_n(x)] = \frac{d}{dx} \sum_{m=0}^{\infty} \frac{(-1)^m x^{2m}}{m! \, (m + n)! \, 2^{2m+n}}$$

$$= \sum_{m=1}^{\infty} \frac{(-1)^m (2m) x^{2m-1}}{m! \, (m + n)! \, 2^{2m+n}}$$

$$= \sum_{m=1}^{\infty} \frac{(-1)^m x^{2m-1}}{(m - 1)! \, (m + n)! \, 2^{2m+n-1}}$$

$$= \sum_{k=0}^{\infty} \frac{(-1)^{k+1} x^{2k+1}}{k! \, (k + n + 1)! \, 2^{2k+n+1}}$$

$$= -x^{-n} \sum_{k=0}^{\infty} \frac{(-1)^k}{k! \, (k + n + 1)!} \left(\frac{x}{2}\right)^{2k+n+1}$$

$$= -x^{-n} J_{n+1}(x).$$

If we now carry out the differentiation on the left-hand side of the equation above and multiply by $x^n$, we obtain the identity

$$J_n'(x) - \frac{n}{x}J_n(x) = -J_{n+1}(x). \qquad (6.104)$$

The derivation of other identities for Bessel functions of integer order $n$ is left for the exercises.

Bessel functions are useful in solving the following boundary value problem, where $n$ is a specified positive integer and $\lambda^2$ is an unknown parameter:

$$x^2y'' + xy' + (\lambda^2x^2 - n^2)y = 0, \qquad 0 < x < a, \qquad (6.105)$$

$$y(a) = 0. \qquad (6.106)$$

Note that since $x = 0$ is a regular singular point, we do not prescribe the value $y(0)$, but instead require that the solution be bounded for $0 \le x \le a$. If we make the change of variable, $\lambda x = z$, we obtain

$$z^2y'' + zy' + (z^2 - n^2)y = 0,$$

where differentiation is now with respect to $z$. Thus the only solution of (6.105) that is bounded for all $0 \le x \le a$ is a constant multiple of $J_n(\lambda x)$. If (6.106) is to be satisfied, the $\lambda^2$ must be chosen so that

$$J_n(\lambda a) = 0. \qquad (6.107)$$

There will, in fact, be an infinite number of values for $\lambda$ which satisfy (6.107) (e.g., observe the start of the oscillatory behavior in Figure 6.3 and in the solutions to exercises 6 and 7). If these values are called $\lambda_j, j = 1, 2, 3, \ldots$, then corresponding to the eigenvalues $\lambda_j^2, j = 1, 2, 3, \ldots$, of the system (6.105) and (6.106), we have the eigenfunctions $J_n(\lambda_j x), j = 1, 2, 3, \ldots$.

To show that these eigenfunctions form an orthogonal set, we return to (6.105) and write it for $\lambda = \lambda_j$ and corresponding eigenfunction $J_n(\lambda_j x) = u$ (as a shorthand notation) as well as $\lambda = \lambda_k$ and $J_n(\lambda_k x) = v$. This gives

$$x^2u'' + xu' + (\lambda_j^2x^2 - n^2)u = 0,$$

$$x^2v'' + xv' + (\lambda_k^2x^2 - n^2)v = 0.$$

If we now multiply the top equation by $v$ and the bottom by $u$ and subtract the results, we obtain

$$x^2(u''v - uv'') + x(u'v - uv') + (\lambda_j^2 - \lambda_k^2)x^2uv = 0. \qquad (6.108)$$

Now we divide out the common factor, $x$, and notice that

$$\frac{d}{dx}(u'v - uv') = u''v - uv'',$$

so

$$\frac{d}{dx}[x(u'v - uv')] = x(u''v - uv'') + (u'v - uv'). \qquad (6.109)$$

If we use (6.109) in (6.108) and integrate from $x = 0$ to $x = a$, we obtain

$$\int_0^a \frac{d}{dx} [x(u'v - uv')] \, dx + (\lambda_j^2 - \lambda_k^2) \int_0^a xuv \, dx = 0$$

or, replacing $u$ and $v$ by their equivalent terms,

$$\left\{ \left[ \frac{d}{dx} J_n(\lambda_j x) \right] J_n(\lambda_k x) - J_n(\lambda_j x) \frac{d}{dx} J_n(\lambda_k x) \right\} x \Bigg|_0^a = (\lambda_k^2 - \lambda_j^2) \int_0^a J_n(\lambda_j x) J_n(\lambda_k x) x \, dx.$$

Since both the Bessel functions and their derivatives are bounded at $x = 0$, the evaluated expression vanishes at $x = 0$. Also since the eigenvalues are given by the solutions of (6.107), the evaluated expression is zero at $x = a$ and we have the orthogonality condition

$$\int_0^a J_n(\lambda_j x) J_n(\lambda_k x) x \, dx = 0, \qquad j \neq k.$$

To evaluate $\| J_n(\lambda_j x) \|^2$, we rewrite (6.105) as

$$\frac{d}{dx} (xy') + \left( \lambda^2 x - \frac{n^2}{x} \right) y = 0.$$

If we multiply both sides of the equation by $2xy'$ and combine terms, we have

$$2(xy') \frac{d}{dx} (xy') + \lambda^2 x^2 (2yy') - n^2 (2yy') = 0$$

or

$$\frac{d}{dx} (xy')^2 + (\lambda^2 x^2 - n^2) \frac{d}{dx} (y^2) = 0.$$

Now we integrate from $x = 0$ to $x = a$ and integrate the term on the right by parts to obtain

$$(xy')^2 \Bigg|_0^a + (\lambda^2 x^2 - n^2) y^2 \Bigg|_0^a - \int_0^a (2\lambda^2 x) y^2 \, dx = 0.$$

Now use the fact that corresponding to $\lambda = \lambda_j$ we have the eigenfunction $J_n(\lambda_j x)$ to rewrite this equation as

$$2\lambda_j^2 \int_0^a [J_n(\lambda_j x)]^2 x \, dx = \left\{ \left[ x \frac{d}{dx} J_n(\lambda_j x) \right]^2 + (\lambda_j^2 x^2 - n^2)[J_n(\lambda_j x)]^2 \right\} \Bigg|_0^a.$$
$$\tag{6.110}$$

Since $J_n(0) = 0$ for $n \geq 1$ and $J_n(\lambda_j x)$ and its derivative are bounded at $x = 0$, the expression on the right-hand side vanishes at the lower limit. Using the fact that $J_n(\lambda_j a) = 0$ leaves (6.110) as

$$2\lambda_j^2 \int_0^a [J_n(\lambda_j x)]^2 x \, dx = \left[ a \frac{d}{dx} J_n(\lambda_j x) \right]^2 \Bigg|_{x=a}. \tag{6.111}$$

From (6.104) and the chain rule,

$$\frac{d}{dx} J_n(\lambda_j x) = J_n'(\lambda_j x) \lambda_j$$

$$= \lambda_j \left[ \frac{n}{\lambda_j x} J_n(\lambda_j x) - J_{n+1}(\lambda_j x) \right].$$

Evaluating this expression for $x = a$ and using the result in (6.111) gives

$$\int_0^a [J_n(\lambda_j x)]^2 x \, dx = \frac{a^2}{2\lambda_j^2}[-\lambda_j J_{n+1}(\lambda_j a)]^2$$

$$= \frac{a^2[J_{n+1}(\lambda_j a)]^2}{2}, \qquad j = 1, 2, 3, \ldots \qquad (6.112)$$

## EXERCISES

1. Find the solution of (6.99) corresponding to the root $-\mu$ of the indicial equation and show that this solution is undefined at $x = 0$.
2. Show that the solution of (6.99) corresponding to $-\mu = -n$, an integer, is $(-1)^n$ times the solution given by (6.102) with $\mu = n$.
3. Show that

$$\frac{d}{dx}[x^n J_n(x)] = x^n J_{n-1}(x), \qquad n = 1, 2, 3, \ldots .$$

4. Use the result of exercise 3 and (6.104) to prove the following identities.

   (a) $J_n'(x) + \dfrac{n}{x} J_n(x) = J_{n-1}(x), \quad n = 1, 2, 3, \ldots .$

   (b) $J_{n+1}(x) = \dfrac{2n}{x} J_n(x) - J_{n-1}(x).$

   (c) $2J_n'(x) = J_{n-1}(x) - J_{n+1}(x).$

5. Use the result of exercise 3 to show that

   (a) $\displaystyle\int_0^a x^n J_{n-1}(x) \, dx = a^n J_n(a), \quad n = 1, 2, 3, \ldots .$

   (b) $\displaystyle\int_0^a x^n J_{n-1}(\lambda x) \, dx = \lambda^{-1} a^n J_n(\lambda a), \quad n = 1, 2, 3, \ldots .$

6. Show that a second solution of Bessel's differential equation (6.99) for $\mu = 0$ is given by

$$J_0(x) \ln x + \frac{x^2}{4} - \frac{3x^4}{2^7} + \cdots .$$

[Hint: Use (6.56).]

7. Show that a second solution of Bessel's differential equation (6.99) for $\mu = 1$ is given by

$$J_1(x) \ln x - x^{-1} - \frac{x}{4} + \frac{5x^3}{2^6} \cdots .$$

[Hint: Use (6.57) with $C = 1$.]

8. Show that the Bessel function of the second kind of order zero,

$$Y_0(x) = \frac{2}{\pi}\left(\gamma + \log\frac{x}{2}\right) + \frac{2}{\pi}\sum_{k=1}^{\infty} \frac{(-1)^{k+1}}{(k!)^2}\left(\sum_{i=1}^{k}\frac{1}{i}\right)\left(\frac{x}{2}\right)^{2k}$$

satisfies Bessel's equation (6.99) with $n = 0$.

9. Find the general solution of (6.99) with $\mu^2 = \frac{1}{4}$.

10. $J_{1/2}(x)$ and $J_{-1/2}(x)$ are defined by choosing $c_0 = c_1 = \sqrt{2/\pi}$ in the series solutions from exercise 9. Use the Taylor series

$$\sin x = \sum_{k=0}^{\infty} \frac{(-1)^k x^{2k+1}}{(2k+1)!},$$

$$\cos x = \sum_{k=0}^{\infty} \frac{(-1)^k x^{2k}}{(2k)!},$$

to show that

$$J_{-1/2}(x) = \sqrt{\frac{2}{\pi x}} \cos x,$$

$$J_{1/2}(x) = \sqrt{\frac{2}{\pi x}} \sin x.$$

11. Bessel functions of the form $J_{1/2\pm m}(x)$, where $m$ is an integer, are spherical Bessel functions. Use appropriate identities to write the following in terms of algebraic functions, $\sin x$ and $\cos x$.
    (a) $J_{3/2}(x)$       (b) $J_{-3/2}(x)$       (c) $J_{5/2}(x)$       (d) $J_{-5/2}(x)$

12. Make a change of dependent variable in (6.99) for $\mu^2 = \frac{1}{4}$ of $y(x) = u(x)/\sqrt{x}$, find the general solution of the resulting differential equation, and compare the results with exercise 10.

13. Some differential equations may be solved by using a change of dependent and/or independent variable to obtain Bessel's differential equation.
    (a) Show that the Airy equation

$$y'' + xy = 0,$$

    may be changed to a form of Bessel's differential equation by letting

$$x = \left(\frac{3z}{2}\right)^{2/3},$$

$$y = \sqrt{x}\, u(z).$$

    (b) Show that the equation

$$x^2 y'' + \left(\alpha^2 x^{2\alpha} + \frac{1}{4} - n^2 \alpha^2\right) y = 0$$

    may be changed to a form of Bessel's differential equation by letting

$$x = z^{1/\alpha},$$

$$y = \sqrt{x}\, u(z).$$

    (c) Show that the equation

$$x^2 y'' + x(a + 2bx^\alpha)y' + [c + \gamma^2 x^{2\beta} + b(a + \alpha - 1)x^\alpha + b^2 x^{2\alpha}]y = 0$$

    may be changed to a form of Bessel's differential equation by letting

$$x = \left(\frac{\beta z}{\gamma}\right)^{1/\beta},$$

$$y = x^{(1-a)/2}\, e^{-(b/\alpha)x^{\alpha}} u.$$

**14.** Find the solution that is bounded at $x = 0$ of each of the differential equations in exercise 13.

**15.** Use the method of Frobenius to find a solution of

$$x^2 y'' + xy' - (x^2 + \nu^2)y = 0, \qquad (6.113)$$

which is bounded at the origin.

**16.** Replacing $x$ by $ix$ in (6.113) gives Bessel's differential equation. The solution of (6.113) that is bounded at $x = 0$ and given by

$$I_\nu(x) = i^{-\nu} J_\nu(ix)$$

is called the *modified Bessel function of order $\nu$ of the first kind*. Show that

(a) $I_{-n}(x) = I_n(x)$.

(b) $\dfrac{d}{dx}[x^\nu I_\nu(x)] = x^\nu I_{\nu-1}(x)$.

(c) $\dfrac{d}{dx}[x^{-\nu} I_\nu(x)] = x^{-\nu} I_\nu(x)$.

(d) $I_{n-1}(x) - I_{n+1}(x) = \dfrac{2n}{x} I_n(x)$.

**17.** The Gauss hypergeometric differential equation is of the form

$$x(1 - x)y'' + [c - (a + b + 1)x]y' - aby = 0.$$

(a) Show that this equation has regular singular points at $x = 0$ and $1$.

(b) Show that the roots of the indicial equation (for a Frobenius solution about $x = 0$) are $0$ and $1 - c$.

(c) Show that the function

$$F(a, b; c; x) = 1 + \frac{ab}{c}x + \frac{(a)(a + 1)(b)(b + 1)}{(c)(c + 1)2!}x^2$$

$$+ \frac{(a)(a + 1)(a + 2)(b)(b + 1)(b + 2)}{(c)(c + 1)(c + 2)3!}x^3 + \cdots$$

is a solution of the hypergeometric equation above. $F(a, b; c; x)$ is called a hypergeometric function or hypergeometric series. It is sometimes written $_2F_1(a, b; c; x)$.

(d) If $c$ is not an integer, show that a second solution of the hypergeometric equation is

$$x^{1-c} F(a + 1 - c, b + 1 - c; 2 - c; x).$$

(e) Show that $F(a, 1; a; x) = 1/(1 - x), \quad |x| < 1$.

# REVIEW EXERCISES

**1.** Find the first four terms in the Taylor series expansion of the solution to

$$y' - e^{-xy} = 0, \qquad y(0) = 1.$$

2. Find the Taylor series expansion of the solution to

$$y'' + y = 0, \qquad y(0) = A, \qquad y'(0) = B,$$

and show that it is identical with the solution obtained using the method of undetermined coefficients.

Solve the differential equations subject to any specified conditions. If no conditions are specified, find the general solution. If you cannot find a general pattern, compute the first few terms.

3. $(1 + x^2)y'' - 4xy' + 6y = 0.$

4. $(1 + x^2)y'' + 2xy' + 2y = 0.$

5. $y'' - xy' + y = x \cos x, \quad y(0) = 0, \quad y'(0) = 2.$

6. $2y'' + xy' - 4y = 0.$

7. $y'' + \sin(xy) = e^{x^2}.$

8. $x^2y'' + xy' + (x^2 - 49)y = 0, \quad y(0) = 0.$

9. $(x^2 - 1)y'' + 2xy' - 20y = 0, \quad y \text{ bounded for } -1 \le x \le 1.$

10. $xy'' + y = 0.$

11. $x^2y'' + x^2y' - 2y = 0.$

12. $xy'' + (x^2 - 3)y' - 2xy = 0.$

13. $(1 - x^2)y'' - 2xy' + 6y = 0.$

14. $xy'' + y' + xy = 0.$

15. $y' = x + e^y, \quad y(0) = 1.$

16. $2xy'' + y' + xy = 0.$

17. The generating function for the Legendre polynomials is

$$(1 - 2xt + t^2)^{-1/2} = \sum_{k=0}^{\infty} P_k(x)t^k. \tag{6.114}$$

(a) Show by direct calculation that $g(x, t) = (1 - 2xt + t^2)^{-1/2}$ satisfies the partial differential equation

$$(1 - 2xt + t^2)\frac{\partial g}{\partial t} - (x - t)g = 0. \tag{6.115}$$

(b) Substitute the series expression for $g$ from (6.114) into (6.115) to derive the identity

$$(n + 1)P_{n+1}(x) + nP_{n-1}(x) = (2n + 1)xP_n(x), \qquad n = 1, 2, 3, \ldots .$$

(c) Derive the identity

$$nP_{n-1}(x) = P_n'(x) - xP_{n-1}'(x), \qquad n = 1, 2, 3, \ldots ,$$

by first showing that $g(x, t)$ satisfies

$$t\frac{\partial}{\partial t}(tg) - (1 - tx)\frac{\partial g}{\partial x} = 0.$$

18. Derive the following identities for the Bessel functions.

(a) $J_0'(x) = -J_1(x).$

(b) $\displaystyle\int_0^a xJ_0(x)\, dx = aJ_1(a).$

(c) $\displaystyle\int_0^a x^3J_0(x)\, dx = (a^3 - 4a)J_1(a) + 2a^2J_0(a).$

# CHAPTER 7

# Laplace Transforms

The Laplace transform of a function $f$ is defined for all $f(t)$ which are piecewise continuous in every subinterval of $[0, \infty)$ and of exponential order as $t \to \infty$. In this chapter the fundamental operational property and a number of addition properties of the transform are developed. The convolution theorem is stated and its use in the inversion of the transform and solving integral equations is demonstrated. Finally, the use of the transform in solving ordinary differential equations and systems of ordinary differential equations is presented.

## 7.1 INTRODUCTION

We are already familiar with linear operators that transform functions into other functions. In particular we have the operator $D = d/dx$, which operates upon differentiable functions via

$$Df(x) = \frac{df}{dx}.$$

We may also use the operation of integration to define transforms of functions. In the general case we assume a function $K(s, t)$, defined for $a < t < b$ with $s$ a parameter, and consider all functions, $f(t)$, defined for $a < t < b$, such that

$$\int_a^b K(s, t)f(t)\, dt \tag{7.1}$$

exists. In this text, $s$ is taken as a real variable, while in more advanced texts $s$ may be complex. For complex $s$, any rule in this book that requires "$s > 0$" will then be replaced by "real part of $s > 0$".

For the particular choices $[a, b) = [0, \infty)$ and $K(s, t) = e^{-st}$, we obtain

$$F(s) = \int_0^\infty e^{-st} f(t) \, dt. \tag{7.2}$$

The transform (7.2), if it exists, is called the *Laplace transform* of $f(t)$ and will be denoted by $F(s)$ or $\mathscr{L}\{f(t)\}$. The first thing we want to do is to show that the Laplace transform is linear. Thus we assume that $f(t)$ and $g(t)$ have Laplace transforms and compute

$$\mathscr{L}\{af(t) + bg(t)\} = \int_0^\infty e^{-st}[af(t) + bg(t)] \, dt.$$

If we now use the properties of improper integrals, we see that the right-hand side of the equation above may be written as

$$a \int_0^\infty e^{-st} f(t) \, dt + b \int_0^\infty e^{-st} g(t) \, dt = a\mathscr{L}\{f(t)\} + b\mathscr{L}\{g(t)\}.$$

Thus

$$\mathscr{L}\{af(t) + bg(t)\} = a\mathscr{L}\{f(t)\} + b\mathscr{L}\{g(t)\}$$

and indeed the Laplace transform is linear.

Setting aside the question of existence, to which we shall return in Section 7.2, we proceed to calculate a few transforms. Consider the function $f(t) = 1$ for all $t$. Then, from (7.2),

$$\mathscr{L}\{1\} = \int_0^\infty e^{-st} dt = \lim_{b \to \infty} -\frac{1}{s} e^{-st} \Big|_0^b,$$

so that if $s$ is positive, $\mathscr{L}\{1\} = 1/s$.

From this we see that for $f(t) = C$, a constant, we have from the fact that the Laplace transform is linear,

$$\mathscr{L}\{C\} = C\mathscr{L}\{1\} = C\left(\frac{1}{s}\right) = \frac{C}{s}. \tag{7.3}$$

Now let $f(t) = t$, $\quad t \geq 0$. Then from (7.2)

$$\mathscr{L}\{t\} = \int_0^\infty e^{-st} t \, dt, \tag{7.4}$$

and an elementary integration by parts in (7.4) shows that

$$\mathscr{L}\{t\} = \frac{1}{s^2}, \qquad s > 0.$$

The Laplace transform of $t^n$ may be established by repeated integrations by parts or by an induction proof. We also may use more elegant methods (see exercise 4, Section 7.3, and exercise 25, Section 7.4) to establish that when $n$ is a positive integer,

$$\mathscr{L}\{t^n\} = \frac{n!}{s^{n+1}}.$$

A slightly more interesting function is

$$f(t) = e^{at},\tag{7.5}$$

where $a$ is a constant. For (7.5) we have

$$\mathcal{L}\{e^{at}\} = \int_0^\infty e^{-st} e^{at} \, dt = \int_0^\infty e^{-(s-a)t} \, dt$$

and it is easy to see that

$$\mathcal{L}\{e^{at}\} = \frac{1}{s-a}, \qquad s - a > 0.\tag{7.6}$$

Thus Laplace transforms of elementary functions may be obtained using the usual techniques of integral calculus.

Another method for calculating transforms uses (7.6) and the fact that the Laplace transform is linear. For example, let

$$f(t) = \cosh t,$$

so that

$$f(t) = \frac{1}{2}(e^t + e^{-t}).$$

Then we have

$$\mathcal{L}\{\cosh t\} = \mathcal{L}\left\{\frac{e^t + e^{-t}}{2}\right\}$$

and since the Laplace transform is linear,

$$\mathcal{L}\{\cosh t\} = \frac{1}{2}\,\mathcal{L}\{e^t\} + \frac{1}{2}\,\mathcal{L}\{e^{-t}\}$$

$$= \frac{1}{2}\left(\frac{1}{s-1}\right) + \frac{1}{2}\left(\frac{1}{s+1}\right) = \frac{s}{s^2 - 1}.\tag{7.7}$$

To obtain the Laplace transform of the sine and cosine functions, we may use Euler's formula,

$$e^{it} = \cos t + i \sin t.$$

Using the above expression and the corresponding expression for $e^{-it}$, we may show that

$$\cos t = \frac{e^{it} + e^{-it}}{2}\tag{7.8}$$

and

$$\sin t = \frac{e^{it} - e^{-it}}{2i}.$$

From (7.8) we have

$$\mathcal{L}\{\cos t\} = \frac{1}{2}\,\mathcal{L}\{e^{it}\} + \frac{1}{2}\,\mathcal{L}\{e^{-it}\}$$

and using (7.6) we have

$$\mathcal{L}\{\cos t\} = \frac{1}{2}\left(\frac{1}{s-i}\right) + \frac{1}{2}\left(\frac{1}{s+i}\right) = \frac{s}{s^2+1}, \qquad s > 0.$$

Once we have the Laplace transform of a function, the following theorem allows us to calculate the transform of functions obtained from the original one by a change of scale (see exercise 1 for a proof).

**Theorem 7.1.**    If $\mathcal{L}\{f(t)\} = F(s)$,

$$\mathcal{L}\{f(at)\} = \frac{1}{a}F\left(\frac{s}{a}\right), \qquad a > 0.$$

Using this theorem, we may obtain the Laplace transform of cos $at$ by the following method. If $f(t) = \cos t$, $F(s) = s/(s^2 + 1)$, so $f(at) = \cos at$ and

$$\mathcal{L}\{\cos at\} = \frac{1}{a}\left[\frac{s/a}{(s/a)^2 + 1}\right] = \frac{s}{s^2 + a^2}.$$

Recall from calculus that if we know the formula for the derivative of a function, we also know the formula for an antiderivative of this resulting derivative. That is, if

$$f'(t) = g(t),$$

then an antiderivative of $g(t)$ is $f(t)$. We may ask a similar question about Laplace transforms, namely, if

$$\mathcal{L}\{f(t)\} = F(s), \tag{7.9}$$

do we know a formula for the inverse Laplace transform of $F(s)$? For functions, $f$, which are continuous on $[0, \infty)$ we may state the following theorem.

**Theorem 7.2.**    If $F(s)$ is defined by (7.9) and $f(t)$ is continuous on $[0, \infty)$, then the inverse Laplace transform of $F(s)$, $\mathcal{L}^{-1}\{F(s)\}$, is given by

$$\mathcal{L}^{-1}\{F(s)\} = f(t).$$

(Note that by definition the Laplace transform of a function involves calculating a definite integral. Thus from calculus, if the Laplace transform of $t$ exists, it is unique.)

Now, since $e^{at}$ is continuous on $[0, \infty)$, we have from (7.6)

$$\mathcal{L}^{-1}\left\{\frac{1}{s-a}\right\} = e^{at},$$

and since cosh $t$ and cos $t$ are continuous on $[0, \infty)$, from (7.7) and (7.8) we have

$$\mathcal{L}^{-1}\left\{\frac{s}{s^2 - 1}\right\} = \cosh t \qquad \text{and} \qquad \mathcal{L}^{-1}\left\{\frac{s}{s^2 + 1}\right\} = \cos t.$$

## EXERCISES

**1.** Prove Theorem 7.1. (*Hint:* Change the variable of integration in $\mathcal{L}\{f(at)\}$.)

Evaluate the following transforms.

**2.** $\mathcal{L}\{\sin t\}$

**3.** $\mathcal{L}\{a \sin t + b \cos t\}$

**4.** $\mathcal{L}\{a + bt\}$

**5.** $\mathcal{L}\{e^{at} - e^{bt}\}$

**6.** $\mathcal{L}\{\sinh t\}$

**7.** $\mathcal{L}\{a \cosh t + b \sinh t\}$

**8.** $\mathcal{L}\{\sin at\}$

**9.** $\mathcal{L}\{\sinh at\}$

**10.** $\mathcal{L}\{\cosh at\}$

**11.** $\mathcal{L}\{f(t)\}$  if  $f(t) = \begin{cases} t - 2, & 0 \le t < 2 \\ 0, & 2 \le t \end{cases}$

**12.** $\mathcal{L}\{f(t)\}$  if  $f(t) = \begin{cases} e^{2t}, & 0 \le t \le 10 \\ e^{20}, & t > 10 \end{cases}$

Evaluate the following inverse transforms (note that the inverse Laplace transform is also a linear operator if all the inverse transforms involved exist).

**13.** $\mathcal{L}^{-1}\left\{\dfrac{3}{s^3}\right\}$

**14.** $\mathcal{L}^{-1}\left\{\dfrac{7}{s^2}\right\}$

**15.** $\mathcal{L}^{-1}\left\{\dfrac{5}{s - 2}\right\}$

**16.** $\mathcal{L}^{-1}\left\{\dfrac{1}{s} + \dfrac{2}{s^2}\right\}$

**17.** $\mathcal{L}^{-1}\left\{\dfrac{2}{s - 3} - \dfrac{3}{s - 2}\right\}$

**18.** $\mathcal{L}^{-1}\left\{\dfrac{3s}{s^2 - 1} + \dfrac{8s}{s^2 + 1}\right\}$

**19.** $\mathcal{L}^{-1}\left\{\dfrac{14}{s^2 - 3s - 10}\right\}$   (*Hint:* Recall partial fractions.)

**20.** $\mathcal{L}^{-1}\left\{\dfrac{3s - 8}{s^2 - 3s - 10}\right\}$

**21.** Prove that $\mathcal{L}\{t^n\} = \dfrac{n!}{s^{n+1}}$ using induction.

## 7.2 EXISTENCE OF THE TRANSFORM

We must now address the question: When does the integral in (7.2) exist? To answer this we will have to restrict the class of functions to which we apply the transform in such a way as to guarantee the existence of the integral in (7.2). To this end we introduce the following definitions:

**Definition 7.1.**    The function $f(t)$ is said to be of *exponential order* as $t \to \infty$ if there exists a constant $\alpha$ such that

$$e^{-\alpha t}|f(t)|$$

is bounded for all $t$ greater than some $T$.
Thus if this is true there also exists a constant $M$ such that

$$|f(t)| < Me^{\alpha t}, \qquad t > T.$$

[An alternative statement is that $f(t)$ is of the order of $e^{\alpha t}$ as $t \to \infty$.]

**Definition 7.2.**    The function $f(t)$ is said to be piecewise continuous on $[a, b]$ if this interval may be subdivided into a finite number of subintervals such that

(i)  $f(t)$ is continuous in each subinterval;
(ii) $f(t)$ has finite limits as $t$ approaches either endpoint of each subinterval from the interior.

An example of a piecewise continuous function is

$$u_a(t) = \begin{cases} 0, & 0 \le t < a \\ 1, & t > a \end{cases},$$

which is shown in Figure 7.1.  [$u_a(t)$ is called the *unit step function*.]

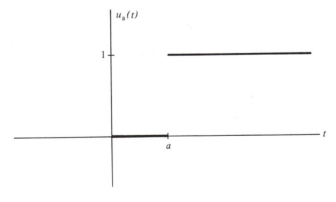

**Figure 7.1**   Graph of the unit step function.

We note that the only discontinuities of piecewise continuous functions are jump discontinuities as in Figure 7.1; that is, there are no "infinite discontinuities" as in $f(t) = 1/(t - 1)$ at $t = 1$. We state, without proof, the following existence theorem.

**Theorem 7.3.**   Let $f(t)$ be piecewise continuous on every finite subinterval of the domain $t \geq 0$ and be of exponential order as $t \to \infty$. Then the Laplace transform of $f(t)$ exists.

All the functions treated in Section 7.1 are continuous on every finite subinterval of $[0, \infty)$, hence piecewise continuous. They also are of exponential order as $t \to \infty$, so the transforms do exist and, in fact, are as calculated.

For the unit step function of Figure 7.1, we may choose $\alpha = 0$ and $M = 2$ in Definition 7.1, since $|u_a(t)| < 2$ for all $t > 1$, so $u_a(t)$ is the order $e^0$. $u_a(t)$ is also piecewise continuous, so that by Theorem 7.3 $\mathcal{L}\{u_a(t)\}$ exists. It is computed using the definition as

$$\mathcal{L}\{u_a(t)\} = \int_0^\infty e^{-st} u_a(t) \; dt$$

$$= \int_0^a e^{-st}(0) \; dt + \int_a^\infty e^{-st}(1) \; dt.$$

Hence

$$\mathcal{L}\{u_a(t)\} = \int_a^\infty e^{-st} \; dt = \frac{e^{-as}}{s}.$$

Similarly, let $f(t)$ be defined by

$$f(t) = \begin{cases} -2, & 0 \leq t < 1 \\ 1, & 1 \leq t < 3 \\ \dfrac{e^{2t}}{400}, & t \geq 3 \end{cases}$$

as in Figure 7.2. $f(t)$ is piecewise continuous on every finite subinterval of $[0, \infty)$ and of exponential order 2 for $t > 3$ [note that $e^{-2t}f(t) = \frac{1}{400}$ for $t > 3$]. Thus $\mathcal{L}\{f(t)\}$ exists and

$$\mathcal{L}\{f(t)\} = -2\int_0^1 e^{-st} \; dt + \int_1^3 e^{-st} \; dt + \frac{1}{400}\int_3^\infty e^{-st}e^{2t} dt$$

$$= \frac{2}{s}e^{-st}\Big|_0^1 - \frac{1}{s}e^{-st}\Big|_1^3 + \frac{e^{(2-s)t}}{400(2-s)}\Big|_3^\infty$$

$$= \frac{2e^{-s}}{s} - \frac{2}{s} - \frac{e^{-3s}}{s} + \frac{e^{-s}}{s} - \frac{e^{(2-s)3}}{400(2-s)}, \qquad s > 2.$$

Thus

$$\mathcal{L}\{f(t)\} = -\frac{e^{-3s}}{s} + \frac{3e^{-s}}{s} - \frac{2}{s} + \frac{e^6 e^{-3s}}{400(s-2)}, \qquad s > 2.$$

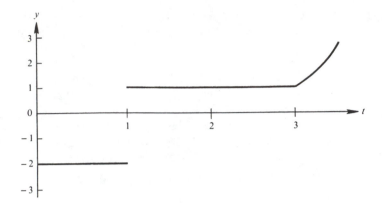

**Figure 7.2**  Graph of $f(t)$

## EXERCISES

1. Let $f(t)$ and $g(t)$ be of exponential order as $t \to \infty$. Show that their sum, $f(t) + g(t)$, also is of exponential order as $t \to \infty$.

2. Let $f(t)$ and $g(t)$ be piecewise continuous on $[a, b]$. Show that $f(t) + g(t)$ also is piecewise continuous on $[a, b]$.

3. Using the results of exercises 1 and 2 show that $\mathcal{L}\{f + g\} = \mathcal{L}\{f\} + \mathcal{L}\{g\}$.

Find the Laplace transforms of the following.

4. $f(t) = -1[u_0(t) - u_2(t)] + u_2(t) = -u_0(t) + 2u_2(t) = \begin{cases} -1, & 0 \leq t < 2 \\ +1, & t \geq 2 \end{cases}$.

5. $f(t) = u_0(t) - u_1(t) + t[u_1(t) - u_2(t)] + 2u_2(t) = \begin{cases} 1, & 0 \leq t < 1 \\ t, & 1 \leq t < 2 \\ 2, & t \geq 2 \end{cases}$.

6. Which of the following functions are piecewise continuous for $0 \leq t \leq 100$?
   (a) $f(t) = t^{-1}$.     (b) $f(t) = (t + 1)^{-1}$.     (c) $f(t) = (t - 2)^{-1}$.
   (d) $f(t) = [\cos(t) + 1]^{-1}$.     (e) $f(t) = [\cos(t) - 2]^{-1}$.
   (f) $f(t) = \tan t$.     (g) $f(t) = \begin{cases} t^2, & 0 \leq t \leq 1 \\ 2t^{-1}, & 1 \leq t \leq 100 \end{cases}$
   (h) $f(t) = \begin{cases} \tan t, & 0 \leq t \leq 1 \\ (2t - 1)^{-1} & 1 < t \leq 100 \end{cases}$

7. Find the values for $\alpha$, $M$, and $T$ in Definition 7.1 for the following functions.
   (a) $f(t) = e^{3t} \cos 2t$.     (b) $f(t) = \begin{cases} 16, & 0 \leq t \leq 10 \\ 100t^{-1}, & 10 < t \end{cases}$
   (c) $f(t) = e^{7t}$.     (d) $f(t) = 2t$.

8. Show that $e^{t^2}$ is not of exponential order.

9. (a) Show that the function $f(t) = 3 \sin(e^{t^2})$ is of exponential order but its derivative $f'(t)$ is not.
   (b) List at least two other functions with this same property.

## 7.3 THE FUNDAMENTAL OPERATIONAL PROPERTY OF THE LAPLACE TRANSFORM

We now turn to the question of how we calculate the Laplace transform of a derivative. Using the definition and integrating by parts gives

$$\mathcal{L}\{f'(t)\} = \int_0^\infty e^{-st} f'(t)\, dt$$
$$= \lim_{b \to \infty} \left[ e^{-st} f(t) \Big|_0^b + s \int_0^b e^{-st} f(t)\, dt \right]. \tag{7.10}$$

In the limit, the integral appearing on the right-hand side of (7.10) is just $\mathcal{L}\{f(t)\}$, so we concentrate on evaluating the other term,

$$\lim_{b \to \infty} e^{-sb} f(b) - e^0 f(0). \tag{7.11}$$

By the properties of absolute values and limits we have

$$\lim_{b \to \infty} \left| e^{-sb} f(b) \right| \le \lim_{b \to \infty} \left| e^{-sb} \right| |f(b)|.$$

Now since $f(b)$ is of the order $e^{\alpha b}$ as $b \to \infty$, for $b > T$ in Definition 7.1,

$$\lim_{b \to \infty} \left| e^{-sb} \right| |f(b)| < \lim_{b \to \infty} e^{-sb} M e^{\alpha b} = \lim_{b \to \infty} M e^{-(s-\alpha)b}.$$

If $s > \alpha$, then $s - \alpha > 0$, and we have that

$$\lim_{b \to \infty} M e^{-(s-\alpha)b} = 0,$$

so $\lim_{b \to \infty} e^{-sb} f(b) = 0$ also. In this case we have

$$\mathcal{L}\{f'(t)\} = s \mathcal{L}\{f(t)\} - f(0). \tag{7.12}$$

We collect these results in the following theorem.

**Theorem 7.4**   If

(i) $f(t)$ is continuous and possesses a piecewise continuous derivative, and
(ii) $f(t)$ and $f'(t)$ are of the order $e^{\alpha t}$ as $t \to \infty$, then $\mathcal{L}\{f'(t)\} = s \mathcal{L}\{f(t)\} - f(0)$ for all $s > \alpha$.

This property is the fundamental operational property of the Laplace transform. It may be applied recursively to yield the transforms of higher order derivatives [see (7.13)] and may be used to calculate the transforms of functions as in the following examples.

**EXAMPLE 7.1**   Let $f(t) = \cos^2 t$, so $f'(t) = -2 \cos t \sin t$ and $f(0) = 1$. If we now use the double-angle formula, $\sin 2t = 2 \sin t \cos t$, along with (7.12), we have

$$\mathcal{L}\{-\sin 2t\} = s\, \mathcal{L}\{\cos^2 t\} - 1.$$

Thus

$$\mathscr{L}\{\cos^2 t\} = \frac{1}{s}[-\mathscr{L}\{\sin 2t\} + 1]$$

$$= \frac{1}{s}\left[\frac{-2}{s^2 + 4} + 1\right] \qquad \text{(see exercise 8, Section 7.1)}$$

$$= \frac{s^2 + 2}{s(s^2 + 4)}.$$

If we formally use (7.12) to evaluate $\mathscr{L}\{f''(t)\}$, we get

$$\mathscr{L}\{f''(t)\} = s\mathscr{L}\{f'(t)\} - f'(0)$$
$$= s[s\mathscr{L}\{f(t)\} - f(0)] - f'(0)$$
$$= s^2\mathscr{L}\{f(t)\} - sf(0) - f'(0).$$

Repeated application of (7.12) yields for $\mathscr{L}\{f^{(n)}(t)\}$:

$$\mathscr{L}\{f^{(n)}(t)\} = s^n\mathscr{L}\{f(t)\} - s^{n-1}f(0) - s^{n-2}f'(0) \qquad (7.13)$$
$$- \cdots - sf^{(n-2)}(0) - f^{(n-1)}(0).$$

We may also prove this by induction. The property listed as (7.13) is useful in solving ordinary differential equations, as it converts a differential equation to an algebraic equation in the variable $s$.

**EXAMPLE 7.2**    Let $f(t) = t \sin t$ and use (7.13) to evaluate $\mathscr{L}\{t \sin t\}$. Since $f''(t)$ will contain the term $t \sin t$, we calculate

$$f'(t) = \sin t + t \cos t$$
$$f''(t) = 2 \cos t - t \sin t$$

and note that $f(0) = f'(0)$. Using (7.13) for $n = 2$ gives

$$\mathscr{L}\{2 \cos t - t \sin t\} = s^2\mathscr{L}\{t \sin t\}.$$

Thus using the linear property of the Laplace transform and rearranging gives

$$(1 + s^2)\mathscr{L}\{t \sin t\} = 2\mathscr{L}\{\cos t\} = \frac{2s}{s^2 + 1}$$

and

$$\mathscr{L}\{t \sin t\} = \frac{2s}{(s^2 + 1)^2}.$$

## EXERCISES

1.  Use (7.13) (with $n = 2$) to find the Laplace transform of
    (a) $\sin \omega t$.
    (b) $\cos \omega t$.

   (c)  sinh $\omega t$.
   (d)  cosh $\omega t$.
   (e)  $t^2$.
   (f)  $t \sin \omega t$.
   (g)  $t \cos \omega t$.
   (h)  $t \sinh t$.
   (i)  $t \cosh t$.
2.  Use Theorem 7.4 to
   (a)  Derive $\mathscr{L}\{\cos \omega t\}$ from $\mathscr{L}\{\sin \omega t\}$.
   (b)  Derive $\mathscr{L}\{\sinh \omega t\}$ from $\mathscr{L}\{\cosh \omega t\}$.
3.  Following the procedure of Example 7.1, find
   (a)  $\mathscr{L}\{\sin^2 t\}$.
   (b)  $\mathscr{L}\{\sinh^2 t\}$. (*Hint:* $2 \sinh^2 t + 1 = \cosh 2t$.)
   (c)  $\mathscr{L}\{\cosh^2 t\}$. (*Hint:* $\cosh^2 t - \sinh^2 t = 1$.)
4.  Let $f(t) = t^n$ in (7.13) to show that $\mathscr{L}\{t^n\} = n!/s^{n+1}$.


## 7.4 MORE PROPERTIES OF THE LAPLACE TRANSFORM

We will now develop some more properties of the Laplace transform so that we may simplify the use of the transform. Throughout this section we assume that the functions $f(t)$ and $g(t)$ satisfy the conditions for the existence of the transform (Theorem 7.3) whenever they are used.

The first property has already been used but is repeated here for the sake of completeness.

   **Property 1.**    $\mathscr{L}$ is a linear operator.

This property, as previously noted, is a direct consequence of the linear properties of the integral appearing in the definition. In terms of the operator $\mathscr{L}$, it takes the form

$$\mathscr{L}\{af(t) + bg(t)\} = a\mathscr{L}\{f(t)\} + b\mathscr{L}\{g(t)\},$$

where $a$ and $b$ are aribtrary constants.

   **Property 2.**    The shifting property. This property tells us that

$$\mathscr{L}\{e^{at}f(t)\} = F(s - a),$$

where

$$F(s) = \mathscr{L}\{f(t)\}. \tag{7.14}$$

Thus multiplication of a function by $e^{at}$ shifts its transform to $F(s - a)$.

This property is easily obtained from the definition, since for $s > a$,

$$\mathscr{L}\{e^{at}f(t)\} = \int_0^\infty e^{-st} e^{at} f(t)\ dt$$

$$= \int_0^\infty e^{-(s-a)t} f(t)\ dt. \tag{7.15}$$

In (7.15) we are free to replace $s - a$ by $\sigma$, obtaining

$$\mathcal{L}\{e^{at}f(t)\} = \int_0^\infty e^{-\sigma t}f(t)\ dt = F(\sigma), \qquad (7.16)$$

but $F(\sigma) = F(s - a)$, so the result is established.

**Property 3.**   If $\mathcal{L}\{f(t)\} = F(s)$, then for $n$ an integer, $n \geq 0$, we have

$$\mathcal{L}\{t^n f(t)\} = (-1)^n F^{(n)}(s), \qquad (7.17)$$

where $F^{(n)}(s)$ denotes the $n$th derivative of $F(s)$ with respect to $s$.

To establish this property we note that if it is legal to interchange the order of integration and differentiation, we may differentiate the expression

$$F(s) = \mathcal{L}\{f(t)\} = \int_0^\infty e^{-st}f(t)\ dt$$

to obtain

$$F'(s) = \int_0^\infty (-t)e^{-st}f(t)\ dt,$$

where the prime denotes differentiation with respect to $s$.
    Now

$$\int_0^\infty (-t)e^{-st}f(t)\ dt = -\mathcal{L}\{tf(t)\},$$

so that

$$\mathcal{L}\{tf(t)\} = -F'(s) = -\frac{d}{ds}[\mathcal{L}\{f(t)\}].$$

If a second differentiation is legal, we obtain

$$F''(s) = \int_0^\infty t^2 e^{-st}f(t)\ dt$$

$$= \mathcal{L}\{t^2 f(t)\}.$$

It is easy to see that repeated differentiations formally lead to (7.17) (or you may prove this using mathematical induction).

**Property 4.**   If $\mathcal{L}\{f(t)\} = F(s)$ for $s > \alpha$, and $f(t)/t$ is bounded for $t > 0$, then

$$\mathcal{L}\{\frac{1}{t}f(t)\} = \int_s^\infty F(u)\ du, \qquad s > \alpha. \qquad (7.18)$$

To establish (7.18) we write

$$F(u) = \int_0^\infty e^{-ut}f(t)\ dt$$

and integrate both sides from $s$ to $\infty$. Thus we have

$$\int_s^\infty F(u)\,du = \int_s^\infty \int_0^\infty e^{-ut} f(t)\,dt\,du. \tag{7.19}$$

On the right-hand side of (7.19) if we may interchange the order of integration, we obtain

$$\int_s^\infty F(u)\,du = \int_0^\infty \left( \int_s^\infty e^{-ut}\,du \right) f(t)\,dt$$

or, upon carrying out the $u$ integration,

$$\int_s^\infty F(u)\,du = \int_0^\infty e^{-st}\frac{1}{t}f(t)\,dt. \tag{7.20}$$

Since the right-hand side of (7.20) is just $\mathcal{L}\{f(t)/t\}$, (7.18) is formally established.

**Property 5.**    If $\mathcal{L}\{f(t)\} = F(s)$, then

$$\mathcal{L}\left\{ \int_0^t f(\tau)\,d\tau \right\} = \frac{1}{s}F(s). \tag{7.21}$$

This is established using the fundamental operational property as applied to the function $g(t)$ defined by

$$g(t) = \int_0^t f(\tau)\,d\tau. \tag{7.22}$$

Thus

$$\mathcal{L}\{g'(t)\} = s\mathcal{L}\{g(t)\} - g(0) = s\mathcal{L}\{g(t)\}. \tag{7.23}$$

Differentiation of (7.22) yields

$$g'(t) = f(t),$$

so that (7.23) becomes

$$\mathcal{L}\{f(t)\} = s\mathcal{L}\{g(t)\},$$

from which (7.21) follows trivially.

**Property 6.**    If $\mathcal{L}\{f(t)\} = F(s)$, then $\mathcal{L}\{u_a(t)f(t-a)\} = e^{-as}F(s)$.

To derive this property we use the definition of the unit step function to write

$$u_a(t)f(t-a) = \begin{cases} 0, & 0 < t < a \\ f(t-a), & a < t \end{cases}.$$

Thus

$$\mathcal{L}\{u_a(t)f(t-a)\} = \int_0^a e^{-st}(0)\,dt + \int_a^\infty e^{-st}f(t-a)\,dt$$

and the change of variable $t - a = \tau$ in the second integral gives

$$\mathscr{L}\{u_a(t)f(t - a)\} = \int_0^\infty e^{-s(\tau + a)}f(\tau)\, d\tau$$

$$= e^{-sa} \int_0^\infty e^{-s\tau}f(\tau)\, d\tau = e^{-sa}F(s).$$

The use of Properties 1 through 6 is illustrated in the following examples.

**EXAMPLE 7.3**    Let $f(t) = 3te^{2t}$. Then by Property 1, $\mathscr{L}\{f(t)\} = 3\mathscr{L}\{te^{2t}\}$. Since $\mathscr{L}\{t\} = 1/s^2$, the shifting property, Property 2, gives $\mathscr{L}\{te^{2t}\} = 1/(s - 2)^2$. We thus have $\mathscr{L}\{3te^{2t}\} = 3/(s - 2)^2$.

**EXAMPLE 7.4**    Let $f(t) = e^{at} \cos \omega t$. Since

$$\mathscr{L}\{\cos \omega t\} = \frac{s}{s^2 + \omega^2},$$

from Property 2 we have

$$\mathscr{L}\{e^{at} \cos \omega t\} = \frac{s - a}{(s - a)^2 + \omega^2}.$$

**EXAMPLE 7.5**    Since $\mathscr{L}\{\sin kt\} = k/(s^2 + k^2)$ and

$$\frac{d^2}{ds^2}\left[\frac{k}{s^2 + k^2}\right] = \frac{2k(3s^2 - k^2)}{(s^2 + k^2)^3},$$

Property 3 gives

$$\mathscr{L}\{t^2 \sin kt\} = (-1)^2 \frac{d^2}{ds^2} \mathscr{L}(\sin kt)$$

$$= \frac{2k(3s^2 - k^2)}{(s^2 + k^2)^3}. \tag{7.25}$$

**EXAMPLE 7.6**    Let $f(t) = \cosh kt \cos kt$. Since $\cosh kt = (e^{kt} + e^{-kt})/2$ we have

$$f(t) = \frac{1}{2} e^{kt} \cos kt + \frac{1}{2} e^{-kt} \cos kt.$$

Using the fact that the Laplace transform is linear (Property 1) we have

$$\mathscr{L}\{f(t)\} = \frac{1}{2} \mathscr{L}\{e^{kt} \cos kt\} + \frac{1}{2} \mathscr{L}\{e^{-kt} \cos kt\}. \tag{7.26}$$

Now $\mathscr{L}\{\cos kt\} = s/(s^2 + k^2)$, so applying the shifting property (Property 2) to the transforms appearing on the right-hand side of (7.26) yields

$$\mathscr{L}\{f(t)\} = \frac{1}{2} \frac{s - k}{(s - k)^2 + k^2} + \frac{1}{2} \frac{s + k}{(s + k)^2 + k^2} = \frac{s^3}{s^4 + 4k^4}.$$

**EXAMPLE 7.7**    Let $f(t) = 3e^{-4t}\{\cos 4t - t \sin 4t\}$. Then

$$\mathcal{L}\{f(t)\} = 3\mathcal{L}\{e^{-4t} \cos 4t\} - 3\mathcal{L}\{e^{-4t}t \sin 4t\}. \tag{7.27}$$

For the first transform on the right-hand side of (7.27), we have from (7.24)

$$\mathcal{L}\{e^{-4t} \cos 4t\} = \frac{s + 4}{(s + 4)^2 + 4^2}. \tag{7.28}$$

The second transform, $\mathcal{L}\{e^{-4t}t \sin 4t\}$, may be approached in different ways. We may apply the shifting property to $\mathcal{L}\{t \sin 4t\}$ or we may apply Property 3 to $\mathcal{L}\{e^{-4t} \sin t\}$. Note that we are using Properties 2 and 3 in either case; only the order in which they are applied is changed.
Since

$$\mathcal{L}\{e^{-4t} \sin 4t\} = \frac{4}{(s + 4)^2 + 4^2}$$

by the shifting property, using Property 3 gives

$$\mathcal{L}\{te^{-4t} \sin 4t\} = -\frac{d}{ds}\left(\frac{4}{(s + 4)^2 + 4^2}\right)$$

$$= \frac{8(s + 4)}{[(s + 4)^2 + 4^2]^2}. \tag{7.29}$$

Substituting (7.28) and (7.29) into (7.27) gives

$$\mathcal{L}\{te^{-4t} \sin 4t\} = \frac{3(s + 4)}{(s + 4)^2 + 4^2} - \frac{24(s + 4)}{[(s + 4)^2 + 4^2]^2}.$$

**EXAMPLE 7.8**    Let $f(t) = (e^t - e^{-t})/t$; then

$$\mathcal{L}\{f(t)\} = \mathcal{L}\left\{\frac{e^t - e^{-t}}{t}\right\} = \mathcal{L}\left\{\frac{2 \sinh t}{t}\right\}.$$

By Property 4, we have

$$\mathcal{L}\left\{\frac{2 \sinh t}{t}\right\} = \int_s^\infty \frac{2}{u^2 - 1} \, du$$

$$= \lim_{b \to \infty} \int_s^b \left[\frac{1}{u - 1} - \frac{1}{u + 1}\right] du.$$

Thus

$$\mathcal{L}\left\{\frac{e^t - e^{-t}}{t}\right\} = \lim_{b \to \infty} \left[\ln|u - 1| - \ln|u + 1|\right]\Big|_s^b$$

$$= \lim_{b \to \infty} \ln\left|\frac{u - 1}{u + 1}\right|\Big|_s^b = \ln\left|\frac{s + 1}{s - 1}\right|.$$

**EXAMPLE 7.9**    Let $\mathcal{L}\{f(t)\} = k/[s(s^2 + k^2)]$. We will use Property 5 to invert this transform. Since $k/(s^2 + k^2) = \mathcal{L}(\sin kt)$ we have

$$\frac{k}{s(s^2 + k^2)} = \frac{1}{s}\,\mathcal{L}\{\sin kt\}. \tag{7.30}$$

According to Property 5,

$$\frac{1}{s}\,\mathcal{L}\{f(t)\} = \mathcal{L}\{\int_0^t F(\tau)\,d\tau\},$$

so that (7.30) implies that

$$\frac{k}{s(s^2 + k^2)} = \mathcal{L}\{\int_0^t \sin k\tau\,d\tau\}.$$

Now

$$\int_0^t \sin k\tau\,d\tau = -\frac{1}{k}\cos k\tau\,\Big|_0^t$$

$$= -\frac{1}{k}(\cos kt - 1) = \frac{1}{k}(1 - \cos kt).$$

Thus

$$\mathcal{L}^{-1}\left\{\frac{k}{s(s^2 + k^2)}\right\} = \frac{1}{k}(1 - \cos kt).$$

**EXAMPLE 7.10**    Here we see that the application of property 5 to the inversion of transforms may be applied recursively (see also exercise 28). We seek

$$\mathcal{L}^{-1}\left\{\frac{k}{s^2(s^2 + k^2)}\right\} = \mathcal{L}^{-1}\left\{\frac{1}{s}\frac{k}{s(s^2 + k^2)}\right\}.$$

In Example 7.9 we found that

$$\mathcal{L}^{-1}\left\{\frac{k}{s(s^2 + k^2)}\right\} = \frac{1}{k}(1 - \cos kt).$$

Thus, by Property 5,

$$\mathcal{L}^{-1}\left\{\frac{k}{s^2(s^2 + k^2)}\right\} = \int_0^t \frac{1}{k}(1 - \cos k\tau)d\tau$$

$$= \frac{kt - \sin kt}{k^2}.$$

**EXAMPLE 7.11**    Find the Laplace transform of the function shown in Figure 7.3 and given by

$$f(t) = u_{\pi/2}(t)\cos t + u_{5\pi/2}(t)\left(t - \frac{5\pi}{2}\right).$$

Property 6 requires functions of the form $u_a(t)f(t - a)$, so we use properties of

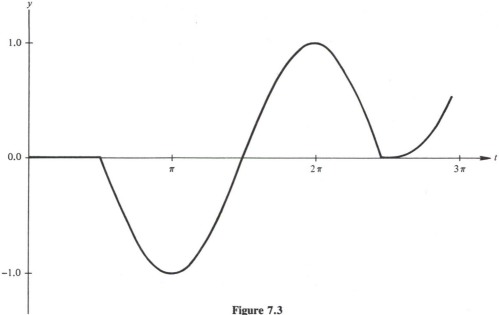

**Figure 7.3**

the trigonometric functions to write

$$\cos t = \sin\left(\frac{\pi}{2} - t\right) = -\sin\left(t - \frac{\pi}{2}\right).$$

Thus combining the linear property (Property 1) with Property 6 gives

$$\mathcal{L}\{f(t)\} = -\mathcal{L}\left\{u_{\pi/2}(t) \sin\left(t - \frac{\pi}{2}\right)\right\} + \mathcal{L}\left\{u_{5/2}(t)\left(t - \frac{5\pi}{2}\right)\right\}$$

$$= -e^{-s\pi/2}\frac{1}{s^2 + 1} + e^{-s5\pi/2}\frac{1}{s^2}.$$

As an aid in working the exercises for this section, we collect the properties from this section and Section 7.3 in Table 7.1.

**TABLE 7.1**  SOME PROPERTIES OF THE LAPLACE TRANSFORM

1. $\mathcal{L}\{af(t) + bg(t)\} = a\mathcal{L}\{f(t)\} + b\mathcal{L}\{g(t)\}$

2. $\mathcal{L}\{e^{at}f(t)\} = F(s - a)$

3. $\mathcal{L}\{t^n f(t)\} = (-1)^n F^{(n)}(s)$

4. $\mathcal{L}\left\{\frac{1}{t}f(t)\right\} = \int_s^\infty F(u)\, du$

5. $\mathcal{L}\{\int_0^t f(\tau)\, d\tau\} = \frac{1}{s}F(s)$

6. $\mathcal{L}\{u_a(t)f(t - a)\} = e^{-as}F(s)$

7. $\mathcal{L}\{f^{(n)}(t)\} = s^n \mathcal{L}\{f(t)\} - \sum_{j=0}^{n-1} s^{(n-j-1)}f^{(j)}(0)$

## EXERCISES

Find the Laplace transform of each of the following functions.

**1.** $e^{at} t^2$

**2.** $e^{at} \sin \omega t$

**3.** $e^{at} \sinh kt$

**4.** $e^{at} \cosh kt$

**5.** $t^2 \sinh kt$

**6.** $t^2 \cosh kt$

**7.** $e^{at} t^2 \sinh kt$

**8.** $e^{at} t^2 \cosh kt$

**9.** $\cos^2 kt$

**10.** $e^{kt} \sin^2 kt$

**11.** $t^2 \cos kt$

**12.** $\sin^3 t \left[ Hint: \sin^3 t = \dfrac{3 \sin t - \sin 3t}{4} \right]$

**13.** $\cos^3 t \left[ Hint: \cos^3 t = \dfrac{\cos 3t + 3 \cos t}{4} \right]$

**14.** $\sin^2 t \cos t$

**15.** $\cos^2 t \sin t$

**16.** $u_3(t) \cos(t - 3)$

**17.** $u_\pi(t) \sin t = \begin{cases} 0, & 0 < t < \pi \\ \sin t, & \pi < t \end{cases}$.

**18.** $u_\pi(t) \cos t$

**19.** $u_1(t)(t - 1)$

**20.** $u_2(t)(t + 3)$

**21.** $u_2(t)(t^2 + 2t - 4)$ [*Hint:* Write $t^2 + 2t - 4$ as a polynomial in powers of $(t - 2)$.]

**22.** $u_1(t)(t^3 - 1)$

**23.** Given that $f''(t) + f'(t) + f(t) = 0$ and that $f(0) = f'(0) = 1$, find $f(t)$.

**24. (a)** Given that $f'(t) + \int_0^t f(\tau) \, d\tau = 0$ and that $f(0) = 0$, find $f(t)$.
   **(b)** Given that $f(t) + \int_0^t f(\tau) \, d\tau = e^{-t}$, find $f(t)$.

**25.** In equation (7.17) choose $f(t) = 1$ to show that

$$\mathcal{L}\{t^n\} = \frac{n!}{s^{n+1}}.$$

**26. (a)** Show that $\mathcal{L}\{f(t) \cosh at\} = \frac{1}{2}[F(s + a) + F(s - a)]$.
   **(b)** Use the result of part (a) to find $\mathcal{L}\{\sinh bt \cosh at\}$.

**27.** Use Property 4 to evaluate
   **(a)** $\mathcal{L}\left\{ \dfrac{\sin \omega t}{t} \right\}$.
   **(b)** $\mathcal{L}\left\{ \dfrac{e^{at} - e^{bt}}{t} \right\}$.

**28.** Derive the result that $\mathcal{L}\{\int_0^t \int_0^\tau f(u) \, du \, d\tau\} = (1/s^2)F(s)$. (*Hint:* Use Property 5.)

**29.** Prove that if $f(t)$ is piecewise continuous and of exponential order that $\lim_{s \to \infty} F(s) = 0$.

(*Hint:* Write

$$\mathcal{L}\{f(t)\} = \int_0^T e^{-st}f(t)\ dt + \int_T^\infty e^{-st}f(t)\ dt,$$

where $T$ is from Definition 7.1, and use the facts that $\left|e^{-st}f(t)\right|$ is bounded for $t > T$ and that the absolute value of an integral is less than or equal to the integral of the absolute value of the integrand.)

**30.** Show that $\mathcal{L}\{tf'(t)\} = -sF'(s) - F(s)$ and $\mathcal{L}\{t^2f''(t)\} = s^2F''(s) + 4sF'(s) + 2F(s)$.

**31.** Solve the initial value problem $f''(t) + tf'(t) - 2f(t) = 1, f(0) = f'(0) = 0$ in the following manner.

   **(a)** Show that taking the Laplace transform of the differential equation and using the results of exercise 30 gives

$$F'(s) + \left(\frac{3}{s} - s\right)F(s) = \frac{-1}{s^2}.$$

   **(b)** Show that this differential equation for $F(s)$ has an integrating factor of $s^3 e^{-s^2/2}$ and a solution of

$$F(s) = \frac{1}{s^3} + Cs^{-3}e^{s^2/2}.$$

   **(c)** Evaluate the arbitrary constant in part (b) using exercise 29 and find the resulting $f(t)$. Verify that your answer is indeed a solution.

**32.** Solve the initial value problem $f''(t) - 2tf'(t) + 2f(t) = 0, f(0) = 0, f'(0) = 1$.

**33.** For what value of the constant $a$ will the differential equation resulting from the Laplace transform of

$$t^2f''(t) - atf'(t) + 3f(t) = 0$$

be of the same form as the original differential equation?

**34.** Property 3 (with $n = 1$) may also be used to invert Laplace transforms by writing it as $tf(t) = -\mathcal{L}^{-1}\{F'(s)\}$. Thus if we know the inverse transform of $F'(s)$, we also know the inverse transform of $F(s)$. Use this property to find

   **(a)** $\mathcal{L}^{-1}\left\{\ln\left|\frac{s-a}{s+b}\right|\right\}$.

   **(b)** $\mathcal{L}^{-1}\left\{\ln\left(1 + \frac{1}{s^2}\right)\right\}$.

   **(c)** $\mathcal{L}^{-1}\left\{\frac{\pi}{2} - \tan^{-1}\frac{s}{2}\right\}$.

## 7.5  THE CONVOLUTION THEOREM

The convolution of two functions $f(t)$ and $g(t)$, denoted $f(t) * g(t)$, is given by

$$f(t) * g(t) = \int_0^t f(\tau)g(t - \tau)\ d\tau$$

whenever the integral is defined. The product of Laplace transforms is related to the convolution as follows.

**Theorem 7.5.**　　Let $F(s) = \mathcal{L}\{f(t)\}$ and $G(s) = \mathcal{L}\{g(t)\}$, where $f(t)$ and $g(t)$ are piecewise continuous and of exponential order $\alpha$ as $t \to \infty$. Then for $s > \alpha$ we have

$$\mathcal{L}^{-1}\{F(s)\,G(s)\} = f(t) * g(t).$$

This is referred to as the convolution theorem.

We may use this theorem to find the inverse transform of

$$\frac{s}{(s^2 - a^2)(s - b)}, \qquad a \neq b$$

as follows. Since $s/(s^2 - a^2) = \mathcal{L}\{\cosh at\}$ and $1/(s - b) = \mathcal{L}\{e^{bt}\}$, we may take $F(s) = s/(s^2 - a^2)$, $f(t) = \cosh at$, $G(s) = 1/(s - b)$, $g(t) = e^{bt}$. Then

$$
\begin{aligned}
\mathcal{L}^{-1}\left\{\frac{s}{(s^2 - a^2)(s - b)}\right\} &= (\cosh at) * e^{bt} \\
&= \int_0^t \cosh a\tau \, e^{b(t-\tau)} \, d\tau \\
&= e^{bt} \int_0^t \left(\frac{e^{a\tau} + e^{-a\tau}}{2}\right) e^{-b\tau} \, d\tau \\
&= \frac{e^{bt}}{2} \int_0^t \left[e^{(a-b)\tau} + e^{-(a+b)\tau}\right] d\tau \\
&= \frac{e^{bt}}{2} \left[\frac{e^{(a-b)t}}{a - b} - \frac{e^{-(a+b)t}}{a + b} - \frac{2b}{a^2 - b^2}\right] \\
&= \frac{1}{2}\left(\frac{e^{at}}{a - b} - \frac{e^{-at}}{a + b} - \frac{2be^{bt}}{a^2 - b^2}\right).
\end{aligned}
$$

[An alternative way is to use the fact that

$$\int e^{-b\tau} \cosh a\tau \, d\tau = \frac{e^{-b\tau}}{a^2 - b^2}(a \sinh a\tau + b \cosh a\tau).]$$

The substitution $\lambda = t - \tau$ when made in the defining integral itself gives

$$f(t) * g(t) = \int_0^t f(\tau)g(t - \tau) \, d\tau = \int_0^t f(t - \lambda)g(\lambda) \, d\lambda$$

$$= \int_0^t g(\lambda)f(t - \lambda) \, d\lambda = g(t) * f(t),$$

so that the convolution operation is commutative. It is also distributive with respect to addition, and associative. Further examples of the use of the convolution theorem in the inversion of Laplace transforms follow.

**EXAMPLE 7.12**　　Let

$$F(s) = \frac{1}{s^2 + 4s + 4}.$$

We factor the denominator to obtain

$$F(s) = \frac{1}{(s + 2)(s + 2)} = \frac{1}{s + 2} \cdot \frac{1}{s + 2}.$$

Now $\mathcal{L}^{-1}\{1/(s + 2)\} = e^{-2t}$. Hence

$$\mathcal{L}^{-1}\left\{\frac{1}{(s + 2)^2}\right\} = e^{-2t} * e^{-2t}$$

$$= \int_0^t e^{-2\tau} e^{-2(t-\tau)} \, d\tau$$

$$= \int_0^t e^{-2t} \, d\tau = te^{-2t}.$$

**EXAMPLE 7.13**    Let

$$F(s) = \frac{1}{s^2 + 4s - 5}.$$

Again we factor the denominator to obtain

$$F(s) = \frac{1}{(s + 5)(s - 1)} = \frac{1}{s + 5} \cdot \frac{1}{s - 1}.$$

Now

$$\mathcal{L}^{-1}\left\{\frac{1}{s + 5}\right\} = e^{-5t} \quad \text{and} \quad \mathcal{L}^{-1}\left\{\frac{1}{s - 1}\right\} = e^{t};$$

hence

$$\mathcal{L}^{-1}\left\{\frac{1}{s^2 + 4s - 5}\right\} = e^{-5t} * e^{t}.$$

Since

$$e^{-5t} * e^{t} = \int_0^t e^{-5\tau} e^{t-\tau} \, d\tau$$

$$= \int_0^t e^{t} e^{-6\tau} \, d\tau = e^{t}\left[\frac{1}{-6}e^{-6\tau}\right]\Bigg|_0^t$$

$$= \frac{-1}{6}e^{-5t} + \frac{1}{6}e^{t},$$

we have that

$$\mathcal{L}^{-1}\left\{\frac{1}{s^2 + 4s - 5}\right\} = \frac{-1}{6}e^{-5t} + \frac{1}{6}e^{t}.$$

**EXAMPLE 7.14**    Let

$$F(s) = \frac{k}{(s - 1)(s^2 + k^2)}.$$

We may write

$$F(s) = \frac{1}{s-1} \cdot \frac{k}{s^2+k^2},$$

and since

$$\frac{1}{s-1} = \mathscr{L}\{e^t\} \quad \text{and} \quad \frac{k}{s^2+k^2} = \mathscr{L}\{\sin kt\},$$

we have

$$\mathscr{L}^{-1}\left\{\frac{k}{(s-1)(s^2+k^2)}\right\} = e^t * \sin kt$$

$$= \sin kt * e^t$$

$$= \int_0^t \sin k\tau \, e^{t-\tau} \, d\tau$$

$$= e^t \int_0^t e^{-\tau} \sin k\tau \, d\tau.$$

The last integral of Example 7.14 needs a simple integration by parts and is left as exercise 9.

Theorem 7.2 gives conditions under which we can be sure that the Laplace transform of a function exists. However, there are functions which have Laplace transforms that do not satisfy these conditions. One such function is $f(t) = t^n$, where $-1 < n < 0$. If we use the definition to calculate its Laplace transform, we obtain

$$\mathscr{L}\{t^n\} = \int_0^\infty e^{-st} t^n \, dt$$

$$= \frac{1}{s^{n+1}} \int_0^\infty e^{-z} z^n \, dz,$$

with the change of the variable of integration $st = z$. Using the definition of the gamma function (see exercise 18) we may write

$$\mathscr{L}\{t^n\} = \frac{\Gamma(n+1)}{s^{n+1}}, \quad \text{for any } n > -1.$$

For $n = -\frac{1}{2}$ we have $\mathscr{L}\{t^{-1/2}\} = \Gamma(\frac{1}{2})s^{-1/2} = \sqrt{\pi/s}$ (see exercise 19) and for $n = \frac{1}{2}$, $\mathscr{L}\{t^{1/2}\} = \Gamma(\frac{3}{2})s^{-3/2} = (\sqrt{\pi}/2)s^{-3/2}$ (see exercise 20).

**EXAMPLE 7.15**    Here we use the result of the preceding paragraph together with the convolution theorem to evaluate $\mathscr{L}^{-1}\{1/[\sqrt{s}(s-1)]\}$. Taking $F(s) = 1/\sqrt{s}$, $G(s) = 1/(s-1)$, we have $f(t) = 1/\sqrt{\pi t}$ and $g(t) = e^t$, so

$$\mathscr{L}^{-1}\left\{\frac{1}{\sqrt{s}(s-1)}\right\} = \frac{1}{\sqrt{\pi t}} * e^t = \frac{1}{\sqrt{\pi}} \int_0^t \frac{1}{\sqrt{\tau}} e^{t-\tau} \, d\tau = \frac{e^t}{\sqrt{\pi}} \int_0^t \frac{e^{-\tau}}{\sqrt{\tau}} \, d\tau.$$

The change of variable $z = \sqrt{\tau}$ shows that

$$\int_0^t \frac{e^{-\tau}}{\sqrt{\tau}} \, d\tau = 2 \int_0^{\sqrt{t}} e^{-z^2} \, dz = \sqrt{\pi} \, \text{erf}(\sqrt{t}),$$

where $\text{erf}(\sqrt{t})$ is the error function with argument $\sqrt{t}$ (see exercise 21). Thus we have

$$\mathcal{L}^{-1}\left\{ \frac{1}{\sqrt{s}(s-1)} \right\} = e^t \, \text{erf}(\sqrt{t})$$

and by use of Property 2,

$$\mathcal{L}^{-1}\left\{ \frac{1}{\sqrt{s+1}\,s} \right\} = e^{-t}[e^t \, \text{erf}(\sqrt{t})] = \text{erf}(\sqrt{t}).$$

The Laplace transform may also be used to solve certain types of integral equations. One type, called a *Volterra integral equation*, has the form

$$y(t) = f(t) + \int_0^t K(t-\tau)y(\tau) \, d\tau,$$

where $f$ and $K$ are known and $y$ is to be determined. Applying the Laplace transform to the above equation gives

$$\mathcal{L}\{y(t)\} = \mathcal{L}\{f(t)\} + \mathcal{L}\{K(t) * y(t)\}$$

or by using the convolution theorem,

$$\mathcal{L}\{y(t)\} = \mathcal{L}\{f(t)\} + \mathcal{L}\{K(t)\}\mathcal{L}\{y(t)\}.$$

Solving for $\mathcal{L}\{y(t)]$ gives

$$\mathcal{L}\{y(t)\} = \frac{\mathcal{L}\{f(t)\}}{1 - \mathcal{L}\{K(t)\}}.$$

To find $y(t)$, for given $f$ and $K$, we need to evaluate the right-hand side of the above equation and then find the inverse Laplace transform of the result. Specific examples are given in the exercises.

## EXERCISES

Invert the following transforms.

**1.** $\dfrac{1}{s^2 - 3s + 2}$

**2.** $\dfrac{1}{s^2 + 2s + 1}$

**3.** $\dfrac{1}{s^3 + 2s^2 - s - 2}$

**4.** $\dfrac{1}{s^2(s+1)}$

**5.** $\dfrac{s}{(s-1)(s^2+k^2)}$

**6.** $\dfrac{1}{(s+1)(s^2+1)}$

**7.** $\dfrac{1}{s(s^2+4s+13)}$

**8.** $\dfrac{1}{s(s^2+16)}$

**9.** $\dfrac{k}{(s-1)(s^2+k^2)}$ (i.e., complete Example 7.14)

**10.** $\dfrac{1}{(s-b)(s^2-a^2)}$

Solve the following equations for $y(t)$.

**11.** $y(t) = 4t - 3 - \int_0^t \sin(t - \tau)y(\tau)\, d\tau.$
**12.** $y(t) = 3 \sin t - 2 \int_0^t \cos(t - \tau)y(\tau)\, d\tau.$
**13.** $y(t) = \sin t + 4e^{-t} - 2 \int_0^t \cos(t - \tau)y(\tau)\, d\tau.$
**14.** $y(t) = t - \int_0^t (t - \tau)y(\tau)\, d\tau.$
**15.** $y(t) = t - \int_0^t e^{t-\tau}y(\tau)\, d\tau.$
**16.** $y(t) = \cos t + \int_0^t e^{t-\tau}y(\tau)\, d\tau.$

**17. (a)** The behavior of an *RLC* circuit (as shown in Figure 7.4) can be described by an integrodifferential equation with the current as the dependent variable. (We used a second order differential equation for charge to describe this circuit in Chapter 4.) This integrodifferential equation is

$$L\frac{dI}{dt} + RI + \frac{1}{C}\int_0^t I(\tau)\, d\tau = E(t).$$

Take the Laplace transform of this equation and solve for $\mathcal{L}\{I(t)\}$ assuming that $I(0) = 0$.

**(b)** If $E(t)$ from part (a) is a constant, $E_0$, show that the inverse transform of your solution in part (a) gives

$$I(t) = \frac{E_0}{\omega L}e^{-Rt/(2L)} \sin \omega t, \qquad \omega^2 = \frac{1}{LC} - \left(\frac{R}{2L}\right)^2.$$

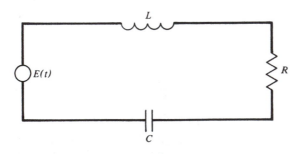

**Figure 7.4**

**18.** The gamma function, $\Gamma(n)$, is defined for $n > 0$ as

$$\Gamma(n) = \int_0^\infty t^{n-1} e^{-t}\, dt.$$

    **(a)** Show that $\Gamma(n + 1) = n\Gamma(n)$. (*Hint:* Use integration by parts.)
    **(b)** If $n$ is a positive integer, show that $\Gamma(1) = 1$ and $\Gamma(n + 1) = n!$.

**19.** Evaluate $\Gamma(\frac{1}{2})$ using the following steps.
    **(a)** Make a change of variable $x = \sqrt{t}$ in the integral expression for $\Gamma(\frac{1}{2})$ to find
$\Gamma(\frac{1}{2}) = 2 \int_0^\infty e^{-x^2}\, dx$.
    **(b)** Write $[\Gamma(\frac{1}{2})]^2$ as

$$\left[\Gamma\!\left(\frac{1}{2}\right)\right]^2 = \Gamma\!\left(\frac{1}{2}\right)\Gamma\!\left(\frac{1}{2}\right) = 4\int_0^\infty e^{-x^2}\, dx \int_0^\infty e^{-y^2}\, dy = 4\int_0^\infty\!\!\int_0^\infty e^{-x^2-y^2}\, dx\, dy.$$

    **(c)** Evaluate the double integral in part (b) using polar coordinates to obtain

$$\left[\Gamma\!\left(\frac{1}{2}\right)\right]^2 = \pi.$$

**20.** Use results of the previous exercises to show that
    **(a)** $\Gamma\!\left(\dfrac{3}{2}\right) = \dfrac{\sqrt{\pi}}{2}$, $\Gamma\!\left(\dfrac{5}{2}\right) = \dfrac{3\sqrt{\pi}}{4}$.

    **(b)** $\Gamma\!\left(\dfrac{2n + 1}{2}\right) = \dfrac{(2n - 1)\,!\sqrt{\pi}}{2^{2n-1}(n - 1)!}$.

**21.** The error function is defined by erf $x = (2/\sqrt{\pi}) \int_0^x e^{-z^2}\, dz$.
    **(a)** Use the results of exercise 19 to show that $\lim\limits_{x\to\infty} \text{erf } x = (1/\sqrt{\pi})\Gamma(\frac{1}{2}) = 1$.
    **(b)** Prove that erf $x$ is an odd function.
    **(c)** Prove that $\text{erf}(0) = 0$ and erf $x$ is an increasing function that is concave down for $x > 0$. Sketch a graph of $y = \text{erf } x$.
    **(d)** Show that the complementary error function, erfc $x = 1 - \text{erf } x$, is a decreasing function which is concave up. Find $\text{erfc}(0)$, $\lim_{x\to\infty} \text{erfc } x$ and sketch a graph of $y = \text{erfc } x$.

Since the Laplace transform is defined as an improper integral, we have all the techniques of calculus at our disposal when we compute such transforms. We use some of these techniques now as we calculate the Laplace transform of a function needed in Chapter 11.

**22.** Compute the Laplace transform of $f(t) = (1/\sqrt{t})e^{-1/t}$ by the following steps.
    **(a)** Show that

$$\mathscr{L}\{f(t)\} = \int_0^\infty \frac{1}{\sqrt{t}} e^{-st-1/t}\, dt = 2e^{-2\sqrt{s}} \int_0^\infty e^{-(\sqrt{s}z-1/z)^2}\, dz$$

    by letting $\sqrt{t} = z$ and completing the square.
    **(b)** Show that a second expression for $\mathscr{L}\{f(t)\}$ [obtained by making the change of variable $\sqrt{s}\,z = y^{-1}$ in the result of part (a)] is

$$\mathscr{L}\{f(t)\} = 2e^{-2\sqrt{s}} \int_0^\infty \frac{1}{\sqrt{s}\,y^2} e^{-(\sqrt{s}y-1/y)^2}\, dy.$$

    **(c)** Add the results of parts (a) and (b) to find that

$$\mathscr{L}\{f(t)\} = e^{-2\sqrt{s}} \int_0^\infty e^{-(\sqrt{s}y-1/y)^2}\left(1 + \frac{1}{\sqrt{s}\,y^2}\right) dy.$$

(d) Make the change of variable $\sqrt{s}\,y - 1/y = u$ to show that

$$\mathcal{L}\{f(t)\} = e^{-2\sqrt{s}} \int_{-\infty}^{\infty} e^{-u^2} \frac{1}{\sqrt{s}}\, du = \sqrt{\frac{\pi}{s}}\, e^{-2\sqrt{s}}.$$

**23.** Use Theorem 7.1 to show that

$$\mathcal{L}\left\{\frac{1}{\sqrt{\pi t}}\, e^{-a^2/(4t)}\right\} = \frac{1}{\sqrt{s}}\, e^{-a\sqrt{s}}.$$

**24.** Show that

$$\mathcal{L}^{-1}\left\{\frac{1}{s}\, e^{-a\sqrt{s}}\right\} = \operatorname{erfc}\left(\frac{a}{2\sqrt{t}}\right).$$

**25.** An outline of the proof of the convolution theorem uses the following steps.
  (a) Show that $\mathcal{L}\{f(t) * g(t)\} = \iint_R e^{-st} f(\tau) g(t - \tau)\, d\tau\, dt$, where $R$ is the region shown in Figure 7.5.

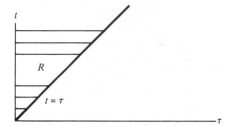

$t = \tau$

Figure 7.5

  (b) Make the change of variables

$$u = t - \tau,$$

$$v = \tau$$

  to show that

$$\mathcal{L}\{f(t) * g(t)\} = \int_0^\infty \int_0^\infty e^{-s(u+v)} f(v) g(u)\, du\, dv.$$

  (c) Write the integral in part (b) as the product of two integrals and note that this product contains the Laplace transforms of $f$ and $g$.
**26.** Obtain Property 5 of Section 7.4 by computing $\mathcal{L}\{f(t) * 1\}$.


## 7.6 SOLUTION OF SIMPLE ODEs

We now take our elementary knowledge of Laplace transforms and see if we can figure out how to solve simple differential equations. Actually, we shall restrict our attention at once to initial value problems, where the transform is most useful. We could solve many of these problems using the techniques from Chapter 4. The advantages of using Laplace transforms are that they handle nonhomogeneous equations and initial conditions automatically and are very convenient for use with discontinuous forcing functions.

**EXAMPLE 7.16**    Suppose that we are given the simple initial value problem which models the motion of a spring–mass system or a pendulum,

$$x''(t) + \omega^2 x(t) = 0, \qquad x(0) = x_0, \quad x'(0) = v_0, \qquad (7.31)$$

where $\omega^2$ is taken to be constant. $\omega^2 = k/m$ for a spring–mass system or $g/\ell$ for a pendulum. If we take the Laplace transform of (7.31), we obtain

$$\mathscr{L}\{x''(t)\} + \omega^2 \mathscr{L}\{x(t)\} = 0. \qquad (7.32)$$

By the fundamental operational property [see (7.13)],

$$\begin{aligned} \mathscr{L}\{x''(t)\} &= s^2 \mathscr{L}\{x(t)\} - sx(0) - x'(0) \\ &= s^2 \mathscr{L}\{x(t)\} - x_0 s - v_0. \end{aligned} \qquad (7.33)$$

Substitution from (7.33) into (7.32) yields

$$s^2 \mathscr{L}\{x(t)\} - x_0 s - v_0 + \omega^2 \mathscr{L}\{x(t)\} = 0. \qquad (7.34)$$

We now solve (7.34) for $\mathscr{L}\{x(t)\}$, obtaining

$$\mathscr{L}\{x(t)\} = x_0 \frac{s}{s^2 + \omega^2} + v_0 \frac{1}{s^2 + \omega^2}. \qquad (7.35)$$

Taking the inverse transform of (7.35), using results from Table 7.2, we have

$$x(t) = x_0 \cos \omega t + \frac{v_0}{\omega} \sin \omega t.$$

Here we see an advantage of the Laplace transform in the solution of an initial value problem, namely, that the solution is obtained directly without having first to find a general solution and then evaluate the integration constants.

**EXAMPLE 7.17**    In Section 3.2 we solved the following example involving the unit step function

$$\frac{dx}{dt} + \frac{x}{a} = \frac{1}{a} u_1(t) = \begin{cases} 0, & 0 < t < 1 \\ 1/a, & 1 < t \end{cases}, \quad x(0) = A. \qquad (7.36)$$

The technique there involved finding general solutions of the differential equation in two separate regions ($0 < t < 1$, $t > 1$) and then requiring that the solution be continuous at $t = 1$. Notice that such difficulties do not occur when the problem is solved using the Laplace transform. Let $X(s) = \mathscr{L}\{x(t)\}$.

The Laplace transform of the differential equation in (7.36) gives

$$sX(s) - A + \frac{1}{a}X(s) = \frac{1}{a}\frac{e^{-s}}{s}$$

or

$$X(s) = \frac{A}{s + 1/a} + \frac{1}{a}\frac{e^{-s}}{s(s + 1/a)}.$$

Since $\mathscr{L}^{-1}\{1/(s + 1/a)\} = e^{-t/a}$,

$$\mathscr{L}^{-1}\left\{\frac{1}{s(s + 1/a)}\right\} = \frac{1 - e^{-t/a}}{1/a} \qquad \text{(by Property 5)}$$

**TABLE 7.2**  LAPLACE TRANSFORMS OF SELECTED FUNCTIONS

| $f(t)$ | $\mathcal{L}\{f(t)\}$ |
| --- | --- |
| 1. $t^n$ | $\dfrac{n!}{s^{n+1}} \left[ \text{or } \dfrac{\Gamma(n+1)}{s^{n+1}} \text{ if } n \neq \text{integer}, n > -1 \right]$ |
| 2. $u_a(t)$, unit step at $t = a$ | $\dfrac{e^{-as}}{s}$ |
| 3. $\sin at$ | $\dfrac{a}{s^2 + a^2}$ |
| 4. $\cos at$ | $\dfrac{s}{s^2 + a^2}$ |
| 5. $\sinh at$ | $\dfrac{a}{s^2 - a^2}$ |
| 6. $\cosh at$ | $\dfrac{s}{s^2 - a^2}$ |
| 7. $e^{at}$ | $\dfrac{1}{s - a}$ |
| 8. $te^{at}$ | $\dfrac{1}{(s - a)^2}$ |
| 9. $e^{at}(1 + at)$ | $\dfrac{s}{(s - a)^2}$ |
| 10. $\dfrac{e^{at} - e^{bt}}{a - b}$ | $\dfrac{1}{(s - a)(s - b)}, \; a \neq b$ |
| 11. $\dfrac{ae^{at} - be^{bt}}{a - b}$ | $\dfrac{s}{(s - a)(s - b)}, \; a \neq b$ |
| 12. $\dfrac{1}{2}\left( \dfrac{e^{at}}{a - b} - \dfrac{e^{-at}}{a + b} - \dfrac{2be^{bt}}{a^2 - b^2} \right)$ | $\dfrac{s}{(s^2 - a^2)(s - b)}, \; a \neq b$ |
| 13. $t \sin at$ | $\dfrac{2as}{(s^2 + a^2)^2}$ |
| 14. $t \cos at$ | $\dfrac{s^2 - a^2}{(s^2 + a^2)^2}$ |
| 15. $\dfrac{\sin at - at \cos at}{2a^3}$ | $\dfrac{1}{(s^2 + a^2)^2}$ |
| 16. $\dfrac{\sin at + at \cos at}{2a}$ | $\dfrac{s^2}{(s^2 + a^2)^2}$ |
| 17. $t \sinh at$ | $\dfrac{2as}{(s^2 - a^2)^2}$ |
| 18. $t \cosh at$ | $\dfrac{s^2 + a^2}{(s^2 - a^2)^2}$ |
| 19. $\dfrac{at \cosh at - \sinh at}{2a^3}$ | $\dfrac{1}{(s^2 - a^2)^2}$ |

and the solution is given by (see Property 6, page 277)

$$x(t) = Ae^{-t/a} + u_1(t)[1 - e^{-(t-1)/a}]. \tag{7.37}$$

Note that this example models the charge, $x(t)$, in an $RC$ circuit where $R = 1$, $C = a$, the initial charge is $A$ and a voltage of $1/a$ is applied after $t = 1$ (Figure 7.6).

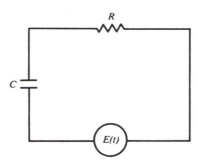

**Figure 7.6**  RC circuit.

**EXAMPLE 7.18**    As a third example we look at the initial value problem

$$x''(t) - x'(t) - 6x(t) = -2,$$
$$x(0) = 1, \qquad x'(0) = 0.$$

(7.38)

Taking Laplace transforms we have

$$s^2\mathcal{L}\{x(t)\} - sx(0) - x'(0) - s\mathcal{L}\{x(t)\} + x(0) - 6\mathcal{L}\{x(t)\} = \frac{-2}{s}$$

or

$$(s^2 - s - 6)\mathcal{L}\{x(t)\} - s + 1 = \frac{-2}{s}.$$

Thus

$$(s^2 - s - 6)\mathcal{L}\{x(t)\} = \frac{s^2 - s - 2}{s}$$

and

$$\mathcal{L}\{x(t)\} = \frac{s^2 - s - 2}{s(s^2 - s - 6)}.$$

(7.39)

The solution is obtained by taking the inverse transform of both sides of (7.39). To do this, since the right-hand side of (7.39) is not recognized as a known transform, we may rewrite this function of $s$ in terms of partial fractions. (See exercise 11 for an alternative method.)

Applying a partial fraction expansion, we seek constants $A$, $B$, and $C$ such that

$$\frac{s^2 - s - 2}{s(s^2 - s - 6)} = \frac{A}{s} + \frac{B}{s - 3} + \frac{C}{s + 2}.$$

The usual procedure yields

$$A = \frac{1}{3}, \qquad B = \frac{4}{15}, \qquad C = \frac{2}{5},$$

and

$$\frac{s^2 - s - 2}{s(s^2 - s - 6)} = \frac{1}{3}\left(\frac{1}{s}\right) + \frac{4}{15}\left(\frac{1}{s - 3}\right) + \frac{2}{5}\left(\frac{1}{s + 2}\right).$$

Substituting this expression into (7.39) and taking the inverse transform gives

$$x(t) = \mathcal{L}^{-1}\left\{\frac{1}{3}\left(\frac{1}{s}\right)\right\} + \mathcal{L}^{-1}\left\{\frac{4}{15}\left(\frac{1}{s - 3}\right)\right\} + \mathcal{L}^{-1}\left\{\frac{2}{5}\left(\frac{1}{s + 2}\right)\right\}.$$

Thus

$$x(t) = \frac{1}{3} + \frac{4}{15}e^{3t} + \frac{2}{5}e^{-2t}$$

is the solution to our initial value problem.

Example 7.18 points out the fact that it may not be immediately clear how the equation

$$\mathcal{L}\{x(t)\} = X(s)$$

is to be inverted. Generally, the procedure is to rewrite $X(s)$ so that recognizable transforms exist or in a form one or more of Properties 1 through 6 of Section 7.4 or the convolution theorem may be applied. Typical of the techniques often used for the inversion of transforms which may occur in solving differential equations are those contained in the following examples.

**EXAMPLE 7.19**    Find

$$\mathcal{L}^{-1}\left\{\frac{2}{(s - 3)^5}\right\}.$$

Here

$$\mathcal{L}^{-1}\left\{\frac{2}{(s - 3)^5}\right\} = 2\mathcal{L}^{-1}\left\{\frac{1}{(s - 3)^5}\right\}.$$

Now $4!/s^5 = \mathcal{L}\{t^4\}$, and by the shifting theorem (Property 2),

$$\mathcal{L}\{e^{3t}t^4\} = \frac{4!}{(s - 3)^5}, \qquad \text{so that} \qquad \frac{1}{(s - 3)^5} = \frac{1}{4!}\mathcal{L}\{e^{3t}t^4\}.$$

Hence

$$\mathcal{L}^{-1}\left\{\frac{1}{(s - 3)^5}\right\} = \frac{t^4}{4!}e^{3t} \qquad \text{and} \qquad \mathcal{L}^{-1}\left\{\frac{2}{(s - 3)^5}\right\} = \frac{1}{12}t^4e^{3t}.$$

**EXAMPLE 7.20**    Consider the *RLC* circuit of Figure 7.4, page 284, where $L = 1$, $R = 6$, $C = \frac{1}{10}$, $q(0) = 0$, $q'(0) = 1$, and $E(t) = V_0[u_1(t) - u_2(t)]$ (Figure 7.7). Taking the Laplace transform of the appropriate differential equation,

$$q'' + 6q'(t) + 10q(t) = E(t) \tag{7.40}$$

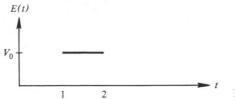

**Figure 7.7**   Graph of $E(t)$.

gives

$$s^2 Q(s) - 1 + 6sQ(s) + 10Q(s) = V_0\left(\frac{e^{-s}}{s} - \frac{e^{-2s}}{s}\right).$$

Solving for $Q(s)$ gives

$$Q(s) = \frac{1}{s^2 + 6s + 10} + V_0\left[\frac{e^{-s}}{s(s^2 + 6s + 10)} - \frac{e^{-2s}}{s(s^2 + 6s + 10)}\right].$$

Since

$$\frac{1}{s^2 + 6s + 10} = \frac{1}{(s^2 + 6s + 9) + 1} = \frac{1}{(s + 3)^2 + 1}$$

and $1/(s^2 + 1) = \mathcal{L}\{\sin t\}$, by Property 2

$$\mathcal{L}\{e^{-3t} \sin t\} = \frac{1}{(s + 3)^2 + 1}.$$

Hence

$$\mathcal{L}^{-1}\left\{\frac{1}{s^2 + 6s + 10}\right\} = e^{-3t} \sin t,$$

and using Property 5, we have

$$\mathcal{L}^{-1}\left\{\frac{1}{s(s^2 + 6s + 10)}\right\} = \int_0^t e^{-3\tau} \sin \tau \, d\tau$$

$$= -\frac{e^{-3t}}{10}(3 \sin t + \cos t) + \frac{1}{10} = h(t). \qquad (7.41)$$

Thus combining the results above with Property 6 allows us to write our solution as

$$q(t) = e^{-3t} \sin t + V_0[u_1(t)h(t - 1) - u_2(t)h(t - 2)],$$

where $h(t)$ is given by (7.41). The graph of $q(t)$ for $0 \le t \le 3$ and $V_0 = 2$ is given in Figure 7.8.

Notice that we may use the convolution theorem to write the solution of (7.40) for any $E(t)$ as an integral. By taking the Laplace transform of (7.40) we obtain

$$s^2 Q(s) - 1 + 6sQ(s) + 10Q(s) = \mathcal{L}\{E(t)\}$$

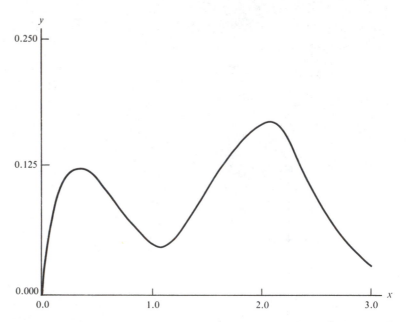

**Figure 7.8**   Graph of the charge resulting from the input voltage $V_0[u_1(t) - u_2(t)]$.

or

$$Q(s) = \frac{1}{s^2 + 6s + 10} + \frac{1}{s^2 + 6s + 10}\mathcal{L}\{E(t)\}.$$

Thus taking the inverse Laplace transform gives

$$q(t) = e^{-3t}\sin t + \int_0^t e^{-3\tau}\sin \tau\, E(t - \tau)\, d\tau.$$

There are many situations where the forcing function in a differential equation is a periodic function which is not expressible in a simple formula. Consider the full-wave rectification of $\sin t$ as shown in Figure 7.9 or the square wave of Figure 7.10. Laplace transforms are particularly useful in solving differential equations with such forcing functions. A nice way to evaluate the Laplace transform of a periodic function is given in the following theorem.

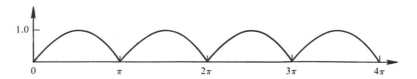

**Figure 7.9**   Full-wave rectification of $\sin t$.

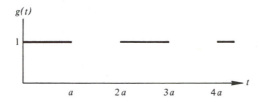

g(t)

**Figure 7.10**  Square wave.

**Theorem 7.6.**    If $f(t)$ is piecewise continuous for $0 \le t \le p$ and periodic with period $p$,

$$\mathcal{L}\{f(t)\} = \frac{1}{1 - e^{-sp}} \int_0^p e^{-st} f(t) \, dt.$$

(See exercise 15 for an outline of the proof.) Thus for the function of Figure 7.9,

$$\mathcal{L}\{h(t)\} = \frac{1}{1 - e^{-\pi s}} \int_0^\pi e^{-st} \sin t \, dt$$

$$= \frac{1}{1 - e^{-\pi s}} \left( \frac{1}{1 + s^2} \right) (e^{-s\pi} + 1) = \frac{1 + e^{-s\pi}}{1 - e^{-s\pi}} \left( \frac{1}{1 + s^2} \right), \qquad (7.42)$$

while for the function of Figure 7.10,

$$\mathcal{L}\{g(t)\} = \frac{1}{1 - e^{-2as}} \left[ \int_0^a e^{-st} (1) \, dt + \int_a^{2a} e^{-st} (0) \, dt \right]$$

$$= \frac{1}{1 - e^{-2as}} \left( \frac{1 - e^{-as}}{s} \right). \qquad (7.43)$$

**EXAMPLE 7.21**    Find the solution of

$$x'' + \omega^2 x = A g(t), \qquad x(0) = x_0, \qquad x'(0) = v_0, \qquad (7.44)$$

where $g(t)$ is the square wave of Figure 7.10. This initial value problem models undamped oscillations of a

$$\begin{Bmatrix} \text{spring–mass system} \\ \text{pendulum} \\ LC \text{ circuit} \end{Bmatrix}$$

depending on the interpretation of $x$, $\omega$, $A g(t)$, $x_0$, and $v_0$. Regardless of the physical situation we find the solution of this initial value problem by taking its Laplace transform to obtain

$$s^2 \mathcal{L}\{x(t)\} - sx_0 - v_0 + \omega^2 \mathcal{L}\{x(t)\} = A \mathcal{L}\{g(t)\}$$

or

$$\mathcal{L}\{x(t)\} = \frac{sx_0}{s^2 + \omega^2} + \frac{v_0}{s^2 + \omega^2} + \frac{A\,\mathcal{L}\{g(t)\}}{s^2 + \omega^2}. \tag{7.45}$$

The inverse Laplace transform of the first two terms on the right-hand side of (7.45) are obvious, but in order to invert the third term, we write it as

$$\mathcal{L}\{g(t)\} = \frac{1}{s}\frac{1 - e^{-as}}{1 - e^{-2as}} = \frac{1}{s}\frac{1 - e^{-as}}{(1 - e^{-as})(1 + e^{-as})} = \frac{1}{s(1 + e^{-as})}.$$

If we now expand $1/(1 + e^{-as})$ as an infinite series in terms of powers of $e^{-as}$, we find

$$\frac{1}{1 + e^{-as}} = 1 - e^{-as} + e^{-2as} - e^{-3as} \cdots = \sum_{n=0}^{\infty} (-1)^n e^{-nas}.$$

[Recall that $\dfrac{1}{1 + x} = \displaystyle\sum_{n=0}^{\infty}(-1)^n x^n$, for $|x| < 1$.] Also note that by Property 5 (or the convolution theorem)

$$\mathcal{L}^{-1}\left\{\frac{1}{s(s^2 + \omega^2)}\right\} = \int_0^t \frac{\sin \omega\tau}{\omega}\,d\tau = \frac{1 - \cos \omega t}{\omega^2}.$$

If we define $r(t)$ by

$$r(t) = \frac{1 - \cos \omega t}{\omega^2},$$

we may write the solution of our original problem as

$$x(t) = x_0 \cos \omega t + \frac{v_0}{\omega} \sin \omega t + A\,\mathcal{L}^{-1}\left\{\mathcal{L}\{r(t)\} \sum_{n=0}^{\infty} (-1)^n e^{-nas}\right\}$$

$$= x_0 \cos \omega t + \frac{v_0}{\omega} \sin \omega t + A \sum_{n=0}^{\infty} (-1)^n u_{na}(t)r(t - na). \tag{7.46}$$

Figure 7.11 shows the effect of a square-wave forcing function on this solution. The graph on the left is the solution of (7.44) with $A = 0$ and $x_0 = v_0 = \omega = 1$, while the one on the right has $A = a = x_0 = v_0 = \omega = 1$.


## EXERCISES

Solve the following initial value problems.

**1.** $x'' + x = e^{-2t} \sin t$, $\quad x(0) = x'(0) = 0$.

**2.** $x'' + 4x' + 4x = f(t)$, $\quad x(0) = 1$, $\quad x'(0) = 0$.

**3.** $x'' + 2x' + 2x = 0$, $\quad x(0) = 1$, $\quad x'(0) = 0$.

**4.** $x'' + x' - 2x = e^{-t} \sin t$, $\quad x(0) = x'(0) = 0$.

**5.** $x' - x = [u_1(t) - u_3(t)]e^t$, $\quad x(0) = 4$.

**6.** $x' + x = [1 - u_\pi(t)] \sin t$, $\quad x(0) = -2$.

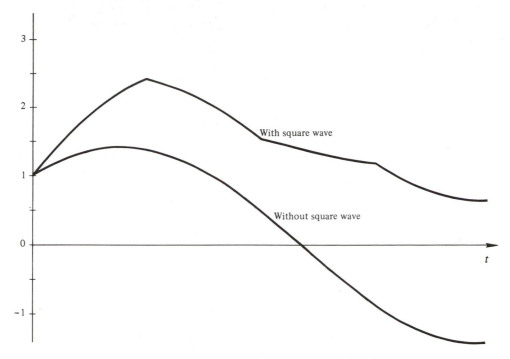

**Figure 7.11**   Effect of the square wave on the solution of (7.44).

**7.** $x'' + \omega^2 x = f(t), \quad x(0) = 10, \quad x'(0) = 0.$
**8.** $x'' + \omega^2 x = f(t), \quad x(0) = 0, \quad x'(0) = 1.$
**9.** $x'' + 3x' + 2x = t^2, \quad x(0) = x'(0) = 0.$
**10.** $x'' + 4x' + 4x = 4 \cos 2t, \quad x'(0) = x(0) = 1.$
**11.** Solve for $x(t)$ in (7.39) by the following procedure.
   **(a)** Show that

$$\mathcal{L}\{x(t)\} = \frac{s^2 - s - 6 + 4}{s(s^2 - s - 6)} = \frac{1}{s} + \frac{4}{s(s - 3)(s + 2)}.$$

   **(b)** Use Table 7.2 to compute $\mathcal{L}^{-1}\left\{\dfrac{4}{(s - 3)(s + 2)}\right\}$ and then use Property 5 to find $x(t)$.

**12.** The differential equation for the current, $I(t)$, in a series circuit with inductance $L$ and resistance $R$ is

$$L\frac{dI}{dt} + RI = E(t), \tag{7.47}$$

where $E(t)$ is the externally applied voltage. Solve (7.47) subject to the initial condition $I(0) = A$ if
   **(a)** $E(t) = E_0[u_1(t) - u_2(t)]$ (a square pulse).
   **(b)** $E(t) = E_0 \sin t[1 - u_\pi(t)]$ (a single pulse of a sine wave).
   **(c)** $E(t) = E_0 h(t)$ (the full-wave rectification of $\sin(t)$ from Figure 7.9).
   **(d)** $E(t) = E_0 g(t)$ (the square wave of Figure 7.10).

**13.** Solve the initial value problem (7.44) for a harmonic oscillator if the right-hand side,
$Ag(t)$, is given by
 (a) $E_0[u_1(t) - u_2(t)]$.
 (b) $E_0 \sin t[1 - u_\pi(t)]$.
 (c) $E_0 h(t)$ (the full-wave rectification of $\sin(t)$ from Figure 7.9).

**14.** The differential equation

$$V\frac{d}{dt}(x(t)) = r_i f(t) - r_o x(t)$$

describes a compartmental model of the absorption of a drug by a body organ of volume
$V$. $x(t)$ is a concentration of the drug in the body fluid at time $t$ in the organ, $r_i$ and $r_o$ are
the respective rates of fluid into and out of the organ, and $f(t)$ is the concentration of the
drug entering the organ.
 (a) Find $x(t)$ if $x(0) = 0$, $r_i = r_o$ and $f(t) = f_0 t[1 - u_a(t)]$.
 (b) Find $x(t)$ if $x(0) = x_0$, $r_i \neq r_o$ and $f(t) = f_0 t[1 - u_a(t)]$.

## 7.7 LAPLACE TRANSFORMS AND SYSTEMS OF EQUATIONS

Initial value problems involving systems of equations may also be solved using the
Laplace transform by techniques illustrated in the following examples. The advantages
of this method over those of Chapter 5 include automatic satisfaction of initial values
and the fact no special techniques are needed to find either particular solutions or a
fundamental matrix (as in the case of repeated eigenvalues).

**EXAMPLE 7.22**    Here we consider the system of equations

$$\begin{aligned}
x_1' &= x_1 - 2x_2, \\
x_2' &= 5x_1 - x_2,
\end{aligned}$$

(7.48)

subject to initial conditions

$$x_1(0) = -1, \qquad x_2(0) = 2.$$

Transforming (7.48) we have

$$sX_1(s) - x_1(0) = X_1(s) - 2X_2(s),$$
$$sX_2(s) - x_2(0) = 5X_1(s) - X_2(s)$$

or

$$(s - 1)X_1(s) + 2X_2(s) = x_1(0),$$
$$-5X_1(s) + (s + 1)X_2(s) = x_2(0).$$

Thus

$$\begin{bmatrix} s - 1 & 2 \\ -5 & s + 1 \end{bmatrix}\begin{bmatrix} X_1(s) \\ X_2(s) \end{bmatrix} = \begin{bmatrix} -1 \\ 2 \end{bmatrix}$$

and by Cramer's rule

$$X_1(s) = \frac{-(s + 1) - 4}{(s - 1)(s + 1) + 10}.$$

Thus

$$X_1(s) = \frac{-s - 5}{s^2 + 9} = -\frac{s}{s^2 + 9} - \frac{5}{3}\frac{3}{s^2 + 9}$$

and taking the inverse Laplace transform gives

$$x_1(t) = -\cos 3t - \frac{5}{3} \sin 3t.$$

Similarly,

$$X_2(s) = \frac{2(s - 1) - 5}{s^2 + 9} = 2\frac{s}{s^2 + 9} - \frac{7}{3}\frac{3}{s^2 + 9}$$

and $x_2(t) = 2 \cos 3t - \frac{7}{3} \sin 3t$.

**EXAMPLE 7.23**    Consider the system of equations

$$\begin{aligned} x_1' &= -x_1 + x_2 + u_1(t), \\ x_2' &= 2x_1, \end{aligned} \tag{7.49}$$

subject to the initial conditions

$$x_1(0) = 0, \qquad x_2(0) = 1.$$

Applying the Laplace transform to both of the equations of the system (7.49) we have, denoting transforms by capital letters,

$$sX_1(s) - x_1(0) = -X_1(s) + X_2(s) + \frac{e^{-s}}{s},$$

$$sX_2(s) - x_2(0) = 2X_1(s).$$

Using the initial conditions gives

$$sX_1(s) = -X_1(s) + X_2(s) + \frac{e^{-s}}{s},$$

$$sX_2(s) - 1 = 2X_1(s), \tag{7.50}$$

which in matrix form looks like

$$\begin{bmatrix} s + 1 & -1 \\ -2 & s \end{bmatrix} \begin{bmatrix} X_1(s) \\ X_2(s) \end{bmatrix} = \begin{bmatrix} \dfrac{e^{-s}}{s} \\ 1 \end{bmatrix}. \tag{7.50}$$

Solving (7.50), say by Cramer's rule or by using the inverse of the coefficient matrix, gives

$$X_1(s) = \frac{1 + e^{-s}}{s(s + 1) - 2},$$

$$X_2(s) = \frac{s + 1 + 2e^{-s}/s}{s(s + 1) - 2}.$$

Now

$$\mathcal{L}^{-1}\left\{\frac{1}{s^2 + s - 2}\right\} = \mathcal{L}^{-1}\left\{\frac{1}{(s + 2)(s - 1)}\right\} = \frac{1}{3}(e^t - e^{-2t})$$

(use a partial fraction decomposition, the convolution theorem, or results from Table 7.2.) Thus from Property 6, page 277,

$$x_1(t) = \frac{1}{3}(e^t - e^{-2t}) + u_1(t)\frac{e^{(t-1)} - e^{-2(t-1)}}{3}. \tag{7.51}$$

Now consider

$$X_2(s) = \frac{s + 1 + 2e^{-s}/s}{(s - 1)(s + 2)}$$

and note that

$$\mathcal{L}^{-1}\left\{\frac{s + 1}{(s - 1)(s + 2)}\right\} = \frac{1}{3}\mathcal{L}^{-1}\left\{\frac{2}{s - 1} + \frac{1}{s + 2}\right\} = \frac{1}{3}(2e^t + e^{-2t})$$

so by Property 5, page 277,

$$\mathcal{L}^{-1}\left\{\frac{s + 1}{s(s - 1)(s + 2)}\right\} = \frac{1}{3}\int_0^t (2e^\tau + e^{-2\tau})\, d\tau = \frac{1}{3}\left(2e^t - \frac{e^{-2t}}{2} - \frac{3}{2}\right).$$

Thus, by the above results and Property 6, we have

$$x_2(t) = \frac{1}{3}(2e^t + e^{-2t}) + u_1(t)\frac{4e^{t-1} - e^{-2(t-1)} - 3}{6} \tag{7.52}$$

and the solution of (7.49) is given by (7.51) and (7.52).

**EXAMPLE 7.24**    Let us see how the Laplace transform handles a system of equations from Section 5.3, page 178, where finding a fundamental matrix led to multiple eigenvalues. The differential equation under consideration is

$$D\begin{bmatrix} x_1 \\ x_2 \end{bmatrix} = \begin{bmatrix} 1 & 1 \\ -1 & 3 \end{bmatrix}\begin{bmatrix} x_1 \\ x_2 \end{bmatrix}, \tag{5.45}$$

and let the initial conditions be $x_1(0) = A$, $x_2(0) = B$. Taking the Laplace transform of (7.53) and using the initial conditions gives

$$\begin{bmatrix} sX_1(s) - A \\ sX_2(s) - B \end{bmatrix} = \begin{bmatrix} 1 & 1 \\ -1 & 3 \end{bmatrix}\begin{bmatrix} X_1(s) \\ X_2(s) \end{bmatrix}$$

or

$$\begin{bmatrix} s - 1 & -1 \\ 1 & s - 3 \end{bmatrix}\begin{bmatrix} X_1(s) \\ X_2(s) \end{bmatrix} = \begin{bmatrix} A \\ B \end{bmatrix}.$$

Solving for $X_1(s)$ and $X_2(s)$ gives

$$X_1(s) = \frac{A(s-3) + B}{(s-2)^2} = \frac{A(s-2) + B - A}{(s-2)^2},$$

$$X_2(s) = \frac{B(s-1) - A}{(s-2)^2} = \frac{B(s-2) + (B-A)}{(s-2)^2}.$$

(7.53)

Taking the inverse Laplace transform of (7.53), (see Table 7.2, page 288), gives our solution as

$$\begin{bmatrix} x_1(t) \\ x_2(t) \end{bmatrix} = \begin{bmatrix} Ae^{2t} + (B-A)te^{-2t} \\ Be^{-2t} + (B-A)te^{-2t} \end{bmatrix}.$$

(7.54)

[Note that if the general solution for this problem from page 179 is required to satisfy $x_1(0) = A$, $x_2(0) = B$, we get $C_1 = B$, $C_2 = B - A$, so that solution is identical with (7.54). Note also that difficulties with multiple eigenvalues in determining a fundamental matrix do not occur with the Laplace transform solution of systems of equations.]

## EXERCISES

**1–10.** Use the Laplace transforms to solve the initial value problems given in exercises 1 through 10 of Section 5.5, page 182. (You may also use Laplace transform to solve all the problems on applications of systems of equations in Section 5.6, page 190.)

**11.** Another illustration of the straightforward way the Laplace transfrom handles the situation where finding a fundamental matrix requires consideration of a multiple eigenvalue is given by the solution of the system of equations

$$D\begin{bmatrix} x_1 \\ x_2 \\ x_3 \\ x_4 \end{bmatrix} = \begin{bmatrix} 1 & 1 & 0 & 0 \\ 0 & 1 & 2 & 0 \\ 0 & 0 & 1 & 0 \\ 0 & -2 & 0 & 1 \end{bmatrix}\begin{bmatrix} x_1 \\ x_2 \\ x_3 \\ x_4 \end{bmatrix}, \quad \begin{bmatrix} x_1(0) \\ x_2(0) \\ x_3(0) \\ x_4(0) \end{bmatrix} = \begin{bmatrix} a_1 \\ a_2 \\ a_3 \\ a_4 \end{bmatrix}$$

Solve this initial value problem using Laplace transforms. [Note this is the same system of differential equations given by (5.53). Use the fundamental matrix given by (5.54) to solve the initial value problem given above and show your answers for the two methods agree.]

**12.** Use Laplace transforms to solve exercise 12 of Section 5.5.

Use Laplace transforms to solve the following initial value problems.

**13.** $D\begin{bmatrix} x_1 \\ x_2 \\ x_3 \end{bmatrix} = \begin{bmatrix} 3 & -1 & -1 \\ -1 & 3 & -1 \\ -1 & -1 & 3 \end{bmatrix}\begin{bmatrix} x_1 \\ x_2 \\ x_3 \end{bmatrix}, \quad \begin{bmatrix} x_1(0) \\ x_2(0) \\ x_3(0) \end{bmatrix} = \begin{bmatrix} 1 \\ 0 \\ 0 \end{bmatrix}.$

**14.** $D\begin{bmatrix} x_1 \\ x_2 \\ x_3 \end{bmatrix} = \begin{bmatrix} 2 & 0 & 9 \\ 0 & 3 & 0 \\ 1 & 0 & 2 \end{bmatrix}\begin{bmatrix} x_1 \\ x_2 \\ x_3 \end{bmatrix}, \quad \begin{bmatrix} x_1(0) \\ x_2(0) \\ x_3(0) \end{bmatrix} = \begin{bmatrix} 1 \\ 0 \\ 1 \end{bmatrix}.$

**15.** $D\begin{bmatrix} x_1 \\ x_2 \\ x_3 \end{bmatrix} = \begin{bmatrix} 2 & 0 & 5 \\ 0 & 1 & 2 \\ -4 & 5 & 0 \end{bmatrix}\begin{bmatrix} x_1 \\ x_2 \\ x_3 \end{bmatrix}, \quad \begin{bmatrix} x_1(0) \\ x_2(0) \\ x_3(0) \end{bmatrix} = \begin{bmatrix} 0 \\ 1 \\ 1 \end{bmatrix}.$

**16.** $D\begin{bmatrix} x_1 \\ x_2 \\ x_3 \end{bmatrix} = \begin{bmatrix} 0 & 3 & 3 \\ 2 & -1 & -3 \\ 0 & -1 & -1 \end{bmatrix} \begin{bmatrix} x_1 \\ x_2 \\ x_3 \end{bmatrix}$, $\begin{bmatrix} x_1(0) \\ x_2(0) \\ x_3(0) \end{bmatrix} = \begin{bmatrix} 3 \\ 0 \\ 7 \end{bmatrix}$.

**17.** $D\begin{bmatrix} x_1 \\ x_2 \end{bmatrix} = \begin{bmatrix} -2 & 4 \\ 1 & 1 \end{bmatrix} \begin{bmatrix} x_1 \\ x_2 \end{bmatrix} + \begin{bmatrix} 7e^{3t} \\ -2e^{3t} \end{bmatrix}$, $\begin{bmatrix} x_1(0) \\ x_2(0) \end{bmatrix} = \begin{bmatrix} 1 \\ 0 \end{bmatrix}$.

**18.** $D\begin{bmatrix} x_1 \\ x_2 \end{bmatrix} = \begin{bmatrix} 2 & 1 \\ -3 & -2 \end{bmatrix} \begin{bmatrix} x_1 \\ x_2 \end{bmatrix} + \begin{bmatrix} 3e^t \\ -e^t \end{bmatrix}$, $\begin{bmatrix} x_1(0) \\ x_2(0) \end{bmatrix} = \begin{bmatrix} 1 \\ 2 \end{bmatrix}$.

**19.** $D\begin{bmatrix} x_1 \\ x_2 \end{bmatrix} = \begin{bmatrix} -2 & 4 \\ 1 & 1 \end{bmatrix} \begin{bmatrix} x_1 \\ x_2 \end{bmatrix} + \begin{bmatrix} e^{2t} \\ -e^{2t} \end{bmatrix}$, $\begin{bmatrix} x_1(0) \\ x_2(0) \end{bmatrix} = \begin{bmatrix} 0 \\ 1 \end{bmatrix}$.

**20.** $D\begin{bmatrix} x_1 \\ x_2 \end{bmatrix} = \begin{bmatrix} 2 & 1 \\ -3 & -2 \end{bmatrix} \begin{bmatrix} x_1 \\ x_2 \end{bmatrix} + \begin{bmatrix} \cos t \\ 0 \end{bmatrix}$, $\begin{bmatrix} x_1(0) \\ x_2(0) \end{bmatrix} = \begin{bmatrix} 1 \\ 0 \end{bmatrix}$.

Another advantage in using the Laplace transform is that system of equations need not be in the form $D\mathbf{x} = \mathbf{A}\mathbf{x} + \mathbf{b}$.

**21. (a)** Show that if $X_1(s) = \mathcal{L}\{x_1(t)\}$, $X_2(s) = \mathcal{L}\{x_2(t)\}$, using the Laplace transform converts the initial value problem

$$\begin{bmatrix} 2D & D-1 \\ D & D \end{bmatrix} \begin{bmatrix} x_1 \\ x_2 \end{bmatrix} = \begin{bmatrix} t \\ t^2 \end{bmatrix}, \qquad \begin{bmatrix} x_1(0) \\ x_2(0) \end{bmatrix} = \begin{bmatrix} 1 \\ 0 \end{bmatrix}$$

into the algebraic system

$$\begin{bmatrix} 2s & s-1 \\ s & s \end{bmatrix} \begin{bmatrix} X_1(s) \\ X_2(s) \end{bmatrix} = \begin{bmatrix} 2 + \dfrac{1}{s^2} \\ 1 + \dfrac{2}{s^3} \end{bmatrix}.$$

**(b)** Solve the equation above for $X_2(s)$ as

$$X_2(s) = \frac{4-s}{s^3(s+1)} = \frac{4}{s^3(s+1)} - \frac{1}{s^2(s+1)}$$

and use partial fractions or the convolution theorem to show that

$$x_2(t) = 5 - 5t + 2t^2 - 5e^{-t}.$$

**(c)** Show that $X_1(s) + X_2(s) = 1/s + 2/s^4$ and solve for $x_1(t)$.

Solve the following initial value problems using the Laplace transform.

**22.** $\begin{bmatrix} D+2 & D \\ 2D & 3D+5 \end{bmatrix} \begin{bmatrix} x_1 \\ x_2 \end{bmatrix} = \begin{bmatrix} 16e^{-2t} \\ 15 \end{bmatrix}$, $\begin{bmatrix} x_1(0) \\ x_2(0) \end{bmatrix} = \begin{bmatrix} 4 \\ -3 \end{bmatrix}$.

**23.** $\begin{bmatrix} 2D-2 & D \\ D-3 & D-3 \end{bmatrix} \begin{bmatrix} x_1 \\ x_2 \end{bmatrix} = \begin{bmatrix} 1 \\ 2 \end{bmatrix}$, $\begin{bmatrix} x_1(0) \\ x_2(0) \end{bmatrix} = \begin{bmatrix} 0 \\ 0 \end{bmatrix}$.

**24.** $\begin{bmatrix} D & -D \\ 2D & -2D-1 \end{bmatrix} \begin{bmatrix} x_1 \\ x_2 \end{bmatrix} = \begin{bmatrix} -e^t \\ 8 \end{bmatrix}$, $\begin{bmatrix} x_1(0) \\ x_2(0) \end{bmatrix} = \begin{bmatrix} -1 \\ -10 \end{bmatrix}$.

**25.** $\begin{bmatrix} D & -D-6 \\ D-3 & 2D \end{bmatrix} \begin{bmatrix} x_1 \\ x_2 \end{bmatrix} = \begin{bmatrix} 0 \\ 0 \end{bmatrix}$, $\begin{bmatrix} x_1(0) \\ x_2(0) \end{bmatrix} = \begin{bmatrix} 2 \\ 3 \end{bmatrix}$.

**26.** $\begin{bmatrix} D^2+2 & -4D \\ D & D^2-4 \end{bmatrix} \begin{bmatrix} x_1 \\ x_2 \end{bmatrix} = \begin{bmatrix} 0 \\ 0 \end{bmatrix}$, $\begin{bmatrix} x_1(0) \\ x_2(0) \end{bmatrix} = \begin{bmatrix} -4 \\ 1 \end{bmatrix}$, $\begin{bmatrix} x_1'(0) \\ x_2'(0) \end{bmatrix} = \begin{bmatrix} 8 \\ 2 \end{bmatrix}$.

**27.** The differential equations which describe small motions of a compound pendulum from equilibrium (see Figure 7.12) are

$$(m_1 + m_2)\ell_1^2 x_1'' + (m_1 + m_2)\ell_1 g x_1 + m_2\ell_1\ell_2 x_2'' = 0,$$

$$m_2\ell_1\ell_2 x_1'' + m_2\ell_2^2 x_2'' + m_2\ell_2 g x_2 = 0. \qquad (7.55)$$

(a) Solve the equations in (7.55) if $g = 32$, $\ell_1 = \ell_2 = 32$, $m_1 = 3$, $m_2 = 1$, $x_1(0) = x_2(0) = 0$, $x_1'(0) = -1$, $x_2'(0) = 1$.

(b) Solve the equation in (7.55) if $g = 32$, $\ell_1 = \ell_2 = 32$, $m_1 = 7$, $m_2 = 9$, $x_1(0) = 1$, $x_2(0) = -1$, $x_1'(0) = x_2'(0) = 0$.

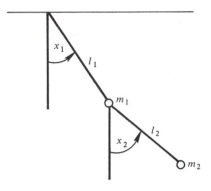

**Figure 7.12**   Compound pendulum.

**28.** The differential equations which describe the electric circuit in Figure 7.13 are

$$LI_1' + RI_2 = E(t),$$

$$RI_2' + \frac{1}{C}(I_2 - I_1) = 0. \qquad (7.56)$$

(a) Solve the equations in (7.56) if $I_1(0) = I_2(0) = 0$, $E(t) = \sin \omega t$, $L = 1$, $R = 5$, $C = \frac{1}{20}$.

(b) Repeat part (a) if the initial conditions are changed to $I_1(0) = 0$, $I_2(0) = 1$.

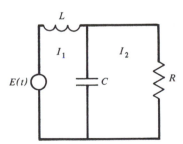

**Figure 7.13**   Electric circuit.

**29.** Show that if **A** is an $n \times n$ matrix of constants, $\mathbf{b}(t)$ an $n \times 1$ matrix of functions of time, $\mathbf{x}(t)$ an $n \times 1$ matrix of unknowns, and $\mathbf{x}_0$ an $n \times 1$ matrix of initial values, the solution of the initial value problem

$$D\mathbf{x} = A\mathbf{x} + \mathbf{b}, \quad \mathbf{x}(0) = \mathbf{x}_0$$

by Laplace transform techniques leads to the algebraic equation $(s\mathbf{I} - \mathbf{A})\mathbf{X}(s) = \mathbf{B}(s) + \mathbf{x}_0$, where $\mathbf{I}$ is the $n \times n$ identity matrix.

## REVIEW EXERCISES

1. Which of the following functions are piecewise continuous for $0 \leq t < 100$?

   (a) $f(t) = \dfrac{\cos \omega t}{t}$.

   (b) $f(t) = \dfrac{t}{\cos \omega t}$.

   (c) $f(t) = \dfrac{\sin \omega t}{t}$.

   (d) $f(t) = \dfrac{t}{\sin \omega t}$.

   (e) $f(t) = \dfrac{1}{t^2 + 1}$.

   (f) $f(t) = \dfrac{1}{t^2 - 1}$.

   (g) $f(t) = u_1(t)t + u_2(t)t^{-1}$.

2. Find values of $\alpha$, $M$, and $T$ in Definition 7.1 for the following functions.

   (a) $f(t) = e^{4t} \sin 6t$.

   (b) $f(t) = [u_0(t) - u_2(t)]t^{40} + u_1(t)e^{3t}$.

   (c) $f(t) = t^2$.

   (d) $f(t) = te^{2t}$.

Evaluate the Laplace transform of the following functions.

3. $f(t) = [1 - u_1(t)]t + u_2(t)e^t = \begin{cases} t, & 0 < t < 1 \\ 0, & 1 < t < 2 \\ e^t, & 2 < t \end{cases}$.

4. $g(t) = \cos^2 \omega t$ and $h(t) = t \cos^2 \omega t$.

5. $f(t) = t^2(1 + t)^2$.

6. $h(t) = u_2(t) - u_3(t)$.

7. $g(t) = \cosh bt \cosh at$.

8. $f(t) = \dfrac{\cos \omega t - 1}{t}$.

9. $h(t) = e^{-2t}t(t + 1)(t - 1)$.

10. $g(t) = u_3(t)(t^2 + 2)$.

11. $f(t) = t$, $0 < t < \pi$, $f(t + \pi) = f(t)$, $t > \pi$. (Also, sketch a graph of $y = f(t)$, $-\pi \leq t \leq 3\pi$.]

12. $f(t)$ given by the graph of Figure 7.14.

13. $f(t)$ given by the graph of Figure 7.15.

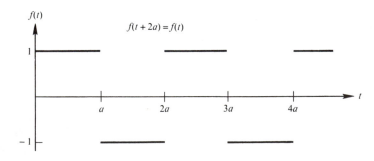

**Figure 7.14**

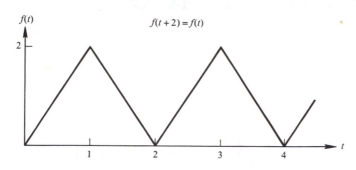

**Figure 7.15**

Evaluate the following inverse transforms.

**14.** $\mathcal{L}^{-1}\left\{\dfrac{1}{(s+2)^3(s-3)^2}\right\}$    **15.** $\mathcal{L}^{-1}\left\{\dfrac{s-1}{s^2-2s-3}\right\}$    **16.** $\mathcal{L}^{-1}\left\{\dfrac{3+s}{s^3+2s^2+s}\right\}$

**17.** $\mathcal{L}^{-1}\left\{\dfrac{6s-4}{s^2+8s+17}\right\}$    **18.** $\mathcal{L}^{-1}\left\{\dfrac{3s+2}{(s+1)(s^2+4)}\right\}$    **19.** $\mathcal{L}^{-1}\left\{\dfrac{e^{-3s}}{(s+1)^2}\right\}$

**20.** $\mathcal{L}^{-1}\left\{\dfrac{e^{-2s}+e^{-4s}}{s^2-4s+13}\right\}$    **21.** $\mathcal{L}^{-1}\left\{\ln\left(1-\dfrac{1}{s^2}\right)\right\}$    **22.** $\mathcal{L}^{-1}\left\{\tan^{-1}\left(\dfrac{1}{s}\right)\right\}$

Solve the following differential or integral equations.

**23.** $x''(t)+\omega^2 x(t)=f(t)$, $\quad x(0)=3$, $\quad x'(0)=10$.

**24.** $x''(t)+2x'(t)+10x(t)=\sin\omega t$, $\quad x(0)=0$, $\quad x'(0)=-2$. (Leave your answer as an integral.)

**25.** $x''(t)+\omega^2 x(t)=u_2(t)-u_3(t)$, $\quad x(0)=x'(0)=0$.

**26.** $x''(t)+4tx'(t)-4x(t)=0$, $\quad x(0)=0$, $\quad x'(0)=3$.

**27.** $tx''(t)+2(t-1)x'(t)-2x(t)=0$, $\quad x(0)=0$.

**28.** $\displaystyle\int_0^t \frac{x(\tau)\,d\tau}{(t-\tau)^\alpha}=f(t)$, $\quad 0<\alpha<1$. (This is called *Abel's integral equation*.)

**29.** $x(t)=4t^2-\displaystyle\int_0^t x(\tau)e^{(-t+\tau)}\,d\tau$.

**30.** $x'(t)=t+\displaystyle\int_0^t x(\tau)\cos(t-\tau)\,d\tau$, $\quad x(0)=2$.

**31.** $2\displaystyle\int_0^t x(\tau)x(t-\tau)\,d\tau=\sin t-t\cos t$.

**32.** The forced vibrations of a mass $m$ at the end of a vertical spring, with spring constant $k$, are described by

$$mx''(t)+kx(t)=f(t)=f_0+f_0 u_1(t)-2f_0 u_2(t).$$

Solve this differential equation if $k/m=4$, $f_0/m=1$, $x(0)=x'(0)=0$. Plot the solution for $0\le t\le 3$ and compare it with the graph of $f(t)$.

**33.** Prove that $(f(t)*g(t))*h(t)=f(t)*(g(t)*h(t))$.

**34.** Solve $x_1'(t)-3x_1(t)+x_2(t)=0,$
$\qquad x_1(t)+x_2'(t)-x_2(t)=0,$ $\quad x_1(0)=0$, $\quad x_2(0)=2$.

**35.** Solve $x_1''(t)+x_2''(t)=t^2$, $\quad x_1(0)=-2$,
$\qquad x_1''(t)-x_2''(t)=4t$, $\quad x_2(0)=x_1'(0)=x_2'(0)=0$.

# CHAPTER 8

# Numerical Methods

Often it is necessary to solve a differential equation for which there is no closed-form or series solution. It is then necessary to turn to the computer to attempt a numerical solution. These solutions can be quite good if the "real" solution that we are approximating is not too "wild." In this chapter we discuss some general ideas about numerical solutions and give some simple methods for generating them. The reader should review isoclines, direction fields, and the method of Picard as well as the method of Taylor series in Section 6.1.

## 8.1 EULER'S METHOD

This method is very simple. In essence, we will try to obtain a numerical solution by using the tangents to the solution curve as an approximation to the curve itself.

Suppose we have an equation of the form

$$y' = F(x, y) \qquad (8.1)$$

subject to the initial condition

$$y(x_0) = y_0. \qquad (8.2)$$

Written in the form

$$\frac{dy}{dx} = F(x, y) \qquad (8.3)$$

(8.1) tells us that at the given initial point $(x_0, y_0)$ the solution curve has slope $F(x_0, y_0)$. Thus we know a point on the curve and the slope of the tangent line at the point.

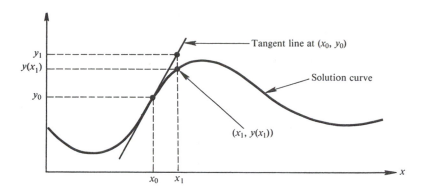

**Figure 8.1**

Now, from Figure 8.1 we see that the tangent line may be a good approximation to the solution curve for values of $x$ near $x_0$. We let $x_1 = x_0 + \Delta x$, where $\Delta x$ is small, and approximate $y(x_1)$ by the value of $y$ along the tangent line when $x = x_1$. We label this value of $y$ as $y_1$, and use the point-slope form of the equation of a line to get

$$y_1 - y_0 = F(x_0, y_0)(x_1 - x_0).$$

But $x_1 - x_0 = \Delta x$, so

$$y_1 = y_0 + F(x_0, y_0)\, \Delta x$$

is the approximation to $y(x_1)$. In a similar manner we calculate $F(x_1, y_1)$ and obtain an approximation, $y_2$, to the solution curve for $x = x_2$,

$$x_2 = x_0 + 2\, \Delta x,$$

as

$$y_2 = y_1 + F(x_1, y_1)\, \Delta x.$$

Continuing in this manner gives a sequence $y_1, y_2, y_3, \ldots, y_n$, as an approximation to the values of the exact solution of our differential equation, $y(x_1)$, $y(x_2)$, $y(x_3)$, $\ldots, y(x_n)$. If we choose $\Delta x$ small enough and if the solution curve is not too wild, we will get a reasonable solution for small values of $n$ (see Figure 8.2).

Euler's method, although conceptually simple and easy to use, is seldom used. It may be summarized as the iterated evaluation of the formulas

$$x_n = x_0 + n\, \Delta x,$$
$$y_{n+1} = y_n + F(x_n, y_n)\, \Delta x.$$
(8.4)

Table 8.1 is a tabulated solution of the very simple initial value problem

$$y' = xy,$$
$$y(1) = 1,$$

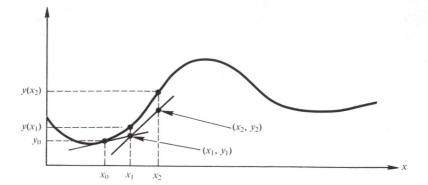

**Figure 8.2**

**TABLE 8.1**  EULER'S METHOD WITH $\Delta x = 0.1$

| $n$ | $x_n$ | $y_n$ | Exact |
|-----|-------|-------|-------|
| 0 | 1.0 | 1.0 | 1.0 |
| 1 | 1.1 | 1.1 | 1.1107 |
| 2 | 1.2 | 1.221 | 1.2461 |
| 3 | 1.3 | 1.3675 | 1.4120 |
| 4 | 1.4 | 1.5453 | 1.6161 |
| 5 | 1.5 | 1.7616 | 1.8682 |

on the interval $[1, 1.5]$. Since this problem has an easily obtained solution,

$$y(x) = e^{0.5(x^2-1)},$$

we may compare the numerical result to the exact one.

The steps we go through to generate Table 8.1, or for which we program some computer, are:

1. Choose $x_0$ and $y_0$. (Here $x_0 = y_0 = 1$.)
2. Choose a $\Delta x$, in this case $\Delta x = 0.1$.
3. Set $n = 0$.
4. Evaluate $x_n$ and $y_{n+1}$ from (8.4). Here we have, the first time through,

$$x_0 = x_0 + 0 \cdot \Delta x = x_0 = 1.0$$

$$y_{0+1} = y_1 = y_0 + F(x_0, y_0)\,\Delta x$$

$$= 1.0 + x_0 y_0\,\Delta x$$

$$= 1 + 0.1 = 1.1$$

5. Increase $n$ by 1. If all desired values of $n$ have been processed, print results and quit; otherwise, go back to step 4.

A simple BASIC language routine to generate Table 8.1 using steps 1 through 5 is given in Listing 8.1. The results, for $\Delta x = 0.1$, are given in Table 8.1.

Listing 8.1

```
10   X(0) = 1
20   Y(0) = 1
30   D = 0.1
40   REM D IS DELTA-X
50   N1 = 5
60   REM N1 IS LAST VALUE OF N ( < 10)
70   FOR N = 0 TO N1
80   X(N) = X(0) + N * D
90   M = N + 1
100  Y(M) = Y(N) + X(N) * Y(N) * D
110 NEXT N
120 PRINT "N", "X(N)", "Y(N)"
130 FOR N = 0 TO N1
140 PRINT
150 PRINT N, X(N), Y(N)
160 NEXT N
170 END
```

Recall that the Taylor expansion of a function $y(x)$ about the point $x = a$ is given by

$$y(x) = \sum_{k=0}^{\infty} y^{(k)}(a) \frac{(x-a)^k}{k!}. \qquad (8.5)$$

If, in (8.5) we replace $a$ by $x_n$ and $x$ by $x_n + \Delta x$ and keep only the first two terms we obtain

$$y_{n+1} = y_n + F(x_n, y_n)\,\Delta x$$

of Euler's method.

We say, because of the above, that Euler's method is a first order method. In general, a method is an $n$th order one if it agrees with a Taylor expansion through the $n$th order.

## 8.2 HEUN'S METHOD

We may obtain an improvement on Euler's method by noting that at each step of the method we have two points approximating the solution curve and the slopes at each point. We denote these points by $(x_n, y_n)$ and $(x_{n+1}, \bar{y}_{n+1})$ and note that Euler's method gives $\bar{y}_{n+1}$ as

$$\bar{y}_{n+1} = y_n + F(x_n, y_n)\,\Delta x$$

(see Figure 8.3). The slopes at these two points are given by $F(x_n, y_n)$ and

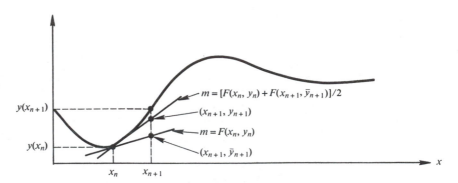

**Figure 8.3**

$F(x_{n+1}, \bar{y}_{n+1})$, and we may compute an average slope, $\bar{F}$, on the interval $[x_n, x_{n+1}]$ as

$$\bar{F} = \frac{F(x_n, y_n) + F(x_{n+1}, \bar{y}_{n+1})}{2}. \qquad (8.6)$$

If we use $\bar{F}$ in place of $f$ in the second of equations (8.4), we obtain a method of solving (8.3) called Heun's method. Thus Heun's method consists in the iterated solution of

$$x_n = x_0 + n\,\Delta x,$$

$$y_{n+1} = y_n + \bar{F}\,\Delta x,$$

with $\bar{F}$ given by (8.6).

A simple BASIC language program for the solution of the problem of Section 8.1 is given in Listing 8.2, with results in Table 8.2.

Listing 8.2

```
10   X(0) = 1
20   Y(0) = 1
30   D = 0.1
40   REM D IS DELTA-X
50   N1 = 5
60   REM N1 IS LAST VALUE OF N( < 10)
70   FOR N = 0 TO N1
80   X(N) = X(0) + N * D
90   NEXT N
100  FOR N = 0 TO N1
110  F = X(N) * Y(N)
120  Y1 = Y(N) + F * D
130  M = N + 1
140  F1 = 0.5 * (F + X(M) * Y1)
```

150 Y(M) = Y(N) + D * F1
160 NEXT N
170 PRINT "N", "X(N)", "Y(N)"
180 FOR N = 0 TO N1
190 PRINT
200 PRINT N, X(N), Y(N)
210 NEXT N
220 END

**TABLE 8.2   HEUN'S METHOD FOR $\Delta x = 0.1$**

| $n$ | $x_n$ | $y_n$ | Exact |
|---|---|---|---|
| 0 | 1.0 | 1.0 | 1.0 |
| 1 | 1.1 | 1.1105 | 1.1107 |
| 2 | 1.2 | 1.2455 | 1.2461 |
| 3 | 1.3 | 1.4109 | 1.4120 |
| 4 | 1.4 | 1.6143 | 1.6161 |
| 5 | 1.5 | 1.8653 | 1.8682 |

This method is one of a class of methods known as *predictor-corrector methods*. They are so called because the equation

$$\bar{y}_{n+1} = y_n + F(x_n, y_n)\, \Delta x$$

predicts a new value of $y$ and the equation

$$y_{n+1} = y_n + \bar{F}\, \Delta x,$$

with $\bar{F}$ as in (8.5) corrects the predicted value. The predicted value is just the value expected when using Euler's method. the corrected value is an improvement over the Euler method, and Heun's method is often called the *improved Euler method*. It can be shown that this method is a second order method.

## 8.3 THE RUNGE–KUTTA METHOD

The Runge–Kutta method is one of the most popular methods of obtaining numerical solutions of ordinary differential equations. It is easily programmed and gives generally good results. This method consists of the iterated solution of the equations

$$x_n = x_0 + n\, \Delta x,$$

$$y_{n+1} = y_n + \frac{1}{6}(k_1 + 2k_2 + 2k_3 + k_4),$$

where

$$k_1 = F(x_n, y_n)\, \Delta x,$$

$$k_2 = F\left(x_n + \frac{\Delta x}{2}, y_n + \frac{k_1}{2}\right) \Delta x,$$

$$k_3 = F\left(x_n + \frac{\Delta x}{2}, \, y_n + \frac{k_2}{2}\right) \Delta x,$$

$$k_4 = F(x_n + \Delta x, \, y_n + k_3) \, \Delta x.$$

The equations given above represent a fourth order Runge–Kutta method.

Again we solve the initial value problem

$$y' = xy, \qquad y(1) = 1.$$

The results are given in Table 8.3 for $\Delta x = 0.1$ using the BASIC language routine for a fourth order Runge–Kutta scheme as given in Listing 8.3.

Listing 8.3

```
10  X(0) = 1
20  Y(0) = 1
30  D = 0.1
40  REM D IS DELTA-X
50  N1 = 5
60  REM N1 IS LAST VALUE OF N ( < 10)
70  FOR N = 0 TO N1
80  X(N) = X(0) + N * D
90  K1 = X(N) * Y(N) * D
100 Y1 = Y(N) + 0.5 * K1
110 X1 = X(N) + 0.5 * D
120 K2 = X1 * Y1 * D
130 Y2 = Y(N) + 0.5 * K2
140 K3 = X1 * Y2 * D
150 Y3 = Y(N) + K3
160 K4 = (X(N) + D) * Y3 * D
170 M = N + 1
180 Y(M) = Y(N) + (K1 + 2 * K2 + 2 * K3 + K4)/6
190 NEXT N
200 PRINT "N", "X(N)", "Y(N)"
210 FOR N = 0 TO N1
220 PRINT N, X(N), Y(N)
230 NEXT N
240 END
```

The three techniques covered in this chapter were presented to give you a brief introduction to the methods of finding numerical solutions of first order differential equations. These basic ideas may also be carried over to numerical solutions of systems of ordinary differential equations. For example, the Euler formulas for the

**TABLE 8.3**  RUNGE–KUTTA METHOD FOR
$\Delta x = 0.1$

| $n$ | $x_n$ | $y_n$ | Exact |
|---|---|---|---|
| 0 | 1.0 | 1.0 | 1.0 |
| 1 | 1.1 | 1.1107 | 1.1107 |
| 2 | 1.2 | 1.2461 | 1.2461 |
| 3 | 1.3 | 1.4120 | 1.4120 |
| 4 | 1.4 | 1.6161 | 1.6161 |
| 5 | 1.5 | 1.8682 | 1.8682 |

numerical approximation to the solution of the nonlinear initial value problem

$$\frac{dy}{dx} = F(y, z, x),$$

$$\frac{dz}{dx} = G(y, z, x),$$

$$y(x_0) = y_0, \qquad z(x_0) = z_0,$$

are

$$y_{n+1} = y_n + F(y_n, z_n, x_n) \, \Delta x,$$

$$z_{n+1} = z_n + G(y_n, z_n, x_n) \, \Delta x.$$

The availability of public domain and commercial software such as RKF4S or others in IMSL has decreased the need for a person who needs a numerical solution of a differential equation to actually write the computer program. However this availability has increased the need for the user to pay close attention to the existence and uniqueness of the solution to the equations under consideration as well as the accuracy of the method being used. The detailed discussion of the various types of errors, including machine round-off errors, that occur in numerical methods for solutions of differential equations are beyond the scope of this book. Such discussions occur in elementary books or courses on numerical analysis or numerical methods.

## EXERCISES

1.  Fill in the details necessary to show that the first two terms of the Taylor series (8.5) gives the expression for $y_{n+1}$ in the Euler method.

2.  Let $y' = F(x, y)$, $y(x_0) = y_0$ be a given initial value problem. Assume a formula for $y_{n+1}$ of the form

$$y_{n+1} = y_n + ak_1 + bk_2, \tag{8.7}$$

where

$$k_1 = F(x_n, y_n) \, \Delta x,$$
$$k_2 = F(x_n + \alpha \, \Delta x, y_n + \beta k_1). \tag{8.8}$$

Show that the choice

$$a = \frac{1}{2}, b = \frac{1}{2}, \alpha = 1, \beta = 1$$

reduces (8.7) and (8.8) to the improved Euler method.

Solve the following problems numerically on the indicated interval using Heun's method and the fourth order Runge–Kutta method.

**3.** $y' = 2xy$, $y(1) = 1$, $[1, \frac{3}{2}]$.      **4.** $y' = 1 + y^2$, $y(0) = 0$, $[0, \frac{1}{2}]$.

**5.** $y' = e^{-y}$, $y(0) = 0$, $[0, \frac{1}{2}]$.      **6.** $y' = 2x - 3y + 1$, $y(1) = 1$, $[1, \frac{3}{2}]$.

**7.** $y' = x$, $y(0) = 1$, $[0, 1]$.

# CHAPTER 9

# Systems
# of Autonomous
# Differential Equations

---

The first order linear system, whose solution is known (Chapter 5), is reexamined in order to determine as much as is possible about its solutions in the phase plane. Knowledge of this behavior of the linear system is then used to obtain information about certain types of second order nonlinear systems. A very brief treatment of the Lyapunov direct method is given, following which the Bendixon and Poincare–Bendixon theorems are stated without proof and illustrated.

## 9.1 INTRODUCTION

The van der Pol equation is the nonlinear ordinary differential equation

$$\frac{d^2x}{dt^2} + \mu(x^2 - 1)\frac{dx}{dt} + x = 0. \tag{9.1}$$

This is an important equation in physics and electrical engineering which first arose as an idealized description of a spontaneously oscillating circuit.

We may rewrite (9.1) as a coupled pair of differential equations by making the change of variables

$$y = \frac{dx}{dt}.$$

Under this change of variables the van der Pol equation becomes the first order system of equations

$$\frac{dx}{dt} = y, \qquad \frac{dy}{dt} = -x - \mu(x^2 - 1)y.$$

Since the van der Pol equation may be rewritten as a first order system of equations we are led to ask if it is possible to do the same thing to other second order nonlinear equations. That we may do so is not difficult to see in the general case, to which we now proceed.

Consider a second order equation of the form

$$\ddot{x}(t) = F[x(t), \dot{x}(t)], \tag{9.2}$$

in which we have written the independent variable as $t$ and will think of it as time, and the dependent variable as $x(t)$ and will think of it as position.

In (9.2) we make the change of variables $y = \dot{x} = dx/dt$ to obtain the equivalent system of first order equations

$$\dot{x} = y,$$
$$\dot{y} = F(x, y). \tag{9.3}$$

If $F(x, y)$ is a linear function, we may use the methods of Chapter 5 to find solutions of (9.3), thus effectively solving (9.2). Even if (9.3) is nonlinear we may be able to obtain from it a linear system, the study of which will give us some information about the solutions to (9.3).

The function $F$ of (9.3) depends, in the case in which $t$ represents time, upon both the position and its time rate of change, that is, the velocity, of a particle whose motion is obtained as a solution to (9.3). Physically, $y(t) = \dot{x}(t)$ represents the momentum of a particle of *unit* mass moving along the trajectory $x = x(t)$. The two dimensional space whose coordinates are $(x(t), p(t))$, $p(t)$ denoting momentum, is called the *phase space* of the particle motion. Since we are dealing with two dimensions this is also called the *phase plane*. Thus the coordinates $(x(t), y(t))$ where $y(t) = \dot{x}(t)$ are the *phase plane coordinates* of a particle of unit mass moving along the trajectory $x = x(t)$.

Note that the equations under consideration, (9.2) or (9.3), do not have an explicit dependence upon $t$. Such equations are called *autonomous*. Equations in which $t$ *does* appear explicitly have the form

$$\ddot{x}(t) = \phi(t, x(t), \dot{x}(t)) \tag{9.4}$$

and are not considered in this text.

Suppose now that we have an equation of the form (9.2) which we do not know how to solve. It may be that by transforming to the system (9.3) we may obtain information about the nature of the solutions in the $(x, y)$ plane, the phase plane, without solving the equation at all. Plots of solutions in the phase plane are called *trajectories*. A study of the form of the trajectories in the phase plane is called *phase plane analysis* and will prove quite useful in studying nonlinear equations. Since we already know how to solve linear systems, however, we will begin by looking at their solutions in the phase plane.

**EXAMPLE 9.1**     As a preliminary example we look at the linear system of first order equations

$$\dot{x} = y,$$
$$\dot{y} = -x, \quad x(0) = 0,$$

which has an obvious solution

$$x(t) = C_1 \sin t,$$
$$y(t) = C_1 \cos t.$$

This solution describes a particle moving in $(x, t)$ space along the curve

$$x(t) = C_1 \sin t$$

with velocity

$$y(t) = C_1 \cos t.$$

In phase space, on the other hand, we have

$$\left(\frac{x}{C_1}\right)^2 + \left(\frac{y}{C_1}\right)^2 = 1$$

and the phase space trajectories are circles centered at the origin with radius $C_1$.

If we plot one period of the motion in $(x, t)$ space, and under it the corresponding circle in the phase plane we obtain the curves of Figure 9.1. As the actual motion of the particle goes through one complete oscillation in $(x, t)$ space about $x(t) = 0$, a moving point in the phase plane goes once around the circle in the clockwise direction, starting from the point $(0, C_1)$. We are thus able to relate phase plane trajectories to those actually followed by the moving particle.

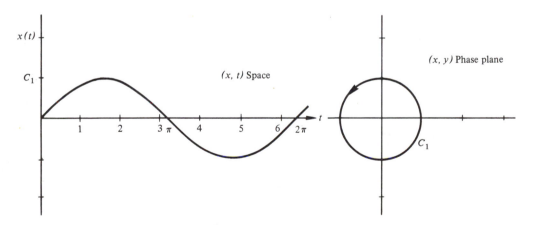

**Figure 9.1**

Note also that each point $(x, y)$ in phase space represents $(x(t), \dot{x}(t))$ for some time $t$. If the hypotheses of the existence and uniqueness theorem (Theorem 1.1) for the equation $\ddot{x} = F(x, \dot{x})$ are satisfied, with $x(t_0)$ and $\dot{x}(t_0)$ specified for some $t = t_0$, then the trajectory passing through the point $(x, y)$ in the phase plane is uniquely given. Thus under these conditions through each point in the phase plane there passes at most one trajectory.

The ordinary differential equation of second order, as has already been noted, with the form

$$\ddot{x} = F(x, \dot{x}), \tag{9.5}$$

may be written as the system of linear equations,

$$\dot{x} = y,$$
$$\dot{y} = F(x, y). \tag{9.6}$$

We may eliminate $t$ between the equations of (9.6) by dividing the second by the first. Thus

$$\frac{\dot{y}}{\dot{x}} = \frac{dy}{dx} = \frac{F(x, y)}{y}. \tag{9.7}$$

In a number of cases (9.7) is separable and may be integrated to obtain $y$ as a function of $x$. The equation $y = y(x)$ then describes the phase plane trajectories of the original equation (9.5), or (9.6).

**EXAMPLE 9.2**    Consider the differential equation which describes a harmonic oscillator,

$$\ddot{x} + \omega^2 x = 0.$$

Here $F(x, y) = -\omega^2 x$ and we have

$$\frac{dy}{dx} = \frac{-\omega^2 x}{y},$$

which separates as

$$y \, dy = -\omega^2 x \, dx$$

with solution

$$y^2 = -\omega^2 x^2 + C^2. \tag{9.8}$$

Equation (9.8) may be put into the form

$$\frac{y^2}{C^2} + \frac{x^2}{C^2/\omega^2} = 1,$$

so we see that the phase plane trajectories are ellipses.

**EXAMPLE 9.3**    If we reverse the sign of $\omega^2$ in Example 9.2, we obtain the equation

$$\ddot{x} - \omega^2 x = 0$$

or

$$\frac{dy}{dx} = \frac{\omega^2 x}{y}. \tag{9.9}$$

Equation (9.9) may also be integrated to give

$$y^2 = \omega^2 x^2 + C. \tag{9.10}$$

The phase plane trajectories, obtained from (9.10), are given by

$$\frac{y^2}{C} - \frac{x^2}{C/\omega^2} = 1$$

and thus are hyperbolas.

This approach may also be effected if the second order equation is non-linear, as the following example shows.

**EXAMPLE 9.4**    Let $x^2\ddot{x} - (\dot{x})^3 = 0$. We may write this in the form

$$\ddot{x} = \frac{(\dot{x})^3}{x^2} \qquad (9.11)$$

and let $y = \dot{x}$ to obtain the system of equations

$$\dot{x} = y,$$

$$\dot{y} = \frac{y^3}{x^2}.$$

Here $F(x, y) = y^3/x^2$, so that

$$\frac{dy}{dx} = \frac{y^2}{x^2}.$$

This last equation is a separable differential equation,

$$\frac{dy}{y^2} = \frac{dx}{x^2},$$

so that

$$-\frac{1}{y} = -\frac{1}{x} + C = \frac{Cx - 1}{x}.$$

Thus $y = x/(1 - Cx)$ gives the phase plane trajectories of (9.11).


# EXERCISES

1. For the system

$$\dot{x} = -y,$$

$$\dot{y} = x,$$

   subject to the conditions $x(0) = y(\pi/2) = 1$, plot the motion in $(x, t)$ space and in $(x, y)$ phase space for one period. Show the direction of motion and starting point in each case.

2. Write the following second order ODEs as a system of first order ODEs and obtain the equations for the phase plane trajectories.
   (a) $\ddot{x} = x \sin x^2$.
   (b) $(4 + x^2)\ddot{x} = 2x - x(\dot{x})^2$.
   (c) $x\ddot{x} = (\dot{x})^2$.
   (d) $y\ddot{x} = -\sin x$.

## 9.2 BASIC DEFINITIONS

We now look at the system

$$\frac{dx}{dt} = \dot{x} = F(x, y),$$

$$\frac{dy}{dt} = \dot{y} = G(x, y) \tag{9.12}$$

and assume that $F(x, y)$, $G(x, y)$ have continuous partial derivatives everywhere. If we think of this pair of equations as defining a particle path in the $(x, t)$ space, then $\dot{x}(t)$, $\dot{y}(t)$ are simply the components of the velocity vector and the slope of the trajectory in the $(x, y)$ plane is given by

$$\frac{dy}{dx} = \frac{\dot{y}}{\dot{x}} = \frac{G(x, y)}{F(x, y)}. \tag{9.13}$$

When both $F(x, y)$ and $G(x, y)$ are zero, the right-hand side of (9.13) is indeterminate. Such points are called *critical points* of (9.12). If $(x_0, y_0)$ is a critical point of (9.12) which has the property that there is a circle of radius $\delta$, $\delta > 0$, with $(x_0, y_0)$ as center which contains no other critical points of (9.12), we call $(x_0, y_0)$ an *isolated critical point*. From here on we assume that we deal only with isolated critical points. In fact, we will always consider the critical points to be at $(0, 0)$, since, if they are not, the change of coordinates

$$\bar{x} = x - x_0, \qquad \bar{y} = y - y_0$$

will move them there (see exercise 5, Section 9.4, page 233).

The curve $\gamma(t)$, given parametrically by $(x(t), y(t))$ in the $xy$-plane, is called a *path* or an *orbit* or a *trajectory* of the system. Once we know where the critical points are we need only determine the behavior of the paths near the critical points to be able to determine the behavior of the paths everywhere. The number of things that may happen at a critical point is, fortunately, limited.

**Definition 9.1.** Let $(0, 0)$ be an isolated critical point of (9.12). If, given $\epsilon > 0$, there exists a $\delta$, $0 < \delta < \epsilon$, such that every path which passes through $(x_1, y_1)$ interior to the circle of radius $\delta$ about $(0, 0)$ at $t = t_1$ remains inside the circle of radius $\epsilon$ about $(0, 0)$ for all $t > t_1$, we say that the critical point is *stable*.

The circumstances of Definition 9.1 are illustrated in Figure 9.2. We see that the path may wander about in the circle of radius $\epsilon$, but once it has passed through $(x_1, y_1)$ at time $t = t_1$ it may never leave the circle of radius $\epsilon$.

**Definition 9.2.** Let $(0, 0)$ be a stable critical point of (9.12). If there exists a $\delta_1 > 0$ such that every path $(x, y)$ which comes within $\delta_1$ of $(0, 0)$ at time $t_1$ has the property that

$$\lim_{t \to \infty} x(t) = 0,$$

$$\lim_{t \to \infty} y(t) = 0,$$

we say that the critical point is *asymptotically stable*.

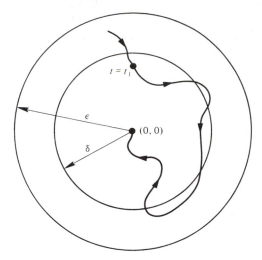

**Figure 9.2**   A stable critical point.

Asymptotic stability is stronger than stability in the sense that all stability requires is that after having gotten as close as we desire to (0, 0), we stay at least that close, while asymptotic stability requires that as $t \to \infty$ we actually get arbitrarily close to (0, 0).

**Definition 9.3.**   Critical points which are not stable are called *unstable*.

Suppose now we say that a path $(x, y)$ *approaches* $(0, 0)$ *as* $t \to \infty$ if $\lim_{t \to \infty} x(t) = 0$ and $\lim_{t \to \infty} y(t) = 0$. Then a stable critical point is asymptotically stable if every path passing within $\delta_1$ of (0, 0) approaches (0, 0) as $t \to \infty$.

## EXERCISES

Find the critical points of the following.

**1.** $\dot{x} = 2x + y,$
    $\dot{y} = -x + 2y.$

**2.** $\dot{x} = -2x + y,$
    $\dot{y} = x - 1.$

**3.** $\dot{x} = 3x - 2y,$
    $\dot{y} = -2x + 2y.$

**4.** $\dot{x} = 1 - y,$
    $\dot{y} = x + y.$

**5.** $\dot{x} = 5x + 2y + 3,$
    $\dot{y} = 6x - 2y.$

**6.** $\dot{x} = x^2 + y^2,$
    $\dot{y} = x^2 - y^2.$

**7.** $\dot{x} = x^2 + y^2 - 1,$
    $\dot{y} = x - y.$

**8.** $\dot{x} = -2x^2 + y^2 + 1,$
    $\dot{y} = x^2 - y.$

**9.** $\dot{x} = y,$
    $\dot{y} = -x.$

**10.** $\dot{x} = xy,$
     $\dot{y} = -x^2.$

## 9.3 THE LINEAR CASE

We begin by studying the linear system of first order equations

$$\dot{x} = ax + by,$$
$$\dot{y} = cx + dy.$$

(9.14)

From Chapter 5 we know that for $a$, $b$, $c$, $d$ real constants the solutions of (9.14) depend upon the nature of the eigenvalues of its coefficient matrix. Obviously (0, 0) is a critical point and we are led to try to determine the properties of the solution near (0, 0) by examining the eigenvalues.

For (9.14) the eigenvalue equation is

$$\begin{vmatrix} a - \lambda & b \\ c & d - \lambda \end{vmatrix} = 0. \tag{9.15}$$

Expanding (9.15) gives

$$\lambda^2 - (a + d)\lambda + (ad - bc) = 0, \tag{9.16}$$

or in a more concise form,

$$\lambda^2 - p\lambda + q = 0, \tag{9.16}$$

where $p = a + d$, $q = ad - bc$. Solutions of (9.16) are

$$\lambda_1, \lambda_2 = \frac{p \pm \sqrt{p^2 - 4q}}{2},$$

so elementary algebra shows that the eigenvalues, $\lambda_1$ and $\lambda_2$, satisfy

$$p = \lambda_1 + \lambda_2$$

$$q = \lambda_1 \lambda_2. \tag{9.17}$$

The nature of these eigenvalues is subdivided into three categories by values of the discriminant $p^2 - 4q$.

If $p^2 - 4q < 0$, the eigenvalues are complex conjugates; if $p^2 - 4q = 0$, there is a single real eigenvalue of multiplicity two; and if $p^2 - 4q > 0$, they are real and distinct. We use these possibilities to divide the possible behaviors at an isolated critical point into cases as follows.

**Case 1a.** $\lambda_1$, $\lambda_2$ real, distinct, opposite signs ($q < 0$). If $q < 0$, then $p^2 - 4q > 0$ and the eigenvalues are real and distinct. Furthermore, the second of equations (9.17) shows that they are of opposite sign and we may assume that $\lambda_1 < 0$, so that $\lambda_2 > 0$.

Denote the eigenvectors corresponding to $\lambda_1$ and $\lambda_2$ by

$$\begin{bmatrix} a_1 \\ a_2 \end{bmatrix} \quad \text{and} \quad \begin{bmatrix} b_1 \\ b_2 \end{bmatrix},$$

respectively. From Chapter 5 we know that (9.14) has solutions of the form

$$\begin{bmatrix} x \\ y \end{bmatrix} = C_1 \begin{bmatrix} a_1 \\ a_2 \end{bmatrix} e^{\lambda_1 t} + C_2 \begin{bmatrix} b_1 \\ b_2 \end{bmatrix} e^{\lambda_2 t} \tag{9.18}$$

and from (9.18) we may plot representative curves in the $xy$-plane.

Suppose that in (9.18), $C_2 = 0$, $C_1 = \pm 1$. Then we have

$$x = \pm a_1 e^{\lambda_1 t},$$

$$y = \pm a_2 e^{\lambda_1 t},$$

or

$$y = \frac{a_2}{a_1}x. \tag{9.19}$$

Now (9.19) represents straight lines in the $xy$-plane which pass through the origin with slope $m = a_2/a_1$. Since $\lambda_1 < 0$ we have

$$\lim_{t \to \infty} x(t) = 0,$$
$$\lim_{t \to \infty} y(t) = 0,$$

and the paths approach (0, 0) as $t \to \infty$. Thus (9.19) represents incoming paths with respect to (0, 0).

Similarly, if $C_1 = 0$, $C_2 = \pm 1$, we obtain

$$x = \pm b_1 e^{\lambda_2 t}$$
$$y = \pm b_2 e^{\lambda_2 t}$$

and

$$y = \frac{b_2}{b_1}x. \tag{9.20}$$

Again the trajectories are straight lines in the $xy$-plane, but as $\lambda_2 > 0$ we have

$$\lim_{t \to -\infty} x(t) = 0,$$
$$\lim_{t \to -\infty} y(t) = 0,$$

so that the paths approach (0, 0) as $t \to -\infty$. These are outgoing paths, since as $t \to +\infty$ they move away from the origin.

For the general case, in which neither $C_1$ nor $C_2$ is zero, we see that

$$\lim_{t \to \infty} \begin{bmatrix} x(t) \\ y(t) \end{bmatrix} = \lim_{t \to \infty} \begin{bmatrix} C_2 b_1 e^{\lambda_2 t} \\ C_2 b_2 e^{\lambda_2 t} \end{bmatrix}$$

and

$$\lim_{t \to -\infty} \begin{bmatrix} x(t) \\ y(t) \end{bmatrix} = \lim_{t \to -\infty} \begin{bmatrix} C_1 a_1 e^{\lambda_1 t} \\ C_1 a_2 e^{\lambda_1 t} \end{bmatrix}.$$

Thus the lines (9.19) and (9.20) are asymptotes for the solution curves, as $t$ becomes unbounded.

A typical configuration for this situation is shown in Figure 9.3, the arrowheads indicating the direction of travel of a particle moving along one of the paths as $t \to +\infty$. This critical point at (0, 0) is called a *saddle point* because the paths are reminiscent of the level curves of the hyperbolic paraboloid in $(x, y, z)$ space. Since all paths move away from (0, 0) except the incoming straight lines for the situation where $C_2 = 0$, (0, 0) is an unstable critical point.

We may summarize this state of affairs in the following theorem.

**Theorem 9.1.**    Let (0, 0) be an isolated critical point of (9.14). If the eigenvalues of (9.14) are real, distinct, and of opposite sign, (0, 0) is a saddle point.

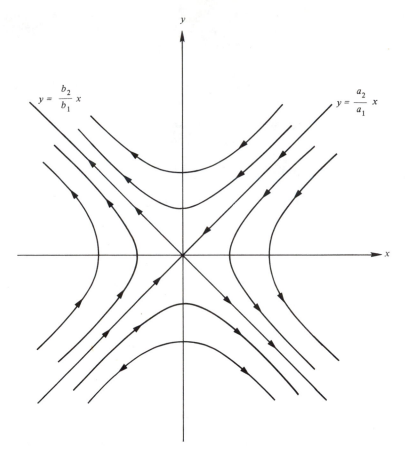

**Figure 9.3**

**EXAMPLE 9.5**   Consider the system

$$\dot{x} = y,$$

$$\dot{y} = x.$$

Here $a = 0$, $b = 1$, $c = 1$, $d = 0$, so the eigenvalue equation is

$$\begin{vmatrix} -\lambda & 1 \\ 1 & -\lambda \end{vmatrix} = 0$$

or

$$\lambda^2 - 1 = 0.$$

The eigenvalues are $\lambda_1 = -1$, $\lambda_2 = +1$, and by Theorem 9.1 there is a saddle point at $(0, 0)$.

The eigenvector belonging to $\lambda_1$ has the form $\alpha \begin{bmatrix} 1 \\ -1 \end{bmatrix}$, $\alpha$ arbitrary, and that belonging to $\lambda_2$ has the form $\beta\begin{bmatrix} 1 \\ 1 \end{bmatrix}$, $\beta$ arbitrary, so that (9.18) becomes

$$\begin{bmatrix} x \\ y \end{bmatrix} = C_1\alpha\begin{bmatrix} 1 \\ -1 \end{bmatrix}e^{-t} + C_2\beta\begin{bmatrix} 1 \\ 1 \end{bmatrix}e^{t}. \qquad (9.21)$$

The asymptotes for incoming paths are given by (9.19), which becomes

$$y = -x,$$

while those for the outgoing paths are given by (9.20), which becomes

$$y = x.$$

**Case 1b.**    $\lambda_1$, $\lambda_2$ real, distinct and have the same sign.    In this case we assume first that $\lambda_1 < \lambda_2 < 0$; that is, both eigenvalues are negative and $\lambda_1$ has the larger absolute value. The general solution is still given by (9.18). In this case all paths are incoming as both $x(t)$ and $y(t)$ approach $(0, 0)$ as $t \to \infty$. In fact, the critical point is asymptotically stable and the paths are as shown in Figure 9.4. The critical point of Figure 9.4 is called a *stable node*.

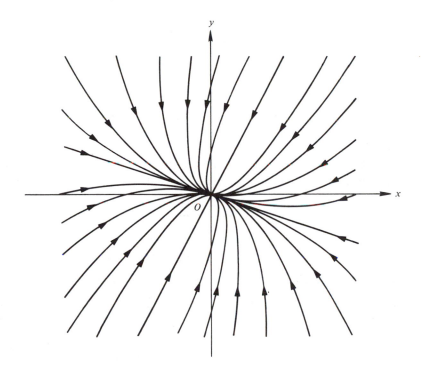

**Figure 9.4**    Stable node.

Next assume that $\lambda_1$ and $\lambda_2$ are both positive. Then $x(t)$, $y(t)$ approach $(0, 0)$ as $t \to -\infty$, so all paths are outgoing. In this case the critical point is called an *unstable node*. An unstable node is shown in Figure 9.5.

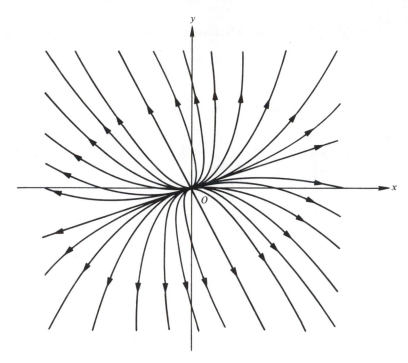

**Figure 9.5** Unstable node.

We summarize this case in the following theorem.

**Theorem 9.2.** Let $(0, 0)$ be an isolated critical point of $(9.7)$. If the eigenvalues of $(9.7)$ are real, distinct, and of the same sign, the critical point is a node. If the eigenvalues are negative, the node is stable. If they are positive, it is unstable.

**EXAMPLE 9.6** Consider the system

$$\dot{x} = 2y,$$
$$\dot{y} = -x - 3y. \qquad (9.22)$$

The eigenvalue equation becomes

$$\begin{vmatrix} -\lambda & 2 \\ -1 & -3 - \lambda \end{vmatrix} = 0$$

or

$$\lambda^2 + 3\lambda + 2 = (\lambda + 2)(\lambda + 1) = 0.$$

Thus $\lambda_1 = -2$, $\lambda_2 = -1$ and $(0, 0)$ is a stable node.

On the other hand, the system

$$\dot{x} = 2y,$$
$$\dot{y} = -x + 3y,$$

which is almost identical to (9.22), has

$$\lambda^2 - 3\lambda + 2 = (\lambda - 2)(\lambda - 1) = 0$$

as its eigenvalue equation and has eigenvalues

$$\lambda_1 = 1, \lambda_2 = 2.$$

Thus for this system $(0, 0)$ is an unstable node.

*Case 2.*    $\lambda_1 = \lambda_2 = \lambda$, $\lambda$ real and nonzero.    If the eigenvalues are equal, then (9.18) no longer gives the solution to our system of differential equations. The theory of Chapter 5, however, tells us that solutions are given by

$$\begin{bmatrix} x \\ y \end{bmatrix} = C_1 \begin{bmatrix} a_1 \\ a_2 \end{bmatrix} e^{\lambda t} + C_2 \left\{ \begin{bmatrix} b_1 \\ b_2 \end{bmatrix} t + \begin{bmatrix} d_1 \\ d_2 \end{bmatrix} \right\} e^{\lambda t}. \tag{9.23}$$

Assume first that $\lambda_1 = \lambda_2 = \lambda < 0$. If $C_2 = 0$, we have

$$x = C_1 a_1 e^{\lambda t},$$
$$y = C_1 a_2 e^{\lambda t}$$

or

$$y = \frac{a_2}{a_1} x \tag{9.24}$$

and the paths are lines of slope $a_2/a_1$ passing through $(0, 0)$. Since $\lambda < 0$, $x(t)$ and $y(t)$ approach $(0, 0)$ as $t \to +\infty$ and the lines are incoming. Just as in case 1b if $C_1$ and $C_2$ are both nonzero, all paths approach $(0, 0)$ as $t \to +\infty$ and $(0, 0)$ is an asymptotically stable node.

Now assume that $\lambda > 0$. Just as in case 1b, we have an unstable node. As a special case we consider (9.14) when $a = d$, $b = c = 0$. Thus we have

$$\dot{x} = ax,$$
$$\dot{y} = ay.$$

The characteristic equation (9.16) becomes

$$\lambda^2 - 2a\lambda + a^2 = 0$$

and has solutions

$$\lambda_1 = \lambda_2 = a,$$

so that the eigenvalues are again real, and equal. The solutions are easily obtained as

$$x = C_1 e^{at},$$
$$y = C_2 e^{at}$$

or

$$\frac{y}{x} = \frac{C_2}{C_1},$$

which we may take as $y = Cx$. Thus all trajectories are lines. They are incoming if

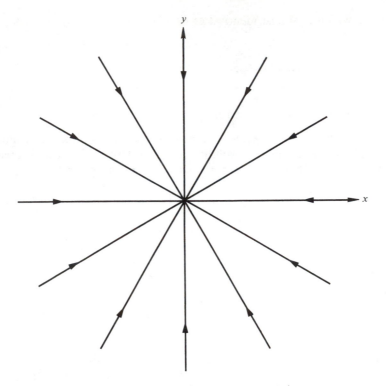

**Figure 9.6**   A star (incoming trajectories).

$a < 0$ and outgoing if $a > 0$. Because of the appearance of the trajectories, as may be observed in Figure 9.6, these nodes are called *stars*.

We summarize these results in the following theorem.

**Theorem 9.3.**   Let $(0, 0)$ be an isolated critical point of (9.14). If the eigenvalues of (9.14) are real, equal, and nonzero, the critical point is a node. If the eigenvalues are negative, the node is stable. If the eigenvalues are positive, the node is unstable.

We must now consider the cases in which the eigenvalues of (9.14) are complex.

*Case 3a.*   $\lambda = \rho \pm i\omega$, $\rho \neq 0$, $\omega \neq 0$. In this case the eigenvalues are complex conjugates and, because $\rho \neq 0$, are not pure imaginary. We may write the solutions in the form

$$\begin{bmatrix} x \\ y \end{bmatrix} = C_1 \begin{bmatrix} \alpha_1 \\ \beta_1 \end{bmatrix} e^{(\rho - i\omega)t} + C_2 \begin{bmatrix} \alpha_2 \\ \beta_2 \end{bmatrix} e^{(\rho + i\omega)t}, \tag{9.25}$$

or, after factoring,

$$\begin{bmatrix} x \\ y \end{bmatrix} = e^{\rho t} \{ C_1 \begin{bmatrix} \alpha_1 \\ \beta_1 \end{bmatrix} e^{-i\omega t} + C_2 \begin{bmatrix} \alpha_2 \\ \beta_2 \end{bmatrix} e^{i\omega t} \}. \tag{9.26}$$

Suppose now that $\rho < 0$. Then, because the terms in braces are periodic, the behavior for large values of $t$ of the solutions is governed by $e^{\rho t}$, which goes to zero as $t \to \infty$. Thus *all* paths approach $(0, 0)$ as $t \to \infty$. To see how they do so, we apply Euler's formula to (9.26) obtaining the solution in the form

$$\begin{bmatrix} x \\ y \end{bmatrix} = e^{\rho t} \begin{bmatrix} u_1 \cos \omega t + v_1 \sin \omega t \\ u_2 \cos \omega t + v_2 \sin \omega t \end{bmatrix}, \tag{9.27}$$

in which, to obtain real solutions, we take $u_1$, $u_2$, $v_1$, $v_2$ to be real constants. The paths in the $xy$-plane are spirals and approach $(0, 0)$ as $t \to \infty$ (see Figure 9.7). If on the other hand, $\rho > 0$, the analysis proceeds as above but the solutions approach $(0, 0)$ as $t \to -\infty$ and so spiral out instead of in. These critical points are called *foci*, and we have the following theorem.

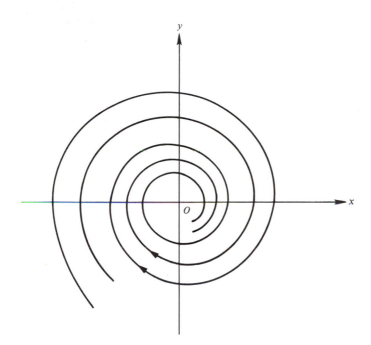

**Figure 9.7**  A focus.

**Theorem 9.4.**    Let $(0, 0)$ be an isolated singular point of (9.14). If the eigenvalues of (9.14) are complex with nonzero real part, the critical point is a focus.

***Case 3b.***    $\lambda = \rho \pm i\omega$, $\rho = 0$, $\omega \neq 0$.    In this case the eigenvalues are pure imaginary and (9.27) reduces to the pair of equations

$$\begin{aligned} x &= u_1 \cos \omega t + v_1 \sin \omega t, \\ y &= u_2 \cos \omega t + v_2 \sin \omega t, \end{aligned} \tag{9.28}$$

with $u_1$, $u_2$, $v_1$, $v_2$ real constants.

Consider the special case of (9.28) in which

$$u_2 = v_1 = 0,$$

so that

$$\frac{x}{u_1} = \cos \omega t,$$

$$\frac{y}{v_2} = \sin \omega t.$$

From (9.29) we have

$$\left(\frac{x}{u_1}\right)^2 + \left(\frac{y}{v_2}\right)^2 = 1,$$

which is the equation of an ellipse in the $xy$-plane. The general case of (9.28) also yields paths that are ellipses but whose axes are rotated away from the $x$ and $y$ axes. In this case the paths are closed and the critical point is obviously stable. The paths do not approach (0, 0) as $t \rightarrow \infty$, so the critical point is not asymptotically stable. Such a critical point is called a *center* and we have the following theorem.

**Theorem 9.5.**    Let $(0, 0)$ be an isolated critical point of (9.14). If the eigenvalues of (9.14) are pure imaginary, the critical point is a center.

### 9.3.1 Summary of Results for Linear Systems

Let $(0, 0)$ be an isolated singular point of

$$\dot{x} = ax + by,$$
$$\dot{y} = cx + dy,$$

whose eigenvalues satisfy

$$\lambda_2 - (a + d)\lambda + (ad - bc) = 0.$$

Then the nature of the critical point is as given in Table 9.1.

**TABLE 9.1**    CLASSIFICATION OF CRITICAL POINTS AND STABILITY

| Eigenvalues, $\lambda_1, \lambda_2$ | Type of critical point | Stability |
|---|---|---|
| 1a. Real, distinct, opposite signs | Saddle | Unstable |
| 1b. Real, unequal, same sign | Node | Asymptotically stable if eigenvalues are negative; unstable if eigenvalues are positive |
| 2. Real, equal | Node (star) | |
| 3a. Complex, $\lambda = \rho \pm i\omega$; $\rho \neq 0, \omega \neq 0$ | Focus | Asymptotically stable if $\rho < 0$; unstable if $\rho > 0$ |
| 3b. Pure imaginary, $\lambda = \rho \pm i\omega$; $\rho = 0, \omega \neq 0$ | Center | Stable but not asymptotically stable |

## EXERCISES

Classify the critical points of the following.

1. $\dot{x} = -2x + 4y,$
   $\dot{y} = x + y,$
2. $\dot{x} = 2x + y,$
   $\dot{y} = -3x - 2y,$
3. $\dot{x} = 2x + 5y,$
   $\dot{y} = x + 6y,$
4. $\dot{x} = -2x + 5y,$
   $\dot{y} = \quad x - 2y,$
5. $\dot{x} = \quad y,$
   $\dot{y} = -x,$
6. $\dot{x} = -x + y,$
   $\dot{y} = -x,$
7. $\dot{x} = x + y,$
   $\dot{y} = -x,$
8. $\dot{x} = x,$
   $\dot{y} = x + y,$
9. $\dot{x} = -y,$
   $\dot{y} = x,$
10. $\dot{x} = y,$
    $\dot{y} = x,$

### 9.3.2 Application

We reconsider (Chapter 4) the motion of a particle of mass $m$ in the presence of a restoring force proportional to the displacement and a damping force proportional to the velocity. According to Newton's second law, the force equation is

$$m\ddot{x} = -kx - c\dot{x}. \tag{9.30}$$

If we let $y = \dot{x}$ in (9.30), we have the equivalent system

$$\dot{x} = y,$$
$$\dot{y} = -\frac{k}{m}x - \frac{c}{m}y. \tag{9.31}$$

The eigenvalue equation for (9.31) is

$$\begin{vmatrix} -\lambda & 1 \\ -\dfrac{k}{m} & -\dfrac{c}{m} - \lambda \end{vmatrix} = 0$$

or

$$\lambda\left(\lambda + \frac{c}{m}\right) + \frac{k}{m} = 0. \tag{9.32}$$

The solutions to (9.32) are easily obtained as

$$\lambda = \frac{1}{2m}(-c \pm \sqrt{c^2 - 4mk}), \tag{9.33}$$

the nature of which depends on the discriminant

$$c^2 - 4mk.$$

We consider the following situations:

   1.   *c = 0, no damping.*   Since $m$ and $k$ are positive, this yields case 3b of

Table 9.1. Physically, this says that the system undergoes stable sinusoidal oscillation about its equilibrium point.

   **2.   $c^2 - 4mk = 0$.**   In this case $\lambda = -c/2m$ and we have case 2 of Table 9.1, where the critical point is asymptotically stable. Physically, the system is critically damped and the motion is not oscillatory but does return to equilibrium.

   **3.   $c^2 - 4mk < 0$, $c \neq 0$.**   In this case the eigenvalues are complex conjugates with negative real parts, so we have case 3a of Table 9.1. Since $\rho = -c/(2m) < 0$, the critical point is asymptotically stable. The motion of the system is a damped oscillation which returns to the equilibrium position as $t \rightarrow \infty$.

   **4.   $c^2 - 4mk > 0$.**   In this case the roots are real, unequal, and negative, so we have case 1b of Table 9.1 and the critical point is asymptotically stable. Physically, the displacement and velocity of the system point go to zero without oscillation. This is the overdamped case.

## EXERCISES

For the cases just considered, assume the following values for the constants $c$, $m$, $k$ and sketch the trajectories in the phase plane.

   **1.** $c = 0$,   $m = k = 1$.
   **2.** $c = 2$,   $m = k = 1$.
   **3.** $c = 1$,   $m = k = 1$.
   **4.** $c = 3$,   $m = k = 1$.
   **5.** Show that the results of the example in this section agree with those obtained by solving $m\ddot{x} = -kx - c\dot{x}$ as a second order differential equation (see Section 4.6.2).

## 9.4 NONLINEAR SYSTEMS: COMPARISON METHOD

We now return to the equations

$$\dot{x} = F(x, y),$$
$$\dot{y} = G(x, y) \tag{9.12}$$

and see if we may use our knowledge of the behavior of linear systems to study nonlinear ones. As it turns out, in many instances this is possible. The conditions and circumstances under which we may will be stated rigorously later in this section. For now let's see what would happen if the following conditions held:

   1. Equation (9.12) has an isolated critical point at the origin.
   2. The functions $F(x, y)$, $G(x, y)$ have Taylor's series expansions about $(0, 0)$.

Since the critical point is isolated we know that there is some circle of radius $\epsilon$, which we may choose as small as we wish, about $(0, 0)$, which contains no other critical points. Throughout this circle both $F$ and $G$ have series expansions that look like

$$F(x, y) = ax + by + f(x, y),$$
$$G(x, y) = cx + dy + g(x, y),$$

(9.34)

where $f(x, y)$ and $g(x, y)$ contains terms of degree 2 or higher. If we stay close to $(0, 0)$, then $x$ and $y$ are small quantities and their higher powers are smaller still. What we would hope would happen is that by staying close enough to $(0, 0)$ we could make the terms $f(x, y)$ and $g(x, y)$ small enough to be neglected. If we can, then using (9.34) in (9.12) will give us

$$\dot{x} = ax + by + f(x, y),$$
$$\dot{y} = cx + dy + g(x, y),$$

(9.35)

and neglecting higher order terms in (9.35) reduces it to

$$\dot{x} = ax + by,$$
$$\dot{y} = cx + dy.$$

(9.36)

Now (9.36) is the linear system we have already studied. We would expect that the behavior of the system (9.35) would be the same as that of (9.36) for $(x, y)$ sufficiently close to $(0, 0)$. Thus, if $F(x, y)$ and $G(x, y)$ are expandable as in (9.34), we may *linearize* the system (9.35) and use the *linearized equation*, (9.36), to determine the behavior of the solutions at the origin for situations described by the following theorem. We must be sure, however, that $f(x, y)$ and $g(x, y)$ go to zero faster than the linear terms so that the linear terms will dominate near the critical point.

**Theorem 9.6.**    Let the system (9.12) be given with $F(x, y)$ and $G(x, y)$ as in (9.34). Then (9.12) has an isolated singular point at $(0,0)$ and the linear terms of (9.34) dominate near $(0, 0)$ if the following three conditions hold.

i) $\begin{vmatrix} a & b \\ c & d \end{vmatrix} \neq 0.$

ii) $f(x, y)$, $g(x, y)$ have continuous first partial derivatives for all $(x, y)$.

iii) $\lim\limits_{(x, y) \to (0, 0)} \dfrac{f(x, y)}{\sqrt{x^2 + y^2}} = \lim\limits_{(x, y) \to (0, 0)} \dfrac{g(x, y)}{\sqrt{x^2 + y^2}} = 0.$

**EXAMPLE 9.7**    Consider the system

$$\dot{x} = 2x + y + x^2,$$
$$\dot{y} = x + 2y + y^2.$$

Here

$$\begin{vmatrix} a & b \\ c & d \end{vmatrix} = \begin{vmatrix} 2 & 1 \\ 1 & 2 \end{vmatrix} = 4 - 1 = 3 \neq 0$$

and $f(x, y) = x^2$, $g(x, y) = y^2$ certainly have continuous first partial derivatives for all $(x, y)$. Thus, to apply Theorem 9.7, we need examine the limits (with the aid of polar coordinates):

$$\lim_{(x,y)\to(0,0)} \frac{x^2}{\sqrt{x^2 + y^2}} = \lim_{r\to 0} r\cos^2\theta = 0, \qquad \text{for all } \theta,$$

$$\lim_{(x,y)\to(0,0)} \frac{y^2}{\sqrt{x^2 + y^2}} = \lim_{r\to 0} r\sin^2\theta = 0, \qquad \text{for all } \theta.$$

Thus the conditions of the theorem are satisfied.

We now have the question: Is the classification of the critical points for the linearized system the same as for the original nonlinear system? The answer is yes in most, but not all cases, as given in the following theorems.

**Theorem 9.7.**   Let Theorem 9.6 hold and consider the linearized system (9.36) associated with the nonlinear system (9.35). In the following cases the nature of the critical point of both systems is the same. $\lambda_1$, $\lambda_2$ are eigenvalues of (9.36).

i) $\lambda_1 \neq \lambda_2$, both real and of the same sign. $(0, 0)$ is a node of both systems.

ii) $\lambda_1 \neq \lambda_2$, both real but of opposite sign. $(0, 0)$ is a saddle point of both systems.

iii) $\lambda_1 = \lambda_2$, both real but $\begin{bmatrix} a & b \\ c & d \end{bmatrix}$ is not a scalar multiple of $\begin{bmatrix} 1 & 0 \\ 0 & 1 \end{bmatrix}$. $(0, 0)$ is a node of both systems.

iv) $\lambda_1 = \bar{\lambda}_2$, $\text{Re}(\lambda_1) \neq 0$, i.e., $\lambda_1$ is the complex conjugate of $\lambda_2$. $(0,0)$ is a focus of both systems.

If $\lambda_1 = \bar{\lambda}_2$, $\text{Re}(\lambda_1) < 0$, the critical point is asymptotically stable for both systems.

If $\lambda_1 = \bar{\lambda}_2$, $\text{Re}(\lambda_1) > 0$, the critical point is unstable for both systems.

**Theorem 9.8.**   Under the hypothesis of Theorem 9.7, for the following cases the nature of the critical points of both systems may be different.

i) $\lambda_1 = \lambda_2$ and $\begin{bmatrix} a & b \\ c & d \end{bmatrix}$ is a scalar multiple of $\begin{bmatrix} 1 & 0 \\ 0 & 1 \end{bmatrix}$. $(0, 0)$ is a node of the linearized system but may be either a node or a focus of the nonlinear system.

ii) $\lambda_1 = \bar{\lambda}_2$, $\text{Re}(\lambda_1) = 0$, i.e., $\lambda_1$ is pure imaginery. $(0, 0)$ is a center of the linearized system but may be either a center or a focus of the nonlinear system. The critical point of the linearized system is stable, but the critical point of the nonlinear system may be asymptotically stable, stable, or unstable.

**EXAMPLE 9.8**   The van der Pol equation, equation (9.1), may, as we have seen, be written as the system of equations

$$\frac{dx}{dt} = y,$$

$$\frac{dy}{dt} = \mu y - x - \mu y x^2. \tag{9.37}$$

The nonlinear term in the second of equations (9.37) is $g(x, y) = \mu y x^2$ and since

$$\lim_{(x, y) \to (0, 0)} \frac{\mu y x^2}{\sqrt{x^2 + y^2}} = 0$$

the system may be linearized. The linearized system is

$$\frac{dx}{dt} = y,$$

$$\frac{dy}{dt} = \mu y - x. \tag{9.38}$$

The eigenvalues of (9.38) are obtained from

$$\begin{vmatrix} -\lambda & 1 \\ -1 & \mu - \lambda \end{vmatrix} = 0.$$

Expanding yields

$$\lambda(\lambda - \mu) + 1 = 0,$$

or

$$\lambda^2 - \mu\lambda + 1 = 0,$$

which has roots

$$\lambda = \frac{\mu \pm \sqrt{\mu^2 - 4}}{2} = \frac{1}{2}\mu \pm i\sqrt{1 - \left(\frac{1}{2}\mu\right)^2}.$$

If the constant $\mu$ is positive and less than 2, the roots are complex congugates with positive real part and the critical point is unstable.

**EXAMPLE 9.9**    In Section 5.7.4, page 197, we considered a predator–prey model of two interacting species which resulted in a linear system of two first order differential equations. We noted then that the model was limited to very small values of time. A more realistic model will now be developed which will not have this limitation.

Let

$$x = \text{prey population at time } t,$$

$$y = \text{predator population at time } t.$$

The model equations from Section 5.7.4 were

$$\dot{x} = a_{11}x - a_{12}y,$$

$$\dot{y} = a_{21}x - a_{22}y,$$

with the $a_{ij}$, $i, j, = 1, 2$, taken to be constants. We now assume that the rate of decrease in the prey population is not simply proportional to the number of predators, but is proportional to the number of encounters between predators and prey. This means that we may replace the term $a_{12}y$ by $a_{12}xy$ in the above equations. We also replace the term $a_{21}x$ by $a_{21}xy$ to account for a rate of increase in the predator population that is likewise proportional to the number of encounters between the predators and prey. Thus we now consider the system of nonlinear differential equations

$$\dot{x} = a_{11}x - a_{12}xy,$$

$$\dot{y} = a_{21}xy - a_{22}y,$$

which is known as a *Volterra–Lotka model* for a predator–prey problem.

To see if we may use Theorem 9.6 to analyze this nonlinear system, we first note that both $(0, 0)$ and $(a_{22}/a_{21}, a_{11}/a_{12})$ are critical points. For the critical point at $(0, 0)$, we note that all the three hypotheses of Theorem 9.6 are fulfilled since i) The determinant of the associated linear system is $-a_{11}a_{22} \neq 0$, ii) $-a_{12}xy$ and $a_{21}xy$ both have continuous first partial derivatives, and iii) the required limits are both zero (it is easy to see this with the help of polar coordinates). Thus the linear terms dominate near $(0, 0)$. To use Theorem 9.7 to classify the critical point at $(0, 0)$, we need the eigenvalues of the linearized system. These are given by the solutions of the equation

$$\begin{vmatrix} a_{11} - \lambda & 0 \\ 0 & -a_{22} - \lambda \end{vmatrix} = 0.$$

Since $a_{11}$ and $a_{22}$ are both positive, the eigenvalues, given by $a_{11}$ and $-a_{22}$, are real and of opposite sign, so by case ii) of Theorem 9.7 we see that $(0, 0)$ is a saddle point for both systems. [Note that in a predator–prey model this equilibrium point at $(0, 0)$ is not of physical interest.]

To establish the nature of the critical point at $(a_{22}/a_{21}, a_{11}/a_{12})$, we make the change of variables

$$\bar{x} = x - \frac{a_{22}}{a_{21}},$$

$$\bar{y} = y - \frac{a_{11}}{a_{12}},$$

and discover that the resulting differential equations are

$$\dot{\bar{x}} = -\left(\frac{a_{12}a_{22}}{a_{21}}\right)\bar{y} - a_{12}\bar{x}\,\bar{y},$$

$$\dot{\bar{y}} = \left(\frac{a_{21}a_{11}}{a_{12}}\right)\bar{x} + a_{21}\bar{x}\,\bar{y}.$$

It is straightforward calculation to show that the hypotheses of Theorem 9.6 are satisfied (see exercise 5), and that the eigenvalues of the associated linear system are given by

$$\lambda = \pm i\sqrt{a_{11}a_{22}}.$$

Thus from Theorem 9.8 we see that (case ii) $\bar{x} = \bar{y} = 0$ is a stable critical point of the linearized system. We may plot the trajectories in the phase plane by noting that they are given by the solution of

$$\frac{d\bar{y}}{d\bar{x}} = -\frac{(a_{21}a_{11}/a_{12})\bar{x}}{(a_{12}a_{22}/a_{21})\,\bar{y}} \qquad [\text{see } (9.13)]$$

as

$$(a_{12}a_{22}/a_{21})\,\bar{y}^2 + (a_{21}a_{11}/a_{12})\bar{x}^2 = C \qquad \text{(an arbitrary constant).}$$

(Note that the equation above is separable.) These trajectories are shown in Figure 9.7 in the original $x$–$y$ coordinates.

Theorem 9.8 tells us that the critical point at $(a_{22}/a_{21},\, a_{11}/a_{12})$ may be either a center or a focus of the nonlinear system. This theorem says nothing about the stability of this critical point. However, in this case we are fortunate in that the differential equation giving the trajectories in the phase plane for this nonlinear system,

$$\frac{dy}{dx} = \frac{y(a_{21}x - a_{22})}{x(a_{11} - a_{12}y)},$$

is separable. If we write it as

$$\frac{a_{11} - a_{12}y}{y}\,dy - \frac{a_{21}x - a_{22}}{x}\,dx = 0,$$

we may integrate to obtain an implicit solution as

$$a_{11}\ln y - a_{12}y - a_{21}x + a_{22}\ln x = \ln C,$$

with $C$ an arbitrary constant. Note that the left-hand side of the above equation has a minimum at $(a_{22}/a_{21},\, a_{11}/a_{12})$, horizontal tangents when $x = a_{22}/a_{21}$, and vertical tangents when $y = a_{11}/a_{12}$. The fact that the equation has a minimum assures us that the curves are closed. The curves shown in Figure 9.8 typify the

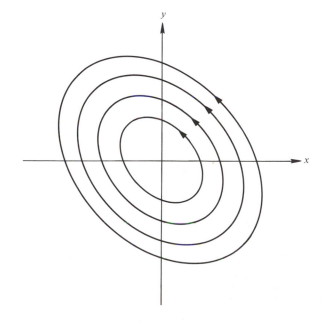

**Figure 9.8**  A center.

trajectories in this phase plane. Thus from these trajectories we see that the critical point at $(a_{22}/a_{21}, a_{11}/a_{12})$ is a stable center. Notice from Figure 9.8 that the motion is counterclockwise and that a maximum (minimum) in the predator population, $y$, occurs about a quarter period after a maximum (minimum) in the prey population. The actual levels of each population depend on which trajectory is traced; there is no situation where either population tends to an equilibrium value.

## EXERCISES

Linearize the following systems and classify their critical points.

**1.** $\dot{x} = x + 4y - x^2,$
$\dot{y} = 6x - y + 2xy.$

**2.** $\dot{x} = \sin x - y,$
$\dot{y} = \sin 2x - 2y.$

**3.** $\dot{x} = e^{-x-3y} - 1,$
$\dot{y} = -x + xy^2.$

**4.** $\dot{x} = 4 - 4x^2 - y^2,$
$\dot{y} = 3xy.$

**5.** Consider the system of equations

$$\dot{x} = 2x - y^2$$

$$\dot{y} = -y + x^2.$$

(a) Linearize this system.
(b) Find and classify all critical points of the linear system.
(c) Find all critical points of the nonlinear system.
(d) Use the transformation

$$\bar{x} = x - x_0$$

$$\bar{y} = y - y_0$$

to translate critical points not at the origin to the origin.
(e) Rewrite the original system in terms of ($\bar{x}$ and $\bar{y}$) for each critical point not at $(0, 0)$: linearize and classify the critical points.

**6.** Show that the system of differential equations for $\bar{x}$ and $\bar{y}$ satisfies the hypotheses of Theorem 9.6 for the singular point $\bar{x} = \bar{y} = 0$.

**7.** Redo Example 9.9 if $a_{12}$ is a negative number and the other three constants remain positive.

**8.** If the growth rate of the prey is inhibited by an increase in its population, the appropriate differential equation is

$$\dot{x} = a_{11}x - a_{12}xy - cx^2 \quad \text{(see Section 2.6.4, page 62)}.$$

If we retain the second differential equation in the predator–prey model as

$$\dot{y} = -a_{21}y + a_{22}xy,$$

determine the critical points of this system and discover what conclusions may be drawn from Theorems 9.6, 9.7, and 9.8.

**9.** Locate and classify the critical points of the system

$$\dot{x} = a_{11}x - a_{12}xy - cx^2,$$

$$\dot{y} = -a_{21}y + a_{22}xy - by^2.$$

**10.** Locate and classify the critical points of the system

$$\dot{x} = -a_{11}x + a_{12}(1 - e^{-by})x,$$

$$\dot{y} = -a_{21}y + a_{22}(1 - e^{-by})x$$

If all the constants are positive and $a_{11} \neq a_{12}$.


## 9.5 NONLINEAR SYSTEMS: DIRECT METHOD

The preceding method of determining the stability at an isolated singular point depends on our being able to linearize the system. Even then we are required to remain close to the singular point. There is a method, due to the Russian mathematician Lyapunov, which gets at the question without first approximating the system by a linear one. The method consists of finding a function $L(x, y)$, called a Lyapunov function, having certain properties to be discussed below. Properties of the derivative with respect to $t$ of $L(x, y)$ will give us information about stability.

**Definition 9.4.**    Let $L(x, y)$ be such that for some region $R$ of the $xy$-plane containing $(0, 0)$,
  i) $L(0, 0) = 0$;
  ii) $L(x, y) > 0$, $(x, y) \neq (0, 0)$, $(x, y) \in R$.
Then $L(x, y)$ is said to be *positive definite* in $R$.
Similarly, let
  i) $L(0, 0) = 0$;
  ii) $L(x, y) < 0$, $(x, y) \neq (0, 0)$, $(x, y) \in R$.
Then $L(x, y)$ is said to be *negative definite* in $R$.

If we replace $>$ by $\geq$ and $<$ by $\leq$ in the definitions above, $L(x, y)$ is said to be positive semidefinite or negative semidefinite, respectively.
To illustrate this definition, consider the following functions.

1. The function $L(x, y) = x^2 + y^2$ is positive definite for any open region in the $xy$-plane.
2. The function $L(x, y) = x - \ln|x + 1| + y^4$ is positive definite in the half plane $x \geq 0$.
3. The function $L(x, y) = (x - y)^2$ fails to be positive definite in any horizontal or vertical half plane because it vanishes not only at $(0, 0)$ but also at every point of the line $y = x$.

Suppose now that we have our standard system (9.12) with an isolated critical point at $(0, 0)$. Let $L(x, y)$ be positive definite in a region $R$ containing $(0, 0)$. If $\gamma(t)$ is a solution curve to (9.12) in $R$, given paramatically by

$$x = x(t),$$

$$y = y(t),$$

we may evaluate $dL(x, y)/dt$ along $\gamma(t)$.

By the chain rule we have

$$\frac{dL}{dt} = \frac{\partial L}{\partial x}\dot{x} + \frac{\partial L}{\partial y}\dot{y}, \tag{9.39}$$

and on $\gamma(t)$ we have

$$\dot{x} = F(x, y),$$
$$\dot{y} = G(x, y), \tag{9.12}$$

so that (9.39) becomes

$$\frac{dL}{dt} = \frac{\partial L}{\partial x}F(x, y) + \frac{\partial L}{\partial y}G(x, y). \tag{9.40}$$

Note that to evaluate the right-hand side of (9.40), we do not have to solve (9.12). If we know $L(x, y)$, we may take $F(x, y)$ and $G(x, y)$ from (9.12) and we have everything we need. (We must also have a certain amount of differentiability, obviously.)

We are now ready to state a theorem.

**Theorem 9.9.**     Let $(0, 0)$ be an isolated critical point of (9.12) and $L(x, y)$ a function which is positive definite in some region $R$ which includes $(0, 0)$. Let $\gamma(t)$ be a solution curve in $R$ and construct the derivative (9.40). Then

   i) if $dL/dt$ is negative semidefinite along $\gamma$ in $R$, the critical point is stable;
   ii) if $dL/dt$ is negative definite along $\gamma$ in $R - \{(0, 0)\}$, the critical point is asymptotically stable.

All we have to do to apply Theorem 9.8 is find a function $L(x, y)$ which satisfies the hypotheses and look at its derivative. That, of course, is the catch. For a given system there are no hard-and-fast rules for constructing a Lyapunov function $L(x, y)$. Often we may make use of the fact that $L(x, y) = Ax^2 + Bxy + Cy^2$ is positive definite if $A > 0$, $C > 0$, and $B^2 - 4AC < 0$.

**EXAMPLE 9.10**     Consider the system

$$\dot{x} = -x + xy,$$
$$\dot{y} = -y - x^2,$$

where $F(x, y) = -x + xy$ and $G(x, y) = -y - x^2$. [Note that $(0, 0)$ is an isolated critical point.] Take the Lyapunov function as $L(x, y) = x^2 + y^2$, which is obviously positive definite for all $x$ and $y$. Equation (9.40) becomes

$$\frac{dL}{dt} = 2x(-x + xy) + 2y(-y - x^2) = -2x^2 - 2y^2.$$

Thus $\partial L/\partial t < 0$ along any curve which does not pass through $(0, 0)$, so the critical point is asymptotically stable.

Since we have made no suppositions about the degree of the functions $F(x, y)$ and $G(x, y)$, the Lyapunov method may be used for linear systems also, as the next example shows.

**EXAMPLE 9.11**    The system

$$\dot{x} = -x + y,$$

$$\dot{y} = -x - y,$$

has $(0, 0)$ as an isolated critical point. If we again use $L(x, y) = x^2 + y^2$, from (9.40), we have

$$\frac{dL}{dt} = 2x(-x + y) + 2y(-x - y) = -2x^2 - 2y^2,$$

which is negative for all $(x, y)$ except $(0, 0)$, so again the critical point is asymptotically stable.

## EXERCISES

1. Show that the system

$$\dot{x} = -x - xy^2,$$

$$\dot{y} = -y - x^2y$$

has an asymptotically stable critical point at $(0, 0)$.

2. Using a Lyapunov function of the form $L(x, y) = Ax^2 + Cy^2$, examine the stability of the critical point at $(0, 0)$ for the following systems.
   (a) $\dot{x} = -x^3 + xy^2,$
       $\dot{y} = -y^3 - 2x^2y.$
   (b) $\dot{x} = -x^3 + 2y^3,$
       $\dot{y} = -2xy^2.$
   (c) $\dot{x} = -x + 2x^2 + y^2,$
       $\dot{y} = -y + xy.$
   (d) $\dot{x} = -x - y - x^3,$
       $\dot{y} = x - y - y^3.$

## 9.6 PERIODIC SOLUTIONS AND LIMIT CYCLES

Suppose the system (9.12) has a center at $(0, 0)$. Then trajectories in the phase plane look like those of Figure 9.8. These trajectories are closed and we imagine some particle following the path. If it is at $(x_0, y_0)$ at time $t_0$, let $T$ be the smallest time for which it will return to $(x_0, y_0)$. Thus we have

$$x(t_0) = x(t_0 + T)$$

$$y(t_0) = y(t_0 + T),$$

so the functions $x(t)$, $y(t)$ are periodic with period $T$. Conversely, every periodic solution must be represented by a closed path in the phase plane.

Since only one trajectory may pass through any point in the phase plane a closed trajectory separates the set of all trajectories into the set of those outside the closed trajectory and those inside. Outside curves may approach the closed trajectory but may

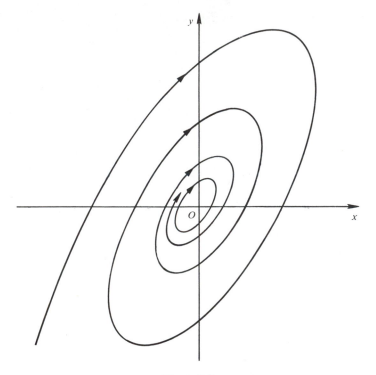

**Figure 9.9**

not cross. In fact, they may spiral in, winding about the closed trajectory in even tighter paths, as in Figure 9.9.

Similarly, trajectories inside may spiral out to the closed trajectory but may not cross. The closed trajectories in these cases are called *limit cycles* of the system (9.12).

**EXAMPLE 9.12**    Consider the system

$$\dot{x} = x - y - x(x^2 + y^2),$$
$$\dot{y} = x + y - y(x^2 + y^2) \tag{9.41}$$

and transform to polar coordinates

$$x = r \cos \theta,$$
$$y = r \sin \theta$$

in the $xy$-plane. From this coordinate transformation we have

$$\dot{x} = \dot{r} \cos \theta - r\dot{\theta} \sin \theta,$$
$$\dot{y} = \dot{r} \sin \theta + r\dot{\theta} \cos \theta. \tag{9.42}$$

Multiplying the first of equations (9.42) by $x$, the second by $y$, and adding gives

$$x\dot{x} + y\dot{y} = r\dot{r}. \tag{9.43}$$

Similarly, multiplying the first of equations (9.42) by $(-y)$, the second by $x$, and adding gives

$$x\dot{y} - y\dot{x} = r^2\dot{\theta}. \tag{9.44}$$

The left-hand sides of (9.43) and (9.44) may be obtained from (9.41) as

$$x\dot{x} = -xy + x^2(1 - x^2 - y^2)$$
$$= -r^2 \sin \theta \cos \theta + r^2 \cos^2 \theta (1 - r^2), \qquad (9.45)$$

$$y\dot{y} = xy + y^2(1 - x^2 - y^2)$$
$$= r^2 \sin \theta \cos \theta + r^2 \sin^2 \theta (1 - r^2). \qquad (9.46)$$

Adding these two equations gives

$$x\dot{x} + y\dot{y} = x^2 + y^2 - x^2(x^2 + y^2) - y^2(x^2 + y^2).$$

Thus (9.43) becomes

$$r\dot{r} = r^2(1 - r^2),$$

or since $r \neq 0$,

$$\dot{r} = r(1 - r^2). \qquad (9.47)$$

To obtain the $\theta$ equation, we multiply the first of equations (9.41) by $-y$ and the second by $x$ and add to get

$$x\dot{y} - y\dot{x} = x^2 + y^2 = r^2. \qquad (9.48)$$

Using (9.44) in (9.48), we have

$$\dot{\theta} = 1. \qquad (9.49)$$

Thus (9.41) in polar coordinates becomes

$$\dot{r} = r(1 - r^2).$$
$$\dot{\theta} = 1. \qquad (9.50)$$

From (9.50) we see that $(0, 0)$ is an isolated critical point.

The equations in (9.50) have the $r$ and $\theta$ dependence separated and may be solved separately. Thus

$$\theta(t) = t + t_0. \qquad (9.51)$$

From

$$\frac{dr}{dt} = r(1 - r^2)$$

we have

$$\left( \frac{1}{r} + \frac{r}{1 - r^2} \right) dr = dt. \qquad (9.52)$$

Integrating both sides of (9.52), we have

$$\ln|r| - \frac{1}{2}\ln|1 - r^2| = t + C_0$$

or

$$\frac{r^2}{1 - r^2} = C_1 e^{2t}.$$

Solving for $r^2$, we have

$$r^2 = \frac{1}{1 + Ce^{-2t}},$$

where $C = 1/C_1$. Thus

$$r = \frac{1}{\sqrt{1 + Ce^{-2t}}}. \tag{9.53}$$

Going back to cartesian coordinates in (9.51) and (9.53) yields

$$x = \frac{\cos t}{\sqrt{1 + Ce^{-2t}}}$$
$$y = \frac{\sin t}{\sqrt{1 + Ce^{-2t}}}. \tag{9.54}$$

If we set $C = 0$ in (9.54) we obtain the circle $x^2 + y^2 = 1$. To see that this circle represents a limit cycle, we note that for $C \neq 0$,

$$x^2 + y^2 = \frac{1}{1 + Ce^{-2t}}. \tag{9.55}$$

If $C > 0$, then $1/(1 + Ce^{-2t}) < 1$ and this fraction approaches 1 as $t \to \infty$. Thus (9.55) represents spirals which approach the unit circle from the inside as $t \to \infty$.

If $C < 0$, $1/(1 + Ce^{-2t}) > 1$ and this fraction approaches 1 as $t \to \infty$. Thus, in this case, (9.55) represents spirals which approach the unit circle from the outside as $t \to \infty$.

In this case, the unit circle is a limit cycle of (9.42).

It is not always possible, in fact it usually is not possible, to solve explicitly for limit cycles. We can, under certain conditions, tell when one exists.

To state an upcoming theorem which gives conditions under which we may have a limit cycle, we need to make the following definition.

**Definition 9.5.**    Let $\gamma(t)$ denote a path in the phase plane. Let $\gamma(t_0)$ be a point on the path and define the *positive half-path* $\gamma^+(t)$ through $t_0$ by $\gamma^+(t) = \{\gamma(t) \,|\, t \geq t_0\}$.

Thus, according to the definition, a positive half-path through $t_0$ is simply that part of the path beginning at $t_0$ and containing all the path for $t$ greater than $t_0$.

We now give the following theorem:

**Theorem 9.10.**    (Poincaré–Bendixon) Let $R$ denote a bounded region of the $xy$-plane for which the system (9.12) is given. Suppose that $F$ and $G$ have continuous partial derivatives in $R$. Let $\bar{R}$ denote a subregion of $R$ which, with its boundary, is wholly contained in $R$. Suppose that $\bar{R}$ contains a positive half-path of (9.12). Then, if there are no critical points of (9.12) in $\bar{R}$, $\bar{R}$ contains a closed path.

The tricky part of applying the Poincaré–Bendixon theorem in this form, however, is in finding the required positive half-path. For Example 9.12 where we discovered the unit circle to be a limit cycle, all the Poincaré–Bendixon theorem would tell us is that in any bounded region which excludes a disk of radius $\frac{1}{2}$, say, and contains a positive half-path, there is a closed path. So one sees that the requirement that $\overline{R}$ contain a positive half-path is crucial.

The use of polar coordinates may ease the task. In Example 9.12 we had

$$r = (1 + Ce^{-2t})^{-1/2},$$

so that

$$\dot{r} = (1 + Ce^{-2t})^{-3/2}(Ce^{-2t})$$

and for $C > 0$, $\dot{r}$ is always positive. Thus for $C > 0$, for each $\theta$ there is a line along which $r$ is increasing and which passes into the region $\frac{1}{2} < r \le 1$ and as $\lim_{t\to\infty} r(t) = 1$ for $C > 0$, the curve never leaves the region. Choosing the region $R$ of the theorem to be the annulus $\frac{1}{2} \le r < \frac{3}{2}$, and the region $\overline{R}$ the annulus $\frac{1}{2} \le r \le 1$ allows the hypotheses of the Poincaré–Bendixon theorem to be satisfied. The closed curve is, in fact, the outer circle of the annulus $\overline{R}$.

The following theorem is also useful, as it tells us where not to look for limit cycles. Before we state it, however, we need the following definition.

**Definition 9.6.**    A region $R$ of the $xy$-plane is *simply connected* if every closed curve in $R$ encloses only points of $R$.

Thus simply connected regions do not have holes in them. The annulus $r_0 < r < r_1$, for example, is not simply connected, while the circle $r = $ constant is.

**Theorem 9.11.    (Bendixon).** Let $R$ be a simply connected region of the $xy$-plane and let the system (9.12) be given in $R$, where
   i)   $F$ and $G$ have continuous partial derivatives in $R$;
   ii)  $\partial F/\partial x + \partial G/\partial y$ does not change sign in $R$.
   Then there are no limit cycles lying entirely in $R$.

One should note that this theorem tells us nothing about regions in which $\partial F/\partial x + \partial G/\partial y$ does change sign. For example, the system (9.41) has

$$\frac{\partial F}{\partial x} + \frac{\partial G}{\partial y} = 2 - 4(x^2 + y^2)$$

and changes sign when $x^2 + y^2 = \frac{1}{2}$, that is, in the circle of radius $1/\sqrt{2}$ about $(0, 0)$. Thus, by Bendixon's negative criterion there is no limit cycle interior to this circle. If we consider the region $R$ to be a circle centered at $(\frac{1}{2}, \frac{1}{2})$ and of radius $\frac{1}{4}$, then an arc of the circle $x^2 + y^2 = \frac{1}{2}$ lies in this region. Thus $\partial F/\partial x + \partial G/\partial y$ changes sign in the region, but there is no limit cycle in the region. The only limit cycle is the circle $x^2 + y^2 = 1$. Again we stress that the negative criterion tells us where not to look for limit cycles; it does not tell us where they may be or even if there are any.

## EXERCISES

1. Show that the equation

$$\ddot{x} + f(x)\dot{x} + g(x)x = 0$$

has no limit cycle in a region of the phase plane in which $f(x)$ does not change sign.
Show that the following systems have no limit cycles.

2. $\dot{x} = y + x^3$,
   $\dot{y} = x + y + y^3$.

3. $\dot{x} = y$,
   $\dot{y} = 1 + x^2 - (1 - x)y$.

4. $\dot{x} = 2xy + y^3$,
   $\dot{y} = -(x^2 + y^2) + y + y^3$.

5. $\dot{x} = y$,
   $\dot{y} = -1 - x^3$.

6. Apply the Poincaré–Bendixon theorem to show that the system

$$\dot{x} = 4x - 4y - x(x^2 + y^2),$$
$$\dot{y} = 4x + 4y - y(x^2 + y^2)$$

has a limit cycle in the annulus bounded by $x^2 + y^2 = 1$ and $x^2 + y^2 = 16$.

# CHAPTER 10

# Orthogonal Functions
# and Fourier Series

We are familiar with situations in which it is useful to replace a function by its Taylor series representation. In this chapter we develop techniques for finding other types of series expansions for functions. These expansions are needed in solving partial differential equations. A crucial concept for these new series expansions is that of a set of orthogonal functions. Such functions have already been briefly encountered as eigenfunctions in Sections 4.6.4, 6.5, and 6.5.1.

## 10.1 SETS OF ORTHOGONAL FUNCTIONS

In Section 4.6.4 we discovered that the eigenfunctions associated with

$$y'' + \lambda y = 0, \qquad y(0) = 0, \quad y(4) = 0, \qquad (4.97)$$

were

$$y_n(x) = \sin \frac{n\pi x}{4}, \qquad n = 1, 2, 3, \ldots$$

(here $\lambda = n^2\pi^2/16$, $n = 1, 2, 3, \ldots$). Note that they have the property that if $m \neq n$,

$$
\begin{aligned}
\int_0^4 y_n(x)y_m(x)\, dx &= \int_0^4 \sin \frac{n\pi x}{4} \sin \frac{m\pi x}{4}\, dx \\
&= \frac{1}{2} \int_0^4 \left[ \cos \frac{(n-m)\pi x}{4} - \cos \frac{(n+m)\pi x}{4} \right] dx \\
&= \frac{1}{2} \left\{ \frac{\sin[(n-m)\pi x/4]}{(n-m)\pi/4} - \frac{\sin[(n+m)\pi x/4]}{(n+m)\pi/4} \right\} \Bigg|_0^4 \\
&= 0.
\end{aligned}
$$

Similarly, from Section 6.5 we had

$$\int_{-1}^{1} P_n(x)P_m(x) \, dx = 0, \qquad n \neq m$$

for the Legendre polynomials and, from Section 6.5.1,

$$\int_{0}^{a} J_n(\lambda_j x)J_n(\lambda_k x)x \, dx = 0, \qquad j \neq k \qquad (10.1)$$

for Bessel functions. Note that if $n = m$ (or $j = k$ in the last example), the associated integral is not zero. These are three special cases of orthogonal functions, which are defined as follows.

**Definition 10.1.**   A set of real-valued functions $\{\phi_i(x), i = 0, 1, 2, \ldots\}$ is said to be *orthogonal* on the interval $[a, b]$ if

$$\int_{a}^{b} \phi_n(x)\phi_m(x) \, dx = 0, \qquad n \neq m.$$

The number $\sqrt{\int_a^b \phi_n^2(x) \, dx} = \|\phi_n\|$ is called the *norm* of the function $\phi_n(x)$. An orthogonal set of functions with $\|\phi_n\| = 1$, $n = 0, 1, 2, \ldots$, is called an *orthonormal* set.

**EXAMPLE 10.1**   For the orthogonal set associated with (4.97) we have

$$\|\phi_n\|^2 = \left\| \sin \frac{n\pi x}{4} \right\|^2 = \int_0^4 \sin^2 \frac{n\pi x}{4} \, dx$$

$$= \int_0^4 \frac{1 - \cos(n\pi x/2)}{2} \, dx$$

$$= 2.$$

Since   $\|\sin(n\pi x/4)\| = \sqrt{2}$,   $\|(1/\sqrt{2}) \sin(n\pi x/4)\| = 1$   and   $\{(1/\sqrt{2}) \sin(n\pi x/4), n = 1, 2, 3, \ldots\}$ is an orthonormal set.

Another example of an orthonormal set is, from page 247, the normalized Legendre polynomials given by

$$\left\{ \sqrt{\frac{2n + 1}{2}} \, P_n(x), n = 0, 1, 2, \ldots \right\}, \qquad \text{for } -1 \leq x \leq 1.$$

One of the useful features of an orthogonal set of functions is its role in providing infinite series expansions for functions. The formal procedure in these expansions is outlined as follows: We assume that $\{\phi_n(x), n = 0, 1, 2, \ldots\}$ is an orthogonal set for $[a, b]$ and that for $a < x < b$ we may write

$$f(x) = \sum_{n=0}^{\infty} c_n \phi_n(x). \qquad (10.2)$$

An expression for the coefficients $c_n$, in (10.2), may be formally obtained by multiplying both sides of (10.2) by $\phi_m(x)$ and integrating the result over the interval

$a < x < b$. This gives

$$\int_a^b f(x)\phi_m(x)\ dx = \int_a^b \left[ \sum_{n=0}^{\infty} c_n \phi_n(x) \right] \phi_m(x)\ dx$$

or

$$\int_a^b f(x)\phi_m(x)\ dx = \sum_{n=0}^{\infty} c_n \int_a^b \phi_n(x)\phi_m(x)\ dx, \qquad (10.3)$$

if we assume that the orders of integration and summation may be interchanged. Since the set of $\phi_n$'s forms an orthogonal set for $[a, b]$, the integral on the right-hand side of (10.3) is zero for $n \neq m$ and we have

$$\int_a^b f(x)\phi_m(x)\ dx = c_m \int_a^b \phi_m^2(x)\ dx = c_m \| \phi_m \|^2. \qquad (10.4)$$

If we now solve for $c_m$ and change the index from $m$ to $n$, we see that if $f(x)$ is represented by (10.2), the coefficients $c_n$ are given by

$$c_n = \frac{\int_a^b f(x)\phi_n(x)\ dx}{\| \phi_n \|^2}. \qquad (10.5)$$

Equation (10.2) is called an expansion of $f$ in terms of the set of orthogonal functions $\{\phi_n(x)\}$ or a *generalized Fourier series*. In this chapter we will not worry too much about what conditions on $f(x)$ are required for such expansions but concentrate on techniques used in making these expansions.

**EXAMPLE 10.2**    To find an expansion for $f(x) = x$ in terms of the orthogonal set of functions $\{\sin(n\pi x/4),\ n = 1, 2, 3, \ldots\}$ for $0 < x < 4$, we simply compute

$$c_n = \frac{1}{2} \int_0^4 x \sin \frac{n\pi x}{4}\ dx$$

$$= \frac{-x \cos(n\pi x/4)}{2n\pi/4} \Big|_0^4 + \frac{2}{n\pi} \int_0^4 \cos \frac{n\pi x}{4}\ dx$$

$$= \frac{-8 \cos n\pi}{n\pi} + \frac{8}{n^2 \pi^2} \sin \frac{n\pi x}{4} \Big|_0^4$$

$$= \frac{-8(-1)^n}{n\pi} = \frac{8(-1)^{n+1}}{n\pi}, \qquad n = 1, 2, 3, \ldots.$$

Thus the desired expansion is

$$x = \sum_{n=1}^{\infty} \frac{8(-1)^{n+1}}{n\pi} \sin \frac{n\pi x}{4}.$$

**EXAMPLE 10.3**    To find an expansion for $f(x) = x^2 - 2$ in terms of the orthogonal set of functions $\{P_n(x),\ n = 0, 1, 2, \ldots\}$ for $-1 < x < 1$, we would compute

$$c_n = \frac{\int_{-1}^{1} (x^2 - 2)P_n(x)\, dx}{\|P_n(x)\|^2}, \qquad n = 0, 1, 2, \ldots, \tag{10.6}$$

where $P_n(x)$ is the Legendre polynomial of degree $n$ (see Section 6.5). We could use the specific polynomial form for $P_n(x)$ for $n = 0, 1, 2, \ldots$ to integrate (10.6), but it is easier to use the properties of the Legendre polynomials. The strategy is to write $x^2 - 2$ as a linear combination of the Legendre polynomials and use the fact that they form an orthogonal set for $-1 < x < 1$. From page 247 we have $P_0(x) = 1$ and $P_2(x) = (3x^2 - 1)/2$, so that

$$x^2 - 2 = \frac{2}{3}\left(\frac{3x^2 - 1}{2}\right) - \frac{5}{3} = \frac{2}{3}P_2(x) - \frac{5}{3}P_0(x). \tag{10.7}$$

Thus

$$\int_{-1}^{1} (x^2 - 2)P_n(x)\, dx = \int_{-1}^{1}\left[\frac{2}{3}P_2(x) - \frac{5}{3}P_0(x)\right]P_n(x)\, dx$$

$$= \frac{2}{3}\int_{-1}^{1} P_2(x)P_n(x)\, dx - \frac{5}{3}\int_{-1}^{1} P_0(x)P_n(x)\, dx. \tag{10.8}$$

Since the Legendre polynomials form an orthogonal set on $-1 < x < 1$, the first integral on the right-hand side of (10.8) is zero for $n \neq 2$ and the second integral is zero for $n \neq 0$. Thus

$$\int_{-1}^{1} (x^2 - 2)P_n(x)\, dx = \begin{cases} \dfrac{2}{3}\displaystyle\int_{-1}^{1} P_2^2(x)\, dx, & n = 2 \\[2ex] -\dfrac{5}{3}\displaystyle\int_{-1}^{1} P_0^2(x)\, dx, & n = 0 \\[2ex] 0, & n \neq 0 \quad \text{or} \quad 2. \end{cases}$$

Since $\|P_n\|^2 = \int_{-1}^{1} P_n^2(x)\, dx$, we know that the expansion for $x^2 - 2$ in terms of Legendre polynomials is simply

$$x^2 - 2 = \frac{2}{3}P_2(x) - \frac{5}{3}P_0(x).$$

Notice that this was already given as (10.7). This is an example of the fact that expansion of a linear combination of members of an orthogonal set in terms of that set is simply the original linear combination (see exercise 12). This is analogous to the fact that the Taylor series expansion of a polynomial is simply the original polynomial.

Note that there was an extra $x$ in the integrand in the integral giving the orthogonality of Bessel functions [see (10.1)]. This is an example of the following:

**Definition 10.2.** A set of real-valued functions $\{\phi_i(x), i = 0, 1, 2, \ldots\}$ is said to be *orthogonal* with respect to the weight function $p(x)$ on the interval $[a, b]$

if

$$\int_a^b \phi_n(x)\phi_m(x)p(x)\ dx = 0, \qquad n \neq m. \tag{10.9}$$

The weighted norm of these functions is defined by

$$\|\phi_n\| = \left( \int_a^b \phi_n^2(x)p(x)\ dx \right)^{1/2} \tag{10.10}$$

Thus if the $\lambda_j$ and $\lambda_k$ in (10.1) are values of $\lambda$ such that

$$J_n(\lambda a) = 0,$$

$\{J_n(\lambda_j x), j = 1, 2, 3, \ldots \}$ forms an orthogonal set with respect to the weight function $x$ for $0 \leq x \leq a$. From (6.112) we have

$$\|J_n(\lambda_j x)\|^2 = \frac{a^2[J_{n+1}(\lambda_j a)]^2}{2}.$$

## EXERCISES

Show that the following sets of functions are orthogonal on the given interval and find the norm of each function in the set.

1. $\{\sin nx, n = 1, 2, 3, \ldots\}, \quad 0 \leq x \leq \pi.$
2. $\{1, \sin nx, n = 1, 2, 3, \ldots\}, \quad -\pi \leq x \leq \pi.$
3. $\{\cos nx, n = 0, 1, 2, \ldots\}, \quad 0 \leq x \leq \pi.$
4. $\{\cos nx, n = 0, 1, 2, \ldots\}, \quad -\pi \leq x \leq \pi.$
5. $\left\{\sin \dfrac{n\pi x}{L}, n = 1, 2, 3, \ldots \right\}, \quad 0 \leq x \leq L.$
6. $\left\{\cos \dfrac{n\pi x}{L}, n = 0, 1, 2, \ldots \right\}, \quad 0 \leq x \leq L.$
7. $\left\{1, \cos \dfrac{n\pi x}{L}, \sin \dfrac{n\pi x}{L}, n = 1, 2, 3, \ldots \right\}, \quad -L \leq x \leq L.$
8. Find the generalized Fourier series for $f(x) = 1, \quad 0 < x < \pi$
   (a) In terms of the functions in exercise 1.
   (b) In terms of the function in exercise 3.
9. Repeat exercise 8 for $f(x) = 1 + x$.
10. Find the generalized Fourier series for $f(x) = \sin^2 x, \; -\pi < x < \pi$ in terms of the functions of exercise 4. (*Hint:* You will not need to compute any integrals if you use a trigonometric identity for $\sin^2 x$.)
11. Find the series expansions for the following functions on $-1 \leq x \leq 1$ in terms of the Legendre polynomials
    (a) $f(x) = \begin{cases} -1, & -1 < x < 0 \\ 1, & 0 < x < 1 \end{cases}$  [*Hint:* Use (6.85).]
    (b) $f(x) = \begin{cases} 0, & -1 < x < 0 \\ 1, & 0 < x < 1 \end{cases}$
    (c) $f(x) = 1 + x + x^2$. (*Hint:* See Example 10.3.)

Exercises 12 through 15 concern the orthogonal set $\{\phi_n(x), n = 0, 1, 2, \ldots\}$ on $a \leq x \leq b$.

**12.** Show that the generalized Fourier series for $\alpha\phi_3(x) + \beta\phi_7(x)$ is simply $\alpha\phi_3(x) + \beta\phi_7(x)$; $\alpha$ and $\beta$ are constants.

**13.** Show that if

$$f(x) = \sum_{n=0}^{\infty} a_n\phi_n(x) \quad \text{and} \quad g(x) = \sum_{n=0}^{\infty} b_n\phi_n(x),$$

the generalized Fourier series for $f(x) + g(x)$ is

$$\sum_{n=0}^{\infty} (a_n + b_n)\phi_n(x).$$

**14.** Show that if a function has two expansions in terms of $\{\phi_n\}$, they are identical; that is, if $\sum_{n=0}^{\infty} a_n\phi_n(x)$ and $\sum_{n=0}^{\infty} a'_n\phi_n(x)$ are expansions of $f(x)$, then $a_n = a'_n$, $n = 0, 1, 2, \ldots$.

**15.** Show that $\| \alpha\phi_i(x) + \beta\phi_j(x) \|^2 = \alpha^2\| \phi_i \|^2 + \beta^2\| \phi_j \|^2$ for any constants $\alpha$, $\beta$ and integers $i, j, i \neq j$.

**16.** Show that $\{P_{2n}(x), n = 0, 1, 2, \ldots\}$ forms an orthogonal set for $0 \leq x \leq 1$ and find $\| P_{2n}(x) \|$. (*Hint:* Use calculus techniques with the limits of integration.)

**17.** On page 251 we discussed Hermite polynomials given by

$$H_0(x) = 1, \quad H_1(x) = 2x, \quad H_2(x) = 4x^2 - 2, \quad H_3(x) = 8x^3 - 12x, \quad \ldots$$

These may be determined by the Rodrigues' formula

$$H_n(x) = (-1)^n e^{x^2} \frac{d^n}{dx^n} e^{-x^2}.$$

Use this formula, along with basic calculus techniques, to show that $\{H_0(x), H_1(x), H_2(x)\}$ form an orthogonal set with respect to the weight function $e^{-x^2}$ on $(-\infty, \infty)$.

**18.** By squaring both sides of (10.2) and integrating the result from $x = a$ to $x = b$, formally derive Parseval's equality

$$\int_a^b (f(x))^2 \, dx = \sum_{n=0}^{\infty} c_n^2,$$

where the $c_n$, $n = 0, 1, 2, \ldots$, are given by (10.5).

**19. (a)** Expand $\cos x$ in terms of the orthogonal set in exercise 2.
   **(b)** The reason for your result in part (a) is that the orthogonal set from exercise 2 was not complete. A set of orthogonal functions is said to be complete if the only continuous function satisfying $\int_a^b f(x)\phi_n(x) \, dx = 0$, for all $n$, is $f(x) = 0$, $a \leq x \leq b$. Find another nonzero function (besides $\cos x$) such that $\int_{-\pi}^{\pi} f(x)\phi_n(x) \, dx = 0$ for the orthogonal functions of exercise 2.

## 10.2 FOURIER SERIES

We now concentrate on the set of functions $\{1, \cos(n\pi x/L), \sin(n\pi x/L), n = 1, 2, 3, \ldots\}$ which form an orthogonal set on $-L \leq x \leq L$. Notice that

$$\| 1 \|^2 = \int_{-L}^{L} 1 \, dx = 2L,$$

$$\left\| \cos \frac{n \pi x}{L} \right\|^2 = \int_{-L}^{L} \cos^2 \frac{n \pi x}{L} \, dx = L,$$

$$\left\| \sin \frac{n \pi x}{L} \right\|^2 = \int_{-L}^{L} \sin^2 \frac{n \pi x}{L} \, dx = L,$$

so by the results of Section 10.1, if $f(x)$, $-L < x < L$, is expanded in a series like (10.2), coefficients are given, from (10.5), by

$$\frac{1}{2L} \int_{-L}^{L} f(x) \, dx \qquad \text{(for } \phi_n = 1\text{)},$$

$$\frac{1}{L} \int_{-L}^{L} f(x) \cos \frac{n \pi x}{L} \, dx \qquad \left( \text{for } \phi_n = \cos \frac{n \pi x}{L} \right),$$

$$\frac{1}{L} \int_{-L}^{L} f(x) \sin \frac{n \pi x}{L} \, dx \qquad \left( \text{for } \phi_n = \sin \frac{n \pi x}{L} \right).$$

Such an expansion was first considered by the French scientist Fourier. The standard form for such an expansion is

$$\frac{a_0}{2} + \sum_{n=1}^{\infty} \left( a_n \cos \frac{n \pi x}{L} + b_n \sin \frac{n \pi x}{L} \right), \tag{10.11}$$

where

$$a_n = \frac{1}{L} \int_{-L}^{L} f(x) \cos \frac{n \pi x}{L} \, dx, \qquad n = 0, 1, 2, \ldots, \tag{10.12}$$

$$b_n = \frac{1}{L} \int_{-L}^{L} f(x) \sin \frac{n \pi x}{L} \, dx, \qquad n = 1, 2, 3, \ldots. \tag{10.13}$$

The series expression in (10.11) is called the *Fourier series* of $f(x)$ on $(-L, L)$, while (10.12) and (10.13) are called the *Fourier coefficients* of $f(x)$ on $(-L, L)$. [Note that writing the constant in (10.11) as $a_0/2$ allowed us to use (10.12) as the formula for $a_0$ as well as $a_n$, $n \geq 1$. The factor of $2^{-1}$ occurs because the norm of 1 is different from the norm of $\cos(n \pi x/L)$, $n = 1, 2, 3, \ldots$.]

**EXAMPLE 10.4**    Find the Fourier series of $f(x) = x - L$, $-L < x < L$. We first calculate

$$a_0 = \frac{1}{L} \int_{-L}^{L} (x - L) \, dx = \frac{(x - L)^2}{2L} \bigg|_{-L}^{L} = -2L,$$

$$a_n = \frac{1}{L} \int_{-L}^{L} (x - L) \cos \frac{n \pi x}{L} \, dx$$

$$= \frac{1}{L} \left[ \frac{(x - L) \sin(n \pi x/L)}{n \pi/L} \bigg|_{-L}^{L} - \frac{L}{n \pi} \int_{-L}^{L} \sin \frac{n \pi x}{L} \, dx \right]$$

$$= \frac{1}{L} \left[ \frac{L^2}{n^2 \pi^2} \cos \frac{n \pi x}{L} \bigg|_{-L}^{L} \right] = 0, \qquad n = 1, 2, 3, \ldots,$$

$$b_n = \frac{1}{L} \int_{-L}^{L} (x - L) \sin \frac{n\pi x}{L} \, dx$$

$$= \frac{1}{L} \left[ \frac{-(x - L) \cos(n\pi x/L)}{n\pi/L} \Big|_{-L}^{L} + \frac{L}{n\pi} \int_{-L}^{L} \cos \frac{n\pi x}{L} \, dx \right]$$

$$= \frac{-2L}{n\pi} \cos(-n\pi) = \frac{-2L(-1)^n}{n\pi}, \qquad n = 1, 2, 3, \ldots,$$

and then write the Fourier series for $f(x) = x - L$, $-L < x < L$ as

$$-L - \frac{2L}{\pi} \sum_{n=1}^{\infty} \frac{(-1)^n \sin(n\pi x/L)}{n}. \tag{10.14}$$

**EXAMPLE 10.5**    Find the Fourier series for the "square wave" given by

$$f(x) = \begin{cases} 0, & -\pi < x < 0 \\ 1, & 0 \le x < \pi \end{cases}.$$

For this function we calculate

$$a_0 = \frac{1}{\pi} \int_{-\pi}^{\pi} f(x) \, dx = \frac{1}{\pi} \int_{0}^{\pi} 1 \, dx = 1,$$

$$a_n = \frac{1}{\pi} \int_{0}^{\pi} \cos nx \, dx = \frac{\sin nx}{n\pi} \Big|_{0}^{\pi} = 0, \qquad n = 1, 2, 3, \ldots,$$

$$b_n = \frac{1}{\pi} \int_{0}^{\pi} \sin nx \, dx = \frac{-\cos nx}{n\pi} \Big|_{0}^{\pi} = \frac{-(\cos n\pi - 1)}{n\pi}$$

$$= \frac{-[(-1)^n - 1]}{n\pi}, \qquad n = 1, 2, 3, \ldots.$$

Thus the Fourier series for this function is

$$\frac{1}{2} - \sum_{n=1}^{\infty} \frac{(-1)^n - 1}{n\pi} \sin nx. \tag{10.15}$$

Note two things about (10.15):

1. Since the coefficient of $\sin nx$ is zero for $n$ an even integer we could also write (10.15) as

$$\frac{1}{2} + \sum_{n=0}^{\infty} \frac{2}{(2n + 1)\pi} \sin(2n + 1)x.$$

2. The value of the Fourier series of this function for $x = 0$ is $\frac{1}{2}$, the mean of 0 and 1 (0 and 1 being the values of $f$ on either side of 0).

The fact that the Fourier series for a function need not converge to that function for all values of $x$ is given by the following theorem.

**Theorem 10.1.**    If $f(x)$ and $f'(x)$ are piecewise continuous for $-L < x < L$, then the Fourier series for $f$, (10.11) with coefficients given by (10.12) and (10.13), converges to

$$\frac{f(x+) + f(x-)}{2}, \qquad -L < x < L,$$

and to a periodic function, with period 2L, for all $x$.

Recall that

$$f(x+) = \lim_{h \to 0^+} f(x + h), \quad f(x-) = \lim_{h \to 0^+} f(x - h)$$

are the right- and left-hand limits of $f$ at $x$. Theorem 10.1 says that the Fourier series for $f$ converges to its value at any point where $f$ is continuous, and to the average value of the left- and right-hand limits of $f$ at any point of discontinuity.

Two other definitions are needed in Theorem 10.1. One of them, piecewise continuity, has already been given as Definition 7.2 on page 266. Piecewise continuity means that a function may have only a finite number of discontinuities and no infinite discontinuities. The definition of a periodic function is given as

**Definition 10.3.**    A function $g(x)$ is said to be *periodic,* or to be a *periodic function,* if it is defined for all $x$ and there exists a positive number $p$ such that

$$g(x + p) = g(x) \tag{10.16}$$

for all $x$. The smallest $p$ for which (10.16) is valid is called the period of $f$ (some authors use "fundamental period" instead of "period").

If we know the behavior of a function for $x$ ranging over a complete period of $f$, we know its behavior for all $x$ (see Figure 10.1).

**Figure 10.1**    A periodic function.

The set of functions $\{\cos(n\pi x/L), \sin(n\pi x/L), n = 1, 2, 3, \ldots\}$, used in construction of the Fourier series for $f$, consists of periodic functions,

$$\cos\left[\frac{n\pi}{L}(x + 2L)\right] = \cos\left(\frac{n\pi x}{L} + 2n\pi\right) = \cos\frac{n\pi x}{L},$$

$$\sin\left[\frac{n\pi}{L}(x + 2L)\right] = \sin\left(\frac{n\pi x}{L} + 2n\pi\right) = \sin\frac{n\pi x}{L},$$

so both these sine and cosine functions have the common period $2L$. Since a constant also satisfies (10.16), for any $p$, we see that the Fourier series for $f$, given by (10.11), is also a periodic function of period $2L$. Thus given a function, $f$, defined for $-L \leq x \leq L$, the Fourier series for $f$ gives a periodic function of period $2L$ and converges to $[f(L-) + f(-L+)]/2$ for $x = L + 2mL$, $m = 0, \pm1, \pm2, \ldots$. This is illustrated in Figures 10.2 and 10.3. Figure 10.2 shows to what the Fourier series for

$$f(x) = \begin{cases} 0, & -\pi < x < 0 \\ 3, & 0 < x < \pi \end{cases} \tag{10.17}$$

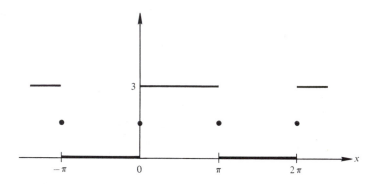

**Figure 10.2**   Graph of Fourier series of (10.17).

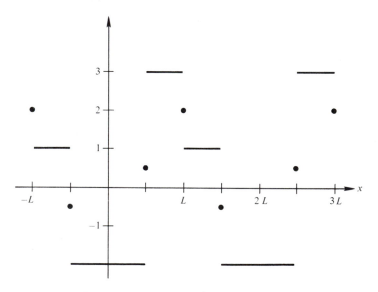

**Figure 10.3**   Graph of Fourier series of $g(x)$.

converges (see Example 10.5) while Figure 10.3 applies to the function

$$g(x) = \begin{cases} 1, & -L < x < \dfrac{-L}{2} \\[2mm] -2, & \dfrac{-L}{2} < x < \dfrac{L}{2} \\[2mm] 3, & \dfrac{L}{2} < x < L \end{cases}.$$

Note that the Fourier series for the function in Example 10.4 converges to $-L$ at $x = L + 2n\pi$, $n = 0, \pm 1, \pm 2, \ldots$.

## EXERCISES

Determine whether or not the following functions are periodic. Give the period for those that are periodic.

**1.** $f(x) = \sin^2 x$.

**2.** $f(x) = \cos \dfrac{x}{2}$.

**3.** $f(x) = \sinh 2x$.
**4.** $f(x) = \sinh 2x + \sin x$.
**5.** $f(x) = \sin 2x + \cos x$.
**6.** $f(x) = x \sin 3x$.
**7.** $f(x) = \sin^3 x$.
**8.** $f(x) = \cos^2 2x$.

**9.** $f(x) = \begin{cases} 1, & 2n \le x \le 2n + 1 \\ -1, & 2n - 1 < x < 2n \end{cases}, \quad n = 0, \pm 1, \pm 2, \ldots$.

**10.** $f(x) = \begin{cases} |x|, & 2n \le x \le 2n + 1 \\ -|x|, & 2n - 1 < x < 2n \end{cases}, \quad n = 0, \pm 1, \pm 2, \ldots$.

**11.** $f(x) = \begin{cases} \sin \pi x, & 2n \le x \le 2n + 1 \\ -\sin \pi x, & 2n - 1 < x < 2n \end{cases}, \quad n = 0, \pm 1, \pm 2, \ldots$.

**12.** $f(x) = \begin{cases} \sin 2\pi x, & 2n \le x \le 2n + 1 \\ -\sin 2\pi x, & 2n - 1 < x < 2n \end{cases}, \quad n = 0, \pm 1, \pm 2, \ldots$.

In each of exercises 13 to 21, a function is defined on an interval $[-L, L]$. In each exercise determine the associated Fourier series and give the value of this series at all points where it does not converge to the value of its original function at that point.

**13.** $f(x) = x$, $\quad -L \le x \le L$.
**14.** $f(x) = |x|$, $\quad -L \le x \le L$.

**15.** $f(x) = \begin{cases} 1, & -\pi \le x \le 0 \\ 0, & 0 < x \le \pi \end{cases}$.

**16.** $f(x) = \begin{cases} x + \pi, & -\pi \le x < 0 \\ x, & 0 \le x \le \pi \end{cases}$.

**17.** $f(x) = -x, \quad -L \le x \le L.$

**18.** $f(x) = \begin{cases} 1, & -1 \le x \le 0 \\ -1, & 0 < x \le 1 \end{cases}.$

**19.** $f(x) = \begin{cases} 1, & -\pi \le x \le 0 \\ -3, & 0 < x \le \pi \end{cases}.$

**20.** $f(x) = \begin{cases} -1, & -\pi \le x \le \dfrac{-\pi}{2} \\[2mm] 0, & \dfrac{-\pi}{2} < x < \dfrac{\pi}{2} \\[2mm] 2, & \dfrac{\pi}{2} \le x \le \pi \end{cases}.$

**21.** $f(x) = x^2, \quad -1 \le x \le 1.$

If we define $S_n(x)$ as the sum of the first $n$ nonzero terms in a Fourier series, for Example 10.5 we have $S_2(x) = \frac{1}{2} + (2/\pi)\sin x$, $S_3(x) = \frac{1}{2} + (2/\pi)\sin x + (2/(3\pi))\sin 3x$, $S_4(x) = S_3(x) + (2/(5\pi))\sin 5x$, and so on. Figure 10.4 shows the comparison of the given function,

$$f(x) = \begin{cases} 0, & -\pi < x < 0 \\ 1, & 0 \le x < \pi \end{cases}$$

with several such sums.

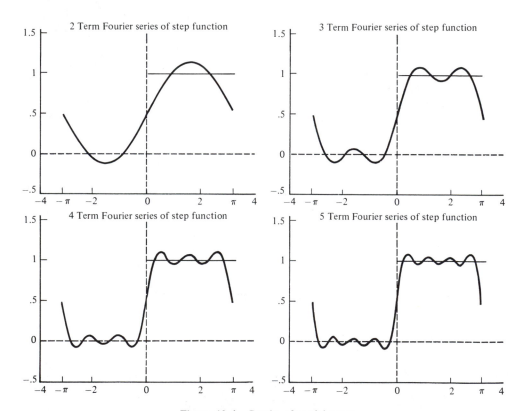

**Figure 10.4**  Graphs of partial sums.

**22.** For each of exercises 13 through 21, graph the original function as well as the corresponding $S_1(x)$, $S_2(x)$, and $S_3(x)$. (If you have a computer available, observe what happens as you increase the number of terms in each Fourier series.)

## 10.3 FOURIER COSINE AND SINE SERIES

You may have noticed that in working the exercises in Section 10.2 that you sometimes computed a Fourier series that consisted only of cosine terms and sometimes it had only sine terms. There are two types of functions, even functions and odd functions, whose properties are useful in simplifying calculations in finding Fourier series. Recall that $f$ is an *even function* if

$$f(x) = f(-x)$$

for all $x$ in the domain of $f$, while it is an *odd function* if

$$f(x) = -f(-x)$$

for all $x$ in the domain of $f$. Table 10.1 gives examples of functions which are even, odd, or neither even nor odd. Figure 10.5 shows the graphs of typical even functions and Figure 10.6 shows a typical odd function.

**TABLE 10.1**   CLASSIFICATION
OF TYPICAL FUNCTIONS

| Function | Classification |
| --- | --- |
| $x^2$ | Even |
| $x + 4x^3$ | Odd |
| $\lvert x \rvert$ | Even |
| $2x + 3$ | Neither |
| $x^2 + x \sin x$ | Even |
| $\sin x + \cos x$ | Neither |
| $x \cos x + 1/x$ | Odd |

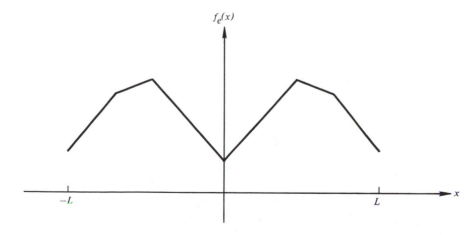

**Figure 10.5**   Graph of the even extension of f.

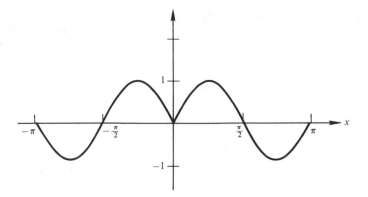

**Figure 10.6**   Graph of the even extension of sin $2x$, $0 < x < \pi$.

One of the important properties of even functions is found in Theorem 10.2.

**Theorem 10.2.**    If $f$ is an integrable *even* function on $[-L, L]$, then

$$\int_{-L}^{L} f(x)\, dx = 2 \int_{0}^{L} f(x)\, dx.$$

The details of the proof of this theorem and their reasons are given as follows:

$$\int_{-L}^{L} f(x)\, dx = \int_{-L}^{0} f(x)\, dx + \int_{0}^{L} f(x)\, dx \quad \text{(property of definite integrals)}$$

$$= - \int_{L}^{0} f(-y)\, dy + \int_{0}^{L} f(x)\, dx \quad \text{(change of variables } x = -y)$$

$$= \int_{0}^{L} f(-y)\, dy + \int_{0}^{L} f(x)\, dx \quad \text{(property of definite integrals)}$$

$$= \int_{0}^{L} f(y)\, dy + \int_{0}^{L} f(x)\, dx \quad \text{(property of even functions)}$$

$$= 2 \int_{0}^{L} f(x)\, dx.$$

The proof of the following companion theorem concerning odd functions is similar to that of Theorem 10.2, and is left for the exercises.

**Theorem 10.3.**    If $f$ is an integrable *odd* function on $[-L, L]$, then

$$\int_{-L}^{L} f(x)\, dx = 0.$$

Two other properties of even and odd functions which are useful in computing Fourier series are:

1. *The product of two even functions or two odd functions is an even function.*
2. *The product of an even function and an odd function is an odd function.*

Since the Fourier coefficients of $f(x)$, $-L \le x \le L$, are given by

$$a_n = \frac{1}{L} \int_{-L}^{L} f(x) \cos \frac{n\pi x}{L} \, dx, \qquad\qquad (10.12)$$

$$b_n = \frac{1}{L} \int_{-L}^{L} f(x) \sin \frac{n\pi x}{L} \, dx, \qquad\qquad (10.13)$$

we have the following results:

1. The Fourier series of an even function consists only of cosine terms with coefficients

$$a_n = \frac{2}{L} \int_{0}^{L} f(x) \cos \frac{n\pi x}{L} \, dx, \qquad n = 0, 1, 2, \dots . \qquad (10.18a)$$

2. The Fourier series of an odd function consists only of sine terms with coefficients given by

$$b_n = \frac{2}{L} \int_{0}^{L} f(x) \sin \frac{n\pi x}{L} \, dx, \qquad n = 1, 2, 3, \dots . \qquad (10.18b)$$

**EXAMPLE 10.6**    The Fourier series of the even function

$$f(x) = 4 - x^2, \qquad -2 \le x \le 2,$$

has coefficients given by

$b_n = 0$, $n = 1, 2, 3, \dots$    (since $f$ is an even function),

$$a_0 = \frac{2}{2} \int_{0}^{2} (4 - x^2) \, dx = 4x - \left. \frac{x^3}{3} \right|_{0}^{2} = \frac{16}{3}$$

$$a_n = \frac{2}{2} \int_{0}^{2} (4 - x^2) \cos \frac{n\pi x}{2} \, dx$$

$$= \left. \frac{(4 - x^2) \sin(n\pi x/2)}{n\pi/2} \right|_{0}^{2} + \frac{2}{n\pi} \int_{0}^{2} (2x) \sin \frac{n\pi x}{2} \, dx$$

$$= \frac{2}{n\pi} \left[ \left. \frac{2x[-\cos(n\pi x/2)]}{n\pi/2} \right|_{0}^{2} + \frac{2}{n\pi} \int_{0}^{2} 2 \cos \frac{n\pi x}{2} \, dx \right]$$

$$= \frac{-16}{n^2 \pi^2} \cos n\pi + \frac{8}{n^2 \pi^2} \int_{0}^{2} \cos \frac{n\pi x}{2} \, dx = \frac{-16}{n^2 \pi^2} (-1)^n, \quad n = 1, 2, 3, \dots .$$

**EXAMPLE 10.7**    The Fourier series of the odd function

$$f(x) = \begin{cases} 1 + x, & -1 \le x < 0 \\ -1 + x, & 0 \le x \le 1, \end{cases}$$

has coefficients given by

$$a_n = 0, \qquad n = 0, 1, 2, \ldots,$$

$$b_n = \frac{2}{1} \int_0^1 (-1 + x) \sin n\pi x \, dx$$

$$= 2 \left[ \frac{-(-1 + x) \cos n\pi x}{n\pi} \Big|_0^1 + \frac{1}{n\pi} \int_0^1 \cos n\pi x \, dx \right]$$

$$= \frac{-2}{n\pi} + \frac{2}{n^2\pi^2} \sin n\pi x \Big|_0^1 = \frac{-2}{n\pi}, \qquad n = 1, 2, 3, \ldots.$$

In solving partial differential equations we are sometimes concerned with a function which describes a physical quantity defined over an interval $[0, L]$. In constructing a Fourier series for such a function, we have a choice in how we extend the definition to the interval $[-L, 0)$. One choice is to define a new function, $f_e$, by

$$f_e(x) = \begin{cases} f(x), & 0 \le x \le L \\ f(-x), & -L \le x < 0 \end{cases}. \tag{10.19}$$

$f_e$ is called the *even extension of $f$* to $[-L, L]$ and will have the Fourier series

$$\frac{a_0}{2} + \sum_{n=1}^{\infty} a_n \cos \frac{n\pi x}{L},$$

with the $a_n$ given by

$$a_n = \frac{2}{L} \int_0^L f(x) \cos \frac{n\pi x}{L} \, dx, \qquad n = 0, 1, 2, \ldots.$$

Thus the Fourier series of $f_e$ converges to a periodic even function with period $2L$. The Fourier series of the even extension of $f$ is called *the half-range cosine expansion* or simply the *Fourier cosine series*.

**EXAMPLE 10.8**     The function

$$f(x) = \sin 2x, \qquad 0 \le x \le \pi,$$

has the even extension

$$f_e(x) = \begin{cases} \sin 2x, & 0 \le x \le \pi \\ -\sin 2x, & -\pi \le x < 0 \end{cases},$$

with a resulting Fourier series

$$\frac{8}{3\pi} \cos x + \frac{4}{\pi} \sum_{n=3}^{\infty} \frac{(-1)^n - 1}{n^2 - 4} \cos nx.$$

The graph of $f_e(x)$ is shown in Figure 10.6.

A second useful manner of extending the definition of a function $f$ defined on $[0, L]$ to one on $[-L, L]$ is by

$$f_o(x) = \begin{cases} f(x), & 0 \le x \le L \\ -f(-x), & -L \le x < 0 \end{cases}. \tag{10.20}$$

Such an extension is called the *odd extension of f* to $[-L, L]$, with an associated Fourier series given by

$$\sum_{n=1}^{\infty} b_n \sin \frac{n\pi x}{L},$$

with the $b_n$ given by

$$b_n = \frac{2}{L} \int_0^L f(x) \sin \frac{n\pi x}{L} \, dx, \qquad n = 1, 2, 3, \ldots .$$

Thus the Fourier series of this odd extension of $f$ converges to a periodic odd function with period $2L$ (see Figure 10.7). This series is also called *the half-range sine expansion* or *the Fourier sine series for f*. For the $f$ defined in Example 10.8, $f_o(x) = f(x)$ and the Fourier series for $f_o$ is simply $\sin 2x$. Example 10.9 is a more challenging example of constructing an odd extension.

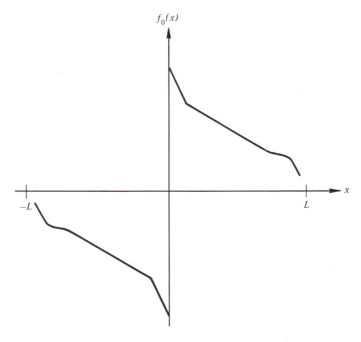

**Figure 10.7**    Graph of the odd extension of f.

**EXAMPLE 10.9**    If

$$f(x) = -1 + x, \qquad 0 \le x \le 1$$

we define its odd extension by

$$f_o(x) = \begin{cases} -1 + x, & 0 \le x \le 1 \\ -(-1 - x), & -1 \le x < 0 \end{cases}. \tag{10.21}$$

Note that the function defined by (10.21) has already been used in Example 10.7, where its Fourier series is given as

$$\frac{-2}{\pi} \sum_{n=1}^{\infty} \frac{\sin n\pi x}{n}.$$

On the other hand, the even extension of this function is

$$f_e(x) = \begin{cases} -1 + x, & 0 \le x \le 1 \\ -1 - x, & -1 \le x < 0 \end{cases},$$

with Fourier coefficients given by

$$a_n = 2 \int_0^1 (-1 + x) \cos n\pi x \, dx, \qquad n = 0, 1, 2, \ldots$$

$$a_n = \frac{2[(-1)^n - 1]}{n^2 \pi^2}, \qquad n = 1, 2, 3, \ldots,$$

$$a_0 = -1.$$

Thus the resulting Fourier cosine series is

$$-\frac{1}{2} - \frac{4}{\pi^2} \sum_{n=1}^{\infty} \frac{\cos(2n - 1)\pi x}{(2n - 1)^2}.$$

Graphs of the even and odd extensions of this function appear in Figure 10.8.

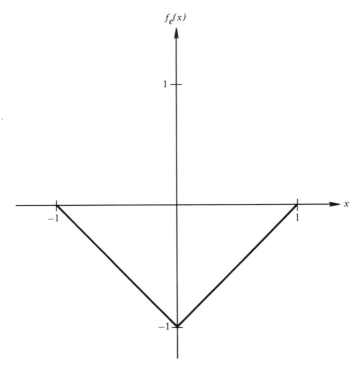

**Figure 10.8** Graphs of the even and odd extensions of $f(x) = -1 + x$, $0 < x < 1$.

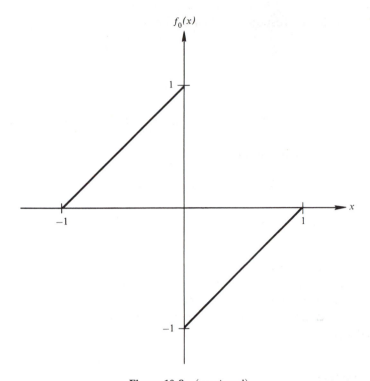

**Figure 10.8**   (*continued*)

We summarize the results of this section by restating Theorem 10.1 for half-range expansions.

**Theorem 10.4.**    Given $f(x)$ and $f'(x)$ as piecewise continuous functions for $0 \leq x \leq L$. Then

(a) the even extension of $f$ to $[-L, 0)$ is

$$f_e(x) = \begin{cases} f(x), & 0 \leq x \leq L \\ f(-x), & -L \leq x < 0, \end{cases}$$

with a corresponding Fourier series

$$\frac{a_0}{2} + \sum_{n=1}^{\infty} a_n \cos \frac{n\pi x}{L}, \tag{10.22}$$

where

$$a_n = \frac{2}{L} \int_0^L f(x) \cos \frac{n\pi x}{L} \, dx, \qquad n = 0, 1, 2, \ldots; \tag{10.23}$$

(b) the odd extension of $f$ to $[-L, 0)$ is

$$f_o(x) = \begin{cases} f(x), & 0 \leq x \leq L \\ -f(-x), & -L \leq x < 0, \end{cases}$$

with a corresponding Fourier series

$$\sum_{n=1}^{\infty} b_n \sin \frac{n\pi x}{L},$$     (10.24)

where

$$b_n = \frac{2}{L} \int_0^L f(x) \sin \frac{n\pi x}{L} dx, \qquad n = 1, 2, 3, \ldots;$$     (10.25)

(c) the series in (10.22) and (10.24) converge to

$$\frac{f(x+) + f(x-)}{2}, \qquad 0 \le x \le L,$$

and to a periodic function with period $2L$ for all $x$.

The Fourier cosine series for $f$ on $[0, L]$ [equation (10.22)] converges to a continuous function if $f$ is continuous. The Fourier sine series for $f$ on $[0, L]$ [equation (10.24)] converges to a continuous function if $f$ is continuous and $f(0) = f(L) = 0$. (Note that the Fourier sine series for any function is 0 for $x = 0$ and $L$.) Typical graphs showing the Fourier series for an even function and an odd function are given in Figures 10.5 and 10.7.

## EXERCISES

Find the half-range cosine expansion and half-range sine expansion of the stated function. Sketch a graph of the function determined by the resulting convergent series.

**1.** $f(x) = 1, \quad 0 \le x \le L.$

**2.** $f(x) = x, \quad 0 \le x \le L.$

**3.** $f(x) = x^2, \quad 0 \le x \le \pi.$

**4.** $f(x) = \sin x, \quad 0 \le x \le \pi.$ (The even extension of this function gives the "fully rectified" sine wave.)

**5.** $f(x) = \cos x, \quad 0 \le x \le \pi.$

**6.** $f(x) = \begin{cases} 1, & 0 \le x < \dfrac{\pi}{2} \\ 0, & \dfrac{\pi}{2} \le x \le \pi \end{cases}.$

**7.** $f(x) = \begin{cases} 1, & 0 \le x < \dfrac{\pi}{2} \\ 2, & \dfrac{\pi}{2} \le x < \pi \end{cases}.$

**8.** $f(x) = \begin{cases} \sin x, & 0 \le x < \dfrac{\pi}{2} \\ \cos x, & \dfrac{\pi}{2} \le x < \pi \end{cases}.$

**9.** $f(x) = 1 + 3x, \quad 0 \le x \le \pi$.

**10.** $f(x) = 1 - x, \quad 0 \le x \le 2$.

**11.** $f(x) = 1 + x, \quad 0 \le x \le 2$.

**12.** $f(x) = e^x, \quad 0 \le x \le 1$.

**13.** $f(x) = e^{-x}, \quad 0 \le x \le 1$.

**14.** $f(x) = \sin^2 x, \quad 0 \le x \le \pi$. (*Hint:* Use trigonometric identities.)

**15.** $f(x) = \cos^2 x, \quad 0 \le x \le \pi$.

**16.** $f(x) = 1 + x^2, \quad 0 \le x \le 1$.

**17.** $f(x) = 1 - x^2, \quad 0 \le x \le 1$.

**18.** $f(x) = \begin{cases} x, & 0 \le x \le \dfrac{\pi}{2} \\ x - \dfrac{\pi}{2}, & \dfrac{\pi}{2} < x \le \pi \end{cases}$ . (The even extension of this function gives a type of "sawtooth wave.")

**19.** $f(x) = \begin{cases} \pi - x, & 0 \le x \le \dfrac{\pi}{2} \\ x, & \dfrac{\pi}{2} < x \le \pi. \end{cases}$

**20.** Prove that
   **(a)** The product of two even functions is an even function.
   **(b)** The product of two odd functions is an even function.
   **(c)** The product of an even function and an odd function is an odd function.
   **(d)** The sum of two even functions is an even function.
   **(e)** The sum of two odd functions is an odd function.
   **(f)** Any function may be written as the sum of an even function and an odd function.

**21.** Prove Theorem 10.3.

**22.** Use Theorem 10.1 and the results of the exercise listed below to obtain a numerical value for the following infinite series.
   **(a)** Exercise 2, $\displaystyle\sum_{n=1}^{\infty} \frac{(-1)^n}{2n - 1}, \sum_{n=1}^{\infty} \frac{1}{(2n - 1)^2}$.
   **(b)** Exercise 3, $\displaystyle\sum_{n=1}^{\infty} \frac{(-1)^n}{n^2}$.
   **(c)** Exercise 4, $\displaystyle\sum_{n=1}^{\infty} \frac{(-1)^n}{4n^2 - 1}$.

**23.** Use Parseval's equality (exercise 18, page 350) together with the results of exercise 3 (this set) to show that

$$\sum_{n=1}^{\infty} \frac{1}{n^4} = \frac{\pi^4}{90}.$$

**24.** Use Parseval's equality together with the result of exercise 4 (this set) to obtain an expression for

$$\sum_{n=1}^{\infty} \frac{1}{(4n^2 - 1)^2}.$$

**25.** Fourier expansions are sometimes useful in solving nonhomogeneous differential equations. Consider the case of a second order harmonic oscillator where the forcing function has the form of a Fourier sine series, namely

$$\frac{d^2y}{dx^2} + \omega^2 y = f(x) = \sum_{n=1}^{\infty} b_n \sin nx.$$

Solve the equation above subject to the initial conditions $y(0) = y'(0) = 0$ by first seeking a particular solution of the form $\sum_{n=1}^{\infty} \alpha_n \sin nx$. (Assume that $\omega$ is positive but not a positive integer.)

**26.** Use the technique of exercise 25 to solve the initial value problem

$$\frac{d^2y}{dx^2} + 16y = f(x), \qquad y(0) = y'(0) = 0,$$

if

$$f(x) = \begin{cases} 1, & 0 < x < \dfrac{\pi}{2} \\ -1, & \dfrac{\pi}{2} < x < \pi \end{cases}$$

and $f(x + \pi) = f(x)$ for $x > 0$.

**27.** If damping is added to the system above, we may need to assume that the particular solution may be expanded as a full Fourier series. Use this idea to solve the initial value problem

$$\frac{d^2y}{dx^2} + 8\frac{dy}{dx} + 15y = f(x), \qquad y(0) = y'(0) = 0,$$

where $f$ is the function from exercise 26.

## 10.4 STURM–LIOUVILLE BOUNDARY VALUE PROBLEMS

After seeing examples of several boundary value problems with second order differential equations which lead to orthogonal functions, we now turn to a more general case of

$$y'' + a_2(x)y' + [a_1(x) + \lambda a_0(x)]y = 0, \qquad L_1 < x < L_2, \qquad (10.26)$$

$$A_1 y(L_1) + A_2 y'(L_1) = 0, \qquad (10.27)$$

$$B_1 y(L_2) + B_2 y'(L_2) = 0. \qquad (10.28)$$

Equation (10.26) may be converted to a more convenient form if it is multiplied by the integrating factor $\exp[\int a_2(x)\, dx]$. This allows the first two terms in (10.26) to be written as the derivative of one term as

$$\frac{d}{dx}\left[e^{\int a_2(x)\, dx} y'\right] + e^{\int a_2(x)\, dx}[a_1(x) + \lambda a_0(x)]y = 0,$$

or

$$\frac{d}{dx}\left[r(x)\frac{dy}{dx}\right] + [q(x) + \lambda p(x)]y = 0, \qquad (10.29)$$

with the definitions of $r(x)$, $q(x)$, and $p(x)$ readily apparent.

Since the integrating factor is never zero, (10.26) and (10.29) will have the same set of solutions. Equation (10.29) together with boundary conditions (10.27) and (10.28) is called a Sturm–Liouville boundary value problem. Such problems often arise in using the method of separation of variables to find solutions of partial differential equations. (This will be considered in Chapter 11.) A useful theorem regarding Sturm–Liouville boundary value problems is as follows:

**Theorem 10.5.**    Given the Sturm–Liouville problem

$$[r(x)y']' + [q(x) + \lambda p(x)]y = 0, \qquad (10.29)$$

$$A_1 y(L_1) + A_2 y'(L_1) = 0, \qquad (10.27)$$

$$B_1 y(L_2) + B_2 y'(L_2) = 0, \qquad (10.28)$$

where $r(x)$, $r'(x)$, $q(x)$, and $p(x)$ are continuous real-valued functions for $L_1 \le x \le L_2$, $p(x)$ and $r(x)$ are positive for $L_1 < x < L_2$, and the constants $A_1$, $A_2$, $B_1$, and $B_2$ are real. ($r$, $q$, $p$, $A_1$, $A_2$, $B_1$, and $B_2$ are independent of $\lambda$, and $|A_1| + |A_2| \ne 0$, $|B_1| + |B_2| \ne 0$.)

If $\lambda_m$ and $\lambda_n$ are two distinct eigenvalues of the Sturm–Liouville problem with corresponding eigenfunctions $y_m$ and $y_n$, then $y_m$ and $y_n$ are orthogonal with respect to the weight function $p(x)$ on $(L_1, L_2)$. This orthogonality also holds if

(a) $r(L_1) = 0$ and instead of (10.27) we have $y(L_1)$ and $y'(L_1)$ bounded;
(b) $r(L_2) = 0$ and instead of (10.28) we have $y(L_2)$ and $y'(L_2)$ bounded;
(c) $r(L_1) = r(L_2)$ if (10.27) and (10.28) are replaced by the periodic boundary conditions

$$\begin{aligned} y(L_1) &= y(L_2), \\ y'(L_1) &= y'(L_2). \end{aligned} \qquad (10.30)$$

To prove this theorem, we rewrite (10.29) as

$$Ly \equiv [r(x)y']' + q(x)y = -\lambda p(x)y. \qquad (10.31)$$

If $\lambda_m$, $y_m$, $\lambda_n$, $y_n$ are as in the hypothesis, we multiply

$$Ly_m = -\lambda_m p(x)y_m$$

by $y_n$, and

$$Ly_n = -\lambda_n p(x)y_n$$

by $y_m$ and subtract the results to obtain

$$y_n Ly_m - y_m Ly_n = (\lambda_n - \lambda_m)p(x)y_n y_m. \qquad (10.32)$$

The left-hand side of (10.32) may be expressed as

$$y_n Ly_m - y_m Ly_n = \frac{d}{dx}[r(x)(y_n y'_m - y_m y'_n)] = \frac{d}{dx}[r(x)\,\Delta(x)], \qquad (10.33)$$

where $\Delta(x) = W[y_n, y_m]$ is the Wronskian of $y_n$ and $y_m$. If we use (10.33) in (10.32) and integrate the result over $[L_1, L_2]$ we obtain

$$(\lambda_n - \lambda_m) \int_{L_1}^{L_2} y_n y_m p(x)\, dx = r(x)\, \Delta(x) \Big|_{L_1}^{L_2} = r(L_2)\, \Delta(L_2) - r(L_1)\, \Delta(L_1). \quad (10.34)$$

If we write the boundary conditions satisfied at $x = L_2$, we have

$$B_1 y_n(L_2) + B_2 y_n'(L_2) = 0,$$
$$B_1 y_m(L_2) + B_2 y_m'(L_2) = 0. \quad (10.35)$$

Now for the homogeneous $2 \times 2$ algebraic system (10.35) to have a nontrivial solution for $B_1$ and $B_2$, we need the determinant of the coefficients to be zero. Since we have $|B_1| + |B_2| \neq 0$, we know that not both $B_1$ and $B_2$ are zero. Thus we also know that this determinant of coefficients, or $\Delta(L_2)$, must equal 0. A similar line of reasoning for the boundary conditions at $x = L_1$ shows that $\Delta(L_1) = 0$, so the right-hand side of (10.34) equals zero. Since $\lambda_n \neq \lambda_m$, the orthogonality of $y_m$ and $y_n$ is established. The proofs for cases (a) to (c) are left to the exercises.

Another useful fact about Sturm–Liouville boundary values problems is that all the resulting eigenvalues are real. That is the result of the next theorem.

**Theorem 10.6.**    The eigenvalues of the Sturm–Liouville problem in Theorem 10.5 are all real.

To prove this theorem, we suppose that an eigenvalue has the form $\lambda = \alpha + i\beta$, $\alpha$ and $\beta$ real. If $y$ is the eigenfunction associated with $\lambda$, (10.29), (10.27), and (10.28) are all satisfied. Since $r$, $q$, and $p$ are real-valued functions and $A_1$, $A_2$, $B_1$, and $B_2$ are real numbers, if we take the complex conjugate of (10.29), (10.27), and (10.28) we obtain

$$[r(x)\bar{y}']' + [q(x) + \bar{\lambda} p(x)]\bar{y} = 0,$$
$$A_1 \bar{y}(L_1) + A_2 \bar{y}'(L_1) = 0,$$
$$B_1 \bar{y}(L_2) + B_2 \bar{y}'(L_2) = 0.$$

Thus $\bar{y}$ is an eigenfunction associated with $\bar{\lambda}$. Since (10.34) is true for any eigenvalue, eigenfunction pair [with the right-hand side of (10.34) set equal to zero] we write it for the pair $\lambda$, $y$ and $\bar{\lambda}$, $\bar{y}$. This gives

$$(\lambda - \bar{\lambda}) \int_{L_1}^{L_2} y\bar{y} p(x)\, dx = 0. \quad (10.36)$$

Since $y\bar{y} = |y|^2$ and $p(x) > 0$ for $L_1 < x < L_2$, the integrand in (10.36) is positive and therefore the integral is not zero. Thus

$$\lambda - \bar{\lambda} = 0$$

or

$$2i\beta = 0.$$

Since $\beta = 0$, we know that $\lambda$ is a real number.

The advantage this theorem gives us is that we need only search for real eigenvalues in solving Sturm–Liouville boundary value problems. The following theorem tells us when the corresponding eigenfunction is unique, so we only need to seek one of those.

**Theorem 10.7.**    To each eigenvalue of a Sturm–Liouville problem, in all cases except (c) with periodic boundary conditions, there corresponds only one linearly independent eigenfunction.

The proof of this theorem is left to the exercises. Since the eigenfunctions of a Sturm–Liouville problem form an orthogonal set with weight function $p(x)$, we may expand a function as an infinite series

$$\sum_{n=0}^{\infty} c_n \phi_n(x), \tag{10.37}$$

where the coefficients are given by

$$c_n = \frac{\int_{L_1}^{L_2} f(x)\phi_n(x)p(x)\,dx}{\|\phi_n\|^2}, \qquad n = 0, 1, 2, \ldots . \tag{10.38}$$

In (10.38)

$$\|\phi_n\|^2 = \int_{L_1}^{L_2} \phi_n^2(x)p(x)\,dx, \qquad n = 0, 1, 2, \ldots . \tag{10.39}$$

The theorem that allows such expansions follows.

**Theorem 10.8.**    If
(a) $\{\phi_n(x), n = 0, 1, 2, \ldots\}$ is a set of eigenfunctions associated with the Sturm–Liouville boundary problem (10.29), (10.27), and (10.28), and
(b) $f$ and $f'$ are piecewise continuous on $[L_1, L_2]$,
then the series in (10.37), with coefficients given by (10.38), converges to

$$\frac{f(x+) + f(x-)}{2}, \qquad \text{for } L_1 < x < L_2.$$

**EXAMPLE 10.10**
    Expand

$$f(x) = \begin{cases} 0, & 0 < x < 2 \\ 1, & 2 < x < 3 \end{cases}$$

in the eigenfunctions of the Sturm–Liouville boundary value problem

$$y'' + \lambda y = 0, \qquad 0 < x < 3,$$

$$y'(0) = 0, \qquad y(3) = 0.$$

The eigenvalues of this problem come from solutions of

$$\cos(\sqrt{\lambda}\,3) = 0,$$

so

$$\lambda_n = \frac{(2n + 1)^2 \pi^2}{36}, \qquad n = 0, 1, 2, \ldots .$$

The corresponding eigenfunctions are

$$\phi_n = \cos \frac{(2n + 1)\pi x}{6}, \qquad n = 0, 1, 2, \ldots$$

with norm obtained from

$$\| \phi_n \|^2 = \frac{3}{2}, \qquad n = 0, 1, 2, \ldots .$$

The coefficients in the expansion for $f$ are given by

$$c_n = \frac{1}{3/2} \int_2^3 \cos \frac{(2n + 1)\pi x}{6} \, dx, \qquad n = 0, 1, 2, \ldots$$

$$= \frac{2}{3} \frac{6}{\pi(2n + 1)} \left[ \sin \frac{(2n + 1)\pi}{2} - \sin \frac{(2n + 1)\pi}{3} \right]$$

so the expansion for $f$ is

$$\frac{4}{\pi} \sum_{n=0}^{\infty} \frac{(-1)^n - \sin \dfrac{(2n + 1)\pi}{3}}{2n + 1} \cos \frac{(2n + 1)\pi x}{6}. \tag{10.40}$$

Note that the series in (10.40) converges to $\frac{1}{2}$ when $x = 2$.

**EXAMPLE 10.11**

Expand

$$f(x) = \begin{cases} x, & 0 < x \le \frac{1}{2} \\ 1 - x, & \frac{1}{2} < x < 1, \end{cases}$$

in terms of the eigenfunctions of

$$y'' + \lambda y = 0, \qquad 0 < x < 1,$$

$$y(0) = 0, \qquad y(1) - By'(1) = 0.$$

The eigenvalues of this problem come from solutions of

$$\sin \sqrt{\lambda} - B \sqrt{\lambda} \cos \sqrt{\lambda} = 0$$

or after dividing by $\cos \sqrt{\lambda}$,

$$\tan \sqrt{\lambda} = B \sqrt{\lambda}. \tag{10.41}$$

To find solutions of (10.41), we graph each side of (10.41) as a function of $\sqrt{\lambda}$. (See Figure 10.9, where for illustrative purposes we have assumed that $0 < B < 1$.) It is clear from Figure 10.9 that there are an infinite number of solutions of (10.41) for $\sqrt{\lambda} > 0$, which we denote by

$$\sqrt{\lambda_n}, \qquad n = 1, 2, 3, \ldots$$

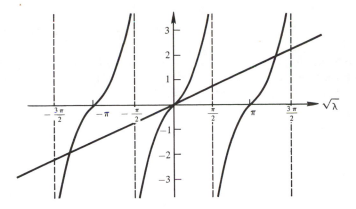

**Figure 10.9** Solutions of $\tan\sqrt{\lambda} = B\sqrt{\lambda}$.

and that they approach $(2n - 1)\pi/2$ for large $n$, that is,

$$\lim_{n\to\infty}\left[\frac{\sqrt{\lambda_n}}{(2n - 1)\pi/2}\right] = 1$$

for $B \neq 0$. The eigenfunction corresponding to $\lambda_n$ is

$$\phi_n = \sin\sqrt{\lambda_n}\, x$$

with a norm obtained from

$$\|\phi_n\|^2 = \int_0^1 \sin^2\sqrt{\lambda_n}\, x\, dx = \frac{1}{2}\int_0^1 (1 - \cos 2\sqrt{\lambda_n}\, x)\, dx$$

$$= \frac{1}{2}\left(1 - \frac{\sin 2\sqrt{\lambda_n}}{2\sqrt{\lambda_n}}\right).$$

Finally, we compute the coefficients in the expansion for $f$ as

$$c_n = \|\phi_n\|^{-1}\left[\int_0^{1/2} x\sin\sqrt{\lambda_n}\, x\, dx + \int_{1/2}^1 (1 - x)\sin\sqrt{\lambda_n}\, x\, dx\right]$$

$$= \frac{\|\phi_n\|^{-1}[2\sin(\sqrt{\lambda_n}/2) - \sin\sqrt{\lambda_n}]}{\sqrt{\lambda_n}}$$

We close this section by showing how eigenfunction expansions may be used to solve nonhomogeneous differential equations. If we seek to solve

$$[r(x)y']' + q(x)y = f(x) \tag{10.42}$$

subject to the boundary conditions

$$A_1 y(L_1) + A_2 y'(L_1) = 0, \tag{10.27}$$

$$B_1 y(L_2) + B_2 y'(L_2) = 0, \tag{10.28}$$

we first consider the associated Sturm–Liouville problem

$$[r(x)y']' + [[q(x) + \lambda p(x)]y = 0, \tag{10.29}$$

with boundary conditions (10.27) and (10.28).

Step one in this procedure is to obtain the eigenvalues and eigenfunctions of this Sturm–Liouville problem as $\lambda_n$ and $\phi_n(x)$, $n = 0, 1, 2, \ldots$.

Step two is to assume that the solution of (10.42) may be expanded in terms of these eigenfunctions as

$$y = \sum_{n=0}^{\infty} a_n \phi_n(x). \tag{10.43}$$

If the series in (10.43) is substituted into (10.42), we obtain

$$\sum_{n=0}^{\infty} a_n(-\lambda_n p(x))\phi_n(x) = f(x). \tag{10.44}$$

[Note we have used the fact that the $\phi_n(x)$ satisfy (10.29) with $\lambda = \lambda_n$, $n = 0, 1, 2, \ldots$.] Since the $\phi_n(x)$ form an orthogonal set with weight function $p(x)$, from (10.44) we conclude that

$$-a_n\lambda_n = \frac{\int_{L_1}^{L_2} f(x)\phi_n(x)\, dx}{\int_{L_1}^{L_2} \phi_n^2(x)p(x)\, dx}, \qquad n = 0, 1, 2, \ldots. \tag{10.45}$$

Thus our solution is given by (10.43) with the $a_n$ obtained from (10.45). As an example of this procedure, consider the boundary value problem

$$y'' + 3y = \begin{cases} x, & 0 < x < \dfrac{\pi}{2} \\[2mm] \pi - x, & \dfrac{\pi}{2} < x < \pi \end{cases} \tag{10.46}$$

$$y(0) = y(\pi) = 0.$$

The associated Sturm–Liouville problem

$$y'' + (3 + \lambda)y = 0, \qquad y(0) = y(\pi) = 0,$$

has eigenvalues $\lambda_n = n^2 - 3$, $n = 1, 2, 3, \ldots$, and eigenfunctions

$$\phi_n(x) = \sin nx, \qquad n = 1, 2, 3, \ldots.$$

Thus assuming a solution of (10.46) as

$$y = \sum_{n=1}^{\infty} a_n \sin nx$$

gives

$$-a_n(n^2 - 3) = \frac{\int_0^{\pi/2} x \sin nx\, dx + \int_{\pi/2}^{\pi}(\pi - x)\sin nx\, dx}{\pi/2}, \qquad n = 1, 2, 3, \ldots.$$

Performing the above integration gives

$$a_n = \frac{-4 \sin(n\pi/2)}{\pi(n^2 - 3)n^2}, \qquad n = 1, 2, 3 \ldots.$$

## EXERCISES

1. Find the eigenvalues, eigenfunctions, and norms of the eigenfunctions for the following Sturm–Liouville problems.

   (a) $y'' + \lambda y = 0$,    $y(0) = 0$,    $y'(1) = 0$.
   (b) $y'' + \lambda y = 0$,    $y'(0) = 0$,    $y'(\pi) = 0$.
   (c) $y'' + \lambda y = 0$,    $y(0) = 0$,    $y(1) + y'(1) = 0$.
   (d) $y'' + \lambda y = 0$,    $y'(0) = 0$,    $y(\pi) + y'(\pi) = 0$.
   (e) $y'' + (1 + \lambda)y = 0$,    $y(0) = 0$,    $y(1) = 0$.
   (f) $y'' + (1 + \lambda)y = 0$,    $y'(0) = 0$,    $y(\pi) = 0$.

2. (a)–(c)  Find the eigenfunction expansion of $f(x) = 1$ using your answers in exercise 1(a), (b), and (c), respectively.

3. (a)–(c)  Find the eigenfunction expansion of $f(x) = x$ using your answers from exercise 1(a), (b), and (c), respectively.

4. Fill in the details in the proof of Theorem 10.5 for
   (a) Case (a).
   (b) Case (b).
   (c) Case (c).

5. Show that the left-hand term in (10.33) is equal to the middle term of (10.33) by expanding each term.

6. Prove Theorem 10.7.

7. In Section 4.5 we had an example of a boundary value associated with the buckling of a long shaft, or column, under axial load. In Chapter 11 we consider the transverse vibrations of a horizontal beam which leads to the following boundary value problems. Find the eigenvalues and eigenfunctions for

   (a) $y'''' - \lambda^4 y = 0$,    $0 < x < L$,    $y(0) = y''(0) = 0$,    $y(L) = y''(L) = 0$.
   (b) $y'''' - \lambda^4 y = 0$,    $0 < x < L$,    $y(0) = y'(L) = 0$,    $y(L) = y'(L) = 0$.
   (c) $y'''' - \lambda^4 y = 0$,    $0 < x < L$,    $y(0) = y'(0) = 0$,    $y''(L) = y'''(L) = 0$.
   (d) $y'''' - \lambda^4 y = 0$,    $0 < x < L$,    $y(0) = y'(0) = 0$,    $y(L) = y''(L) = 0$.

8. Use the techniques developed for solving (10.42) to solve the following boundary value problems.

   (a)  $y'' + 5y = \begin{cases} \sin x, & 0 < x < \dfrac{\pi}{2} \\[2mm] 0, & \dfrac{\pi}{2} < x < \pi \end{cases}$,    $y(0) = y(\pi) = 0$.

   (b)  $y'' + 5y = \begin{cases} -1, & 0 < x < \dfrac{\pi}{2} \\[2mm] 1, & \dfrac{\pi}{2} < x < \pi \end{cases}$,    $y'(0) = y'(\pi) = 0$.

   (c)  $y'' + 5y = \begin{cases} x, & 0 < x < \dfrac{\pi}{2} \\[2mm] \pi - x, & \dfrac{\pi}{2} < x < \pi \end{cases}$,    $y'(0) = y'(\pi) = 0$.

## REVIEW EXERCISES

1. Verify that the first four Legendre polynomials form an orthogonal set on $[-1, 1]$. $[P_0(x) = 1, P_1(x) = x, P_2(x) = (3x^2 - 1)/2, P_3(x) = (5x^3 - 3x)/2.]$

2. Find the series expansion of $f(x) = 3 - 2x + x^3$ in terms of the Legendre polynomials. (*Hint:* See Example 10.3.)

3. Find the series expansion of

$$f(x) = \begin{cases} -2, & -1 < x < 0 \\ 1, & 0 < x < 1 \end{cases}$$

   in terms of the Legendre polynomials.

4. Verify that the first three Laguerre polynomials form an orthogonal set with respect to the weight function $e^{-x}$ on $[0, \infty)$ $[L_0(x) = 1, L_1(x) = x - 1, L_2(x) = x^2 - 4x + 2]$. You may wish to use the Rodrigues' formula,

$$L_n(x) = (-1)^n e^x \frac{d^n}{dx^n}(x^n e^{-x}) \qquad \text{(see Section 6.5.1, page 251).}$$

5. Suppose that you have a set of functions $\{\phi_n(x), n = 0, 1, 2, \ldots\}$ which are orthogonal on $(0, L)$ and you know that

$$\int_0^L x^m \phi_n(x)\, dx = 0, \qquad m \neq n,$$

$$\int_0^L x^n \phi_n(x)\, dx = \frac{1}{n}, \qquad n \neq 0$$

$$\int_0^L \phi_0(x)\, dx = \pi,$$

$$\int_0^L \phi_n^2(x)\, dx = \begin{cases} 2, & n = 0 \\ 4, & n \geq 1 \end{cases}.$$

   (a) If $f(x) = 3 + 7x^2$, find the series expansion of $f(x)$ in terms of this set of orthogonal functions.
   (b) Do likewise for the function $f(x) = 7 + 3x - 5x^4$.

6. (a) Prove that the set $\{\cos nx, n = 1, 2, 3, \ldots\}$ forms an orthogonal set on $[-\pi, 0]$.
   (b) Find the norm of each function in this set.
   (c) Is this set of functions complete? That is, can you find a nonzero function, $f$, such that $\int_{-\pi}^0 f(x) \cos nx\, dx = 0$, $n = 1, 2, 3, \ldots$ ? (See exercise 19, Section 10.1.)

7. Find the Fourier series expansion of

$$f(x) = \begin{cases} -1, & -\pi < x < 0 \\ -2, & 0 < x < \pi \end{cases}$$

   and give the value of this series for all points where it does not converge to the value of the original function at that point.

**8.** Repeat exercise 7 for

$$f(x) = \begin{cases} -2, & -\pi < x < \dfrac{-\pi}{2} \\ 0, & \dfrac{-\pi}{2} < x < \dfrac{\pi}{2} \\ 1, & \dfrac{\pi}{2} < x < \pi \end{cases}.$$

**9.** Find the Fourier series expansion of $f(x) = |x| + x$, $-\pi < x < \pi$.

**10. (a)** Find the Fourier series expansion of

$$f(x) = \begin{cases} 0, & -\pi < x < 0 \\ \sin x, & 0 < x < \pi \end{cases}.$$

**(b)** Find the half-range cosine expansion of $\sin x$, $0 < x < \pi$.
**(c)** Find the half-range sine expansion of $\sin x$, $0 < x < \pi$.
**(d)** Compare the graphs of the resulting functions in parts (a), (b), and (c) for $-2\pi < x < 2\pi$.

**11.** Given the function

$$f(x) = \begin{cases} 1, & 0 < x < 1 \\ -1, & 1 < x < 2 \end{cases}.$$

**(a)** Find the half-range cosine expansion of $f$, and sketch a graph of the function determined by the resulting series for $-3 < x < 3$.
**(b)** Find the half-range sine expansion of $f$, and sketch a graph of the function determined by the resulting series for $-3 < x < 3$.

**12.** Repeat exercise 11 for the function

$$f(x) = \begin{cases} x, & 0 < x < \pi \\ \pi, & \pi < x < 2\pi \end{cases}$$

and sketch the results for $-3\pi < x < 3\pi$.

**13.** Find the eigenvalues and eigenfunctions of the following Sturm–Liouville problems.
**(a)** $y'' + \lambda y = 0$, $y(0) + 3y'(0) = 0$, $y(\pi) = 0$.
**(b)** $y'' + \lambda y = 0$, $y(0) = 0$, $y(1) + 2y'(1) = 0$.

**14.** Find the eigenfunction expansion of $f(x) = 1$ using the answer from exercise 13(a).

**15.** Repeat exercise 14 using the answers from exercise 13(b).

**16.** Verify by direct integration that the eigenfunctions in 13(b) are orthogonal over the interval $(0, 1)$.

**17.** Find the charge in an $LC$ circuit if the applied voltage has the form of a half-rectified sine wave as shown in Figure 10.10. The appropriate differential equation is

$$L\frac{d^2x}{dt^2} + \frac{1}{C}x = E(t).$$

Take $L = 1$, $C = 1/100$, $x(0) = x'(0) = 0$, and $E(t)$ as in Figure 10.10.

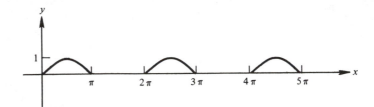

**Figure 10.10**　Half-rectified sine wave.

**18.** Find the Fourier series of $f(x) = x^2$, $-\pi < x < \pi$, and combine the result with Theorem 10.1 to find exact values for the following infinite series:

$$\sum_{n=1}^{\infty} \frac{1}{n^2}, \quad \sum_{n=1}^{\infty} \frac{(-1)^{n+1}}{n^2} \quad \text{and} \quad \sum_{n=0}^{\infty} \frac{1}{(2n+1)^2}.$$

**19.** Solve the following boundary value problems.

**(a)** $y'' + 3y = \begin{cases} \sin x, & 0 < x < \dfrac{\pi}{2} \\ 0, & \dfrac{\pi}{2} < x < \pi \end{cases}$, $\quad y'(0) = y'(\pi) = 0.$

**(b)** $y'' + 7y = \begin{cases} x, & 0 < x < \pi \\ \pi, & \pi < x < 2\pi, \\ 3\pi - x, & 2\pi < x < 3\pi \end{cases}$ $\quad y(0) = y(3\pi) = 0.$

# CHAPTER 11

# Partial Differential Equations

The first 10 chapters of this book have dealt with ordinary differential equations, their solutions, and their applications. We had one or more dependent variables, but only one independent variable. With partial differential equations (PDEs), we have at least two independent variables to deal with and many ways to seek solutions. In this brief chapter we cover only three methods of solution: direct integration, Laplace transform, and separation of variables. All these methods of solution rely heavily on our techniques for solving ordinary differential equations covered in the preceding chapters.

## 11.1 INTRODUCTION

A *partial differential equation* is an equation involving at least one partial derivative. We have already solved some partial differential equations in the section concerning exact ordinary differential equations (Section 2.2). There we discovered that if

$$M(x, y) \, dx + N(x, y) \, dy = 0 \tag{11.1}$$

is exact, its solution is

$$u(x, y) = C,$$

where

$$\frac{\partial u}{\partial x} = M(x, y), \qquad \frac{\partial u}{\partial y} = N(x, y).$$

Example 2.2 asked for a solution of the exact differential equation

$$(-xy \sin xy + \cos xy + e^{2x}) \, dx + (y^2 - x^2 \sin xy) \, dy = 0. \tag{11.2}$$

Here

$$\frac{\partial u}{\partial y} = y^2 - x^2 \sin xy, \tag{11.3}$$

so

$$u = \frac{y^3}{3} + x \cos xy + g(x). \tag{11.4}$$

Thus, in order that our solution in (11.4) satisfy the second condition, $\partial u/\partial x = M$, $g(x)$ must satisfy

$$g'(x) = e^{2x},$$

so

$$g(x) = \frac{e^{2x}}{2}$$

and

$$u(x, y) = \frac{y^3}{3} + x \cos xy + \frac{e^{2x}}{2}$$

is the solution of (11.2). Equation (11.3) is an example of a first order partial differential equation.

**Definition 11.1.** An equation containing a partial derivative of order $n$, but none higher than $n$, is called an $n$th *order partial differential equation*.

**EXAMPLE 11.1**

| *Partial differential equation* | *Order* |
|---|---|
| $\dfrac{\partial^2 u}{\partial x^2} + \dfrac{\partial^2 u}{\partial y^2} = \dfrac{\partial u}{\partial t}$ | 2 |
| $\dfrac{\partial^3 u}{\partial t^3} + \dfrac{\partial^2 u}{\partial x^2} + \dfrac{\partial u}{\partial y} + x^2 yu = 0$ | 3 |
| $\dfrac{\partial^4 u}{\partial x^4} = \dfrac{\partial^2 u}{\partial t^2} + \dfrac{\partial^2 u}{\partial t\, \partial x} + \left(\dfrac{\partial u}{\partial x}\right)^2$ | 4 |

Now consider the second order partial differential equation

$$\frac{\partial^2 u}{\partial x\, \partial y} = 0. \tag{11.5}$$

If we write (11.5) as

$$\frac{\partial}{\partial x}\left(\frac{\partial u}{\partial y}\right) = 0,$$

integration with respect to $x$ gives

$$\frac{\partial u}{\partial y} = G(y), \tag{11.6}$$

where $G(y)$ is an arbitrary function of $y$. Integration of (11.6) gives

$$u = \int G(y)\, dy + f(x)$$
$$= g(y) + f(x), \tag{11.7}$$

where $f(x)$ is an arbitrary function of $x$, and we replace the indefinite integral of the arbitrary function $G(y)$ with the arbitrary function $g(y)$. Direct substitution of (11.7) into (11.5) verifies that (11.7) is indeed a solution.

Notice that our solution (11.4) of a first order partial differential equation contained one arbitrary function, while our solution (11.7) of a second order partial differential equation involved two arbitrary functions. The concept of a general solution of linear partial differential equation is sometimes useful.

**Definition 11.2.**    The *general solution* of an $n$th order linear partial differential equation in two independent variables is one which contains $n$ arbitrary functions.

**EXAMPLE 11.2**    Solve the boundary value problem

$$\frac{\partial^2 u}{\partial x\, \partial y} + \frac{\partial u}{\partial x} = 2x, \qquad x > 0, \quad y > 0, \tag{11.8}$$

$$u(x, 0) = x^2 + \cos x, \, x > 0, \qquad \frac{\partial u}{\partial y}(0, y) = 2y, \quad y > 0.$$

If we write this differential equation as

$$\frac{\partial}{\partial x}\left[\frac{\partial u}{\partial y} + u\right] = 2x$$

and integrate, we obtain

$$\frac{\partial u}{\partial y} + u = x^2 + f(y). \tag{11.9}$$

The left-hand side of (11.9) is a first order linear equation with the integrating factor $e^y$. We multiply (11.9) by $e^y$ and integrate to find

$$\int \frac{\partial}{\partial y}[e^y u]\, dy = \int x^2 e^y\, dy + \int f(y)e^y\, dy + g(x)$$

or

$$u = x^2 + F(y) + e^{-y}g(x), \tag{11.10}$$

where

$$F(y) = e^{-y} \int f(y)e^y\, dy.$$

From the boundary conditions we have

$$u(x, 0) = x^2 + F(0) + g(x) = x^2 + \cos x,$$

$$\frac{\partial u}{\partial y}(0, y) = F'(y) - e^{-y}g(0) = 2y.$$

Thus

$$g(x) = \cos x - F(0)$$

and

$$F(y) = -e^{-y}g(0) + y^2 + C.$$

To evaluate $C$, set $x = 0$ and $y = 0$ in these last two equations to find

$$g(0) = 1 - F(0),$$
$$F(0) = -g(0) + C.$$

This gives

$$C = 1$$

and our solution (11.10) of the original boundary value problem becomes

$$u = x^2 - e^{-y}g(0) + y^2 + C + e^{-y}[\cos x - F(0)]$$
$$= x^2 - e^{-y}[1 - F(0)] + y^2 + 1 + e^{-y}[\cos x - F(0)]$$
$$= x^2 - e^{-y} + y^2 + 1 + e^{-y}\cos x.$$

**EXAMPLE 11.3**   Find the general solution of

$$\frac{\partial^2 u}{\partial y^2} + x^2 u = \pi e^{3y}, \qquad x > 0, \qquad y > 0. \qquad (11.11)$$

Recall that in solving second order linear nonhomogeneous ordinary differential equations we added a particular solution to the general solution of the associated homogeneous differential equation. If we try that procedure on (11.11) we note that

$$\frac{\partial^2 u}{\partial y^2} + x^2 u = 0$$

has solutions of the form

$$u = f(x)\cos xy + g(x)\sin xy.$$

Taking our clue from the method of undetermined coefficients we try a particular solution of (11.11) in the form $k(x)e^{3y}$. Substituting this into (11.11) gives

$$9k(x)e^{3y} + x^2 k(x)e^{3y} = \pi e^{3y}.$$

Solving for $k(x)$ gives

$$k(x) = \frac{\pi}{x^2 + 9}$$

and our general solution of (11.11) is

$$u = f(x) \cos xy + g(x) \sin xy + \frac{\pi e^{3y}}{x^2 + 9}.$$

**EXAMPLE 11.4**    Solve the initial value problem

$$\frac{\partial^2 u}{\partial t^2} = c^2 \frac{\partial^2 u}{\partial x^2}, \qquad t > 0, \quad -\infty < x < \infty, \tag{11.12}$$

$$u(x, 0) = h(x), \qquad \frac{\partial u}{\partial t}(x, 0) = 0. \tag{11.13}$$

Equation (11.12) describes the vibration of an infinitely long elastic string with $c^2$ a positive constant, $c^2 = T/\rho$, $T$ the horizontal component of tension in the string, and $\rho$ the density of the string. From (11.13) we see that at $t = 0$ the string has an initial displacement $h(x)$ and is released from rest. [Note that $u(x, t)$ is the displacement from the equilibrium position at location $x$ along the string and time $t$; see Figure 11.1.]

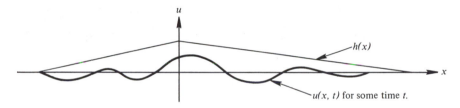

**Figure 11.1**    An infinite string.

The key step in solving this problem is to make a change of independent variables by letting

$$\begin{aligned} r &= x + ct, \\ s &= x - ct. \end{aligned} \tag{11.14}$$

(The reason for this will be explored in the exercises.) Using the chain rule for partial derivatives, we calculate

$$\frac{\partial u}{\partial t} = \frac{\partial u}{\partial r}\frac{\partial r}{\partial t} + \frac{\partial u}{\partial s}\frac{\partial s}{\partial t} = c\left(\frac{\partial u}{\partial r} - \frac{\partial u}{\partial s}\right),$$

$$\frac{\partial^2 u}{\partial t^2} = c\left[\frac{\partial}{\partial r}\left(\frac{\partial u}{\partial r} - \frac{\partial u}{\partial s}\right)\frac{\partial r}{\partial t} + \frac{\partial}{\partial s}\left(\frac{\partial u}{\partial r} - \frac{\partial u}{\partial s}\right)\frac{\partial s}{\partial t}\right]$$

$$= c^2\left(\frac{\partial^2 u}{\partial r^2} - 2\frac{\partial^2 u}{\partial r\, \partial s} + \frac{\partial^2 u}{\partial s^2}\right)$$

and, similarly,

$$\frac{\partial^2 u}{\partial x^2} = \frac{\partial^2 u}{\partial r^2} + 2\frac{\partial^2 u}{\partial r\, \partial s} + \frac{\partial^2 u}{\partial s^2}.$$

Thus in terms of our new independent variables, $r$ and $s$, (11.12) becomes

$$c^2\left(\frac{\partial^2 u}{\partial r^2} - 2\frac{\partial^2 u}{\partial r\,\partial s} + \frac{\partial^2 u}{\partial s^2}\right) = c^2\left(\frac{\partial^2 u}{\partial r^2} + 2\frac{\partial^2 u}{\partial r\,\partial s} + \frac{\partial^2 u}{\partial s^2}\right)$$

or

$$-4c^2\frac{\partial^2 u}{\partial r\,\partial s} = 0. \tag{11.15}$$

Since $-4c^2 \neq 0$, we have the partial differential equation

$$\frac{\partial^2 u}{\partial r\,\partial s} = 0,$$

with the general solution

$$u = g(s) + f(r), \tag{11.16}$$

where $g$ and $f$ are arbitrary functions [see (11.7)]. In terms of our original variables we have the general solution of (11.12) as

$$u = g(x - ct) + f(x + ct). \tag{11.17}$$

To satisfy the second of the initial conditions (11.13), we differentiate (11.17) with respect to $t$ to find

$$\frac{\partial u}{\partial t} = g'(x - ct)(-c) + f'(x + ct)(c),$$

so

$$\frac{\partial u}{\partial t}(x, 0) = [-g'(x) + f'(x)]c = 0.$$

Thus

$$f'(x) = g'(x)$$

and

$$f(x) = g(x) + C. \tag{11.18}$$

Using (11.18) in (11.17) gives

$$u = g(x - ct) + g(x + ct) + C, \tag{11.19}$$

so the first condition in (11.13) requires that

$$g(x) + g(x) + C = h(x)$$

or

$$g(x) = \tfrac{1}{2}[h(x) - C].$$

Using this expression for $g$ in (11.19) gives our solution of the original initial value problem as

$$u = \tfrac{1}{2}[h(x + ct) + h(x - ct)]. \tag{11.20}$$

The form of (11.20) is called the *d'Alembert solution of the wave equation*.

Our concern in this chapter is with linear partial differential equations, primarily those of second order. If we have two independent variables, $x$ and $y$, a *general linear second order partial differential equation* will have the form

$$A(x, y)\frac{\partial^2 u}{\partial x^2} + B(x, y)\frac{\partial^2 u}{\partial x \partial y} + C(x, y)\frac{\partial^2 u}{\partial y^2} + D(x, y)\frac{\partial u}{\partial x} + E(x, y)\frac{\partial u}{\partial y} + F(x, y)u$$

$$= G(x, y). \qquad (11.21)$$

If $G(x, y) = 0$, this equation is said to be *homogeneous;* if not, it is *nonhomogeneous*. This is consistent with our terminology for ordinary differential equations. Three special cases of (11.21) are

$$k\frac{\partial^2 u}{\partial x^2} = \frac{\partial u}{\partial t} \qquad \text{(the heat equation)}, \qquad (11.21a)$$

$$c^2\frac{\partial^2 u}{\partial x^2} = \frac{\partial^2 u}{\partial t^2} \qquad \text{(the wave equation)}, \qquad (11.21b)$$

and

$$\frac{\partial^2 u}{\partial x^2} + \frac{\partial^2 u}{\partial y^2} = 0 \qquad \text{(Laplace's equation)}. \qquad (11.21c)$$

These equations will be considered in considerable detail in the next sections, not only because they describe interesting physical situations, but also because they are prototype models for three classes or types of the general equation (11.21) (see exercises 27 and 28).

## EXERCISES

Find the general solution to the following partial differential equations.

1. $\dfrac{\partial^2 u}{\partial x^2} = y$.

2. $\dfrac{\partial u}{\partial x} + \dfrac{1}{x}u = xy$.

3. $\dfrac{\partial u}{\partial y} + 2yu = y \sin x$.

4. $\dfrac{\partial^2 u}{\partial x^2} + 4y^2 u = 6$.

5. $\dfrac{\partial^2 u}{\partial x^2} + 4y^2 u = y^2 - x + 1$.

6. $\dfrac{\partial^2 u}{\partial x \partial y} = 4x + 4y$.

7. $\dfrac{\partial^2 u}{\partial x \partial y} - 3\dfrac{\partial u}{\partial y} = 2y$.

**8.** $\dfrac{\partial^2 u}{\partial x\, \partial y} + 2\dfrac{\partial u}{\partial x} = \sin x.$

Solve the following partial differential equations subject to the conditions indicated.

**9.** $\dfrac{\partial^2 u}{\partial x\, \partial y} = 0, \quad u(0,y) = \sin y, \quad u(x,0) = xe^x.$

**10.** $\dfrac{\partial^2 u}{\partial y^2} - 12x = 0, \quad u(x,0) = x^2, \quad \dfrac{\partial u}{\partial y}(x,0) = \sin x.$

**11.** $\dfrac{\partial^2 u}{\partial y^2} - 2u = 0, \quad u(x,0) = x^3, \quad \dfrac{\partial u}{\partial y}(x,0) = e^{-x}.$

**12.** $\dfrac{\partial^2 u}{\partial x\, \partial y} - \dfrac{\partial u}{\partial y} = 1, \quad u(x,0) = 0, \quad \dfrac{\partial u}{\partial y}(0,y) = \sin y.$

**13.** Make the change of variables

$$x = r \cos \theta,$$
$$y = r \sin \theta$$

to show that Laplace's equation in rectangular coordinates,

$$\frac{\partial^2 u}{\partial x^2} + \frac{\partial^2 u}{\partial y^2} = 0,$$

becomes

$$\frac{\partial^2 u}{\partial r^2} + \frac{1}{r}\frac{\partial u}{\partial r} + \frac{1}{r^2}\frac{\partial^2 u}{\partial \theta^2} = 0$$

in polar coordinates.

**14.** Show that in spherical coordinates, $x = \rho \sin \phi \sin \theta$, $y = \rho \sin \phi \cos \theta$, $z = \rho \cos \phi$, Laplace's equation,

$$\frac{\partial^2 u}{\partial x^2} + \frac{\partial^2 u}{\partial y^2} + \frac{\partial^2 u}{\partial z^2} = 0,$$

becomes

$$\frac{1}{\rho}\frac{\partial^2(\rho u)}{\partial \rho^2} + \frac{1}{\rho^2}\left(\frac{1}{\sin \phi}\right)\frac{\partial}{\partial \phi}\left(\sin \phi \frac{\partial u}{\partial \phi}\right) + \frac{1}{\rho^2 \sin^2 \phi}\frac{\partial^2 u}{\partial \theta^2} = 0.$$

**15.** One mathematical model of the growth of a spherical tumor (Figure 11.2) uses Laplace's equation as follows. We assume that the tumor is initially the shape of a sphere of radius $a$, and as it grows remains a sphere of radius $R(t)$. Let $p(\rho, t)$ be the pressure within the tumor at distance $\rho$ from its center and at time $t$. Then $p$ satisfies

$$\frac{1}{\rho^2}\frac{\partial^2(\rho p)}{\partial \rho^2} = S, \qquad \rho < R(t),$$

where $S$ is a rate of volume change per unit volume. Let $u(\rho, t)$ be the nutrient concentration outside the tumor and $u_0$ be that inside the tumor. Then $u$ satisfies

$$\frac{1}{\rho^2}\frac{\partial^2(\rho u)}{\partial \rho^2} = 0, \qquad \rho > R(t).$$

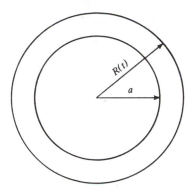

**Figure 11.2**   A spherical tumor.

Additional conditions to be imposed on the system are

(1) $p(0, t)$ is bounded

(2) $p(R(t), t) = \dfrac{\beta}{R(t)}$

(3) $\dfrac{\partial u}{\partial \rho} - \mu(u - u_0)^{1/2} = 0, \quad$ for $\rho = R(t)$

(4) $\lim\limits_{\rho \to \infty} u(\rho, t) = u_\infty$

In the equations above, $\beta$, $\mu$, $u_\infty$, $S$, and $u_0$ are constants.

(a) Find the general solution for $p(\rho, t)$, $\rho < R(t)$, using conditions 1 and 2 to evaluate the arbitrary functions that appear in your solution.

(b) Find the general solution for $u(\rho, t)$, $\rho > R(t)$, using conditions 3 and 4 to evaluate the arbitrary functions that appear in your solution. [When you use the quadratic equation to solve for one of your functions, you may determine the proper sign by noting that $u(\rho, t)$ must be a decreasing function of $\rho$.]

We now have expressions for $u(\rho, t)$ and $p(\rho, t)$ in terms of the radius of the tumor, $R(t)$. However, this expanding radius is not known, but may be determined from the last requirements,

(5) $R'(t) = -\dfrac{\partial p}{\partial \rho} + \lambda \sqrt{u - u_0}, \quad$ for $\rho = R(t)$,

   $R(0) = a.$

Although this separable ordinary differential equation may be solved, it is more instructive to find the limiting size of this tumor. To this end set $R'(t) = 0$ (the radius does not change in time) and solve for the resulting $R$. Problems such as this where one of the boundaries is not known and changes with time are known as moving boundary value problems.

16. (a) Solve the initial value problem

$$\frac{\partial^2 u}{\partial t^2} = c^2 \frac{\partial^2 u}{\partial x^2}$$

$$u(x, 0) = 0, \qquad \frac{\partial u}{\partial t}(x, 0) = H(x).$$

[*Hint:* Look at Example 11.4 and start with the solution of the partial differential equation as given in (11.17).]

**(b)** Use the results of part (a) and Example 11.4 to write down the solution to

$$\frac{\partial^2 u}{\partial t^2} = c^2 \frac{\partial^2 u}{\partial x^2}$$

$$u(x, 0) = h(x), \qquad \frac{\partial u}{\partial t}(x, 0) = H(x).$$

17. Show by direct substitution that if $h$ has two continuous derivatives (11.20) is indeed a solution of (11.12) satisfying initial conditions (11.13).

Show that the given function satisfies the given differential equation.

18. $\dfrac{\partial^2 u}{\partial x^2} + \dfrac{\partial^2 u}{\partial y^2} = 0$,   $e^{ay} \sin ax$,   $a$ is constant.

19. $\dfrac{\partial^2 u}{\partial x^2} + \dfrac{\partial^2 u}{\partial y^2} = 0$,   $e^{ay} \cos ax$,   $a$ is constant.

20. $\dfrac{\partial^2 u}{\partial x^2} + \dfrac{\partial^2 u}{\partial y^2} = 0$,   $ax + by + cxy + d$,   $a, b, c, d$ are constants.

21. $\dfrac{\partial^2 u}{\partial x^2} = \dfrac{\partial^2 u}{\partial t^2}$,   $\sin ax \cos at$.

22. $a^2 \dfrac{\partial^2 u}{\partial x^2} = \dfrac{\partial^2 u}{\partial t^2}$,   $\cos bx \sin abt$.

23. $k \dfrac{\partial^2 u}{\partial x^2} = \dfrac{\partial u}{\partial t}$,   $e^{-ka^2 t} \sin ax$.

24. $\dfrac{\partial^2 u}{\partial x^2} + \dfrac{\partial^2 u}{\partial y^2} + \dfrac{\partial^2 u}{\partial z^2} = 0$,   $\sin ax \sin by \sinh(\sqrt{a^2 + b^2}\, z)$.

25. $k\left(\dfrac{\partial^2 u}{\partial x^2} + \dfrac{\partial^2 u}{\partial y^2}\right) = \dfrac{\partial u}{\partial t}$,   $e^{-2ka^2 t} \sin ax \cos ay$.

26. $a^2\left(\dfrac{\partial^2 u}{\partial x^2} + \dfrac{\partial^2 u}{\partial y^2}\right) = \dfrac{\partial^2 u}{\partial t^2}$,   $\cos \alpha x \cos \beta y \cos(\sqrt{\alpha^2 + \beta^2}\, at)$.

The following table gives a classification of linear second order partial differential equations (11.21) according to the value of $B^2 = 4AC$ in portions of the $xy$-plane.

$$A(x, y)\frac{\partial^2 u}{\partial x^2} + B(x, y)\frac{\partial^2 u}{\partial x\, \partial y} + C(x, y)\frac{\partial^2 u}{\partial y^2} + D(x, y)\frac{\partial u}{\partial y} + E(x, y)\frac{\partial u}{\partial y}$$

$$+F(x, y)u = G(x, y) \qquad (11.21)$$

| Value of $B^2 - 4AC$ in a domain of the $xy$ plane | Classification |
|---|---|
| $< 0$ | Elliptic |
| $> 0$ | Hyperbolic |
| $= 0$ | Parabolic |

27. Classify (11.5), (11.8), (11.11), (11.12), (11.21a–c), and the differential equations in exercises 1, 5, 6, and 7.

Note that as far as classification is concerned, the forms of $D$, $E$, $F$, and $G$ are immaterial. Thus we let them all equal zero in (11.21) and in addition consider the case where $A$, $B$, and $C$ are constant.

$$A\frac{\partial^2 u}{\partial x^2} + B\frac{\partial^2 u}{\partial x\,\partial y} + C\frac{\partial^2 u}{\partial y^2} = 0.\tag{11.21d}$$

**28. (a)** Make the change of independent variables

$$r = x + ay,\qquad s = x + by$$

in (11.21d) and show that the new equation is

$$(A + Ba + Ca^2)\frac{\partial^2 u}{\partial r^2} + (2A + B(a + b) + 2Cab)\frac{\partial^2 u}{\partial r\,\partial s}$$

$$+ (A + Bb + Cb^2)\frac{\partial^2 u}{\partial s^2} = 0.\tag{11.21e}$$

**(b)** If $B^2 - 4AC > 0$, let

$$a = \frac{-B + \sqrt{B^2 - 4AC}}{2C},\qquad b = \frac{-B - \sqrt{B^2 - 4AC}}{2C},\qquad AC \neq 0$$

and show that (11.21e) reduces to

$$-C^{-1}(B^2 - 4AC)\frac{\partial^2 u}{\partial r\,\partial s} = 0.$$

[Thus show that all hyperbolic equations of the form (11.21d), with $AC \neq 0$, have the general solution

$$u = g(x + by) + f(x + ay);$$

see Example 11.4 and especially (11.15).]

**(c)** If $B^2 - 4AC = 0$, let

$$r = x + ay,$$

$$s = x,$$

where $a = -B/(2C)$, to show that (11.21e) reduces to

$$A\frac{\partial^2 u}{\partial s^2} = 0.$$

**(d)** Integrate the differential equation in part (c) to show the general solution of all parabolic equations of the form (11.21d) may be written as

$$u = xf(x + ay) + g(x + ay).$$

The case where $B^2 - 4AC < 0$, the elliptic case, does not give a simple partial differential which easily yields a general solution. Thus we will not treat this case in this exercise set.

## 11.2 SEPARATION OF VARIABLES

The partial differential equations that yielded general solutions were those we could put in a form to use techniques of solution for ordinary differential equations. A much more straightforward technique for solving *linear* partial differential equations is presented in this section and used throughout the remainder of the chapter. Although this method is not universally applicable, it works on many partial differential equations encountered in elementary applications. Its successful application leads to the types of boundary value problems encountered in Chapters 4 and 10.

In Section 2.1 we discovered that an ordinary differential equation of the form

$$P_1(x)P_2(y)\, dx + Q_1(x)Q_2(y)\, dy = 0 \tag{11.22}$$

could be solved by direct integration after division by $P_2(y)Q_1(x)$. The reason this worked is that the division isolated the functions of $x$ from those of $y$. To solve partial differential equations using *separation of variables* we assume that the solution may be expressed as a product of functions of each independent variable. That is, if $u(x, y)$ is our solution, assume that

$$u(x, y) = P(x)Q(y) \tag{11.23}$$

and substitute this expression into the partial differential equation. The success of this method then depends upon our manipulating the resulting equation to obtain ordinary differential equations. The subsequent details for finding $P(x)$ and $Q(y)$ are illustrated by the following three examples.

**EXAMPLE 11.5**　　Find all solutions of

$$\frac{\partial u}{\partial x} + \frac{\partial u}{\partial y} = 3u \tag{11.24}$$

which have the form given in (11.23).

Direct substitution of (11.23) into (11.24) yields

$$P'(x)Q(y) + P(x)Q'(y) = 3P(x)Q(y),$$

which becomes

$$\frac{P'(x)}{P(x)} + \frac{Q'(y)}{Q(y)} = 3 \tag{11.25}$$

if we divide by $P(x)Q(y)$. If we rearrange (11.25) as

$$\frac{P'(x)}{P(x)} = 3 - \frac{Q'(y)}{Q(y)}, \tag{11.26}$$

we see that the left-hand side of (11.26) is a function of $x$ only, while the right-hand side is a function of $y$ only. Since this is valid for all values of $x$ and $y$, and $x$ and $y$ are independent varibles, both sides of (11.26) must equal the same constant. If we call this constant $\lambda$, we have

$$\frac{P'(x)}{P(x)} = \lambda \qquad \text{and} \qquad 3 - \frac{Q'(y)}{Q(y)} = \lambda \tag{11.27a}$$

or

$$P'(x) - \lambda P(x) = 0, \qquad Q'(y) + (\lambda - 3)Q(y) = 0. \tag{11.27b}$$

Either by integrating (11.27a) directly or finding integrating factors of (11.27b) we find their solution as

$$P(x) = C_1 e^{\lambda x}, \qquad Q(y) = C_2 e^{(3-\lambda)y}.$$

(Note that the solutions are still valid if $\lambda = 0$ or $\lambda = 3$.)

Thus solutions of (11.24) may be written as

$$u(x, y) = C e^{\lambda x} e^{(3-\lambda)y},$$

where $\lambda$ and $C$ are arbitrary constants. These constants are determined by additional constraints on the function $u$, as illustrated by the next example.

**EXAMPLE 11.6**    The second order partial differential equation,

$$k \frac{\partial^2 u}{\partial x^2} = \frac{\partial u}{\partial t}, \qquad k > 0 \tag{11.28}$$

is often referred to as the *one-dimensional heat equation* or *one-dimensional diffusion equation*. In this example we consider a thin, homogeneous rod of length $L$ (Figure 11.3) with the following assumptions:

1. No heat is being generated in the rod.
2. Heat may only flow along the $x$ direction.
3. No heat crosses the lateral surfaces of the rod.
4. The specific heat and thermal conductivity of the rod are constant (i.e., $k$ is a constant, $k =$ (thermal conductivity)/[(rod density)(specific heat)]).

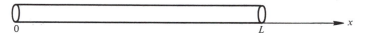

**Figure 11.3**   A heat conducting rod.

If $u(x, t)$ denotes the temperature at point $x$ in the rod at time $t$, (11.28) describes the behavior of $u$. In addition, we assume an initial temperature distribution in the rod given by

$$u(x, 0) = f(x), \qquad 0 < x < L. \tag{11.29}$$

We also need to specify boundary conditions at $x = 0$ and $x = L$. If the temperatures at $x = 0$ and $x = L$ are kept at zero degrees, we have

$$u(0, t) = 0, \qquad t > 0$$
$$u(L, t) = 0, \qquad t > 0. \tag{11.30}$$

[The boundary condition

$$\frac{\partial u}{\partial x}(L, t) = 0, \qquad t > 0 \tag{11.31}$$

is appropriate if the end at $x = L$ is insulated. At an insulated face, there is no movement of heat across the boundary, so the partial derivative of $u$ with respect to $x$ must vanish.] We assume that the temperature $u(x, t)$ may be written as

$$u(x, t) = P(x)Q(t) \tag{11.32}$$

in (11.28). This gives

$$kP''(x)Q(t) = P(x)Q'(t)$$

or

$$\frac{P''(x)}{P(x)} = \frac{Q'(t)}{kQ(t)} = \lambda. \tag{11.33}$$

Thus the partial differential equation is "separated" into two ordinary differential equations,

$$P''(x) - \lambda P(x) = 0, \tag{11.34a}$$

$$Q'(t) - \lambda k Q(t) = 0. \tag{11.34b}$$

The general solution of (11.34a) depends on whether the constant $\lambda$ is positive, negative, or zero. Thus we have

$$P = \begin{cases} C_1 \cosh \sqrt{\lambda}\, x + C_2 \sinh \sqrt{\lambda}\, x, & \text{if } \lambda > 0 \\ C_1 + C_2 x, & \text{if } \lambda = 0 \\ C_1 \cos \mu x + C_2 \sin \mu x, & \text{if } \lambda < 0, \ \lambda = -\mu^2. \end{cases}$$

To determine the value of $\lambda$, we note that the boundary conditions in (11.30) only involve the $x$ variable. That is, by our assumption $u = P(x)Q(t)$, we have

$$u(0, t) = P(0)Q(t) = 0,$$

$$u(L, t) = P(L)Q(t) = 0.$$

Since $u(x, t) = 0$ if $Q(t) = 0$ (and we do not want a solution that is identically zero), this means that we need

$$P(0) = P(L) = 0 \tag{11.35}$$

in order for (11.30) to be satisfied. Note that (11.34a) plus (11.35) constitutes a Sturm–Liouville boundary value problem. (See Section 10.4.)

Now, if $\lambda > 0$, from (11.35) we need

$$C_1 \cosh 0 + C_2 \sinh 0 = 0,$$

$$C_1 \cosh \sqrt{\lambda}L + C_2 \sinh \sqrt{\lambda}L = 0. \tag{11.36}$$

Since

$$\cosh x = \frac{e^x + e^{-x}}{2} \quad \text{and} \quad \sinh x = \frac{e^x - e^{-x}}{2}, \quad \cosh 0 = 1, \quad \sinh 0 = 0,$$

the first equation in (11.36) requires $C_1 = 0$. Since $\sinh \sqrt{\lambda}L \neq 0$ for $\sqrt{\lambda}L \neq 0$, the second equation in (11.36) is true only if $C_2 = 0$ and for $\lambda > 0$ we have only the trivial solution, $P(x) = 0$, which results in $u = 0$.

If $\lambda = 0$, $P(x) = C_1 + C_2x$, so equation (11.35) requires that

$$C_1 + C_2 0 = 0,$$

$$C_1 + C_2 L = 0.$$

The only solution of these two equations is $C_1 = C_2 = 0$ and only the trivial solution, $P(x) = 0$, comes from choosing $\lambda = 0$.

For $\lambda < 0$, we let $\lambda = -\mu^2$ for convenience (it avoids using $\sqrt{-\lambda}$) and note that the boundary conditions of (11.35) are satisfied if

$$C_1 \cos 0 + C_2 \sin 0 = 0,$$

$$C_1 \cos \mu L + C_2 \sin \mu L = 0.$$

Thus $C_1 = 0$ and in order for $C_2 \neq 0$ ($C_2 = 0$ gives $u = 0$), we must choose $\mu$ so that

$$\sin \mu L = 0. \tag{11.37}$$

This means that $\mu$ may take on the values

$$\mu = \frac{n\pi}{L}, \qquad n = 1, 2, 3, \ldots, \tag{11.38}$$

with the corresponding solutions for $P(x)$ given by

$$P(x) = C_2 \sin \frac{n\pi x}{L}, \qquad n = 1, 2, 3, \ldots. \tag{11.39}$$

The function of $t$ may now be determined from (11.34b) as

$$Q(t) = C_3 e^{\lambda kt} = C_3 e^{-\mu^2 kt} = C_3 e^{-n^2 \pi^2 kt/L^2}$$

and $u$ is given by

$$u = PQ = C_2 C_3 \sin \frac{n\pi x}{L} e^{-n^2 \pi^2 kt/L^2}. \tag{11.40}$$

Since $n = 1, 2, 3, \ldots$ in (11.40), and we may choose a different multiplicative constant for each $n$, we collect all such solutions as the infinite series

$$u(x, t) = \sum_{n=1}^{\infty} b_n \sin \frac{n\pi x}{L} e^{-n^2 \pi^2 kt/L^2}. \tag{11.41}$$

The last condition we have to satisfy is the initial condition (11.29), which gives

$$u(x, 0) = f(x) = \sum_{n=1}^{\infty} b_n \sin \frac{n\pi x}{L}, \qquad 0 < x < L. \tag{11.42}$$

We now need an expansion of $f(x)$ in terms of a Fourier sine series, and using (10.25) of Theorem 10.4 we know that the coefficients, $b_n$, are given by

$$b_n = \frac{2}{L} \int_0^L f(x) \sin \frac{n\pi x}{L} \, dx. \tag{11.43}$$

Equation (11.41), with the $b_n$ given by (11.43), is the final form of the solution to our original problem. Note that $\lim_{t \to \infty} u(x, t) = 0$ for all $x$. Thus the limiting temperature is that of zero degrees everywhere (the same as that of the ends of the rod).

Note also that the constant $k$ occurs in every term in the solution for $u(x, t)$ as $e^{-n^2 \pi^2 kt/L^2}$. Thus large values of $k$ make the rod's temperature change from the initial value, $f(x)$, to zero more rapidly than do smaller ones. This makes sense since $k$ is directly proportional to the thermal conductivity and inversely proportional to the rod density and specific heat.

One key step in this method of solution was the determination of boundary conditions for $P(x)$ from those on $u(x, t)$ [that is, going from (11.30) to (11.35)]. This was possible only because the boundary conditions for (11.30) were homogeneous. If we start with nonhomogeneous boundary conditions, the first step we perform before using the method of separation of variables is to make a change of dependent variable. The goal of this transformation is to have homogeneous boundary conditions, so the method of separation of variables leads to a Sturm–Liouville boundary value problem. This procedure is illustrated in the next example.

**EXAMPLE 11.7**    Find the temperature distribution in a rod if one end is kept at $g_0(t)$ degrees and the other at $g_1(t)$ degrees (Figure 11.4). Assume that the initial temperature distribution is

$$u(x, 0) = f(x), \qquad 0 < x < L. \tag{11.44}$$

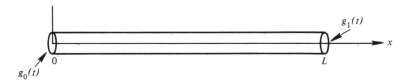

**Figure 11.4**   Rod with nonhomogeneous end conditions.

Here the appropriate differential equation is (11.28), while the boundary conditions are

$$u(0, t) = g_0(t), \qquad t > 0,$$
$$u(L, t) = g_1(t), \qquad t > 0. \tag{11.45}$$

To change this problem to one involving homogeneous boundary conditions, we define a new dependent variable, $v(x, t)$, by

$$v(x, t) = u(x, t) + A(t)x + B(t). \tag{11.46}$$

If we choose $A$ and $B$ so that $v(0, t) = v(L, t) = 0$, we require

$$v(0, t) = u(0, t) + B = g_0(t) + B = 0,$$
$$v(L, t) = u(L, t) + AL + B = g_1(t) + AL + B = 0. \tag{11.47}$$

This gives

$$B = -g_0(t), \qquad A = \frac{g_0(t) - g_1(t)}{L}. \qquad (11.48)$$

The partial differential equation for $v$ is

$$k \frac{\partial^2}{\partial x^2}[v(x, t) - Ax - B] = \frac{\partial}{\partial t}[v(x, t) - Ax - B]$$

or

$$k \frac{\partial^2 v}{\partial x^2} = \frac{\partial v}{\partial t} - x \frac{\partial A}{\partial t} - \frac{\partial B}{\partial t}. \qquad (11.49)$$

If the ends of the rod are held at a *constant temperature*, $A$ and $B$ will be *constants* and (11.49) reduces to

$$k \frac{\partial^2 v}{\partial x^2} = \frac{\partial v}{\partial t}. \qquad (11.50)$$

The appropriate initial condition for $v$ is

$$v(x, 0) = u(x, 0) + Ax + B$$

$$= f(x) + \frac{(g_0 - g_1)x}{L} - g_0. \qquad (11.51)$$

Note that if $g_0$ and $g_1$ are constants, $v(x, t)$ satisfies the same partial differential equation and homogeneous boundary conditions as $u(x, t)$ did in Example 11.6. The difference in our new problem is in the initial condition, (11.51). Thus we may use (11.41) to write the solution for $v(x, t)$ as

$$v(x, t) = \sum_{n=1}^{\infty} b_n \sin \frac{n\pi x}{L} e^{-n^2 \pi^2 kt/L^2},$$

where the coefficients $b_n$ are given by

$$b_n = \frac{2}{L} \int_0^L \left[ f(x) + \frac{(g_0 - g_1)x}{L} - g_0 \right] \sin \frac{n\pi x}{L} dx.$$

In terms of our original dependent variable, we have

$$u(x, t) = \frac{(g_1 - g_0)x}{L} + g_0 + \sum_{n=1}^{\infty} b_n \sin \frac{n\pi x}{L} e^{-n^2 \pi^2 kt/L^2}.$$

Since the contribution of the infinite series to $u(x, t)$ for large $t$ diminishes, and in fact disappears as $t \to \infty$, we see that we may write

$$u(x, t) = \text{steady-state part} + \text{transient part},$$

where $(g_1 - g_0)x/L + g_0$ is the steady-state solution. [Note that it satisfies

$$k \frac{\partial^2 u}{\partial x^2} = 0, \qquad u(0, t) = g_0, \qquad u(L, t) = g_1.]$$

Thus if we made the change of variable as

$$u(x, t) = S(x) + v(x, t),$$

where $S(x)$ was the steady-state solution, we would also have obtained a homogeneous boundary value problem which we could solve using separation of variables.

We close this section with two examples which use Laplace transforms to solve partial differential equations. We may define the Laplace transform of a function of two variables by

$$\mathcal{L}\{u(x, t)\} = \overline{u}(x, s) = \int_0^\infty e^{-st}u(x, t) \, dt.$$

Since $x$ and $t$ are independent variables, $x$ acts like a constant as far as operations with respect to $t$ are concerned. Thus if we are careful, we may use all our previous properties of Laplace transforms. Some of these are collected for you in Table 11.1

**TABLE 11.1   LAPLACE TRANSFORMS**

| Function | Laplace transform |
|---|---|
| $t^n$ | $\dfrac{n!}{s^{n+1}}$ |
| $\sin at$ | $\dfrac{a}{s^2 + a^2}$ |
| $\cos at$ | $\dfrac{s}{s^2 + a^2}$ |
| $\dfrac{1}{\sqrt{\pi t}} e^{-a^2/(4t)}$ | $\dfrac{1}{\sqrt{s}} e^{-a\sqrt{s}}$ |
| $e^{at}f(t)$ | $\begin{cases} 0, & s < a \\ F(s - a), & a < s, \end{cases}$ if $F(s) = \mathcal{L}\{f(t)\}$ |
| $\displaystyle\int_0^t f(\tau)g(t - \tau) \, d\tau$ | $F(s)G(s)$ |
| $\dfrac{\partial u}{\partial t}$ | $s\overline{u}(x, s) - u(x, 0)$ |
| $\dfrac{\partial^2 u}{\partial t^2}$ | $s^2\overline{u}(x, s) - su(x, 0) - \dfrac{\partial u}{\partial t}(x, 0)$ |
| $\dfrac{\partial^2 u}{\partial x^2}$ | $\dfrac{\partial^2 \overline{u}}{\partial x^2}$ |

## EXAMPLE 11.8   Temperature in a semi-infinite rod

Here we consider a very long rod (assumed to be a semi-infinite rod) and replace the boundary condition at the right end of the finite rod with the limit condition

$$\lim_{x \to \infty} u(x, t) = 0.$$

We consider that the initial temperature is zero,

$$u(x, 0) = 0,$$

and that a constant flux of heat is applied at the left end of the rod,

$$\frac{\partial u}{\partial x}(0, t) = -C.$$

To solve this boundary value problem we take the Laplace transform of the partial differential equation

$$k\frac{\partial^2 u}{\partial x^2} = \frac{\partial u}{\partial t}$$

to obtain

$$k\frac{\partial^2 \bar{u}}{\partial x^2} = s\bar{u} - u(x, 0) = s\bar{u}. \tag{11.52}$$

We also take the Laplace transform of both sides of the boundary conditions to obtain

$$\frac{\partial \bar{u}}{\partial x}(0, s) = -\frac{C}{s}, \qquad \lim_{x \to \infty} \bar{u}(x, s) = 0. \tag{11.53}$$

The general solution of (11.52) is

$$\bar{u}(x, s) = C_1(s)e^{-(\sqrt{s/k})x} + C_2(s)e^{(\sqrt{s/k})x}.$$

To satisfy the two conditions in (11.53) we choose

$$C_2(s) = 0, \qquad C_1(s) = \frac{C\sqrt{k}}{s\sqrt{s}}$$

and have

$$\bar{u}(x, s) = \frac{C\sqrt{k}}{s\sqrt{s}}e^{-(\sqrt{s/k})x}.$$

We use the convolution theorem to find the inverse Laplace transform of this function by taking

$$F(s) = \frac{\sqrt{k}}{\sqrt{s}}e^{-(x/\sqrt{k})\sqrt{s}}, \qquad G(s) = \frac{C}{s}.$$

This gives

$$u(x, t) = \mathcal{L}^{-1}\{\bar{u}(x, s)\}$$

$$= \int_0^t f(\tau)g(t - \tau)\,d\tau$$

$$= C\sqrt{k}\int_0^t \frac{1}{\sqrt{\pi\tau}}e^{-x^2/(4k\tau)}(1)\,d\tau. \tag{11.54}$$

We may put this integral in a different form with a change of variable and integration by parts as follows. Let $y = x/(2\sqrt{k\tau})$, so $dy = -(x/(4\sqrt{k}\tau\sqrt{\tau}))\,d\tau$

and the integral becomes

$$u(x,\, t) = \frac{Cx}{\sqrt{\pi}} \int_{x/(2\sqrt{kt})}^{\infty} y^{-2} e^{-y^2}\, dy.$$

This integral may be transformed into one involving the complementary error function (see Section 7.5) by integration by parts ($dv = y^{-2}\, dy$, $u = e^{-y^2}$) as

$$u(x,\, t) = \frac{C}{\sqrt{\pi}} \left[ 2\sqrt{kt}\, e^{-x^2/(4kt)} - 2x \int_{x/(2\sqrt{kt})}^{\infty} e^{-y^2}\, dy \right]$$

$$= C \left[ 2\sqrt{\frac{kt}{\pi}}\, e^{-x^2/(4kt)} - x\, \mathrm{erfc}\!\left( \frac{x}{2\sqrt{kt}} \right) \right].$$

In this form we can see that having the constant flux at $x = 0$ gives rise to a temperature at that end given by

$$u(0,\, t) = 2C \sqrt{\frac{kt}{\pi}}.$$

### EXAMPLE 11.9    Vibration of a semi-infinite string

Example 11.4 considered the vibration of an infinitely long elastic string. We now consider a semi-infinite string where one end (at $x = 0$; Figure 11.5) is subject to a prescribed transverse displacement, $g_0(t)$. The string is initially at rest in its equilibrium position. This gives rise to the differential system

$$\frac{\partial^2 u}{\partial t^2} = c^2 \frac{\partial^2 u}{\partial x^2}, \qquad x > 0, \qquad t > 0,$$

$$u(x,\, 0) = \frac{\partial u}{\partial t}(x,\, 0) = 0, \qquad x > 0,$$

$$u(0,\, t) = g_0(t),$$

$$\lim_{x \to \infty} u(x,\, t) = 0.$$

We may take the Laplace transform of the differential equation and use the initial conditions to find

$$s^2 \bar{u}(x,\, s) = c^2 \frac{\partial^2 \bar{u}}{\partial x^2},$$

with the general solution

$$\bar{u}(x,\, s) = C_1(s) e^{sx/c} + C_2(s) e^{-sx/c}.$$

**Figure 11.5**   A semi-infinite string.

To find values for $C_1$ and $C_2$ we take the Laplace transform of the last two conditions above to find

$$\overline{u}(0,\,s) = \mathcal{L}\{u(0,\,t)\} = \mathcal{L}\{g_0(t)\},$$

$$\lim_{x \to \infty} \overline{u}(x,\,s) \neq 0.$$

This gives $C_1(s) = 0$, $C_2(s) = \mathcal{L}\{g_0(t)\}$, and

$$\overline{u}(x,\,s) = \mathcal{L}\{g_0(t)\}e^{-sx/c}.$$

From Table 11.1 we may write the solution in terms of our original variables as

$$u(x,\,t) = \begin{cases} 0, & t \le \dfrac{x}{c} \\ g_0\!\left(t - \dfrac{x}{c}\right), & t > \dfrac{x}{c} \end{cases}.$$

We may use this solution to illustrate the situation where a person holds one end of a long stretched rope, then shakes it and watches a "wave" progress down the rope. If we model this disturbance as

$$g_0(t) = \begin{cases} \sin t, & 0 < t < \pi \\ 0, & \pi < t \end{cases},$$

we see the solution has the form

$$u(x,\,t) = \begin{cases} 0, & t \le \dfrac{x}{c} \\[2mm] \sin\!\left(t - \dfrac{x}{c}\right), & \dfrac{x}{c} < t < \dfrac{x}{c} + \pi \\[2mm] 0, & \dfrac{x}{c} + \pi < t \end{cases}$$

(see Figure 11.6).

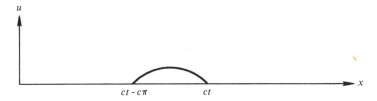

**Figure 11.6**   Travelling wave on a semi-infinite string.

## EXERCISES

Find solutions of the following partial differential equations using separation of variables.

**1.** $\dfrac{\partial u}{\partial x} + 3\dfrac{\partial u}{\partial y} = 0.$

**2.** $2\dfrac{\partial u}{\partial x} + \dfrac{\partial u}{\partial y} = 4u.$

**3.** $2x\dfrac{\partial u}{\partial x} + \dfrac{\partial u}{\partial y} = 0.$

**4.** $\dfrac{\partial u}{\partial x} + 2x\dfrac{\partial u}{\partial y} = 0.$

**5.** $x\dfrac{\partial u}{\partial x} + 3y\dfrac{\partial u}{\partial y} = 0.$

**6.** $y\dfrac{\partial u}{\partial x} + 2x\dfrac{\partial u}{\partial x} = 0.$

**7.** $\dfrac{\partial u}{\partial x} - \dfrac{\partial u}{\partial y} = (x + 2y)u.$

**8.** $x^2\dfrac{\partial u}{\partial x} + 2\dfrac{\partial u}{\partial y} = 0.$

**9.** $\dfrac{\partial u}{\partial x} + 3x^2\dfrac{\partial u}{\partial y} = 0.$

**10.** $3x^3y^3\dfrac{\partial^2 u}{\partial x\, \partial y} + u = 0.$

**11.** $\dfrac{\partial^2 u}{\partial x\, \partial y} + 9x^2y^2u = 0.$

**12.** Find the solution of the heat equation (11.28) for the following initial and boundary conditions.

    **(a)** $u(0, t) = 0,$    $u(L, t) = 0,$    $u(x, 0) = x.$
    **(b)** $u(0, t) = 0,$    $u(L, t) = 0,$    $u(x, 0) = u_0$ (a constant).

    **(c)** $u(0, t) = 0,$   $\dfrac{\partial u}{\partial x}(L, t) = 0,$    $u(x, 0) = x$ (insulated at $x = L$).

    **(d)** $\dfrac{\partial u}{\partial x}(0, t) = 0,$    $u(L, t) = 0,$    $u(x, 0) = f(x)$ (insulated at $x = 0$).

    **(e)** $u(0, t) = 0,$    $Bu(L, t) - \dfrac{\partial u}{\partial x}(L, t) = 0,$    $u(x, 0) = f(x)$

    (an imperfect insulator at $x = L$). If we write the second boundary condition as $\dfrac{\partial u}{\partial x}(L, t) = Bu(L, t)$, we see that the gradient of the temperature at the end of the rod is proportional to the temperature there.

    **(f)** Let $f(x) = x$ in exercise 12(e) and determine the solution. Then take the limit as $B \to 0$ and compare your result with the solution of exercise 12(c).

**13.** A metal rod has $k = 0.25$ and $L = 50$. If both ends of the rod are kept at zero degrees and the initial temperature in the rod is 0 for $0 \le x < 25$ and 10 for $25 < x < 50$, find the resulting temperature for subsequent times.

**14. (a)** A metal rod has $k = 1.0$ and $L = 100$. If both ends of the rod are insulated and the initial temperature in the rod is given by $f(x) = x$, find the resulting temperature subsequent times.

    **(b)** For your answer in part (a), evaluate $\lim\limits_{t \to \infty} u(x, t)$ and compare the result with the solution to the steady-state problem

$$k\dfrac{\partial^2 u}{\partial x^2} = 0, \qquad \dfrac{\partial u}{\partial x}(0) = \dfrac{\partial u}{\partial x}(100) = 0.$$

If the surface of the metal rod is not insulated but radiates heat into the surrounding medium at constant temperature $u_0$ according to Newton's law of cooling, we have the partial differential equation

$$k\frac{\partial^2 u}{\partial x^2} = \frac{\partial u}{\partial t} + k_1(u - u_0), \qquad k_1 > 0.$$

**15. (a)–(d)** Solve this differential equation with $k = 1$, $k_1 = \frac{1}{2}$ for boundary and initial conditions given by exercise 12(a) to (d).

In exercise 16 you need to change the dependent variable (see Example 11.7) before using the method of separation of variables.

**16.** Find the solution of the heat equation (11.28) for the following initial and boundary conditions.

**(a)** $u(0, t) = 6$, $\quad u(1, t) = 4$, $\quad u(x, 0) = x$, $\quad 0 < x < 1$.

**(b)** $\dfrac{\partial u}{\partial x}(0, t) = 3$, $\quad u(2, t) = 0$, $\quad u(x, 0) = \begin{cases} 3, & 0 < x < 1 \\ 0, & 1 < x < 2 \end{cases}$.

**(c)** $u(0, t) = 1$, $\quad \dfrac{\partial u}{\partial x}(\pi, t) = 0$, $\quad u(x, 0) = 2$, $\quad 0 < x < \pi$.

**(d)** $u(0, t) = 1$, $\quad \dfrac{\partial u}{\partial x}(\pi, t) = 3$, $\quad u(x, 0) = x$, $\quad 0 < x < \pi$.

**17.** The differential equation in (11.28) with initial condition (11.29) and boundary conditions (11.45) may be used to describe the diffusion of solutes across a membrane into a cell. For this situation we could consider an idealization of the membrane to one dimension (Figure 11.7). This could be the case for a large cell, or if we were only concerned with a small segment of the membrane. Here $L$ is the width of the cell membrane, $k$ the diffusion constant, $g_0(t)$ the concentration of the solute exterior to the cell, and $g_1(t)$ the concentration of the solute in the interior of the cell. Find the concentration of the solute in the membrane of the cell if $g_0(t) = g_0$ (a constant), $g_1(t) = 0$, and the initial condition is $u(x, 0) = 0$.

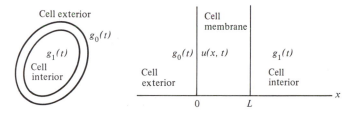

**Figure 11.7**    Cell Geometries.

**18.** Repeat exercise 17 if $g_0(t) = g_0$, $g_1(t) = g_1$ (a constant) and $u(x, 0) = u_0$ (a constant).

**19.** Show that if $v(x, t)$ and $u(x, t)$ satisfy

$$k\frac{\partial^2 v}{\partial x^2} = \frac{\partial v}{\partial t} - F(x, t) \qquad\qquad k\frac{\partial^2 u}{\partial x^2} = \frac{\partial u}{\partial t}$$

$$v(0, t) = v(L, t) = 0 \qquad\qquad u(0, t) = u(L, t) = 0$$

$$v(x, 0) = f(x) \qquad\qquad u(x, 0) = g(x),$$

then $w = u + v$ satisfies

$$k\frac{\partial^2 w}{\partial x^2} = \frac{\partial w}{\partial t} - F(x, t), \qquad w(0, t) = w(L, t) = 0, \qquad w(x, 0) = f(x) + g(x).$$

**20.** The $F(x, t)$ in exercise 19 could represent the loss of heat along the rod. Solve the heat conduction problem

$$k\frac{\partial^2 w}{\partial x^2} = \frac{\partial w}{\partial t} - 2F_0,$$

$$w(0, t) = w(L, t) = 0,$$

$$w(x, 0) = h(x)$$

by the following steps.

**(a)** Solve for $v(x, t)$ from exercise 19 by assuming that $v(x, t) = v(x)$. Then integrate the differential equation and satisfy the boundary conditions at $x = 0$ and $L$. This solution will define the initial function $f(x)$.

**(b)** Solve for $u(x, t)$ from exercise 19 choosing $g(x) = h(x) - f(x)$. [Here $h(x)$ is from the given initial condition and $f(x) = v(x)$ is from part (a).]

**(c)** Note that adding the solution to parts (a) and (b) gives the solution to your original problem.

**21.** Use the procedure from exercise 20 to solve

$$\frac{\partial^2 w}{\partial x^2} = \frac{\partial w}{\partial t} - 4,$$

$$w(0, t) = w(10, t) = 0,$$

$$w(x, 0) = 1 - 2x^2.$$

**22.** Another method of solution of a partial differential equation with a forcing function (i.e., a nonhomogeneous partial differential equation) is outlined as follows. To solve

$$k\frac{\partial^2 w}{\partial x^2} = \frac{\partial w}{\partial t} - F(x, t),$$

$$w(0, t) = w(L, t) = 0,$$

$$w(x, 0) = h(x),$$

**(a)** Assume that $w(x, t) = P(x)Q(t)$ and solve

$$k\frac{\partial^2 w}{\partial x^2} = \frac{\partial w}{\partial t},$$

$$w(0, t) = w(L, t) = 0.$$

**(b)** Let the eigenfunctions and eigenvalues from part (a) be denoted by $\phi_n(x)$, $\lambda_n^2$, $n = 1$, 2, 3, . . . , and expand $F(x, t)$ in terms of these eigenfunctions, treating $t$ as a parameter. If this expansion is

$$F(x, t) = \sum_{n=1}^{\infty} \alpha_n(t)\phi_n(x),$$

and we assume that our unknown function has the expansion

$$w(x, t) = \sum_{n=1}^{\infty} q_n(t)\phi_n(x),$$

show that the $\alpha_n(t)$ and $q_n(t)$ are related by

$$q_n'(t) + \lambda_n^2 k q_n(t) = \alpha_n(t), \qquad n = 1, 2, 3, \ldots .$$

Also show that the initial condition associated with this differential equation is

$$q_n(0) = \frac{\int_0^L h(x)\phi_n(x)\, dx}{\int_0^L \phi_n^2(x)\, dx}.$$

**23.** Use the method of exercise 22 to solve exercise 21.

**24.** A rod which is losing heat via radioactive decay is modeled by the partial differential equation

$$\frac{\partial^2 u}{\partial x^2} = \frac{\partial u}{\partial t} - xe^{-\alpha t}, \qquad 0 < x < \pi, t > 0.$$

If the ends of the rod ($x = 0$ and $x = \pi$) are held at zero degrees and the initial temperature is $u(x, 0) = 100$, find $u(x, t)$ using the method of exercise 22.

**25.** $\dfrac{-\hbar^2}{2m}\{\nabla^2 \psi\} + V(x, y, z)\psi = i\hbar\dfrac{\partial \psi}{\partial t}$ is the Schrödinger equation for a single particle of mass $m$ moving in a potential field $V(x, y, z)$. Show that separating out the $t$ dependence as

$$\psi(x, y, z, t) = \phi(x, y, z)e^{-iEt/\hbar}$$

results in the partial differential equation

$$\nabla^2 \phi + \frac{2m}{\hbar}(E - V)\phi = 0.$$

Here $\hbar = (\text{Planck's constant})/(2\pi)$ and $E$, the total energy, are constants and

$$\nabla^2 = \frac{\partial^2}{\partial x^2} + \frac{\partial^2}{\partial y^2} + \frac{\partial^2}{\partial z^2}$$

is the Laplacian operator.

**26.** Use the Laplace transform to solve the following problem of heat conduction in a semi-infinite bar.

$$k\frac{\partial^2 u}{\partial x^2} = \frac{\partial u}{\partial t}, \qquad 0 < x < \infty, \qquad t > 0,$$

$$u(x, 0) = 0,$$

$$u(0, t) = C, \quad (\text{a constant}),$$

$$\lim_{x \to \infty} u(x, t) = 0.$$

**27.** Solve exercise 26 if the end condition is changed to $u(0, t) = Ce^{-\alpha t}$.

**28.** Use the Laplace transform to derive D'Alembert's solution of the wave equation (see Example 11.4). You will need the additional conditions

$$\lim_{x \to \infty} u(x, t) = \lim_{x \to -\infty} u(x, t) = 0.$$

(*Hint:* It is helpful to write the general solution of $y'' - \alpha^2 y = 0$ as $y = C_1 e^{\alpha|x|} + C_2 e^{-\alpha|x|}$.)

## 11.3 THE WAVE EQUATION

We have already obtained d'Alembert's solution of the wave equation

$$\frac{\partial^2 u}{\partial t^2} = c^2 \frac{\partial^2 u}{\partial x^2},$$  (11.12)

subject to,

$$u(x, 0) = h(x), \qquad \frac{\partial u}{\partial t}(x, 0) = 0$$  (11.13)

(as Example 11.4) by using a change of independent variable. This initial value problem described the vibration of a stretched string released from rest from an initial displacement, $h(x)$. The partial differential equation in (11.12) is called *the wave equation* since it is used to model waves propagating in a continuous medium. Other examples are the motion of water waves, electromagnetic waves, and acoustic waves. We now use separation of variables to find the solution of the problem for a finite string. Thus we require $u(x, t)$ to satisfy (11.12), (11.13), and boundary conditions

$$u(0, t) = u(L, t) = 0,$$  (11.55)

describing the fact that the ends of the string, at $x = 0$ and $x = L$, remain fixed for all time. If we substitute the assumed form of $u$,

$$u(x, t) = P(x)Q(t),$$  (11.56)

into (11.12) and divide the result by $c^2 P(x)Q(t)$, we obtain

$$\frac{Q''(t)}{c^2 Q(t)} = \frac{P''(x)}{P(x)} = \lambda.$$  (11.57)

Since the left-hand side of (11.57) is a function only of $t$ while the right-hand side is a function only of $x$, and since $x$ and $t$ are independent variables, both sides of (11.57) must equal a constant, which we call $\lambda$. This gives ordinary differential equations

$$Q'' - \lambda c^2 Q = 0,$$  (11.58)

$$P'' - \lambda P = 0,$$  (11.59)

for the determination of $P(x)$ and $Q(t)$. The form of (11.56) requires that boundary conditions (11.55) transfer to

$$P(0) = P(L) = 0.$$  (11.60)

Note that (11.59) and (11.60) are the same Sturm–Liouville problem that we encountered in solving the heat equation using separation of variables [see (11.34a) and (11.36)]. Thus we have eigenvalues

$$\lambda = \frac{-n^2 \pi^2}{L^2}, \qquad n = 1, 2, 3, \ldots$$  (11.61)

and corresponding eigenfunctions

$$P(x) = \sin \frac{n \pi x}{L}, \qquad n = 1, 2, 3, \ldots.$$  (11.62)

Using this value of $\lambda$ in (11.58) allows us to see that

$$\sin \frac{n\pi ct}{L} \quad \text{and} \quad \cos \frac{n\pi ct}{L}, \qquad n = 1, 2, 3, \ldots$$

are solutions. Taking a different constant for each $n$ and each function gives us a series for $u(x, t)$ of the form

$$u(x, t) = \sum_{n=1}^{\infty} \left( a_n \sin \frac{n\pi ct}{L} + b_n \cos \frac{n\pi ct}{L} \right) \sin \frac{n\pi x}{L}. \qquad (11.63)$$

If (11.63) is to satisfy (11.13), we need

$$a_n = 0, \qquad n = 1, 2, 3, \ldots$$

and

$$h(x) = \sum_{n=1}^{\infty} b_n \sin \frac{n\pi x}{L}, \qquad 0 < x < L. \qquad (11.64)$$

Thus we need the Fourier sine series expansion for $h(x)$ as given in Theorem 10.4. From (10.25) we obtain

$$b_n = \frac{2}{L} \int_0^L h(x) \sin \frac{n\pi x}{L} \, dx \qquad (11.65)$$

and the solution of our original problem as

$$u(x, t) = \sum_{n=1}^{\infty} b_n \cos \frac{n\pi ct}{L} \sin \frac{n\pi x}{L}. \qquad (11.66)$$

For a specific value of $x$, each term in the series in (11.66) is a product

$$\left( b_n \sin \frac{n\pi x}{L} \right) \cos \frac{n\pi ct}{L}. \qquad (11.67)$$

The behavior of (11.67) is that of a sinusoidal periodic function of time with period $2L/(nc)$ while $b_n \sin(n\pi x/L)$ represents its amplitude. Since frequency $= (2\pi)/$ (period), the frequency of this motion, also called *natural frequency*, is

$$\frac{2\pi}{2L/nc} = \frac{n\pi c}{L}.$$

Since $c^2 = T/\rho = $ tension/density, we see that the natural frequencies are directly proportional to the square root of the tension and inversely proportional to the length of the string and the square root of the density. Thus short strings or those under high tension have higher frequencies than long ones, or those under less tension. Similarly, a more dense string will have lower frequencies than a less dense string. The spatial behavior of this motion is governed by $\sin(n\pi x/L)$, called a *natural mode of vibration*. The natural modes, $\sin(\pi x/L)$ and $\sin(2\pi x/L)$, corresponding to the two lowest frequencies, $\pi c/L$ and $2\pi c/L$, are shown in Figure 11.8. The series form of our solution in (11.66) may be written in the form of d'Alembert's solution by using the trigonometric identity

$$2 \cos a \sin b = \sin(b + a) + \sin(b - a) \qquad (11.68)$$

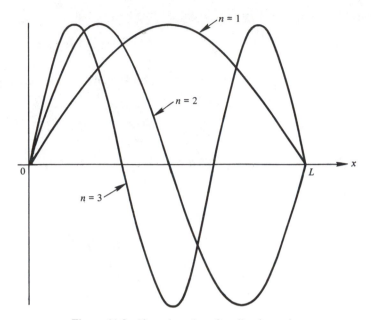

**Figure 11.8**   Natural modes of a vibrating string.

and results from Fourier series. Equation (11.68) allows us to express (11.66) as

$$u(x,\ t) = \frac{1}{2} \sum_{n=1}^{\infty} b_n \left[ \sin \frac{(x + ct)n\pi}{L} + \sin \frac{(x - ct)n\pi}{L} \right].  \qquad (11.69)$$

Theorem 10.4 states that the series [from (11.64)]

$$\sum_{n=1}^{\infty} b_n \sin \frac{n\pi x}{L}$$

converges to $h(x)$ for $0 < x < L$ and to the odd periodic extension of $h$, with period $2L$. If we still denote this extension of $h$ by $h_o$, we may write (11.69) as

$$u(x,\ t) = \frac{1}{2}[h_o(x + ct) + h_o(x - ct)],  \qquad (11.70)$$

which is the same form we had as for the infinite string [see (11.20)].

The method of separation of variables also works if we have more than two independent variables. Thus we consider the two dimensional wave equation

$$\frac{\partial^2 u}{\partial t^2} = c^2 \left( \frac{\partial^2 u}{\partial x^2} + \frac{\partial^2 u}{\partial y^2} \right),  \qquad (11.71)$$

which in polar coordinates has the form

$$\frac{\partial^2 u}{\partial t^2} = c^2 \left( \frac{\partial^2 u}{\partial r^2} + \frac{1}{r} \frac{\partial u}{\partial r} + \frac{1}{r^2} \frac{\partial^2 u}{\partial \theta^2} \right)  \qquad (11.72)$$

(see exercise 13, Section 11.1). To use the method of separation of variables on

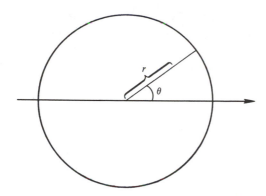

**Figure 11.9**   A circular membrane.

(11.72) to analyze the vibration of a circular elastic membrane (Figure 11.9), we assume that

$$u(r, \theta, t) = R(r)\Theta(\theta)T(t) \tag{11.73}$$

and obtain

$$R(r)\Theta(\theta)T''(t) = c^2\left[R''(r)\Theta(\theta)T(t) + \frac{1}{r}R'(r)\Theta(\theta)T(t) + \frac{1}{r^2}R(r)\Theta''(\theta)T(t)\right].$$

Division by $c^2R(r)\Theta(\theta)T(t)$ gives

$$\frac{T''(t)}{c^2T(t)} = \frac{R''(r) + (1/r)R'(r)}{R(r)} + \frac{\Theta''(\theta)}{r^2\Theta(\theta)} = -\lambda^2. \tag{11.74}$$

Since the left-hand side of (11.74) depends only on $t$ while the right-hand side depends on both $r$ and $\theta$, and since $r$, $\theta$, and $t$ are independent variables, both sides must equal a constant, say $-\lambda^2$. (Later we see that this constant will be negative, so we introduce this notation now for convenience.) Thus we have the two equations

$$T''(t) + \lambda^2c^2T(t) = 0, \tag{11.75}$$

$$\frac{R''(r) + (1/r)R'(r)}{R(r)} + \frac{\Theta''(\theta)}{r^2\Theta(\theta)} = -\lambda^2. \tag{11.76}$$

If we multiply (11.76) by $r^2$ and rearrange, we obtain

$$\frac{r^2R''(r) + rR'(r)}{R(r)} + \lambda^2r^2 = -\frac{\Theta''(\theta)}{\Theta(\theta)} = \mu. \tag{11.77}$$

In (11.77) we again use the idea that $r$ and $\theta$ are independent variables and call the separation constant $\mu$. Rearranging (11.77) gives

$$r^2R''(r) + rR'(r) + (\lambda^2r^2 - \mu)R(r) = 0, \tag{11.78}$$

$$\Theta''(\theta) + \mu\Theta(\theta) = 0. \tag{11.79}$$

To obtain solutions of our three ordinary differential equations with their two unknown separation constants, $\mu$ and $-\lambda^2$, we consider a specific boundary value problem.

**EXAMPLE 11.10**    Find the resulting motion of a circular membrane (such as a drum) of radius $a$, subject to an initial displacement $h(r, \theta)$ and zero initial velocity. The edge of the membrane at $r = a$ remains fixed. If $u(r, \theta, t)$ represents the displacement of the membrane, the facts above translate into

$$\frac{\partial^2 u}{\partial t^2} = c^2 \left( \frac{\partial^2 u}{\partial r^2} + \frac{1}{r} \frac{\partial u}{\partial r} + \frac{1}{r^2} \frac{\partial^2 u}{\partial \theta^2} \right), \tag{11.72}$$

$$u(a, \theta, t) = 0, \tag{11.80}$$

$$u(r, \theta, 0) = h(r, \theta), \quad \frac{\partial u}{\partial t}(r, \theta, 0) = 0. \tag{11.81}$$

We also have

$$u(r, \theta, t) = u(r, \theta + 2\pi, t) \tag{11.82}$$

as a condition on our unknown displacement. Since $u(r, \theta, t) = R(r)\Theta(\theta)T(t)$, (11.82) requires that

$$\Theta(\theta + 2\pi) = \Theta(\theta),$$

that is, the solutions of (11.79) be periodic with period $2\pi$. This will be true only if the separation constant, $\mu$, is the square of an integer, that is,

$$\mu = n^2, \quad n = 0, 1, 2, \ldots, \tag{11.83}$$

and

$$\Theta(\theta) = a_n \cos n\theta + b_n \sin n\theta, \quad n = 0, 1, 2, \ldots. \tag{11.84}$$

Having $\mu = n^2$ means that (11.78) becomes

$$r^2 R''(r) + r R'(r) + (\lambda^2 r^2 - n^2) R(r) = 0. \tag{11.85}$$

This is Bessel's differential equation [see (6.105)] with the general solution

$$R(r) = A J_n(\lambda r) + B Y_n(\lambda r). \tag{11.86}$$

However, since $Y_n(0)$ is not bounded, and $r = 0$ is in our domain (the center of the membrane), we require that

$$B = 0.$$

The boundary condition on $R$ is obtained from (11.80) as

$$R(a) = 0. \tag{11.87}$$

Equations (11.85) and (11.87) define a Sturm–Liouville boundary value problem with orthogonal eigenfunctions

$$J_n(\lambda_j r), \quad j = 1, 2, 3, \ldots.$$

The constants $\lambda_j, j = 1, 2, 3, \ldots$, are determined from

$$J_n(\lambda_j a) = 0. \tag{11.88}$$

[See the text following (6.105) and (6.106), as well as Theorem 10.5.] Knowing the $\lambda$ in (11.75) allows us to write its solution as

$$T(t) = A_j \cos \lambda_j ct + B_j \sin \lambda_j ct, \qquad j = 1, 2, 3, \ldots . \qquad (11.89)$$

We may write our solution as the double infinite series

$$u(r, \theta, t) = \sum_{n=0}^{\infty} \sum_{j=1}^{\infty} [a_n \cos n\theta + b_n \sin n\theta] J_n(\lambda_j r)[A_j \cos \lambda_j ct + B_j \sin \lambda_j ct].$$
$$(11.90)$$

The fact that $\dfrac{\partial u}{\partial t}(r, \theta, 0) = 0$ means that we may set

$$B_j = 0, j = 1, 2, 3, \ldots$$

in (11.90). The remaining coefficients are determined from

$$u(r, \theta, 0) = h(r, \theta) = \sum_{n=0}^{\infty} \sum_{j=1}^{\infty} [A_j a_n \cos n\theta + A_j b_n \sin n\theta] J_n(\lambda_j r).$$
$$(11.91)$$

If we rewrite (11.91) as

$$h(r, \theta) = \left[\sum_{j=1}^{\infty} A_j J_0(\lambda_j r)\right] a_0 + \sum_{n=1}^{\infty} \left\{\left[\sum_{j=1}^{\infty} A_j J_n(\lambda_j r)\right] a_n \cos n\theta\right.$$
$$\left. + \left[\sum_{j=1}^{\infty} A_j J_n(\lambda_j r)\right] b_n \sin n\theta\right\} \qquad (11.92)$$

we see that for $r$ fixed, (11.92) is simply the Fourier series of $h(r, \theta)$. Thus

$$\frac{1}{2\pi} \int_{-\pi}^{\pi} h(r, \theta) \, d\theta = \sum_{j=1}^{\infty} A_j J_0(\lambda_j r) a_0, \qquad (11.93)$$

$$\frac{1}{\pi} \int_{-\pi}^{\pi} h(r, \theta) \cos n\theta \, d\theta = \sum_{j=1}^{\infty} A_j J_n(\lambda_j r) a_n, \qquad (11.94)$$

$$\frac{1}{\pi} \int_{-\pi}^{\pi} h(r, \theta) \sin n\theta \, d\theta = \sum_{j=1}^{\infty} A_j J_n(\lambda_j r) b_n. \qquad (11.95)$$

Now in each of (11.93), (11.94), and (11.95) we may use the orthogonality of the Bessel functions to obtain values of the coefficients $A_j a_0$, $A_j a_n$, and $A_j b_n$ [see (10.1) and (6.112)]. Thus

$$\frac{1}{2\pi} \int_0^a \int_{-\pi}^{\pi} h(r, \theta) J_0(\lambda_k r) r \, d\theta \, dr$$
$$= A_k a_0 \| J_0(\lambda_k r)\|^2, \qquad k = 1, 2, 3, \ldots \qquad (11.96)$$

$$\frac{1}{\pi} \int_0^a \int_{-\pi}^{\pi} h(r, \theta) \cos n\theta \, J_n(\lambda_k r) r \, d\theta \, dr$$
$$= A_k a_n \| J_n(\lambda_k r)\|^2, \qquad \begin{array}{l} k = 1, 2, 3, \ldots \\ n = 1, 2, 3, \ldots \end{array} \qquad (11.97)$$

$$\frac{1}{\pi} \int_0^a \int_{-\pi}^{\pi} h(r, \theta) \sin n\theta \, J_n(\lambda_k r) r \, d\theta \, dr$$

$$= A_k b_n \| J_n(\lambda_k r) \|^2, \qquad \begin{matrix} k = 1, 2, 3, \dots \\ n = 1, 2, 3, \dots \end{matrix} \qquad (11.98)$$

Substituting values of $B_j = 0$, $A_k a_0$, $A_k a_n$, and $A_k b_n$ from (11.96), (11.97), and (11.98) into (11.90) gives the final form of our answer.

The details are considerably simpler if the motion is independent of $\theta$. This happens when $h(r, \theta) = h(r)$, so

$$a_n = b_n = 0, \qquad n = 1, 2, 3, \dots$$

and

$$a_0 A_k = \frac{\int_0^a h(r) J_0(\lambda_k r) r \, dr}{\| J_0(\lambda_k r) \|^2}$$

$$= \frac{\int_0^a h(r) J_0(\lambda_k r) r \, dr}{a^2 J_1^2(\lambda_k a)}.$$

This simpler case has its solution, from (11.90), as

$$u(r, t) = \sum_{j=1}^{\infty} \frac{\int_0^a h(r) J_0(\lambda_j r) r \, dr}{a^2 J_1^2(\lambda_j a)} J_0(\lambda_j r) \cos \lambda_j ct. \qquad (11.99)$$

## EXERCISES

1. Find $u(x, t)$ for a vibrating string that is fixed at its ends, $x = 0$ and $x = \pi$, if $c^2 = 1$, $\frac{\partial u}{\partial t}(x, 0) = 0$, and has an initial deflection given by

   (a) $0.1 \sin x$.
   (b) $0.01 \sin 2x$.
   (c) $0.1 \sin x + 0.01 \sin 2x$.
   (d) $0.05 \sin 2x + 0.2 \sin 3x$.
   (e) $\begin{cases} x, & 0 \le x \le \dfrac{\pi}{2} \\ \pi - x, & \dfrac{\pi}{2} < x \le \pi \end{cases}$.
   (f) $\pi x - x^2$.

2. (a)–(f) Repeat exercise 1 assuming an initial displacement of zero and with the functions in parts (a) through (f) representing the initial velocity.

3. Determine the proper change of dependent variable to change the boundary value problem

   $$\frac{\partial^2 u}{\partial t^2} = \frac{\partial^2 u}{\partial x^2},$$

   $$u(0, t) = 1, \qquad u(L, t) = 3,$$

$$u(x, 0) = f(x), \qquad \frac{\partial u}{\partial t}(x, 0) = 0,$$

to one with homogeneous boundary conditions.

**4.** Find the solution to exercise 3 if $f(x) = 1 + 2x$.

**5.** Solve the boundary value problem in exercise 3 if the initial conditions are changed to

$$u(x, 0) = 0, \qquad \frac{\partial u}{\partial t}(x, 0) = x(L - x).$$

**6.** The partial differential equation describing the forced vibration of an elastic string is

$$\frac{\partial^2 u}{\partial t^2} = c^2 \frac{\partial^2 u}{\partial x^2} + \frac{F(x, t)}{\rho}, \tag{11.100}$$

where $F(x, t)$ is the force per unit length acting at right angles to the string and $\rho$ is the density of the string. If the boundary and initial conditions associated with (11.100) are

$$u(0, t) = u(L, t) = 0, \tag{11.101}$$

$$u(x, 0) = f(x), \qquad \frac{\partial u}{\partial t}(x, 0) = g(x), \tag{11.102}$$

we may find $u(x, t)$ in a manner similar to the method of undetermined coefficients. Note that if (11.100) were homogeneous, i.e., $F(x, t) = 0$, the method of separation of variables would give the solution of (11.100) and (11.101) as

$$\sum_{n=1}^{\infty} T_n(t) \sin \frac{n\pi x}{L},$$

where $T_n(t)$ satisfies

$$T_n'' + \frac{c^2 n^2 \pi^2}{L^2} T_n = 0.$$

We then expand $F(x, t)/\rho$ in a similar series,

$$\frac{F(x, t)}{\rho} = \sum_{n=1}^{\infty} \alpha_n(t) \sin \frac{n\pi x}{L}, \tag{11.103}$$

and write

$$u(x, t) = \sum_{n=1}^{\infty} T_n(t) \sin \frac{n\pi x}{L} \tag{11.104}$$

and choose the $T_n(t)$ so that (11.100) is satisfied.

**(a)** Show that this operation leads to the ordinary differential equations (with $\lambda_n = n\pi/L$)

$$T_n''(t) + c^2 \lambda_n^2 T_n(t) = \alpha_n(t), \qquad n = 1, 2, 3, \ldots . \tag{11.105}$$

**(b)** If $F(x, t)/\rho = F_0$ (a constant), show that the solutions of (11.105) are

$$T_n(t) = a_n \cos \lambda_n ct + b_n \sin \lambda_n ct + \frac{2F_0 L}{c^2 \lambda_n^3}[1 - (-1)^n].$$

**(c)** Using the solution for $T_n(t)$ from part (b) in (11.104), find $a_n$ and $b_n$, $n = 1, 2, 3, \ldots$, such that the initial conditions in (11.102) are satisfied.

7. If the elastic string of exercise 6 is vibrating under the force of gravity, the forcing function in the differential equation is given by $F(x, t) = -g$. Solve the boundary value problem that arises in this situation when the string is released from rest from the initial displacement of

$$u(x, 0) = \begin{cases} x, & 0 < x < \dfrac{L}{2} \\ L - x, & \dfrac{L}{2} < x < L \end{cases}.$$

Thus in addition to the initial condition above, we have

$$\frac{\partial^2 u}{\partial t^2} = c^2 \frac{\partial^2 u}{\partial x^2} - \frac{g}{\rho},$$

$$u(0, t) = u(L, t) = 0, \qquad \frac{\partial u}{\partial t}(x, 0) = 0.$$

8. If the stretched string of this section has a damping force acting on it which is proportional to the velocity, the differential equation becomes

$$\frac{\partial^2 u}{\partial t^2} + \beta \frac{\partial u}{\partial t} = c^2 \frac{\partial^2 u}{\partial x^2}.$$

   (a) If the ends of the above string are fixed (at $x = 0$ and $x = \pi$) and the string is initially at rest with the shape $u(x, 0) = h(x)$, find the resulting damped motion.
   (b) If the initial conditions in part (a) are changed to be

$$u(x, 0) = 0, \qquad \frac{\partial u}{\partial t}(x, 0) = 5,$$

   find the resulting motion.

9. Consider the motion of a circular membrane that is independent of the angle $\theta$ and is at equilibrium at $t = 0$ with a prescribed initial velocity. This motion is governed by

$$\frac{\partial^2 u}{\partial t^2} = c^2 \left( \frac{\partial^2 u}{\partial r^2} + \frac{1}{r} \frac{\partial u}{\partial r} \right),$$

$$u(a, t) = 0, \qquad u(r, 0) = 0, \qquad \frac{\partial u}{\partial t}(r, 0) = h(r).$$

   Find the series expression for $u(r, t)$.

10. Find the series expression for $u(r, \theta, t)$ from Example 11.10 if

$$h(r, \theta) = (1 - r^2) \cos 2\theta.$$

11. Use the method of separation of variables to solve

$$\frac{\partial^2 u}{\partial t^2} = c^2 \left( \frac{\partial^2 u}{\partial r^2} + \frac{1}{r} \frac{\partial u}{\partial r} \right),$$

$$u(a, t) = 0,$$

$$u(r, 0) = f(r), \qquad \frac{\partial u}{\partial t}(r, 0) = g(r).$$

**12.** Small transverse vibrations of a beam are governed by the differential equation

$$\frac{\partial^2 u}{\partial t^2} + c^2 \frac{\partial^4 u}{\partial x^4} = 0.$$

The constant $c^2 = $ (flexural rigidity/(cross-sectional area × density)). Boundary conditions associated with a beam are listed below:

1. A fixed end has the displacement and slope equal to zero, that is,

$$u(a, t) = \frac{\partial u}{\partial x}(a, t) = 0$$

(see Figure 11.10; a fixed end is also called a built-in end or a clamped end).

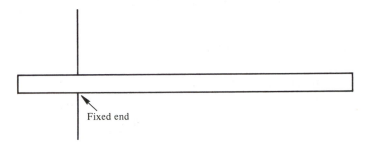

**Figure 11.10**    Fixed end.

2. A simply supported end (Figure 11.11) has displacement and moment equal to zero, that is,

$$u(a, t) = \frac{\partial^2 u}{\partial x^2}(a, t) = 0.$$

Simply supported ends                                   Simply supported ends

**Figure 11.11**    Simply supported end.

3. A free end (Figure 11.12) has zero moment and zero shear, that is,

$$\frac{\partial^2 u}{\partial x^2}(a, t) = \frac{\partial^3 u}{\partial x^3}(a, t) = 0.$$

**(a)** Find the solution for the vibration of a beam that has simply supported ends at $x = 0$ and $x = L$ with initial conditions

$$u(x, 0) = x(L - x), \qquad \frac{\partial u}{\partial t}(x, 0) = 0.$$

**(b)** Redo part (a) if

$$u(x, 0) = 0 \quad \text{and} \quad \frac{\partial u}{\partial t}(x, 0) = x(L - x).$$

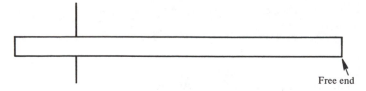

**Figure 11.12**   Free end.

(c) Find an expression for the eigenvalues associated with the vibration of a beam with a fixed end at $x = 0$ and a free end at $x = L$. Show (graphically) that there is an infinite number of eigenvalues.

(d) Find an expression for the eigenvalues associated with the vibration of a beam with fixed ends at $x = 0$ and $x = L$.


## 11.4 LAPLACE'S EQUATION

Another important partial differential equation is Laplace's equation, which in two dimensions is

$$\frac{\partial^2 u}{\partial x^2} + \frac{\partial^2 u}{\partial y^2} = 0. \tag{11.106}$$

This equation describes the steady flow of heat, electric or magnetic field potentials, and is important in the study of certain two dimensional fluid flows, as well as problems in elasticity. Solutions of (11.106) that have continuous second derivatives are called *harmonic* functions. We illustrate the techniques of solving Laplace's equation by working three examples for different coordinate systems.

### EXAMPLE 11.11

Find the temperature distribution in a rectangular plate if three edges of the plate are kept at $0°$ while the fourth edge is kept at $50°$ (see Figure 11.13). These boundary conditions are listed as

$$u(x, 0) = u(x, b) = 0, \tag{11.107}$$

$$u(0, y) = 50, \qquad u(a, y) = 0. \tag{11.108}$$

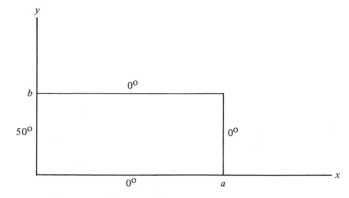

**Figure 11.13**

Assuming that $u(x, y) = P(x)Q(y)$ in (11.106) yields

$$\frac{P''}{P} = -\frac{Q''}{Q} = \lambda$$

or

$$P'' - \lambda P = 0, \qquad Q'' + \lambda Q = 0. \tag{11.109}$$

With this product assumption for $u(x, y)$, the boundary conditions (11.107) and (11.108) become

$$Q(0) = Q(b) = 0, \tag{11.110}$$

$$P(a) = 0. \tag{11.111}$$

Note that the nonhomogeneous boundary condition in (11.108) may not transfer to one involving $x$ only. [This would be more clearly shown if the edge for which $x = 0$ were kept at $f(y)$ degrees instead of 50°.] Since the differential equation for $Q(y)$ in (11.109) plus the two boundary conditions in (11.110) constitute a Sturm–Liouville boundary value problem, we solve it to determine the separation constant, $\lambda$, as well as $Q(y)$. From (11.109), if $\lambda > 0$,

$$Q(y) = A \cos \sqrt{\lambda}\, y + B \sin \sqrt{\lambda}\, y.$$

To satisfy (11.110) we need $A = 0$ and

$$\sqrt{\lambda}\, b = n\pi, \qquad n = 1, 2, 3, \ldots.$$

This means that we have eigenvalues

$$\lambda = \frac{n^2 \pi^2}{b^2}, \qquad n = 1, 2, 3, \ldots$$

and eigenfunctions

$$Q(y) = \sin \frac{n\pi y}{b}, \qquad n = 1, 2, 3, \ldots.$$

Thus the differential equation for $P$ becomes

$$P'' - \left(\frac{n^2 \pi^2}{b^2}\right) P = 0 \tag{11.112}$$

with solution

$$P(x) = A e^{n\pi x/b} + B e^{-n\pi x/b}. \tag{11.113}$$

To satisfy (11.111) we need to choose $A$ and $B$ such that

$$A e^{n\pi a/b} + B e^{-n\pi a/b} = 0$$

or

$$B = -A e^{2n\pi a/b}.$$

Thus (11.113) becomes

$$P(x) = A(e^{n\pi x/b} - e^{2n\pi a/b}e^{-n\pi x/b})$$

$$= Ae^{n\pi a/b}[e^{n\pi(x-a)/b} - e^{-n\pi(x-a)/b}]$$

$$= 2Ae^{n\pi a/b} \sinh \frac{n\pi(x-a)}{b}$$

$$= A_1 \sinh \frac{n\pi(x-a)}{b}. \tag{11.114}$$

[Note that we could have arrived at (11.114) by assuming that the solution of (11.112) was

$$P(x) = A \sinh \frac{n\pi(x-a)}{b} + B \cosh \frac{n\pi(x-a)}{b}.]$$

Collecting all the solutions to this point yields

$$u(x, t) = \sum_{n=1}^{\infty} b_n \sinh \frac{n\pi(x-a)}{b} \sin \frac{n\pi y}{b}, \tag{11.115}$$

where the $b_n$ are chosen to satisfy the last boundary condition,

$$u(0, y) = 50.$$

Setting $x = 0$ in (11.115) gives

$$50 = \sum_{n=1}^{\infty} b_n \sinh \frac{-n\pi a}{b} \sin \frac{n\pi y}{b}.$$

But this means that the $b_n \sinh(-n\pi a/b)$ are simply the Fourier sine coefficients in the expansion of 50 in terms of the set $\{\sin n\pi y/b, n = 1, 2, 3, \ldots\}$. Thus

$$b_n \sinh \frac{-n\pi a}{b} = \frac{2}{b} \int_0^b 50 \sin \frac{n\pi y}{b} \, dy$$

$$= \frac{100}{b} \left[ \frac{-\cos(n\pi y/b)}{n\pi/b} \Big|_0^b \right]$$

$$= \frac{100}{n\pi}[(-1)^{n+1} + 1]$$

and our solution of the original problem is

$$u(x, y) = \sum_{n=1}^{\infty} \frac{100[(-1)^{n+1} + 1]}{n\pi} \frac{\sinh [n\pi(x-a)/b]}{\sinh (-n\pi a/b)} \sin \frac{n\pi y}{b}. \tag{11.116}$$

[Note that as long as we have either both boundary conditions on $x$, or both boundary conditions on $y$, given as homogeneous, we may use separation of variables successfully. If this is not the case, we need to make a change of dependent variable to obtain such homogeneous conditions. See exercise 1(e), page 419.]

## EXAMPLE 11.12

Consider Laplace's equation in polar coordinates (see exercise 13, Section 10.1)

$$\frac{\partial^2 u}{\partial r^2} + \frac{1}{r}\frac{\partial u}{\partial r} + \frac{1}{r^2}\frac{\partial^2 u}{\partial \theta^2} = 0, \qquad \begin{array}{c} r > a, \\ -\pi < \theta \le \pi, \end{array} \qquad (11.117)$$

with

$$u(a, \theta) = f(\theta), \qquad -\pi < \theta < \pi. \qquad (11.118)$$

We also require that $u(r, \theta)$ be periodic in $\theta$ and bounded for large $r$. Such a problem occurs in determining the temperature distribution of an infinite slab with a circular hole (Figure 11.14). The edge of the hole, $r = a$, is kept at the temperature $f(\theta)$. If we use separation of variables, we assume that

$$u(r, \theta) = R(r)\Theta(\theta)$$

and obtain

$$r^2 R''(r) + r R'(r) - \lambda R(r) = 0, \qquad (11.119)$$

$$\Theta''(\theta) + \lambda\Theta(\theta) = 0. \qquad (11.120)$$

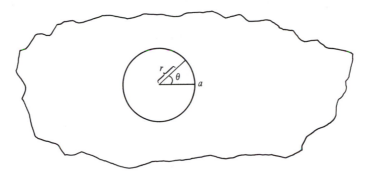

**Figure 11.14**   Infinite slab with a circular hole.

Solutions of (11.120) which satisfy

$$\Theta(\theta + 2\pi) = \Theta(\theta)$$

are

$$\Theta(\theta) = A_n \cos n\theta + B_n \sin n\theta, \qquad n = 0, 1, 2, \ldots, \qquad (11.121)$$

with the separation constant, $\lambda$, determined as

$$\lambda = n^2, \qquad n = 0, 1, 2, \ldots.$$

Equation (11.119) is a Cauchy–Euler equation, so we seek a solution of the form

$$R(r) = r^m$$

and obtain

$$m(m - 1) + m - n^2 = 0$$

or

$$m = \pm n.$$

Thus if $n \neq 0$,

$$R(r) = Ar^{-n} + Br^n, \qquad (11.122)$$

and if $n = 0$,

$$R(r) = A + B \ln r. \qquad (11.123)$$

However, if $u(r, \theta)$ is to be bounded for large $r$, so must $R(r)$. Thus we need $B = 0$ in (11.122) and (11.123) and may then write our solution as

$$u(r, \theta) = A_0 + \sum_{n=1}^{\infty} (A_n \cos n\theta + B_n \sin n\theta)r^{-n}. \qquad (11.124)$$

The boundary condition at $r = a$ is satisfied if

$$f(\theta) = A_0 + \sum_{n=1}^{\infty} (A_n \cos n\theta + B_n \sin n\theta)a^{-n}.$$

We use our results from Fourier series (Section 10.2) to find

$$A_0 = \frac{2}{\pi} \int_{-\pi}^{\pi} f(\theta) \, d\theta,$$

$$a^{-n}A_n = \frac{1}{\pi} \int_{-\pi}^{\pi} f(\theta) \cos n\theta \, d\theta, \qquad n = 1, 2, 3, \ldots$$

$$a^{-n}B_n = \frac{1}{\pi} \int_{-\pi}^{\pi} f(\theta) \sin n\theta \, d\theta, \qquad n = 1, 2, 3, \ldots.$$

The details for solving Laplace's equation for a circular disk are similar and left for the reader as exercise 5.

## EXAMPLE 11.13

We conclude this section with an example involving Laplace's equation in spherical coordinates (Figure 11.15). If we introduce spherical coordinates $(\rho, \phi, \theta)$ by

$$x = \rho \sin \phi \sin \theta,$$

$$y = \rho \sin \phi \cos \theta,$$

$$z = \rho \cos \phi,$$

we discover that Laplace's equation becomes (see Exercise 14, Section 11.1)

$$\frac{1}{\rho} \frac{\partial^2(\rho u)}{\partial \rho^2} + \frac{1}{\rho^2 \sin \phi} \frac{\partial}{\partial \phi}\left(\sin \phi \frac{\partial u}{\partial \phi}\right) + \frac{1}{\rho^2 \sin^2 \phi} \frac{\partial^2 u}{\partial \theta^2} = 0. \qquad (11.125)$$

We take $u$ in (11.125) as representing the electric potential outside a charged sphere of radius $a$. If the potential on the outer surface of the sphere is a function

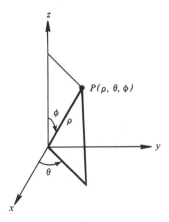

**Figure 11.15**  Spherical coordinates.

only of $\phi$, say

$$u(a, \phi, \theta) = f(\phi), \tag{11.126}$$

and the rest of space is free of charge, we seek to find the resulting $u$ for $\rho > a$. Since the charge on the sphere is independent of $\theta$, so will be the resulting electric potential in space. Thus we assume that

$$u(\rho, \phi, \theta) = R(\rho)\Phi(\phi) \tag{11.127}$$

in (11.125) and obtain

$$\frac{\rho}{R} \frac{d^2}{d\rho^2}(\rho R) = \frac{-1}{\sin \phi\, \Phi} \frac{d}{d\phi}\left(\sin \phi\, \frac{d\Phi}{d\phi}\right) = \lambda. \tag{11.128}$$

This yields two ordinary differential equations

$$\rho^2 R'' + 2\rho R' - \lambda R = 0, \tag{11.129}$$

$$\frac{1}{\sin \phi} \frac{d}{d\phi}\left(\sin \phi\, \frac{d\Phi}{d\phi}\right) + \lambda \Phi = 0. \tag{11.130}$$

Equation (11.130) becomes more familiar if we let

$$v = \cos \phi.$$

Then

$$\frac{d}{d\phi} = \frac{dv}{d\phi} \frac{d}{dv} = -\sin \phi\, \frac{d}{dv}$$

or

$$\frac{-1}{\sin \phi} \frac{d}{d\phi} = \frac{d}{dv}.$$

If we rearrange (11.130) as

$$\frac{1}{\sin \phi} \frac{d}{d\phi}\left(\frac{\sin^2 \phi}{\sin \phi} \frac{d\Phi}{d\phi}\right) + \lambda \Phi = \frac{1}{\sin \phi} \frac{d}{d\phi}\left(\frac{1 - \cos^2 \phi}{\sin \phi} \frac{d\Phi}{d\phi}\right) + \lambda \Phi = 0$$

we see it transforms to

$$-\frac{d}{dv}\left[(1-v^2)\left(-\frac{d\Phi}{dv}\right)\right] + \lambda\Phi = 0$$

or [differentiating and writing the separation constant $\lambda$ as $n(n+1)$]

$$(1-v^2)\frac{d^2\Phi}{dv^2} - 2v\frac{d\Phi}{dv} + n(n+1)\Phi = 0. \qquad (11.131)$$

This is Legendre's differential equation [see (6.94) and the paragraph following (6.94)] with general solution

$$\Phi = AP_n(v) + BQ_n(v). \qquad (11.132)$$

If we want our potential $u(\rho, \phi)$ to be bounded for all $\phi$, $0 \le \phi \le \pi$, i.e., $-1 \le v \le 1$, we must take $B = 0$ and $n$ a nonnegative integer. In terms of $\phi$, this means that we have

$$\Phi(\phi) = P_n(\cos\phi), \qquad n = 0, 1, 2, \ldots. \qquad (11.133)$$

Equation (11.129) [with $\lambda = n(n+1)$, $n$ a nonnegative integer] is of the Cauchy–Euler type and will have solutions of the form

$$R = \rho^m.$$

Inserting this form for $R$ in

$$\rho^2 R'' + 2\rho R' - n(n+1)R = 0$$

gives

$$m(m-1) + 2m - n(n+1) = 0$$

or

$$m(m+1) - n(n+1) = 0.$$

Since this algebraic equation has solutions

$$m = n \qquad \text{and} \qquad m = -n-1,$$

the general solution of (11.129) is

$$R = A\rho^n + B\rho^{-n-1}. \qquad (11.134)$$

(Note that $\rho^n$ and $\rho^{-n-1}$ are linearly independent functions.) In order that $R(\rho)$ be bounded for large $\rho$, we must take $A = 0$ in (11.134). This leaves us with the infinite series for $u$,

$$u(\rho, \phi) = \sum_{n=0}^{\infty} A_n\rho^{-n-1}P_n(\cos\phi), \qquad (11.135)$$

where the $A_n$ are determined from the boundary condition (11.126),

$$f(\phi) = \sum_{n=0}^{\infty} A_n a^{-n-1}P_n(\cos\phi). \qquad (11.136)$$

Since $\{P_n(\cos \phi), \; n = 0, 1, 2, \ldots\}$ forms an orthogonal set on $[0, \pi]$ with weight function $\sin \phi$ (see exercise 11), the $A_n$ are given by

$$A_n = \frac{(2n + 1)a^{n+1}}{2} \int_0^\pi f(\phi)P_n(\cos \phi) \sin \phi \, d\phi.$$

Thus the final form for $u(\rho, \phi)$ is

$$u(\rho, \phi) = \frac{1}{2} \sum_{n=0}^\infty \frac{(2n + 1) \int_0^\pi f(\psi)P_n(\cos \psi) \sin \psi \, d\psi}{(\rho/a)^{n+1}} P_n(\cos \phi). \qquad (11.137)$$

## EXERCISES

1. Solve (11.106) for the interior of a rectangle with the following boundary conditions.
   (a) $u(0, y) = 0, \quad u(a, y) = 50,$
      $u(x, 0) = 0, \quad u(x, b) = 0.$
   (b) $u(0, y) = 20, \quad u(a, y) = 50,$
      $u(x, 0) = 0, \quad u(x, b) = 0.$
   (c) $u(0, y) = 0, \quad u(a, y) = 0,$
      $u(x, 0) = 0, \quad u(x, b) = 50.$
   (d) $u(0, y) = 0, \quad u(a, y) = 0,$
      $u(x, 0) = 0, \quad u(x, b) = x.$
   (e) $u(0, y) = 0, \quad u(a, y) = 10,$
      $u(x, 0) = 5, \quad u(x, b) = 20.$
      (*Hint:* You may want to use a change of variables as in Example 11.7.)
   (f) $u(0, y) = 0, \quad u(a, y) = f(y),$
      $u(x, 0) = 0, \quad u(x, b) = 0.$

2. Show that if $\lambda \leq 0$ in (11.109), the corresponding solution for $Q$ is identically zero if the boundary conditions (11.110) are satisfied.

3. Laplace's equation in cylindrical polar coordinates is

$$\frac{\partial^2 u}{\partial r^2} + \frac{1}{r}\frac{\partial u}{\partial r} + \frac{1}{r^2}\frac{\partial^2 u}{\partial \theta^2} + \frac{\partial^2 u}{\partial z^2} = 0.$$

Assume that $u = R(r)\Theta(\theta)Z(z)$ and use the method of separation of variables to determine the ordinary differential equations satisfied by $R$, $\Theta$, and $Z$.

4. Solve (11.117) subject to $u(r, \theta) = u(r)$,

$$u(a) = u_0,$$

$$u(b) = u_1.$$

This system of equations describes the potential between two coaxial cylinders of radius $a$ and $b$, respectively, $a < b$ (Figure 11.16). The inner cylinder is at potential $u_0$ with the outer one at $u_1$.

5. Solve (11.117) for $r < a$ if

$$u(a, \theta) = f(\theta), \qquad -\pi < \theta < \pi.$$

This models the temperature or electric potential interior to a long right circular cylinder with $f(\theta)$ describing the surface temperature or electric potential.

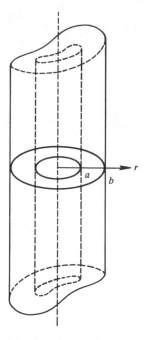

**Figure 11.16** Coaxial cylinder.

6. Assume $u = R(\rho)\Phi(\phi)\Theta(\theta)$ in (11.125) and derive ordinary differential equations for $R$, $\Phi$, and $\Theta$ using the method of separation of variables.

7. Find the solution of (11.125) that is bounded for
   (a) $\rho > a$ and satisfies $u(a, \phi, \theta) = 4$.
   (b) $\rho < a$ and satisfies $u(a, \phi, \theta) = 4$.

8. Find the solution of (11.125) for $a < \rho < b$ that satisfies

$$u(a, \phi, \theta) = u_0 \quad \text{(a constant)},$$

$$u(b, \phi, \theta)\dot{} = u_1 \quad \text{(a constant)}.$$

9. Find the bounded solution of (11.125) for $\rho < a$ if

$$u(a, \phi, \theta) = f(\phi).$$

10. Redo exercise 9 if
    (a) $f(\phi) = \cos \phi$.
    (b) $f(\phi) = 1 + \cos^2 \phi$.

11. Use the fact that the Legendre polynomials, $P_n(x)$, $n = 0, 1, 2, \ldots$, form an orthogonal set on $-1 \le x \le 1$ to show that the set $\{P_n(\cos \phi), n = 0, 1, 2, \ldots\}$, forms an orthogonal set on $0 \le \phi \le \pi$ with weight function $\sin \phi$.

## REVIEW EXERCISES

1. Find the general solution to the following partial differential equations.
   (a) $\dfrac{\partial^2 u}{\partial x\, \partial y} = \sin x + e^y$.
   (b) $\dfrac{\partial^2 u}{\partial x\, \partial y} + \dfrac{\partial u}{\partial x} = 2x$.

**2.** An extension of the example in exercise 29, page 424, which includes a second order reaction term is

$$\frac{\partial w}{\partial t} + c\frac{\partial w}{\partial x} + \lambda w + \beta w^2 = 0. \qquad (11.138)$$

**(a)** Use the change of independent variables $r = x - ct$, $s = t$, to transform (11.138) to

$$\frac{\partial w}{\partial s} + \lambda w + \beta w^2 = 0. \qquad (11.139)$$

**(b)** If $\lambda = \beta = 1$, treat (11.139) as a separable ordinary differential equation and find its general solution as

$$w(r, s) = \frac{f(r)}{e^s - f(r)}.$$

**(c)** Show that if $\lambda = 0$ and $\beta = 1$ in (11.139), the general solution is

$$w(r, s) = \frac{1}{s + f(r)}. \qquad (11.140)$$

**(d)** In terms of the original variables, the solution in (11.140) is

$$w(x, t) = \frac{1}{t + f(x - ct)}. \qquad (11.141)$$

Show that the choice of $f$ as

$$f(r) = \frac{1}{w_1(-r/c)} + \frac{r}{c} + \left[\frac{1}{w_0(r)} - \frac{1}{w_1(-r/c)} - \frac{r}{c}\right]u_0(r),$$

$u_0(r)$ is the unit step function, allows (11.141) to satisfy the initial condition

$$w(x, 0) = w_0(x)$$

and boundary condition

$$w(0, t) = w_1(t).$$

**(e)** Show that the general solution of (11.139) is

$$w(r, s) = \frac{f(r)}{e^{\lambda s} - (\beta/\lambda)f(r)}$$

with $f$ an arbitrary function.

**3.** By making the change of variable

$$r = x + y$$
$$s = -x + y,$$

find the general solution of

$$\frac{\partial w}{\partial x} + \frac{\partial w}{\partial y} - w = 0. \qquad (11.142)$$

**4.** Use the general solution from review exercise 3 to find a solution of (11.142) which satisfies

$$w(x, 0) = w_0(x).$$

**5. (a)** Show that the change of variables

$$r = x \cos \theta + y \sin \theta$$
$$s = -x \sin \theta + y \cos \theta$$

converts the partial differential equation

$$a\frac{\partial w}{\partial x} + b\frac{\partial w}{\partial y} + cw = F(x, y) \qquad (11.143)$$

into a differential equation containing only one derivative.

**(b)** Find the general solution of (11.143) if $F(x, y) = 0$, $abc \neq 0$.

**6.** Show that

$$u(x, y) = e^{ay}(C_1 \sin ax + C_2 \cos ax)$$

satisfies

$$\frac{\partial^2 u}{\partial x^2} + \frac{\partial^2 u}{\partial y^2} = 0$$

for all choices of the constants $a$, $C_1$, and $C_2$.

Use the method of separation of variables to solve the following partial differential equations.

**7.** $\dfrac{\partial u}{\partial x} + 2\dfrac{\partial u}{\partial y} = 0.$

**8.** $3x^2 \dfrac{\partial u}{\partial x} + \dfrac{\partial u}{\partial y} = 0.$

**9.** $2\dfrac{\partial u}{\partial x} + 3\dfrac{\partial u}{\partial y} = u.$

**10.** $\dfrac{\partial u}{\partial x} + \dfrac{\partial u}{\partial y} = (x + 2y)u.$

**11.** $\dfrac{\partial^2 u}{\partial x \partial y} = f(x)g(y), \quad u(0, y) = 0, \quad u(x, 0) = 0.$

**12.** Find the solution of the heat equation (11.28) which satisfies the initial condition

$$u(x, 0) = \begin{cases} x, & 0 < x < \dfrac{L}{2} \\ \\ L - x, & \dfrac{L}{2} < x < L \end{cases}$$

and the boundary conditions

$$u(0, t) = u(L, t) = 0.$$

**13.** Repeat exercise 12 if the boundary conditions are changed to

$$u(0, t) = 0, \qquad \frac{\partial u}{\partial x}(L, t) = 0.$$

**14.** Find the solution of the heat equation (11.28) which satisfies

$$u(x, 0) = 0, \qquad u(0, t) = L, \qquad u(L, t) = 2L.$$

**15.** Use d'Alembert's solution of the wave equation to solve

$$\frac{\partial^2 u}{\partial t^2} = c^2 \frac{\partial^2 u}{\partial x^2}, \qquad -\infty < x < \infty, \quad t > 0,$$

$$u(x, 0) = e^{-|x|} \sin x,$$

$$\frac{\partial u}{\partial t}(x, 0) = 0.$$

**16.** Repeat exercise 15 if the initial conditions are changed to

$$u(x, 0) = 0,$$

$$\frac{\partial u}{\partial t}(x, 0) = xe^{-|x|}.$$

**17.** Find the solution of

$$\frac{\partial^2 u}{\partial t^2} = \frac{\partial^2 u}{\partial x^2}, \qquad 0 < x < L, \qquad t > 0,$$

$$u(0, t) = u(L, t) = 0,$$

$$u(x, 0) = 7 \sin\left(\frac{2\pi x}{L}\right),$$

$$\frac{\partial u}{\partial t}(x, 0) = 0.$$

**18.** Repeat exercise 17 if the initial displacement is changed to

$$u(x, 0) = \begin{cases} x, & 0 < x < \dfrac{L}{2} \\[2mm] L - x, & \dfrac{L}{2} < x < L \end{cases}.$$

**19.** Solve

$$\frac{\partial^2 u}{\partial t^2} = \frac{\partial^2 u}{\partial x^2} + f(x), \qquad 0 < x < \pi, \quad t > 0,$$

$$u(0, t) = u(\pi, t) = 0,$$

$$u(x, 0) = \frac{\partial u}{\partial t}(x, 0) = 0,$$

$$f(x) = \begin{cases} x, & 0 < x < \dfrac{\pi}{2} \\[2mm] \pi - x, & \dfrac{\pi}{2} < x < \pi \end{cases}.$$

(*Hint:* See exercise 6, page 409.)

**20.** Find a solution of Laplace's equation

$$\frac{\partial^2 u}{\partial x^2} + \frac{\partial^2 u}{\partial y^2} = 0, \qquad 0 < x < 2, \quad 0 < y < 1,$$

which satisfies

$$u(x, 0) = u(x, 1) = 0,$$

$$u(0, y) = 0, \qquad u(x, 0) = \begin{cases} x, & 0 < x < 1 \\ 2 - x, & 1 < x < 2 \end{cases}.$$

**21.** Solve

$$\frac{\partial^2 u}{\partial r^2} + \frac{1}{r}\frac{\partial u}{\partial r} + \frac{1}{r^2}\frac{\partial^2 u}{\partial \theta^2} = 0, \qquad \begin{matrix} r > 1 \\ -\pi < \theta < \pi \end{matrix}$$

if $u(r, \theta)$ is bounded and

$$u(1, \theta) = \sin\theta + 2\sin 3\theta.$$

**22.** Repeat exercise 21 if the boundary condition is changed to

$$u(1, \theta) = \begin{cases} 1, & -\pi < \theta < 0 \\ 2, & 0 < \theta < \pi \end{cases}.$$

**23.** The partial differential equation which describes the small displacement, $w$, of a heavy flexible chain, of length $L$, from equilibrium is

$$\frac{\partial^2 w}{\partial t^2} = -g\frac{\partial w}{\partial x} + g(L - x)\frac{\partial^2 w}{\partial x^2}$$

($g$ is the gravitational constant).
**(a)** Transform the above differential equation to

$$\frac{\partial^2 w}{\partial t^2} = g\frac{\partial w}{\partial y} + gy\frac{\partial^2 w}{\partial y^2} \tag{11.144}$$

by the change of variable $y = L - x$, $u(y, t) = w(L - x, t)$.
**(b)** Use separation of variables to show that the solution of (11.144) which is bounded for $0 \le y \le L$ and satisfies

$$w(0, t) = 0$$

is

$$\sum_{n=1}^{\infty} J_0\left(2\lambda_n\sqrt{\frac{y}{g}}\right)(A_n \cos\lambda_n t + B_n \sin\lambda_n t),$$

where the $\lambda_n$ are determined from $J_0(2\lambda_n\sqrt{L/g}) = 0$, $n = 1, 2, 3, \ldots$ . (*Hint*: Once you separate variables, let $z = \sqrt{y}$ in order to convert the differential equation in $y$ to a form of Bessel's equation.)

**29.** A partial differential equation which describes the advective transport of a chemical, $w$, subject to first order reaction is

$$R\frac{\partial w}{\partial t} = -V\frac{\partial w}{\partial x} - Kw, \qquad x > 0, \quad t > 0,$$

where $R$ is a retardation coefficient, $V$ the velocity of the solution carrying the chemical, and $K$ the first-order reaction coefficient.

**(a)** Write this equation as

$$\frac{\partial w}{\partial t} + c\frac{\partial w}{\partial x} + \lambda w = 0, \qquad c = \frac{V}{R}, \qquad \lambda = \frac{K}{R},$$

and make the change of variables

$$r = x - ct, \qquad s = t$$

to derive the following expression for its general solution:

$$w(x, t) = f(x - ct)e^{-\lambda t},$$

where $f$ is an arbitrary function.

**(b)** Show that choosing $f$ so the initial condition

$$w(x, 0) = w_0(x)$$

and the boundary condition

$$w(0, t) = w_1(t)$$

are satisfied leads to the equations

$$f(x) = w_0(x) \qquad \text{and} \qquad f(-ct) = w_1(t)e^{\lambda t}.$$

**(c)** Write the last condition of $f$, in part (b), as $f(t) = w_1(-t/c)e^{-\lambda t/c}$ and show that the choice of $f$ as

$$f(r) = w_1\!\left(\frac{-r}{c}\right)e^{-\lambda r/c} + \left[w_0(r) - w_1\!\left(\frac{-r}{c}\right)e^{-\lambda r/c}\right]u_0(r),$$

where $u_0(r)$ is the unit step function (see page 266), gives a solution of the partial differential equation of part (a) that satisfies the two conditions of part (b).

# Appendix:
# Determinants
# and Systems
# of Linear Equations

## A.1 DETERMINANTS AND CRAMER'S RULE

An $m \times n$ system of linear equations in the $n$ variables $x_1, \ldots, x_n$ is a system of $m$ equations of the form

$$
\begin{aligned}
a_{11}x_1 + a_{12}x_2 + \cdots + a_{1n}x_n &= b_1, \\
a_{21}x_1 + a_{22}x_2 + \cdots + a_{2n}x_n &= b_2, \\
&\ \ \vdots \\
a_{m1}x_1 + a_{m2}x_2 + \cdots + a_{mn}x_n &= b_m,
\end{aligned}
\qquad \text{(the } a_{ij} \text{ and } b_i \text{ are constants)}. \qquad \text{(A.1)}
$$

In the event that the numbers $m$ and $n$ are equal, the system (A.1) has as many equations as unknowns. An example of a system of two equations in two unknowns is

$$
\begin{aligned}
x_1 + 2x_2 &= 3, \\
3x_1 - x_2 &= -5.
\end{aligned}
\qquad \text{(A.2)}
$$

Each equation in the system (A.2) is the equation of a straight line in the $x_1x_2$-plane. Thus the simultaneous solution of the two equations in (A.2) must represent the intersection of two lines. A little thought tells us the following about lines in the plane and their intersections.

1. There is the possibility that they do not intersect (i.e., they are parallel). In this case the system (A.2) has no solution.

2. They may intersect in exactly one point. In this case the system (A.2) has only one solution (i.e., has a unique solution).
3. The two equations may describe one and the same line. In this case (A.2) has an infinite number of solutions.

As we recall from elementary algebra, one technique for solving the system of equations (A.2) is the following. We note that multiplying the second of the two equations by 2 yields the equivalent system

$$x_1 + 2x_2 = \quad 3,$$
$$6x_1 - 2x_2 = -10. \tag{A.3}$$

The principle that "when equals are added to equals the results are equal" permits us to add the two equations of the system (A.3), obtaining

$$7x_1 = -7. \tag{A.4}$$

We note that (A.4) contains only the variable $x_1$ and gives

$$x_1 = -1. \tag{A.5}$$

We now solve either of the members of (A.2) for $x_2$ as a function of $x_1$. As it turns out, the second equation is the easier of the two and solving it yields

$$x_2 = 3x_1 + 5. \tag{A.6}$$

Substitution from (A.5) into (A.6) yields

$$x_2 = -3 + 5 = 2.$$

As a check we substitute the solution pair

$$(x_1, x_2) = (-1, 2) \tag{A.7}$$

into both equations of the sytstem:

$$-1 + 2(2) = \quad 3,$$
$$3(-1) - 2 = -5.$$

Since (A.7) satisfies both of the equations of the system, it is the unique solution of the system.

As a second example consider

$$x_1 - 3x_2 = 4,$$
$$-3x_1 + 9x_2 = 5. \tag{A.8}$$

If the first of these equations is multiplied by 3 and added to the second, we obtain the contradictory statement that

$$0 = 17.$$

Thus (A.8) does not have a solution.

The reason this system of equations does not have a solution may be seen by putting both equations in slope-intercept form as

$$x_2 = \frac{1}{3}x_1 - \frac{4}{3},$$

$$x_2 = \frac{1}{3}x_1 + \frac{5}{9}.$$

We see immediately that (A.8) represents a pair of parallel lines in the $x_1x_2$-plane which do not intersect.

The details of solutions for systems larger than $n = m = 2$ will usually be more complicated than those of these first two examples. Before attempting to solve complicated systems it would be useful to know if the system has a solution. Thus we will develop criteria for establishing when an $m \times n$ system of linear equations has solutions, how many solutions it has, and methods of solution of this system. To this end, we will begin with the case easiest to work with, that in which $n = m$. Thus we will be looking at the sytems of equations of the form

$$
\begin{aligned}
a_{11}x_1 + a_{12}x_2 + \cdots + a_{1n}x_n &= b_1, \\
&\;\;\vdots \\
a_{61}x_1 + a_{62}x_2 + \cdots + a_{6n}x_n &= b_6, \\
&\;\;\vdots \\
a_{n1}x_1 + a_{n2}x_2 + \cdots + a_{nn}x_n &= b_n.
\end{aligned}
\tag{A.9}
$$

Note that the first subscript on the coefficient $a_{ij}$ represents the equation number, while the second subscript is identical to the subscript of the variable it multiplies.

As an alternative notation we rewrite (A.9) in the form

$$
\mathbf{AX} =
\begin{bmatrix}
a_{11} & a_{12} & \cdots & a_{1n} \\
\vdots & \vdots & & \vdots \\
a_{61} & a_{62} & \cdots & a_{6n} \\
\vdots & \vdots & & \vdots \\
a_{n1} & a_{n2} & \cdots & a_{nn}
\end{bmatrix}
\begin{bmatrix}
x_1 \\
\vdots \\
x_6 \\
\vdots \\
x_n
\end{bmatrix}
=
\begin{bmatrix}
b_1 \\
\vdots \\
b_6 \\
\vdots \\
b_n
\end{bmatrix}
= \mathbf{B},
\tag{A.10}
$$

where we let $\mathbf{A}$ represent the left array above, $\mathbf{X}$ the arrays of $x$'s, and $\mathbf{B}$ the array of $b$'s.

This is just a shorthand way of writing (A.9), since we may think of (A.9) as obtained from (A.10) in the following way. To form the $p$th equation of the system (A.9), we multiply each element of the $p$th row of $\mathbf{A}$ by the corresponding entry in $\mathbf{X}$ and add all the products, then set the result equal to $b_p$.

As an example, take $p = 6$. We form the products

$$a_{61}x_1,$$

$$a_{62}x_2,$$

$$\vdots$$

$$a_{6n}x_n,$$

then add them together

$$a_{61}x_1 + a_{62}x_2 + \cdots + a_{6n}x_n,$$

and set the result equal to $b_6$ to obtain

$$a_{61}x_1 + a_{62}x_2 + \cdots + a_{6n}x_n = b_6.$$

This would indeed be the sixth equation of the system (A.9).

In this notation we may write the system of equations in (A.2) as

$$\begin{bmatrix} 1 & 2 \\ 3 & -1 \end{bmatrix} \begin{bmatrix} x_1 \\ x_2 \end{bmatrix} = \begin{bmatrix} 3 \\ -5 \end{bmatrix}$$

and the system of equations in (A.8) as

$$\begin{bmatrix} 1 & -3 \\ -3 & 9 \end{bmatrix} \begin{bmatrix} x_1 \\ x_2 \end{bmatrix} = \begin{bmatrix} 4 \\ 5 \end{bmatrix}.$$

Putting the system of equations in (A.9) in the form (A.10) allows a better understanding of the following definitions.

**Definition A.1.**    The array

$$A = \begin{bmatrix} a_{11} & \cdots & a_{1n} \\ \vdots & & \vdots \\ a_{n1} & \cdots & a_{nn} \end{bmatrix}$$

is called the *matrix,* or the *coefficient matrix* of the system (A.10) [or (A.9)].

**Definition A.2.**    The column

$$X = \begin{bmatrix} x_1 \\ \vdots \\ x_n \end{bmatrix}$$

is called the *vector* (or *column matrix*) of *unknowns* of the system.

**Definition A.3.**    The column

$$B = \begin{bmatrix} b_1 \\ \vdots \\ b_n \end{bmatrix}$$

is called the *constant vector* (or *column matrix*) of the system.

With each $n \times n$ coefficient matrix $A$ we will associate a real number called the *determinant of* $A$. First we will do this for a $2 \times 2$ coefficient matrix and then use the result to obtain a rule for calculating determinants of $n \times n$ matrices. As our simplest possible case, then, we choose the $2 \times 2$ system

$$a_{11}x_1 + a_{12}x_2 = b_1,$$
$$a_{21}x_1 + a_{22}x_2 = b_2,$$

or

$$\begin{bmatrix} a_{11} & a_{12} \\ a_{21} & a_{22} \end{bmatrix} \begin{bmatrix} x_1 \\ x_2 \end{bmatrix} = \begin{bmatrix} b_1 \\ b_2 \end{bmatrix}.$$

The coefficient matrix is

$$\mathbf{A} = \begin{bmatrix} a_{11} & a_{12} \\ a_{21} & a_{22} \end{bmatrix}.$$

For a $2 \times 2$ system we will define the determinant of $\mathbf{A}$ to be

$$\det \mathbf{A} = \begin{vmatrix} a_{11} & a_{12} \\ a_{21} & a_{22} \end{vmatrix} = a_{11}a_{22} - a_{21}a_{12}. \tag{A.11}$$

For the $2 \times 2$ system we considered earlier, (A.2), we have

$$\mathbf{A} = \begin{bmatrix} 1 & 2 \\ 3 & -1 \end{bmatrix}$$

and

$$\det \mathbf{A} = \begin{vmatrix} 1 & 2 \\ 3 & -1 \end{vmatrix} = (1)(-1) - (2)(3) = -1 - 6 = -7.$$

The next higher value of $n$ is $n = 3$. For a $3 \times 3$ system the coefficient matrix $\mathbf{A}$ has the form

$$\mathbf{A} = \begin{bmatrix} a_{11} & a_{12} & a_{13} \\ a_{21} & a_{22} & a_{23} \\ a_{31} & a_{32} & a_{33} \end{bmatrix}$$

and we will define its determinant to be

$$\det \mathbf{A} = a_{11} \begin{vmatrix} a_{22} & a_{23} \\ a_{32} & a_{33} \end{vmatrix} - a_{12} \begin{vmatrix} a_{21} & a_{23} \\ a_{31} & a_{33} \end{vmatrix} + a_{13} \begin{vmatrix} a_{21} & a_{22} \\ a_{31} & a_{32} \end{vmatrix}. \tag{A.12}$$

Since we know how to calculate the $2 \times 2$ determinants in (A.12), we know how to calculate $\det \mathbf{A}$.

As an example, consider the $3 \times 3$ system

$$x_1 - 2x_2 + x_3 = b_1,$$
$$-x_1 + x_2 - 4x_3 = b_2,$$
$$3x_1 + 3x_2 + x_3 = b_3.$$

Here

$$\mathbf{A} = \begin{bmatrix} 1 & -2 & 1 \\ -1 & 1 & -4 \\ 3 & 3 & 1 \end{bmatrix}$$

and

$$\det \mathbf{A} = 1 \begin{vmatrix} 1 & -4 \\ 3 & 1 \end{vmatrix} + 2 \begin{vmatrix} -1 & -4 \\ 3 & 1 \end{vmatrix} + 1 \begin{vmatrix} -1 & 1 \\ 3 & 3 \end{vmatrix}$$

$$= 1 - (3)(-4) + 2[-1 - 3(-4)] + [-3 - (3)(1)]$$

$$= 1 + 12 + 2(11) - 6$$

$$= 13 + 22 - 6 = 29.$$

The generalization of this procedure to larger $n \times n$ systems is easy. As an example let's do a $4 \times 4$ determinant. We have the $4 \times 4$ matrix

$$\mathbf{A} = \begin{bmatrix} a_{11} & a_{12} & a_{13} & a_{14} \\ a_{21} & a_{22} & a_{23} & a_{24} \\ a_{31} & a_{32} & a_{33} & a_{34} \\ a_{41} & a_{42} & a_{43} & a_{44} \end{bmatrix}$$

and we calculate det $\mathbf{A}$ as follows:

$$\det \mathbf{A} = a_{11} \begin{vmatrix} a_{22} & a_{23} & a_{24} \\ a_{32} & a_{33} & a_{34} \\ a_{42} & a_{43} & a_{44} \end{vmatrix} - a_{12} \begin{vmatrix} a_{21} & a_{23} & a_{24} \\ a_{31} & a_{33} & a_{34} \\ a_{41} & a_{43} & a_{44} \end{vmatrix}$$

$$+ a_{13} \begin{vmatrix} a_{21} & a_{22} & a_{24} \\ a_{31} & a_{32} & a_{34} \\ a_{41} & a_{42} & a_{44} \end{vmatrix} - a_{14} \begin{vmatrix} a_{21} & a_{22} & a_{23} \\ a_{31} & a_{32} & a_{33} \\ a_{41} & a_{42} & a_{43} \end{vmatrix}. \qquad (A.13)$$

We now reduce the $3 \times 3$ determinants to $2 \times 2$ determinants, as before. One should carefully note the alternation of signs as we move across the top row. To evaluate the determinant of

$$\mathbf{A} = \begin{bmatrix} -1 & 0 & 2 & 3 \\ 0 & 4 & -1 & 1 \\ 3 & 4 & 2 & 1 \\ 3 & 4 & 2 & 2 \end{bmatrix}$$

we have

$$\det \mathbf{A} = (-1) \begin{vmatrix} 4 & -1 & 1 \\ 4 & 2 & 1 \\ 4 & 2 & 2 \end{vmatrix} - 0 \begin{vmatrix} 0 & -1 & 1 \\ 3 & 2 & 1 \\ 3 & 2 & 2 \end{vmatrix} + 2 \begin{vmatrix} 0 & 4 & 1 \\ 3 & 4 & 1 \\ 3 & 4 & 2 \end{vmatrix} - 3 \begin{vmatrix} 0 & 4 & -1 \\ 3 & 4 & 2 \\ 3 & 4 & 2 \end{vmatrix}.$$

Now expanding these determinants (note that since the second $3 \times 3$ determinant is multiplied by 0, we do not need to evaluate it) gives

$$\begin{vmatrix} 4 & -1 & 1 \\ 4 & 2 & 1 \\ 4 & 2 & 2 \end{vmatrix} = 4 \begin{vmatrix} 2 & 1 \\ 2 & 2 \end{vmatrix} - (-1) \begin{vmatrix} 4 & 1 \\ 4 & 2 \end{vmatrix} + 1 \begin{vmatrix} 4 & 2 \\ 4 & 2 \end{vmatrix}$$

$$= 4(2) + 4 + 0 = 12,$$

$$\begin{vmatrix} 0 & 4 & 1 \\ 3 & 4 & 1 \\ 3 & 4 & 2 \end{vmatrix} = 0 \begin{vmatrix} 4 & 1 \\ 4 & 2 \end{vmatrix} - 4 \begin{vmatrix} 3 & 1 \\ 3 & 2 \end{vmatrix} + 1 \begin{vmatrix} 3 & 4 \\ 3 & 4 \end{vmatrix}$$

$$= 0 - 4(3) + 0 = -12,$$

$$\begin{vmatrix} 0 & 4 & -1 \\ 3 & 4 & 2 \\ 3 & 4 & 2 \end{vmatrix} = 0 \begin{vmatrix} 4 & 2 \\ 4 & 2 \end{vmatrix} - 4 \begin{vmatrix} 3 & 2 \\ 3 & 2 \end{vmatrix} + (-1) \begin{vmatrix} 3 & 4 \\ 3 & 4 \end{vmatrix} = 0.$$

Thus the value of det $\mathbf{A}$ is given by

$$(-1)(12) - 0 + 2(-12) - 3(0) = -36.$$

Now that we know how to calculate determinants we may write our first theorem about solutions to $n \times n$ systems of equations. To this end we give the following definition:

**Definition A.4.** An $n \times n$ matrix, $\mathbf{A}$, for which det $\mathbf{A} = 0$ is said to be *singular*.

Matrices for which det $\mathbf{A} \neq 0$ are said to be *nonsingular* and we have

**Theorem A.1.** The $n \times n$ system of equations $\mathbf{AX} = \mathbf{B}$ has a unique solution if and only if $\mathbf{A}$ is nonsingular.

This theorem tells us that any time we have an $n \times n$ system of equations whose coefficient matrix is nonsingular, we are guaranteed a unique solution. Of course, we are not told how to get the solution, and at this point we have only the technique from elementary algebra already mentioned.

There is another method of solution available to us now that we know how to calculate determinants. This is *Cramer's rule,* and it works in the following way. Consider the $2 \times 2$ system

$$\begin{bmatrix} a_{11} & a_{12} \\ a_{21} & a_{22} \end{bmatrix} \begin{bmatrix} x_1 \\ x_2 \end{bmatrix} = \begin{bmatrix} b_1 \\ b_2 \end{bmatrix}. \tag{A.14}$$

If

$$\det \begin{bmatrix} a_{11} & a_{12} \\ a_{21} & a_{22} \end{bmatrix} = a_{11}a_{22} - a_{21}a_{12} \neq 0,$$

we know that a unique solution exists. Cramer's rule tells us that the solution to (A.14) is given by

$$x_1 = \frac{\begin{vmatrix} b_1 & a_{12} \\ b_2 & a_{22} \end{vmatrix}}{\begin{vmatrix} a_{11} & a_{12} \\ a_{21} & a_{22} \end{vmatrix}}, \qquad x_2 = \frac{\begin{vmatrix} a_{11} & b_1 \\ a_{21} & b_2 \end{vmatrix}}{\begin{vmatrix} a_{11} & a_{12} \\ a_{21} & a_{22} \end{vmatrix}}. \tag{A.15}$$

If we think of $x_1$ as the first variable and $x_2$ as the second variable, the procedure for using Cramer's rule may be explained as follows:

**Step 1.**    Replace the first column of the coefficient matrix by the constant vector.

**Step 2.**    Calculate the determinant of this new matrix.

**Step 3.**    Divide by the determinant of the coefficient matrix (which must not $= 0$) and call the result $x_1$.

As one may easily see from (A.15), the procedure is exactly the same for $x_2$ except that since $x_2$ is the second variable, we replace the second column of the coefficient matrix with the constant vector in step 1 above.

The generalization of Cramer's rule to systems with $n$ larger than 2 is straight-forward and is left to the exercises. It should be carefully noted here that because determinants are defined only for $n \times n$ matrices, Cramer's rule applies only to $n \times n$ systems of equations and then only to systems whose solution is unique.

In using Cramer's rule it may prove useful to have the following facts about determinants.

1. If any two rows of a determinant are proportional, the determinant is identically zero.
2. If any two columns of a determinant are proportional, the determinant is identically zero.
3. If any row or column of a matrix **A** is composed entirely of zeros, the determinant is identically zero.

For example, the solution of

$$3x_1 - 2x_2 = -2,$$
$$-x_1 + 4x_2 = 4,$$

using Cramer's rule is

$$x_1 = \frac{\begin{vmatrix} -2 & -2 \\ 4 & 4 \end{vmatrix}}{\begin{vmatrix} 3 & -2 \\ -1 & 4 \end{vmatrix}} = \frac{0}{10} = 0,$$

$$x_2 = \frac{\begin{vmatrix} 3 & -2 \\ -1 & 4 \end{vmatrix}}{10} = 1.$$

An example of a $3 \times 3$ system of linear equations is

$$3x_1 - x_2 + x_3 = 2,$$
$$2x_1 - x_2 - x_3 = -1,$$
$$-x_1 + x_2 + x_3 = 1.$$

The determinant of the coefficient matrix of this system of equations is

$$\begin{vmatrix} 3 & -1 & 1 \\ 2 & -1 & -1 \\ -1 & 1 & 1 \end{vmatrix} = 3 \begin{vmatrix} -1 & -1 \\ 1 & 1 \end{vmatrix} - (-1) \begin{vmatrix} 2 & -1 \\ -1 & 1 \end{vmatrix} + 1 \begin{vmatrix} 2 & -1 \\ -1 & 1 \end{vmatrix}$$

$$= 3(0) + 1 + 1 = 2.$$

Then using Cramer's rule gives

$$x_1 = \frac{1}{2} \begin{vmatrix} 2 & -1 & 1 \\ -1 & -1 & -1 \\ 1 & 1 & 1 \end{vmatrix} = 0$$

(since the bottom two rows are proportional).

$$x_2 = \frac{1}{2} \begin{vmatrix} 3 & 2 & 1 \\ 2 & -1 & -1 \\ -1 & 1 & 1 \end{vmatrix}$$

$$= \frac{1}{2} \left[ 3 \begin{vmatrix} -1 & -1 \\ 1 & 1 \end{vmatrix} - 2 \begin{vmatrix} 2 & -1 \\ -1 & 1 \end{vmatrix} + 1 \begin{vmatrix} 2 & -1 \\ -1 & 1 \end{vmatrix} \right]$$

$$= \frac{1}{2}[3(0) - 2(1) + 1(1)] = -\frac{1}{2},$$

$$x_3 = \frac{1}{2} \begin{vmatrix} 3 & -1 & 2 \\ 2 & -1 & -1 \\ -1 & 1 & 1 \end{vmatrix}$$

$$= \frac{1}{2} \left[ 3 \begin{vmatrix} -1 & -1 \\ 1 & 1 \end{vmatrix} - (-1) \begin{vmatrix} 2 & -1 \\ -1 & 1 \end{vmatrix} + 2 \begin{vmatrix} 2 & -1 \\ -1 & 1 \end{vmatrix} \right]$$

$$= \frac{1}{2}[3(0) + 1(1) + 2(1)] = \frac{3}{2},$$

and $(0, -\frac{1}{2}, \frac{3}{2})$ is the solution to the original system of equations.

### The Homogeneous n × n System

If in (A.10) we set all the entries in the constant vector to zero, we obtain the system of equations

$$\begin{bmatrix} a_{11} & a_{12} & \cdots & a_{1n} \\ a_{21} & a_{22} & \cdots & a_{2n} \\ \vdots & & & \\ a_{n1} & a_{n2} & \cdots & a_{nn} \end{bmatrix} \begin{bmatrix} x_1 \\ x_2 \\ \vdots \\ x_n \end{bmatrix} = \begin{bmatrix} 0 \\ 0 \\ \vdots \\ 0 \end{bmatrix}. \tag{A.16}$$

An equation of the form (A.16) is called a *homogeneous system of equations,* or simply, a homogeneous system. We see by inspection that (A.16) always has the solution

$$x_1 = x_2 = \cdots = x_n = 0. \tag{A.17}$$

Also, if det **A** $\neq$ 0, fact 3 above and an application of Cramer's rule will quickly convince one that (A.17) is indeed a solution of (A.16).

Now we have a theorem which tells us that if det **A** $\neq$ 0, the system (A.10) has a unique solution. This theorem places no restrictions whatever on the entries $b_i$, so it must apply even when all the $b_i$ are zero, that is, in the homogeneous system. Thus if the determinant of the coefficient matrix in (A.16) is not zero, (A.16) will have a unique solution. But (A.17) is *always* a solution to (A.16), so if the coefficient matrix of (A.16) is nonsingular, (A.17) is the *only* solution. It is not a very interesting solution and is called the *trivial solution*.

We now have the following corollary to Theorem A.1.

**Corollary A.1.**    If det **A** $\neq$ 0 for the homogeneous system (A.16) the trivial solution is the only solution (i.e., is unique).

Obviously, the interesting homogeneous systems will be those with nontrivial solutions, namely, those whose coefficient matrices are singular. However, Cramer's rule is useless in these cases, so we need to have other ways of solving systems of equations.

## EXERCISES

Calculate the following determinants.

**1. (a)** $\begin{vmatrix} 1 & 2 \\ 3 & 4 \end{vmatrix}$    **(b)** $\begin{vmatrix} 3 & 4 \\ 1 & 2 \end{vmatrix}$    **(c)** $\begin{vmatrix} 2 & 1 \\ 4 & 3 \end{vmatrix}$    **(d)** $\begin{vmatrix} 6 & -5 \\ 4 & -3 \end{vmatrix}$

**2. (a)** $\begin{vmatrix} x^2 & x \\ 2x & 1 \end{vmatrix}$    **(b)** $\begin{vmatrix} \sin x & \cos x \\ \cos x & -\sin x \end{vmatrix}$    **(c)** $\begin{vmatrix} e^x & e^{2x} \\ e^x & 2e^{2x} \end{vmatrix}$

**3. (a)** $\begin{vmatrix} 1 & 2 & 3 \\ -1 & 0 & 1 \\ 2 & -2 & 4 \end{vmatrix}$    **(b)** $\begin{vmatrix} -1 & 0 & 1 \\ 1 & 2 & 3 \\ 2 & -2 & 4 \end{vmatrix}$    **(c)** $\begin{vmatrix} 2 & 1 & 0 \\ 1 & -1 & -3 \\ 3 & 1 & 3 \end{vmatrix}$

**4. (a)** $\begin{vmatrix} x^3 & x & x^2 \\ 3x^2 & 1 & 2x \\ 6x & 0 & 2 \end{vmatrix}$    **(b)** $\begin{vmatrix} e^x & e^{2x} & e^{-x} \\ e^x & 2e^{2x} & -e^{-x} \\ e^x & 4e^{2x} & e^{-x} \end{vmatrix}$    **(c)** $\begin{vmatrix} 1 & \sin x & \cos x \\ 0 & \cos x & -\sin x \\ 0 & -\sin x & -\cos x \end{vmatrix}$

If we have the system of equations **Ax** = **B** with

$$\mathbf{A} = \begin{bmatrix} a_{11} & a_{12} & a_{13} \\ a_{21} & a_{22} & a_{23} \\ a_{31} & a_{32} & a_{33} \end{bmatrix}, \qquad \mathbf{X} = \begin{bmatrix} x_1 \\ x_2 \\ x_3 \end{bmatrix}, \qquad \mathbf{B} = \begin{bmatrix} b_1 \\ b_2 \\ b_3 \end{bmatrix},$$

Cramer's rule gives

$$x_1 = \frac{1}{\det \mathbf{A}} \begin{vmatrix} b_1 & a_{12} & a_{13} \\ b_2 & a_{22} & a_{23} \\ b_3 & a_{32} & a_{33} \end{vmatrix}, \qquad x_2 = \frac{1}{\det \mathbf{A}} \begin{vmatrix} a_{11} & b_1 & a_{13} \\ a_{21} & b_2 & a_{23} \\ a_{31} & b_3 & a_{33} \end{vmatrix},$$

$$x_3 = \frac{1}{\det \mathbf{A}} \begin{vmatrix} a_{11} & a_{12} & b_1 \\ a_{21} & a_{22} & b_2 \\ a_{31} & a_{32} & b_3 \end{vmatrix}.$$

Solve the following systems of equations.

**5. (a)** $x_1 + 2x_2 = -4,$
$3x_1 + 4x_2 = 9.$

**(b)** $2x_1 + x_2 = 8,$
$4x_1 + 3x_2 = 1.$

**(c)** $3x_1 - x_2 = 4,$
$x_1 + 3x_2 = -6.$

**6. (a)** $x_1 + 7x_2 = 14,$
$7x_1 - x_2 = 48.$

**(b)** $9x_1 - 7x_2 = 0,$
$x_1 - x_2 = 12.$

**(c)** $6x_1 + 5x_2 = 0,$
$4x_1 - 6x_2 = 0.$

Solve the following systems of equations.

**7. (a)** $4x_1 + 9x_2 = 8,$
$8x_1 + 6x_2 = -1,$
$6x_2 + 6x_3 = -1.$

**(b)** $2x_1 - x_2 + 3x_3 = 5,$
$x_1 + x_2 - x_3 = 2,$
$3x_1 + 2x_2 - 2x_3 = 5.$

**(c)** $x_1 - 2x_2 + 3x_3 = 0,$
$2x_1 + x_2 + 2x_3 = 1,$
$3x_1 + 6x_2 + x_3 = -3.$

**8. (a)** $2x_2 - 3x_2 = -2,$
$-6x_1 + x_2 + 2x_3 = 0,$
$3x_1 - x_2 - 2x_3 = -3.$

**(b)** $2x_1 + 3x_2 - x_3 = -5,$
$-x_1 + x_2 + 2x_3 = -5,$
$3x_1 - 2x_2 + 4x_3 = 3.$

**(c)** $x_1 + x_2 + x_3 = -2,$
$x_1 - x_2 + x_3 = 2,$
$-x_1 + x_2 - x_3 = -2.$

Using Theorem A.1, determine whether or not the following systems of equations have unique solutions (do not solve for the unknowns).

**9. (a)** $x_1 + 2x_2 + 3x_3 = 1,$
$2x_1 - x_2 + 3x_3 = 7,$
$x_1 - 2x_2 + 3x_3 = -6.$

**(b)** $x_1 - 2x_2 + 2x_3 = 6,$
$4x_1 + x_2 - x_3 = 4,$
$2x_1 + 5x_2 - 5x_3 = 1.$

**(c)** $3x_1 + 2x_2 + x_3 = 4,$
$-x_1 + 2x_2 - 2x_3 = 1,$
$x_1 - 3x_2 + 4x_3 = 6.$

**10.** For what values of $r$ will the following systems have a unique solution?

**(a)** $rx_1 + 3x_2 = 4,$
$8x_1 + 2x_2 = 1.$

**(b)** $3x_1 - rx_2 = 8,$
$2rx_1 - 6x_2 = 1.$

**(c)** $x_1 + 3x_3 = 9,$
$2x_1 + 2x_2 + rx_3 = 0,$
$x_1 + x_2 + 3x_3 = 4.$

**11.** Show that the following expressions all give the same value for the determinant as (A.12).

**(a)**
$$\begin{vmatrix} a_{11} & a_{12} & a_{13} \\ a_{21} & a_{22} & a_{23} \\ a_{31} & a_{32} & a_{33} \end{vmatrix} = a_{11}\begin{vmatrix} a_{22} & a_{23} \\ a_{32} & a_{33} \end{vmatrix} - a_{21}\begin{vmatrix} a_{12} & a_{13} \\ a_{32} & a_{33} \end{vmatrix} + a_{31}\begin{vmatrix} a_{12} & a_{13} \\ a_{22} & a_{23} \end{vmatrix}.$$

**(b)**
$$\begin{vmatrix} a_{11} & a_{12} & a_{13} \\ a_{21} & a_{22} & a_{23} \\ a_{31} & a_{32} & a_{33} \end{vmatrix} = -a_{12}\begin{vmatrix} a_{21} & a_{23} \\ a_{31} & a_{33} \end{vmatrix} + a_{22}\begin{vmatrix} a_{11} & a_{13} \\ a_{31} & a_{33} \end{vmatrix} - a_{32}\begin{vmatrix} a_{11} & a_{13} \\ a_{21} & a_{23} \end{vmatrix}.$$

**(c)**
$$\begin{vmatrix} a_{11} & a_{12} & a_{13} \\ a_{21} & a_{22} & a_{23} \\ a_{31} & a_{32} & a_{33} \end{vmatrix} = a_{31}\begin{vmatrix} a_{12} & a_{13} \\ a_{22} & a_{23} \end{vmatrix} - a_{32}\begin{vmatrix} a_{11} & a_{13} \\ a_{21} & a_{23} \end{vmatrix} + a_{33}\begin{vmatrix} a_{11} & a_{12} \\ a_{21} & a_{22} \end{vmatrix}.$$

**(d)**
$$\begin{vmatrix} a_{11} & a_{12} & a_{13} \\ a_{21} & a_{22} & a_{23} \\ a_{31} & a_{32} & a_{33} \end{vmatrix} = -a_{21}\begin{vmatrix} a_{12} & a_{13} \\ a_{32} & a_{33} \end{vmatrix} + a_{22}\begin{vmatrix} a_{11} & a_{13} \\ a_{31} & a_{33} \end{vmatrix} - a_{23}\begin{vmatrix} a_{11} & a_{12} \\ a_{31} & a_{32} \end{vmatrix}.$$

**(e)**
$$\begin{vmatrix} a_{11} & a_{12} & a_{13} \\ a_{21} & a_{22} & a_{23} \\ a_{31} & a_{32} & a_{33} \end{vmatrix} = a_{13}\begin{vmatrix} a_{21} & a_{22} \\ a_{31} & a_{32} \end{vmatrix} - a_{23}\begin{vmatrix} a_{11} & a_{12} \\ a_{31} & a_{32} \end{vmatrix} + a_{33}\begin{vmatrix} a_{11} & a_{12} \\ a_{21} & a_{22} \end{vmatrix}.$$

Evaluate.

**12. (a)** $\begin{vmatrix} a & 0 & 0 \\ d & b & 0 \\ e & f & c \end{vmatrix}$     **(b)** $\begin{vmatrix} a & d & e \\ 0 & b & f \\ 0 & 0 & c \end{vmatrix}$     **(c)** $\begin{vmatrix} 0 & 0 & c \\ 0 & b & f \\ a & d & e \end{vmatrix}$

**(d)** $\begin{vmatrix} 1 & 0 & 0 & 0 \\ a & 0 & 0 & 0 \\ d & b & 0 & 0 \\ e & f & c & 0 \end{vmatrix}$     **(e)** $\begin{vmatrix} 0 & 0 & 0 & 1 \\ 0 & 0 & 0 & a \\ 0 & 0 & b & d \\ 0 & c & e & f \end{vmatrix}$     **(f)** $\begin{vmatrix} 0 & 0 & 0 & a \\ 0 & 0 & b & e \\ 0 & c & f & g \\ d & h & i & j \end{vmatrix}$

**13. (a)** $\begin{vmatrix} 1 & 2 & 10 \\ 1 & 11 & 10 \\ 0 & 0 & 0 \end{vmatrix}$     **(b)** $\begin{vmatrix} 1 & 0 & 1 \\ 1 & 0 & 1 \\ 6 & 9 & 12 \end{vmatrix}$     **(c)** $\begin{vmatrix} 1 & -1 & 1 \\ 1 & -1 & 1 \\ a & b & c \end{vmatrix}$

**(d)** $\begin{vmatrix} 0 & 1 & 2 \\ 0 & 3 & 4 \\ 0 & 5 & 6 \end{vmatrix}$     **(e)** $\begin{vmatrix} 1 & 1 & -1 \\ 2 & 2 & 2 \\ 1 & 1 & -1 \end{vmatrix}$     **(f)** $\begin{vmatrix} a & b & c \\ 1 & 3 & 2 \\ 1 & 3 & 2 \end{vmatrix}$

Cramer's rule for an $n \times n$ system, $\mathbf{AX} = \mathbf{B}$, where

$$\mathbf{A} = \begin{bmatrix} a_{11} & a_{12} & \cdots & a_{1n} \\ a_{21} & a_{22} & \cdots & a_{2n} \\ \vdots & & & \\ a_{n1} & a_{n2} & \cdots & a_{nn} \end{bmatrix}, \quad \mathbf{X} = \begin{bmatrix} x_1 \\ x_2 \\ \vdots \\ x_n \end{bmatrix} \quad \mathbf{B} = \begin{bmatrix} b_1 \\ b_2 \\ \vdots \\ b_n \end{bmatrix}$$

is

$$x_i = \left\{ \frac{1}{\det \mathbf{A}} \right\} \cdot \left\{ \begin{array}{l} \text{determinant of the matrix obtained} \\ \text{by replacing the } i\text{th column } of\ \mathbf{A} \text{ with} \\ \text{the elements in } \mathbf{B} \end{array} \right\}, \quad i = 1, 2, 3, \ldots, n.$$

Although Cramer's rule has great theoretical value, it is not very efficient for solving systems for $n > 3$ unless the determinant has zeros in "convenient" places.

**14.** Use Cramer's rule to solve the following equations.

**(a)** $\begin{bmatrix} 1 & 0 & 0 & -1 \\ 0 & 2 & 0 & -1 \\ 1 & 0 & 2 & 5 \\ 0 & 0 & 2 & 1 \end{bmatrix} \begin{bmatrix} x_1 \\ x_2 \\ x_3 \\ x_4 \end{bmatrix} = \begin{bmatrix} 3 \\ 1 \\ 5 \\ 2 \end{bmatrix}.$

**(b)** $\begin{bmatrix} 1 & 1 & 1 & 1 \\ 1 & 2 & 4 & 2 \\ -1 & 1 & -1 & -1 \\ -1 & 3 & -1 & 1 \end{bmatrix} \begin{bmatrix} x_1 \\ x_2 \\ x_3 \\ x_4 \end{bmatrix} = \begin{bmatrix} 2 \\ 1 \\ -6 \\ -2 \end{bmatrix}.$

## A.2 SOLUTION OF A SYSTEM OF EQUATIONS BY LOWER GAUSS ELIMINATION

To introduce some new terminology for a different method of solving equations, consider the system of equations

$$
\begin{aligned}
3x_1 - 5x_2 + x_3 &= 0, \\
-x_1 - x_2 + x_3 &= 1, \\
2x_1 - 4x_2 - 3x_3 &= -1.
\end{aligned} \tag{A.18}
$$

The coefficient matrix of the system (A.18) is

$$
\begin{bmatrix}
3 & -5 & 1 \\
-1 & -1 & 1 \\
2 & -4 & -3
\end{bmatrix}. \tag{A.19}
$$

The constant vector is

$$
\begin{bmatrix}
0 \\
1 \\
-1
\end{bmatrix}. \tag{A.20}
$$

Using (A.19) and (A.20), we define a new matrix associated with the system. This new matrix is called the *augmented matrix* and is formed by adjoining the column matrix (A.20) onto (A.19) as a fourth column

$$
\begin{bmatrix}
3 & -5 & 1 & 0 \\
-1 & -1 & 1 & 1 \\
2 & -4 & -3 & -1
\end{bmatrix}. \tag{A.21}
$$

Matrix (A.21) is actually a representation of the original system in which the variables have been suppressed and only the coefficients of the unknowns and the constants kept. We may perform certain operations on the rows of (A.21) to effect a solution to the system (A.18). These *elementary row operations* are:

(1) We may multiply or divide any row by a nonzero constant.

(2) We may replace any row by the sum of that row and a constant times another row.     (A.22)

(3) We may interchange any pair of rows.

The reasons that we are permitted to do this are illustrated as follows.

1. The first row, for example, represents the equation

$$
3x_1 - 5x_2 + x_3 = 0
$$

and multiplying or dividing this equation by any nonzero constant is a well-defined legitimate operation. (Here a *legitimate operation* is one that leads to an *equivalent system,* that is, a new system with the same solution set.) Thus the same is true for this equation in the form of an augmented matrix

$$[3 \quad -5 \quad 1 \quad 0].$$

2. Adding two valid equations together is also a legitimate operation; for example, take

$$3x_1 - 5x_2 + x_3 = 0,$$

$$-x_1 - x_2 + x_3 = 1.$$

Replacing the top equation by the result of adding it to 3 times the bottom equation gives the equivalent system

$$-8x_2 + 4x_3 = 3,$$

$$-x_1 - x_2 + x_3 = 1.$$

The corresponding operation in an augmented matrix would be to replace

$$\begin{bmatrix} 3 & -5 & 1 & 0 \\ -1 & -1 & 1 & 1 \end{bmatrix}$$

with

$$\begin{bmatrix} 0 & -8 & 4 & 3 \\ -1 & -1 & 1 & 1 \end{bmatrix}.$$

3. Interchanging rows simply means writing down the equations of the system (A.18) in a different order. This obviously results in an equivalent system.

The Gauss elimination procedure is motivated by the ease of finding solutions of systems of equations in a special form. Consider a new system of equations given by

$$x_1 + 2x_2 - 3x_3 = -4,$$

$$x_2 + 2x_3 = 3, \tag{A.23}$$

$$6x_3 = -6,$$

where some uncoupling of the variables is present. By substituting the value of $x_3$ from the third equation above, $x_3 = -1$, into the second equation, we determine $x_2$ as

$$x_2 = 3 - 2x_3 = 3 - 2(-1) = 5.$$

In a similar manner, we find $x_1$ as

$$x_1 = -4 - 2x_2 + 3x_3$$

$$= -4 - 2(5) + 3(-1) = -17.$$

This is called solving a system by *back substitution*.

The message given by this example is that augmented matrices of the form

$$\begin{bmatrix} 1 & 2 & -3 & -4 \\ 0 & 1 & 2 & 3 \\ 0 & 0 & 6 & -6 \end{bmatrix} \tag{A.24}$$

represent systems which are easy to solve. The reason these equations were easy to solve is that the only unknown occurring in the last row is $x_3$, only $x_2$ and $x_3$ are present in the second row and the coefficient of $x_2$ is 1, and the coefficient of $x_1$ in the first row is 1; or, more concisely stated, because the lower left-hand portion of the augmented matrix looked like

$$\begin{bmatrix} 1 & & \\ 0 & 1 & \cdots \\ 0 & 0 & \end{bmatrix}. \tag{A.25}$$

The process of using the operations listed under (A.22) to convert an arbitrary augmented matrix to one which looks like (A.25) is called the *lower Gauss elimination* procedure or *Gaussian elimination*. We now illustrate this method with the system of equations given by (A.21), with the augmented matrix

$$\begin{bmatrix} 3 & -5 & 1 & 0 \\ -1 & -1 & 1 & 1 \\ 2 & -4 & -3 & -1 \end{bmatrix}. \tag{A.21}$$

We begin by interchanging rows 1 and 2 to obtain

$$\begin{bmatrix} -1 & -1 & 1 & 1 \\ 3 & -5 & 1 & 0 \\ 2 & -4 & -3 & -1 \end{bmatrix}.$$

Then multiply row 1 by $-1$:

$$\begin{bmatrix} 1 & 1 & -1 & -1 \\ 3 & -5 & 1 & 0 \\ 2 & -4 & -3 & -1 \end{bmatrix}. \tag{A.26}$$

We have achieved the first objective of Gaussian elimination; namely, we have rearranged (A.21) to get (A.26), which has a 1 in the upper left-hand corner. Now we subtract 3 times row 1 from row 2, obtaining

$$\begin{bmatrix} 1 & 1 & -1 & -1 \\ 0 & -8 & 4 & 3 \\ 2 & -4 & -3 & -1 \end{bmatrix}.$$

If we now multiply row by 1 by 2 and subtract the result from row 3, we obtain

$$\begin{bmatrix} 1 & 1 & -1 & -1 \\ 0 & -8 & 4 & 3 \\ 0 & -6 & -1 & 1 \end{bmatrix}$$

and we have achieved the second objective of Gaussian elimination, namely, we have eliminated all nonzero entries in the first column of the augmented matrix except the first. In terms of the original system of equations, we have eliminated the first variable, $x_1$, from the second and third equations. The method continues as follows.

Divide row 2 by $-8$ to obtain

$$\begin{bmatrix} 1 & 1 & -1 & -1 \\ 0 & 1 & -\dfrac{1}{2} & -\dfrac{3}{8} \\ 0 & -6 & -1 & 1 \end{bmatrix}.$$

Now multiply row 2 by 6 and add the result to row 3 to obtain

$$\begin{bmatrix} 1 & 1 & -1 & -1 \\ 0 & 1 & -\dfrac{1}{2} & -\dfrac{3}{8} \\ 0 & 0 & -4 & -\dfrac{5}{4} \end{bmatrix}. \qquad (A.27)$$

The last row of the resulting augmented matrix contains an entry only for the variable $x_3$. The second row contains entries only for $x_2$ and $x_3$ and row 1 for $x_1$, $x_2$, and $x_3$. If we write out the system of equations corresponding to (A.27), we obtain

$$x_1 + x_2 - x_3 = -1,$$

$$x_2 - \frac{1}{2}x_3 = -\frac{3}{8}, \qquad (A.28)$$

$$-4x_3 = -\frac{5}{4}.$$

Now solve the third equation in (A.28) obtaining

$$x_3 = \frac{5}{16}$$

and substitute this value into the second equation in (A.28) and obtain

$$x_2 = \frac{1}{2}x_3 - \frac{3}{8} = \frac{1}{2}\left(\frac{5}{16}\right) - \frac{3}{8} = \frac{5}{32} - \frac{12}{32} = -\frac{7}{32}.$$

From the first of equations (A.28) we now obtain

$$x_1 = -1 - x_2 + x_3 = -1 + \frac{7}{32} + \frac{5}{16} = -\frac{15}{32}$$

and the system is solved.

Now let's go back and review what we have done. The first step was to write down the augmented matrix, (A.21). The next step was to get a 1 in the upper left-hand corner, in the $a_{11}$ position. This we did by interchanging two rows. We could also have done it by simply dividing the first row by 3. Once we have a 1 in the $a_{11}$ position, we "obtain zeros" in all the positions below it.

We now turn our attention to the second column. If we do not have a 1 in the $a_{22}$ position, we obtain one if possible. We then "obtain zeros" in all lower positions of the second column using legitimate row operations listed in (A.22). Then we move on to the third column, then the fourth, and so on. In the last nonzero row remaining after this process, the left entry corresponds to the coefficient of a variable, while the right entry is the constant. We now rewrite this system in equation form and solve these resulting equations successively for the variables $x_n, x_{n-1}, \ldots, x_1$.

Up to this point we have been considering only systems of equations in which $m$, the number of equations, is equal to $n$, the number of unknowns. It often happens that this is not the case and we have more equations than unknowns or more unknowns than equations. In neither of these cases does Cramer's rule apply, as we may calculate determinants only for $n \times n$ matrices. We must then fall back on the Gauss elimination procedure to reduce the coefficient matrix to a more convenient form.

Let us consider first the case in which there are more equations than unknowns, for example, the system of equations

$$2x_1 + 8x_2 = 14,$$
$$x_1 - 3x_2 = 0, \tag{A.29}$$
$$4x_1 + 2x_2 = 14.$$

We may begin the elimination method by writing the augmented matrix as

$$\begin{bmatrix} 2 & 8 & 14 \\ 1 & -3 & 0 \\ 4 & 2 & 14 \end{bmatrix}. \tag{A.30}$$

Now divide row 1 by 2 to obtain the equivalent matrix

$$\begin{bmatrix} 1 & 4 & 7 \\ 1 & -3 & 0 \\ 4 & 2 & 14 \end{bmatrix}.$$

Now replace row 2 with the result of subtracting row 1 from it and replace row 3 by the result of subtracting 4 times row 1 from row 3, obtaining

$$\begin{bmatrix} 1 & 4 & 7 \\ 0 & -7 & -7 \\ 0 & -14 & -14 \end{bmatrix}. \tag{A.31}$$

The augmented matrix (A.31) corresponds to the system of equations

$$x_1 + 4x_2 = 7,$$
$$-7x_2 = -7, \tag{A.32}$$
$$-14x_2 = -14.$$

At this point we could stop, having seen that the last two equations of (A.32) are equivalent equations. The solution obtained from (A.32) is easily found to be

$$x_1 = 3, \qquad x_2 = 1. \tag{A.33}$$

We did not, however, have to stop the elimination with (A.31), and a further step is revealing. In (A.31) we divide the second row by $-7$ to get

$$\begin{bmatrix} 1 & 4 & 7 \\ 0 & 1 & 1 \\ 0 & -14 & -14 \end{bmatrix}.$$ (A.34)

If we now replace row 3 by the result of adding it to 14 times row 2, we get

$$\begin{bmatrix} 1 & 4 & 7 \\ 0 & 1 & 1 \\ 0 & 0 & 0 \end{bmatrix}.$$ (A.35)

From (A.35), because the bottom row has all zeros in it, we conclude that the original system of equations is equivalent to the system

$$x_1 + 4x_2 = 7,$$
$$x_2 = 1.$$ (A.36)

The determinant of the coefficient matrix of (A.36) is

$$\begin{vmatrix} 1 & 4 \\ 0 & 1 \end{vmatrix} = 1,$$

and thus is nonsingular, so the solution, (A.33), is unique.

As another example of what may occur when there are more equations than unknowns, consider the system

$$3x_1 - x_2 = 4,$$
$$6x_1 - 2x_2 = 8,$$ (A.37)
$$-9x_1 + 3x_2 = -12.$$

The augmented matrix for (A.37) is

$$\begin{bmatrix} 3 & -1 & 4 \\ 6 & -2 & 8 \\ -9 & 3 & -12 \end{bmatrix}.$$ (A.38)

To strictly follow the Gauss elimination procedure, we should obtain a 1 in the upper left-hand corner of (A.38). However, since the other coefficients of $x_1$ (i.e., 6 and $-9$) are both multiples of 3, and the second step in our procedure is to obtain zeros in the other two positions beneath the 3 in (A.38), we proceed as follows. Replace row 2 by the result of multiplying row 1 by $-2$ and adding it to row 2 of (A.38):

$$\begin{bmatrix} 3 & -1 & 4 \\ 0 & 0 & 0 \\ -9 & 3 & -12 \end{bmatrix}$$ (A.39)

and replace row 3 by the result of multiplying row 1 by 3 and adding to row 3:

$$\begin{bmatrix} 3 & -1 & 4 \\ 0 & 0 & 0 \\ 0 & 0 & 0 \end{bmatrix}.$$ (A.40)

This matrix is equivalent to the single equation

$$3x_1 - x_2 = 4. \tag{A.41}$$

Equation (A.41) does not have a unique solution, since solving for $x_1$ in terms of $x_2$ yields

$$x_1 = \frac{1}{3}(x_2 + 4) \tag{A.42}$$

and in (A.42) we may choose $x_2$ arbitrarily and calculate the $x_1$ corresponding to it.

Another possibility is embodied in a system of equations of the type

$$\begin{aligned}
2x_1 - 4x_2 &= -4, \\
3x_1 - 4x_2 &= -2, \\
-4x_1 + 3x_2 &= 6.
\end{aligned} \tag{A.43}$$

The augmented matrix is

$$\begin{bmatrix} 2 & -4 & -4 \\ 3 & -4 & -2 \\ -4 & 3 & 6 \end{bmatrix} \tag{A.44}$$

which, under the by now familiar procedure, reduces to

$$\begin{bmatrix} 1 & -2 & -2 \\ 0 & 1 & 2 \\ 0 & 0 & 8 \end{bmatrix}. \tag{A.45}$$

The third of equations (A.45), $0x_1 + 0x_2 = 8$, has no solution since there are no values of $x_1$ and $x_2$ which satisfy it. Thus no solution exists for (A.43).

Thus for systems of equations in which there are more equations than unknowns, we have seen the following cases arise:

1. There is a unique solution.
2. There is an infinite number of solutions.
3. There is no solution.

In each of these cases, a careful application of the Gauss elimination method leads us to the right conclusion. These three possibilities are, in fact, the only ones when there are more equations than unknowns.

The next situation we will consider is the case in which there are more unknowns than equations. An example is the system of equations

$$\begin{aligned}
x_1 + 2x_2 - x_3 + 2x_4 &= 0, \\
-2x_1 - 5x_2 + 3x_3 &= 2, \\
x_1 + x_3 + 10x_4 &= 7.
\end{aligned} \tag{A.46}$$

The augmented matrix for this system is

$$\begin{bmatrix} 1 & 2 & -1 & 2 & 0 \\ -2 & -5 & 3 & 0 & 2 \\ 1 & 0 & 1 & 10 & 7 \end{bmatrix} \tag{A.47}$$

which may be brought, by three steps using the Gauss elimination method, to the form

$$\begin{bmatrix} 1 & 2 & -1 & 2 & 0 \\ 0 & -1 & 1 & 4 & 2 \\ 0 & 0 & 0 & 0 & 3 \end{bmatrix}. \tag{A.48}$$

From the last of equations (A.48) we again may conclude that this system of equations has no solution.

Now consider a different system of equations:

$$\begin{aligned} x_1 + 2x_2 - x_3 + 2x_4 &= 0, \\ -2x_1 - 5x_2 + 3x_3 &= 2, \\ x_1 + 3x_2 + x_3 + 6x_4 &= -7, \end{aligned} \tag{A.49}$$

with augmented matrix

$$\begin{bmatrix} 1 & 2 & -1 & 2 & 0 \\ -2 & -5 & 3 & 0 & 2 \\ 1 & 3 & 1 & 6 & -7 \end{bmatrix}. \tag{A.50}$$

Now use the Gauss elimination procedure to reduce (A.50) to

$$\begin{bmatrix} 1 & 2 & -1 & 2 & 0 \\ 0 & 1 & -1 & -4 & -2 \\ 0 & 0 & 3 & 8 & -5 \end{bmatrix}$$

which corresponds to the equivalent system of equations

$$\begin{aligned} x_1 + 2x_2 - x_3 + 2x_4 &= 0, \\ x_2 - x_3 - 4x_4 &= -2, \\ 3x_3 + 8x_4 &= -5. \end{aligned} \tag{A.51}$$

The system (A.51) may be written as

$$\begin{aligned} x_1 &= -2x_2 + x_3 - 2x_4, \\ x_2 &= x_3 + 4x_4 - 2, \\ x_3 &= -\frac{8}{3}x_4 - \frac{5}{3}, \end{aligned} \tag{A.52}$$

and we see from (A.52) that we may choose $x_4$ arbitrarily and that $x_1$, $x_2$, and $x_3$ are determined in terms of this value of $x_4$. Again, we have a system of equations with an infinite number of solutions.

These are the only two possibilities for the case in which we have more un-knowns than equations. As in the case of more equations than unknowns, the Gauss elimination method leads us to the right conclusion. We summarize our findings in the following statement. For the system of equations

$$\mathbf{AX} = \mathbf{B} \tag{A.53}$$

and the situation where there are more unknowns than equations, the Gauss elimi-nation method will either lead to an infinite number of solutions or show that no solution exists. For the situation where there are more equations than unknowns, we may have no solution, a unique solution, or an infinite number of solutions. Whatever the case, Gaussian elimination will give us the proper answer. Notice that if $\mathbf{B} = 0$ in (A.53), we have a homogeneous system which always has at least one solution, namely, the trivial solution.

## EXERCISES

Use Gaussian elimination to find solutions to the following systems of equations.

**1. (a)**
$$\begin{aligned} -x_1 + 2x_2 + 3x_3 &= 5, \\ x_1 + x_2 - x_3 &= 3, \\ 2x_1 - 3x_2 - 2x_3 &= 5. \end{aligned}$$
   **(b)**
$$\begin{aligned} 2x_1 + x_2 + 3x_3 &= 4, \\ 2x_1 - 2x_2 - x_3 &= 1, \\ -2x_1 + 4x_2 + x_3 &= 1. \end{aligned}$$

**2. (a)**
$$\begin{aligned} 3x_1 - 2x_2 - x_3 &= 3, \\ 3x_2 - 2x_3 &= 2, \\ -6x_1 + 2x_2 + x_3 &= 0. \end{aligned}$$
   **(b)**
$$\begin{aligned} 3x_1 + 2x_2 - x_3 &= 1, \\ 2x_1 - 2x_2 + 3x_3 &= 4, \\ x_1 - 6x_2 + 7x_3 &= 5. \end{aligned}$$

**3. (a)**
$$\begin{aligned} 3x_1 + x_2 - 2x_3 &= 11, \\ 3x_1 - 4x_2 - 2x_3 &= -4, \\ -6x_1 + 18x_2 + 18x_3 &= 24. \end{aligned}$$
   **(b)**
$$\begin{aligned} x_1 - 2x_2 - x_3 &= 3, \\ 2x_1 + 3x_2 + x_3 &= 4, \\ -x_1 - x_2 - 2x_3 &= 3. \end{aligned}$$

**4. (a)**
$$\begin{aligned} -x_1 + x_2 + 2x_3 &= -9, \\ 2x_1 - 4x_2 + x_3 &= 17, \\ x_1 + 3x_2 - 2x_3 &= 1. \end{aligned}$$
   **(b)**
$$\begin{aligned} x_1 + x_2 - x_3 &= 0, \\ 2x_1 - x_2 + 2x_3 &= 6, \\ 3x_1 - 2x_3 &= -3. \end{aligned}$$

**5. (a)**
$$\begin{aligned} x_1 - x_2 + x_3 + x_4 &= 6, \\ -x_1 - 3x_2 + x_3 + x_4 &= 2, \\ 2x_1 + 2x_2 + 4x_3 + x_4 &= 1, \\ x_1 + x_2 + x_3 + x_4 &= 2. \end{aligned}$$
   **(b)**
$$\begin{aligned} x_2 + 4x_3 - 3x_4 &= 0, \\ -x_1 + 5x_3 + 2x_4 &= 0, \\ x_1 + 2x_2 + 3x_3 - 8x_4 &= 0, \\ x_1 + x_2 - x_3 - 5x_4 &= 0. \end{aligned}$$

**6. (a)**
$$\begin{aligned} 3x_1 + 2x_2 - 7x_3 &= 2, \\ x_1 - x_2 + x_3 &= 3, \\ x_1 + 4x_2 - 9x_3 &= 4, \\ 2x_1 + 3x_2 - 8x_3 &= 7. \end{aligned}$$
   **(b)**
$$\begin{aligned} x_1 + x_2 - x_3 &= 6, \\ 2x_1 - x_2 + 3x_3 &= 4, \\ 4x_1 + x_2 + x_3 &= 16, \\ 2x_1 + 5x_2 - 7x_3 &= 20. \end{aligned}$$

**7. (a)**
$$\begin{aligned} 2x_1 - 4x_2 + x_3 - 3x_4 &= 6, \\ x_1 - 2x_2 + 3x_3 + 6x_4 &= 2. \end{aligned}$$
   **(b)**
$$\begin{aligned} x_1 + x_2 + x_3 + x_4 &= 6, \\ 2x_1 - x_2 + x_3 - 2x_4 &= 2, \\ 3x_1 + 2x_2 - x_3 + 3x_4 &= 4. \end{aligned}$$

**8. (a)**
$$\begin{aligned} x_1 + x_2 &= 4, \\ 3x_1 - 4x_2 &= 9, \\ 5x_1 - 2x_2 &= 17. \end{aligned}$$
   **(b)**
$$\begin{aligned} 3x_1 - 2x_2 &= 16, \\ -6x_1 + 4x_2 &= -32, \\ -3x_1 + 2x_2 &= -16. \end{aligned}$$

For what value(s) of $r$ will the following systems of equations have a unique solution?

**9. (a)** $5x_1 + rx_2 = 4,$
$4x_1 + 3x_2 = 3,$
$rx_1 - 6x_2 = 3.$

**(b)** $x_1 + x_2 + x_3 = 6,$
$x_1 + rx_2 + rx_3 = 2.$

**10.** For what value(s) of $r$ will the linear equations in exercise 9 have no solution?

**11.** For what value(s) of $r$ will the linear equations in exercise 9 have an infinite number of solutions?

# Answers

## CHAPTER 1

### Section 1.3, page 8

**1.** (a) 2nd order, linear   (c) 3rd order, nonlinear   (e) 4th order, nonlinear
**3.** (a) $r = 3, -2$   (c) $r = 3, -4$   (e) $r = \pm 2i, \pm 2$   (g) $r = 2, -1 \pm \sqrt{3}i$
**4.** (a) $r = 3$   (c) $r = 32$   (e) $r = 0$      **6.** (a) impossible   (c) $C = -\frac{1}{4}$
**7.** (a) $-e^x + e^{3x}$   (c) $9 \cos 4x$   (e) $\frac{1}{3}x^2 + \frac{5}{3}x^{-1}$      **8.** (a) ok   (c) F(1, 3) is not defined

### Section 1.4, page 10

**5.** (a) $r = 0, 1$   (c) $C_1 = C_4 = -C_3, C_2, C_3$ arbitrary

### Section 1.6, page 18

**1.** (a) $3x^2 = k$      **3.** (a) $x^2 + y^2 = k$
**7.** $y_3 = 1 + 2x + 2x^2 + 8x^3/3 + 8x^4/3 + 28x^5/15 + 8x^6/9 + 16x^7/63$

### Section 1.7, page 20

**1.** (a) $e^{-x}$   (c) 0   (e) $2 + 2x + 3x^2 + x^3$   (g) $(-\gamma^2 + \omega^2) \sin \gamma x$   (i) $(a\gamma^2 + b\gamma + c)e^{\gamma x}$
**2.** (a) $(D + 3)(D - 2)$   (c) $(D - 4)(D + 3)$   (e) $(D - 2)(D + 2)(D^2 + 4)$
   (g) $(D - 2)(D + 2)(D - 3)$   (i) $(3D - 1)^2$      **10.** (a) 3, 1   (c) 1, 2

## CHAPTER 2

### Section 2.1, page 30

1. (a) $y = x/(1 - Cx)$ (c) $y = \pm\sqrt{-\cos x^2 + 2C}$ (e) $y = 2\tan\left(\dfrac{2}{x} + C\right)$

 (g) $y = -\ln|-e^x - C|$ (i) $y + \ln|y + 1| = \frac{1}{2}(x + \frac{1}{2}\sin 2x) + C$

2. (a) $y = x/e$ (c) $y^3 = 8 - \int_1^x \exp(-t^2)\,dt$ (e) $y = (1 - \pi + 2\sin^{-1} x)^{-1/2}$

 (g) $y^3 = 27 + \int_0^x \sin t^2\,dt$ (i) $y^2/2 + \ln|y| = \frac{3}{2} - \exp(-\cos x)$

3. (a) no (c) degree 1 (e) degree 0 (g) degree 0

4. (a) $y = x\sqrt{(13x^{-3} - 1)/3}$ (c) $y = xe^{Cx}$ (e) $y^2 = x^2(Cx^{-8/3} - 1)/4$

 (g) $y = (-x + \sqrt{10 - x^2})/2$

### Section 2.2, page 35

1. (a) $x^3/3 - xy^2 = C$ (c) $e^x + xy - x + 3e^y - 7y = C$ (e) $x^4y^3 + x^3/12 - 16y = C$

 (g) $x\sin(x + y) + y^2/2 = C$ (i) $\sin xy + 2y + x = C$ (k) $x^3\ln|y| = C$

2. (a) $r = -2$ (c) any value of $r$

### Section 2.3, page 42

1. $y^2x - y^4/4 = C$  3. $y = x/(x^2 - C)$  5. $y = x/(-1 + Cx)$  7. $y^2x - x^{-1} = C$

9. $x^2y^{-1} + y + \ln|y| = C$  11. $y = 3 + Ce^{-x^2/2}$  13. $y = [x^2/2 + C]e^{2x}$

15. $y = x^3 + (C - x^3/3)/\ln|x|$  17. $x\sin y - y\ln|x| = C$  19. $xy^{-1} + 2\ln|y| = C$

21. $x^2y^3 + y^4 = C$  23. $x^2y^2/2 + x^3y + x^2y = C$

### Section 2.4.1, page 45

1. $y = [\frac{2}{3} + C\exp(-x^{3/2})]^{2/3}$  3. $y = x(\ln|x| + C)^{-1}$  5. $y = [2x^{10} + Cx^8]^{-1/4}$

7. $y^2 = 1 + Ce^{-x^2}$  9. $y = [-(\frac{2}{3})\ln|x| + C(\ln|x|)^{-1/2}]^{-1}$

### Section 2.4.2, page 48

1. $(x - y + 1)^3 = C(x + y - 3)$  3. $(3x - y - 11)^2 = 11(x - 3)^2 + C$

5. $(x + 7)^2y = -x^3/3 - 5x^2 - 21x + C$  7. $(y - 2x - 3)^2 = C(y - x - 2)$

9. $9x^2 - \frac{225}{16} + (4x + 5y - 5)^2 = C$

### Section 2.5, page 52

1. $y = C_1 + C_2\ln|x|$  3. $y^2 = C_1x + C_2$  5. $y = C_1 - (1 - C_2^2x^2)^{1/2}/C_2$

7. $y = \ln|x| + C_1x^{-1} + C_2$  9. $y = \tan(C_1x + C_2)$  11. $y_2 = xe^{2x}$  13. $y_2 = x\ln x$

15. $y_2 = x\cos x$

### Section 2.6, page 64

1. (a) $x = 600[1 - \exp(-t/50)]$ (b) $t = 50\ln 3$ (c) 3  3. $30 - 25/e^3$

5. $t = 10^4\ln(200/199.55) \approx 27.5$

7. (b) $y = (-2 + \sqrt{3})x$ (d) $\tan^{-1}(y/x) + \frac{1}{2}\ln(1 + y^2/x^2) = -\ln|x| + C$

9. $t = 6(\ln 5)/\ln 2$  11. $t = (\ln 2)/\ln(1.25)$

# CHAPTER 3

## Section 3.1, page 71

**1.** $y = x - 1 + Ce^{-x}$     **3.** $y = e^x/2 + Ce^{-x}$     **5.** $y = -3\cos^2 x + C\cos x$
**7.** $y = x \int x^{-1} \cos x \, dx + Cx$     **9.** $y = -\frac{1}{2} + Ce^{-2/x}$     **11.** $y = xe^{-2x} + Cx^{-1}e^{-2x}$
**13.** $y = x^3 e^{-3x} + Ce^{-3x}$     **15.** $y = -x^{-3} + Cx^{-2}$

## Section 3.2, page 79

**1.** $x = Ce^{-t}$     **3.** $x = t - 1 + Ce^{-t}$     **5.** $x = t + 2 + (\sin t - \cos t)/2 + Ce^{-t}$
**7.** $x = 3e^{3t} - 2$     **9.** $x = -2t + (e^{-7t} - 1)(\frac{2}{7})$     **11.** $x = (3e^{-t^2} + 1)/2$
**13.** $x = Ce^{-t} + e^{it}/(1 + i) = Ce^{-t} + (\sin t + \cos t)/2 + i(\sin t - \cos t)/2$

**15.** $x = e^t/2 + Ce^{-t}$     **17.** $x = \begin{cases} 2(1 - e^{-t}), & 0 \le t < 1 \\ 2(e - 1)e^{-t}, & 1 \le t \end{cases}$

**19.** $x = \begin{cases} (t + x_0)e^{-t}, & 0 \le t < 2 \\ e^{-2} + (1 + x_0)e^{-t}, & 2 \le t \end{cases}$     **21.** $x = 2(t - 1) + 1 + Ce^{-t}$

**23.** $x = 3(\sin t + \cos t) + \sin 2t - 2\cos 2t + Ce^{-t}$     **25.** $x = 5 + e^t + Ce^{-t}$

## Section 3.3, page 83

**1.** $t = 5[\ln(\frac{12}{52})]/\ln(\frac{35}{52}) \approx 18.5$     **3.** $i = (1 - e^{-60t})/10$
**5.** transient charge is $4.95e^{-10t}$, steady state is $\frac{1}{20}$
**9.** **(a)** $v = 5 + 95e^{-32t/5}$     **(b)** $t = (\frac{5}{32}) \ln(\frac{19}{9}) \approx 0.117$     **(c)** 5
**11.** **(a)** $\dfrac{dv}{dt} = -32, \; v(0) = 64$     **(b)** $t = 2$     **(c)** 68

# CHAPTER 4

## Section 4.1, page 87

**1.** independent     **3.** dependent     **5.** independent     **7.** independent     **9.** independent
**13.** **(a)** independent   **(c)** dependent

## Section 4.2.1, page 94

**1.** $y = C_1 e^{3x} + C_2 e^{-2x}$     **3.** $y = C_1 e^x + C_2 x e^x$     **5.** $y = C_1 e^{-4x} + C_2 e^{2x}$
**7.** $y = C_1 e^{4x} + C_2 e^{-3x}$     **9.** $y = C_1 e^{3x} + C_2 e^{2x}$     **11.** $y = C_1 \cos 4x + C_2 \sin 4x$
**13.** $y = C_1 e^{x/3} + C_2 e^{-x/2}$     **15.** $y = C_1 e^{-x} \cos \sqrt{3} x + C_2 e^{-x} \sin \sqrt{3} x$
**17.** **(a)** $y = e^x - e^{-2x}$     **(c)** $y = 2e^{-3x} + 6xe^{-3x}$     **(e)** $y = 5 - e^{-3x}$
    **(g)** $y = 15e^{-5x} \cos 5\sqrt{3} x + (\frac{79}{5}\sqrt{3})e^{-5x} \sin 5\sqrt{3} x$     **19.** $y = 4\cos(4x - \pi/4)$

## Section 4.2.2, page 98

**1.** $y = C_1 + C_2 \cos x + 5C_3 \sin x$     **3.** $y = (C_1 + C_2 x) \cos \sqrt{2} x + (C_3 + C_4 x) \sin \sqrt{2} x$
**5.** $y = C_1 e^{\sqrt{2} x} + C_2 e^{-\sqrt{2} x} + C_3 \cos 2x + C_4 \sin 2x$     **7.** $y = C_1 e^x + C_2 e^{-x} + C_3 e^{2x}$
**9.** $y = C_1 e^x + C_2 e^{-x} \cos 2x + C_3 e^{-x} \sin 2x$     **11.** $y = -6 + 4x - 2e^{-2x} + 8e^{-x}$
**13.** $y = e^x - e^{-3x}$     **15.** yes     **17.** no     **19.** yes     **21.** yes     **23.** yes     **25.** yes

## Section 4.3, page 108

**3.** $y = C_1 e^{2x} + C_2 e^{-3x} + (\frac{1}{6})e^{3x}$    **5.** $y = C_1 e^x + C_2 e^{-4x} - x^2/4 - 3x/8 - \frac{21}{32}$
**7.** $y = C_1 \cos 2x + C_2 \sin 2x - (\frac{1}{5}) \cos 3x$
**9.** $y = C_1 \cos 2x + C_2 \sin 2x - (\frac{6}{5}) \cos 3x - 4 \sin 3x$
**11.** $y = C_1 e^{-2x} + C_2 x e^{-2x} + (\frac{1}{8}) \sin 2x$    **13.** $y = C_1 e^x + C_2 e^{-x} + C_3 e^{-2x} - 4 \cos x - 2 \sin x$
**15.** $y = C_3 e^x + e^{3x}(C_1 \cos 2x + C_2 \sin 2x) + e^{2x} \sin x$
**17.** $y = -(\frac{1}{36})e^{-3x} + (\frac{35}{6})xe^{-3x} + (\frac{1}{36})e^{3x}$
**19.** $y = 4\pi \cos 2x + (1 - \pi) \sin 2x - 2x \cos 2x + 2x \sin 2x$

## Section 4.4, page 112

**1.** $y = C_1 x^2 + C_2 x^{-3}$    **3.** $y = C_1 x^{-4+\sqrt{14}} + C_2 x^{-4-\sqrt{14}}$
**5.** $y = C_1 x^3 \cos(4 \ln x) + C_2 x^3 \sin(4 \ln x)$    **7.** $y = C_1 x + C_2 x^{1/2}$    **9.** $y = C_1 x + C_2 x^{-1/2}$
**11.** $y = C_1(x - 1)^{-1} + C_2(x - 1)^{-1} \ln|x - 1|$
**13.** $y = C_1(x - 3) + C_2(x - 3) \ln|x - 3| + C_3(x - 3)^{-1}$
**19.** **(a)** $y = C_1 x^2 + C_2 x^{-3}$    **(c)** $y = C_1 x^{-2} + C_2 x^{-2} \ln x$
   **(e)** $y = C_1 x^2 \cos(2 \ln x) + C_2 x^2 \sin(2 \ln x) + C_3 x$
   **(g)** $y = C_1 x + C_2 x^{-1} + C_3 \sin(\ln x) + C_4 \cos(\ln x)$

## Section 4.5, page 121

**1.** $y = C_1 \cos x + C_2 \sin x - \cos x \ln|\sec x + \tan x|$    **3.** $y = C_1 + C_2 e^x + \ln(\sec x)$
**5.** $y = C_1 e^{-x} + C_2 x e^{-x} - 2e^{-x} \ln|x|$    **7.** $y = C_1 x + C_2 x \ln x + (\ln x)^3 x/6$
**9.** $y = C_1 \cos(\ln x) + C_2 \sin(\ln x) + x^3/10$

## Section 4.6.1, page 124

**1.** $x = 3 \cos 4t + 3 \sin 4t$    **3.** $x = \pi \cos(3t/2) + 2 \sin(3t/2)$    **5.** $x = 2 \cos(t/2)$
**7.** $\sqrt{148}, 8\pi$    **9.** $2, 4\pi/3, 2$    **11.** $x = (\frac{1}{2}) \sin 4t$    **13.** $\frac{1}{600}$    **15.** $2\ell/3g$

## Section 4.6.2, page 130

**1.** **(a)** $c > 12$    **(c)** $c = 12$    **7.** $x = -3te^{-4t}$    **9.** $x = e^{-t/3}(\cos 2t + (\frac{2}{3}) \sin 2t)$
**11.** **(a)** $0 < \alpha < 2$    **13.** $\sim 0.432$

## Section 4.6.3, page 133

**1.** **(a)** $x = [(\sqrt{3} - 1)/4] \sin 4t + \sin t$    **(b)** $2\pi$

**3.** **(a)** $x = -\left[k_0 \cos \omega t + \dfrac{\ell_0 \gamma + k_0 \alpha}{\omega} \sin \omega t\right]e^{-\alpha t} + k_0 \cos \gamma t + \ell_0 \sin \gamma t,$
   $k_0 = -2\alpha\gamma/[(\alpha^2 + \omega^2 - \gamma^2)^2 + (2\alpha\gamma)^2]$
   $\ell_0 = (\alpha^2 + \omega^2 - \gamma^2)/[(\alpha^2 + \omega^2 - \gamma^2)^2 + (2\alpha\gamma)^2]$
   **(c)** $[(\alpha^2 + \omega^2 - \gamma^2)^2 + (2\alpha\gamma)^2]^{-1/2}, \gamma = \omega^2 - \alpha^2$
**5.** $x = k_0 \cos \gamma t + \ell_0 \sin \gamma t - [k_0 \cos(3t/4) + \frac{4}{3}(\gamma\ell_0 + k_0/4) \sin(3t/4)]e^{-t/4}$
   $k_0 = -8\gamma E_0/[(10 - 16\gamma^2)^2 + (8\gamma)^2]; \ell_0 = (10 - 16\gamma^2)E_0/[(10 - 16\gamma^2)^2 + (8\gamma)^2]$
**7.** $x = 2e^{-t} \sin 3t + 2$

### Section 4.6.4, page 140

1. $y = 4 \cos 10x + 50 \sin 10x$     3. $y = (10/\sqrt{47})e^{-3x/2} \sin(\sqrt{47}\,x/2)$
5. (a) no values (0 is always a solution)   (c) $b = (2n + 1)\pi/8$, $n = 0, 1, 2, \ldots$
7. $(n\pi/b)^2$, $\sin(n\pi x/b)$     9. $[(2n + 1)\pi/(2b)]^2$, $n = 0, 1, 2, \ldots$, $\sin[(2n + 1)\pi x/(2b)]$
11. $k/m = (n\pi/3)^2$, $n = 1, 2, 3, \ldots$

## CHAPTER 5

### Section 5.2, page 153

1. (a) $\begin{bmatrix} 5 & 8 \\ 5 & 6 \end{bmatrix}$   (c) $\begin{bmatrix} -2 & 0 \\ 0 & -2 \end{bmatrix}$   (e) $\begin{bmatrix} 0 & 5 \\ 2 & 11 \end{bmatrix}$   (g) $\begin{bmatrix} 0 & 0 \\ 0 & 0 \end{bmatrix}$   (i) $\begin{bmatrix} 3 & 3 \\ 2 & 6 \end{bmatrix}$   (k) $\begin{bmatrix} 8 & 0 \\ -5 & 5 \end{bmatrix}$

(m) $\begin{bmatrix} 0 & 0 \\ 0 & 0 \end{bmatrix}$   (o) $\begin{bmatrix} 4 & -2 \\ -3 & 1 \end{bmatrix}$

2. (a) $\begin{bmatrix} 4 & 2 & 3 \\ -3 & 2 & 4 \\ 0 & -1 & 2 \end{bmatrix}$   (c) $\begin{bmatrix} 0 & 0 & 0 \\ 0 & 0 & 0 \\ 0 & 0 & 0 \end{bmatrix}$   (e) $\begin{bmatrix} 1 & 0 & 1 \\ -1 & 0 & 3 \\ 0 & -2 & 0 \end{bmatrix}$   (g) $\begin{bmatrix} 3 & 3 & 3 \\ -3 & -3 & -3 \\ 3 & 3 & 3 \end{bmatrix}$

(i) $\begin{bmatrix} -3 & 0 & 0 \\ 0 & -3 & 0 \\ 0 & 0 & -3 \end{bmatrix}$   (k) $\begin{bmatrix} -1 & 5 & -1 \\ -1 & 5 & -1 \\ -1 & 5 & -1 \end{bmatrix}$   (m) $\begin{bmatrix} -3 & 0 & 0 \\ 0 & -3 & 0 \\ 0 & 0 & -3 \end{bmatrix}$

3. (a) 4   (c) 0   (e) −3   (g) −8

4. (a) $\begin{bmatrix} 2 & -1 \\ 1 & 2 \end{bmatrix}$   (c) $\begin{bmatrix} 0 & 0 \\ 0 & 0 \end{bmatrix}$   (e) $\begin{bmatrix} 0 & 0 \\ 0 & 0 \end{bmatrix}$   (g) $\dfrac{1}{-2}\begin{bmatrix} 4 & -2 \\ -3 & 1 \end{bmatrix}$   (i) $\dfrac{1}{5}\begin{bmatrix} -2 & 1 \\ -1 & -2 \end{bmatrix}$

(k) $\begin{bmatrix} 1 & 0 \\ 0 & 1 \end{bmatrix}$   (m) $\dfrac{-23}{10}\begin{bmatrix} 1 & 0 \\ 0 & 1 \end{bmatrix}$

5. (a) $\begin{bmatrix} 2 & -1 & 0 \\ 0 & 1 & -2 \\ 1 & 3 & 1 \end{bmatrix}$   (e) $\dfrac{1}{16}\begin{bmatrix} 7 & -2 & -1 \\ 1 & 2 & -7 \\ 2 & 4 & 2 \end{bmatrix}$   (g) doesn't exist

(i) $\dfrac{-1}{32}\begin{bmatrix} 7 & -2 & -1 \\ 1 & 2 & -7 \\ 2 & 4 & 2 \end{bmatrix}$   (k) $\dfrac{1}{4}\begin{bmatrix} 3 & 0 & -1 \\ -1 & 0 & -1 \\ 3 & 2 & -1 \end{bmatrix}$

### Section 5.3.1, page 157

1. (a) independent   (c) dependent   (e) dependent
2. (a) $(1, 0, 2) = -5(1, 0, 0) + 4(1, -1, 1) + 2(1, 2, -1)$   (c) impossible
6. (a) dependent if $k = 1$     7. (a) $k \neq 1$
9. (a) independent   (c) independent   (e) independent     11. (a) $(0, 0, 1)$
12. (a) $(0, 1, 0, 0)$   (c) $(1, 0, 0, 0)$

### Section 5.3.2, page 162

1. (a) $\lambda = 2 \sim a\begin{bmatrix} 1 \\ 1 \end{bmatrix}$, $\lambda = -1 \sim b\begin{bmatrix} 2 \\ -1 \end{bmatrix}$   (c) $\lambda = 2 \sim a\begin{bmatrix} 3 \\ -1 \end{bmatrix}$

(e) $\lambda = 0 \sim a\begin{bmatrix} 1 \\ 0 \end{bmatrix}$, $b\begin{bmatrix} 0 \\ 1 \end{bmatrix}$   (g) $\lambda = 0 \sim a\begin{bmatrix} 1 \\ -1 \end{bmatrix}$, $\lambda = 3 \sim b\begin{bmatrix} 1 \\ 2 \end{bmatrix}$   (i) $\lambda = 2 \sim a\begin{bmatrix} 1 \\ 1 \end{bmatrix}$

**2. (a)** $\lambda = 1 \sim a\begin{bmatrix} 0 \\ 1 \\ 0 \end{bmatrix}$, $\lambda = 0 \sim b\begin{bmatrix} 1 \\ 0 \\ -1 \end{bmatrix}$, $\lambda = 2 \sim c\begin{bmatrix} 1 \\ 0 \\ 1 \end{bmatrix}$

**(c)** $\lambda = 0 \sim a\begin{bmatrix} 1 \\ 0 \\ 0 \end{bmatrix}$, $b\begin{bmatrix} 0 \\ 1 \\ 0 \end{bmatrix}$, $c\begin{bmatrix} 0 \\ 0 \\ 1 \end{bmatrix}$ **(e)** $\lambda = 1 \sim a\begin{bmatrix} 1 \\ 0 \\ 0 \end{bmatrix}$, $\lambda = 2 \sim b\begin{bmatrix} 2 \\ 1 \\ 0 \end{bmatrix}$

**(g)** $\lambda = 2 \sim a\begin{bmatrix} 1 \\ -1 \\ 1 \end{bmatrix}$, $b\begin{bmatrix} 1 \\ -1 \\ 0 \end{bmatrix}$, $\lambda = 4 \sim c\begin{bmatrix} 1 \\ 1 \\ 0 \end{bmatrix}$

**(i)** $\lambda = 2 \sim a\begin{bmatrix} 0 \\ 1 \\ -1 \end{bmatrix}$, $\lambda = \sqrt{3} \sim b\begin{bmatrix} -1 \\ -\sqrt{3} \\ 1 \end{bmatrix}$, $\lambda = -\sqrt{3} \sim b\begin{bmatrix} -1 \\ \sqrt{3} \\ 1 \end{bmatrix}$

**5. (a)** $r > -\frac{1}{2}$ **(c)** $r = -\frac{1}{2}$

## Section 5.4, page 170

**1.** $\begin{bmatrix} x_1 \\ x_2 \end{bmatrix} = C_1 e^{2t}\begin{bmatrix} \cos t \\ -\sin t \end{bmatrix} + C_2 e^{2t}\begin{bmatrix} \sin t \\ \cos t \end{bmatrix}$

**3.** $\begin{bmatrix} x_1 \\ x_2 \end{bmatrix} = C_1\begin{bmatrix} \cos t \\ \cos t + \sin t \end{bmatrix} + C_2\begin{bmatrix} \sin t \\ -\cos t + \sin t \end{bmatrix} + \begin{bmatrix} t \\ t - 1 \end{bmatrix}$

**5.** $\begin{bmatrix} x_1 \\ x_2 \end{bmatrix} = C_1\begin{bmatrix} 2 \\ 1 \end{bmatrix} e^{2t} + C_2\begin{bmatrix} 1 \\ 2 \end{bmatrix} e^{-t}$ **7.** $\begin{bmatrix} x_1 \\ x_2 \\ x_3 \end{bmatrix} = C_1\begin{bmatrix} 2 \\ 2 \\ 1 \end{bmatrix} e^{3t} + C_2\begin{bmatrix} 0 \\ -2 \\ 1 \end{bmatrix} e^{t} + C_3\begin{bmatrix} 1 \\ 1 \\ 0 \end{bmatrix} e^{2t}$

**9.** $\begin{bmatrix} x_1 \\ x_2 \\ x_3 \end{bmatrix} = C_1\begin{bmatrix} 1 \\ 2 \\ -1 \end{bmatrix} + C_2\begin{bmatrix} -2 \\ 1 \\ 0 \end{bmatrix} e^{t} + C_3\begin{bmatrix} 1 \\ 2 \\ 5 \end{bmatrix} e^{6t}$

**10.** $\begin{bmatrix} x_1 \\ x_2 \\ x_3 \end{bmatrix} = C_1\begin{bmatrix} -1 \\ 1 \\ 0 \end{bmatrix} e^{t} + C_2\begin{bmatrix} 0 \\ 1 \\ 1 \end{bmatrix} e^{t} + C_3\begin{bmatrix} 1 \\ 1 \\ 1 \end{bmatrix} e^{-t}$ **13.** $\begin{bmatrix} x_1 \\ x_2 \end{bmatrix} = C_1\begin{bmatrix} 1 \\ 1 \end{bmatrix} e^{-t} + C_2\begin{bmatrix} 5 \\ -4 \end{bmatrix} e^{-10t}$

**15.** $\begin{bmatrix} x_1 \\ x_2 \end{bmatrix} = C_1\begin{bmatrix} t \\ 1 \end{bmatrix}(t^2 + 1)^{-1/2} + C_2\begin{bmatrix} 1 \\ 0 \end{bmatrix} + \begin{bmatrix} 4t \\ 4 \end{bmatrix}$

**17.** $\begin{bmatrix} x_1 \\ x_2 \end{bmatrix} = C_1\begin{bmatrix} 2\cos(\sqrt{3}\,t/2) \\ -\cos(\sqrt{3}\,t/2) + \sqrt{3}\,\sin(\sqrt{3}\,t/2) \end{bmatrix} e^{7t/2}$

$\qquad + C_2\begin{bmatrix} 2\sin(\sqrt{3}\,t/2) \\ -\sqrt{3}\,\cos(\sqrt{3}\,t/2) - \sin(\sqrt{3}\,t/2) \end{bmatrix} e^{7t/2} + \begin{bmatrix} \frac{4}{13} - 2e^{2t}/3 \\ \frac{12}{13} + e^{2t}/3 \end{bmatrix}$

**19.** $\begin{bmatrix} x_1 \\ x_2 \end{bmatrix} = C_1\begin{bmatrix} 1 \\ -1 \end{bmatrix} e^{t} + C_2\begin{bmatrix} 1 \\ 2 \end{bmatrix} e^{2t} + C_3\begin{bmatrix} 1 \\ 1 \end{bmatrix} e^{-t}$

**21.** $\begin{bmatrix} x_1 \\ x_2 \end{bmatrix} = C_1\begin{bmatrix} 1 \\ 3t \end{bmatrix} + C_2\begin{bmatrix} 2 \\ -5 \end{bmatrix} e^{-2t} + C_3\begin{bmatrix} 0 \\ 1 \end{bmatrix} + \begin{bmatrix} -t^2 + 2t^3/3 \\ t + t^2 - 4t^3/3 + t^4/2 \end{bmatrix}$

## Section 5.5, page 182

**1.** $\mathbf{X} = \begin{bmatrix} e^{7t} & 5e^{t} \\ e^{7t} & -e^{t} \end{bmatrix}\begin{bmatrix} -\frac{7}{6} \\ \frac{5}{6} \end{bmatrix}$ **3.** $\mathbf{X} = \begin{bmatrix} 3e^{7t} & -2e^{2t} \\ e^{7t} & e^{2t} \end{bmatrix}\begin{bmatrix} 5 \\ 5 \end{bmatrix}$ **5.** $\mathbf{X} = \begin{bmatrix} 3e^{-t} & (3t+1)e^{-t} \\ -e^{-t} & -te^{-t} \end{bmatrix}\begin{bmatrix} -7 \\ 24 \end{bmatrix}$

**7.** $\mathbf{X} = \begin{bmatrix} e^{t}(\cos t - \sin t) \\ e^{t}(2\cos t - \sin t) \end{bmatrix}$ **9.** $\mathbf{X} = \begin{bmatrix} e^{3t}\sin 5t \\ e^{3t}\cos 5t \end{bmatrix}$

**13.** $\mathbf{X} = \begin{bmatrix} 13 & (3+\sqrt{57})e^{\lambda_1 t} & (3-\sqrt{57})e^{\lambda_2 t} \\ -6 & 12e^{\lambda_1 t} & 12e^{\lambda_2 t} \\ -1 & (3+\sqrt{57})e^{\lambda_1 t} & (3-\sqrt{57})e^{\lambda_2 t} \end{bmatrix}\begin{bmatrix} C_1 \\ C_2 \\ C_3 \end{bmatrix}, \quad \begin{aligned} \lambda_1 &= \dfrac{1+\sqrt{57}}{2} \\ \lambda_2 &= \dfrac{1-\sqrt{57}}{2} \end{aligned}$

**15.** $\mathbf{X} = \begin{bmatrix} e^t & -e^{4t} & 0 \\ e^t & e^{4t} & e^{4t} \\ e^t & 0 & -e^{4t} \end{bmatrix}\begin{bmatrix} C_1 \\ C_2 \\ C_3 \end{bmatrix}$  **16.** $\mathbf{X} = \begin{bmatrix} e^{-2t} & 2e^t & 2e^{-t} \\ 2e^{-2t} & e^t & 3e^{-t} \\ e^{-2t} & -e^t & -e^{-t} \end{bmatrix}\begin{bmatrix} C_1 \\ C_2 \\ C_3 \end{bmatrix}$

**19.** $\mathbf{X} = \begin{bmatrix} 5e^{2t} & e^{2t}\cos 4t & e^{2t}\sin 4t \\ 4e^{2t} & 0 & 0 \\ 0 & -e^{2t}\sin 4t & e^{2t}\cos 4t \end{bmatrix}\begin{bmatrix} C_1 \\ C_2 \\ C_3 \end{bmatrix}$

**21.** #15 with $C_1 = \frac{2}{3}$, $C_2 = C_3 = -\frac{1}{3}$  **23.** #19 with $C_1 = -\frac{1}{20}$, $C_2 = \frac{1}{4}$, $C_3 = 1$

### Section 5.6, page 190

**1.** $\mathbf{X}_p = \begin{bmatrix} 1 \\ -\frac{1}{2} \end{bmatrix}e^{3t}$  **3.** $\mathbf{X}_p = \dfrac{1}{25}\begin{bmatrix} (-35t+12)e^{2t} - 12e^{-3t} \\ (-35t-3)e^{2t} + 3e^{-3t} \end{bmatrix}$

**5.** $\mathbf{X}_p = \dfrac{1}{\pi^2 - 1}\begin{bmatrix} \pi+5 \\ 3\pi-9 \end{bmatrix}e^{\pi t}$  **7.** $\mathbf{X}_p = \dfrac{1}{2}\begin{bmatrix} -2\cos t + \sin t \\ 3\cos t \end{bmatrix}$

### Section 5.7, page 198

**1.** $x_1 = (-42\sin t + 7\sqrt{6}\sin \sqrt{6}t)/15$
$\quad x_2 = (-168\sin t - 7\sqrt{6}\sin \sqrt{6}t)/30$
**3. (a)** $x_1 = C_1\cos 5t + C_2\sin 5t + C_3\cos 3t + C_4\sin 3t$
$\quad x_2 = -C_1\cos 5t - C_2\sin 5t + C_3\cos 3t + C_4\sin 3t$
**7. (b)** $I_1 = 2(1 - e^{-50t}\cos 350t) - (\frac{2}{7})e^{-50t}\sin 350t$  **11. (a)** $t = -12.5\ln(\frac{3}{5})$
$\quad I_2 = 2(1 - e^{-50t}\cos 350t) + (\frac{48}{7})e^{-50t}\sin 350t$
**13. (a)** $x_1 = (1199A/40 - 1.5S_0)e^{-2t} + (401A/10 - 2S_0)te^{-2t} + 3S_0/2$
$\quad\quad - A[(\frac{1199}{40})\cos t/10 + (\frac{2001}{800})\sin t/10]$
$\quad x_2 = (-S_0/2 + 399A/40)e^{-2t} + (-S_0 + 401A/20)te^{-2t} + S_0/2$
$\quad\quad - A[\sin t/10 + (\frac{399}{40})\cos t/10]$

## CHAPTER 6

### Section 6.1, page 209

**1.** $-1 + 2(x+1) - 3(x+1)^2 + 22(x+1)^3/6$
**3. (a)** $1 + x + x^2/2 + x^3/3$  **(b)** radius $\geq 1 - \exp\left(\dfrac{-\sqrt{5}}{10}\right)$
**5. (a)** $1 - (\sin 1)x^2/2 + (\sin 2)x^4/48$  **7.** $1 + 2x + 5x^3/6 - x^4/3$

### Section 6.2, page 214

**5.** $y = 4\displaystyle\sum_{k=0}^{\infty}\dfrac{x^{4k}}{4^k k!} + 3\displaystyle\sum_{k=0}^{\infty}\dfrac{x^{4k+2}}{2^{k+1}(2k+1)(2k-1)\cdots(3)(1)}$  **7.** $y = 4\displaystyle\sum_{k=0}^{\infty}\dfrac{(\frac{2}{3})^k(x-1)^{3k}}{k!}$
**9.** $y = -3 - 3(x - \pi/4) - 9(x - \pi/4)^2/2 - 11(x - \pi/4)^3/2 - 57(x - \pi/4)^4/8 + \cdots$
**11.** $y = 2 - 2(x-1) + 6(x-1)^2 - 26(x-1)^3/3 + 44(x-1)^4/3 + \cdots$

**13.** $y = 3[x - x^4/4 + 3x^7/56 - x^{10}/112 + \cdots]$
**15.** $y = x - x^2/2 + x^3/3 - x^4/6 + x^5/6 + \cdots$

## Section 6.3, page 219

**1.** (a) $1, -2, -3$  (c) $-6, 1$  (e) $3i, -3i, -7$  (g) $2, \pm n\pi, n = 0, 1, 2, \ldots$
**2.** (a) $0$  (c) $2$  (e) $3$  (g) $\pi - 3$
**3.** (a) $y = 3x + 3x^2/2 - x^4/2 + \cdots, h = \infty$
  (c) $y = -2x + x^2/2 - x^3/6 + x^4/3 + \cdots, h = \infty$
  (e) $y = 2(x - 1) + (x - 1)^2 + 2(x - 1)^3/3 + (x - 1)^4/6 + \cdots, h = 1$
**4.** (a) $y = c_0[1 + x^4/12 + x^8/672 + \cdots] + c_1[x + x^5/20 + x^9/1440 + \cdots], h = \infty$
  (c) $y = c_0[1 - 5x^2/2 + 35x^4/24 + \cdots] + c_1[x - 7x^3/6 + 7x^5/24 + \cdots], h = 1$
  (e) $y = (x - 1) - (x - 1)^2/2 + (x - 1)^3/6 - (x - 1)^4 + \cdots, h = 1$

## Section 6.3.1, page 223

**1.** regular $0, -1, 4, -4$, irregular $1$    **3.** regular $-1, \pm n\pi, n = 1, 2, 3, \ldots$, irregular $0$
**5.** regular $-2, 4, -4$, irregular $0$    **7.** regular $1$, irregular $0, 4$

## Section 6.4.1, page 230

**1.** $y = c_0 \sum\limits_{k=0}^{\infty} \dfrac{(-1)^k x^k}{2^k k!} + c_0^* x^{1/2} \sum\limits_{k=0}^{\infty} \dfrac{(-1)^k x^k}{(2k + 1)(2k - 1) \cdots (3)(1)}$

**3.** $y = c_0\left[1 + \sum\limits_{k=1}^{\infty} \dfrac{x^k}{k!(3k - 2)(3k - 5)\cdots (7)(4)(1)}\right]$

$\qquad\qquad\qquad + c_0^* x^{2/3}\left[1 + \sum\limits_{k=1}^{\infty} \dfrac{x^k}{k!(3k + 2)(3k - 1)\cdots (8)(5)}\right]$

**5.** $y = c_0 \sum\limits_{k=0}^{\infty} \dfrac{(-1)^k x^k}{(2k)!} + c_0^* x^{1/2} \sum\limits_{k=0}^{\infty} \dfrac{(-1)^k x^k}{(2k + 1)!}$

**7.** $y = c_0\left[1 + \sum\limits_{k=1}^{\infty} \dfrac{(-1)^k x^k}{(2k - 1)(2k - 3) \cdots (3)(1)}\right] + c_0^* x^{1/2} \sum\limits_{k=0}^{\infty} \dfrac{(-1)^k x^k}{k!}$

**9.** $y = c_0 x^{-1}\left[1 + \sum\limits_{k=1}^{\infty} \dfrac{(-1)^k x^k}{k!(2k - 7)(2k - 9) \cdots (-3)(-5)}\right]$

$\qquad\qquad\qquad + c_0^* x^{5/2}\left[1 + \sum\limits_{k=1}^{\infty} \dfrac{(-1)^k x^k}{k!(2k + 7)(2k + 5) \cdots (11)(9)}\right]$

**11.** $y = c_0[1 + e(x - 1) - (e + e^2)(x - 1)^2/2 - (e + e^2 - e^3)(x - 1)^3/18 + \cdots]$
$\qquad + c_0^*(x - 1)^{3/2}[1 - e(x - 1)/5 + (-5e + e^2)(x - 1)^2/70$
$\qquad - (35e - 19e^2 + e^3)(x - 1)^3/1890 + \cdots]$

**13.** $y = c_0(x + 1)^{1/2} \sum\limits_{k=0}^{\infty} \dfrac{(\frac{3}{2})^k(x + 1)^k}{k!} + c_0^*(x + 1)^2 \sum\limits_{k=0}^{\infty} \dfrac{3^{k+1}(x + 1)^k}{(2k + 3)(2k + 1) \cdots (5)(3)}$

**15.** $y = c_0(x - 1) + c_0^*(x - 1) \displaystyle\int (x - 1)^{-1} e^{-(x-1)^2/2}\, dx$

$\qquad = c_0(x - 1) + c_0^*(x - 1)\left[\ln|x - 1| + \sum\limits_{k=1}^{\infty} \dfrac{(-1)^k(x - 1)^{2k}}{k!\, 2^k(2k)}\right]$

**17.** $y = c_0[y_1(x) \cos \ln|x| - y_2(x) \sin \ln|x|] + c_0^*[y_1(x) \sin \ln|x| + y_2(x) \cos \ln|x|]$
$\qquad y_1(x) = 1 - x/5 - x^2/40 + 3x^3/520 + \cdots$
$\qquad y_2(x) = 2x/5 - 3x^2/40 + 7x^3/1560 + \cdots$

## Section 6.4.2, page 242

**1.** $y = c_0 y_1(x) + c_0^*[y_1(x) \ln x - x^2/4 - 3x^4/128 - 11x^6/13824 + \cdots]$

$y_1(x) = 1 + \sum_{k=1}^{\infty} \frac{x^{2k}}{(k!)^2 2^{2k}}, \ h = \infty$

**3.** $y = c_0(1 + 2x/3 + x^2/3) + c_0^* x^4 \sum_{k=0}^{\infty} (k + 1)x^k, \ h = 1$

**5.** $y = c_0 y_1(x) + c_0^*[y_1(x) \ln x + 2x^{-2}], \ y_1(x) = x^{-4} - x^{-2}, \ h = 1$

**7.** $y = c_0 x + c_0^*\left[x \ln x + \sum_{k=1}^{\infty} \frac{(-1)^k x^{k+1}}{kk!}\right], \ h = \infty$

**9.** $y = c_0 y_1(x) + c_0^*[-y_1(x) \ln x + 1 - x^2 + 3x^3/4 - 7x^4/36 + \cdots], \ h = \infty$

## Section 6.5.1, page 249

**9.** (a) $x^2 = \frac{2}{3}P_2(x) + \frac{1}{3}P_0(x), \ x^3 = \frac{2}{5}P_3(x) + \frac{3}{5}P_1(x)$

**11.** $y = \sum_{k=0}^{n} \frac{(n - k + 1)(n - k + 2) \cdots (n)(n + 1) \cdots (n + k)}{2^k (k!)^2}(x - 1)^k$

**15.** (b) $\lambda = n, \ n = 0, 1, 2, \ldots$

  (d) $L_0(x) = 1, \ L_1(x) = x - 1, \ L_2(x) = x^2 - 4x + 2, \ L_3(x) = x^3 - 9x^2 + 18x - 6$

## Section 6.5.2, page 257

**1.** $y = x^{-\mu}\left[1 + \sum_{k=1}^{\infty} \frac{(-1)^k x^{2k}}{2^{2k} k!(1 - \mu)(2 - \mu) \cdots (k - \mu)}\right]$

**9.** $y = c_0 x^{-1/2} \sin x + c_0^* x^{-1/2} \cos x$

**11.** (a) $J_{3/2}(x) = \sqrt{\frac{2}{\pi x}}\left(\frac{\sin x}{x} - \cos x\right)$   (c) $J_{5/2}(x) = \sqrt{\frac{2}{\pi x}}\left[\left(\frac{3}{x^2} - 1\right) \sin x - \frac{3 \cos x}{x}\right]$

# CHAPTER 7

## Section 7.1, page 265

**3.** $(a + bs)/(s^2 + 1)$    **5.** $s/(s^2 - 1)$    **7.** $(as + b)/(s^2 - 1)$    **9.** $a/(s^2 - a^2)$

**11.** $(1 - 2s - e^{-2s})/s^2$    **13.** $3t^2/2$    **15.** $5e^{2t}$    **17.** $2e^{3t} - 3e^{2t}$    **19.** $2(e^{5t} - e^{-2t})$

## Section 7.2, page 268

**5.** $\mathcal{L}\{f(t)\} = 1/s + e^{-s}/s^2 - e^{-2s}/s^2$    **6.** (a) no  (c) no  (e) yes  (g) yes

**7.** (a) $\alpha = 3, \ M = 2, \ T = 0$

## Section 7.3, page 270

**1.** (a) $\omega/(s^2 + \omega^2)$  (c) $\omega/(s^2 - \omega^2)$  (e) $2/s^3$  (i) $(s^2 + 1)/(s^2 - 1)^2$

**3.** (a) $[1/s - s/(s^2 + 4)]/2$

## Section 7.4, page 278

**1.** $2/(s - a)^3$    **3.** $k/[(s - a)^2 + k^2]$    **5.** $1/(s - k)^3 - 1/(s + k)^3$

**7.** $-2k/[(s - a)^2 - k^2]^2 - 4k(s - a)^2/[(s - a)^2 - k^2]^3$    **9.** $(s^2 + 2)/[s(s^2 + 4)]$

**13.** $[s/(s^2 + 9) + 3s/(s^2 + 1)]/4$    **17.** $-e^{-\pi s}/(s^2 + 1)$    **19.** $e^{-s}/s^2$
**21.** $e^{-2s}[2/s^3 + 6/s^2 + 4/s]$    **23.** $f(t) = e^{-t/2}[\cos \sqrt{3}\,t/2 + \sqrt{3}\,\sin \sqrt{3}\,t/2]$
**27. (a)** $\pi/2 - \tan^{-1} s$

## Section 7.5, page 283

**1.** $e^{2t} - e^t$    **3.** $e^{-2t}/3 - e^{-t}/2 + e^t/6$    **5.** $\dfrac{e^t + k \sin kt - \cos kt}{1 + k^2}$
**7.** $[3 - 3e^{-2t} \cos 3t - 2e^{-2t} \sin 3t]/39$    **9.** $[ke^t - k \cos kt - \sin kt]/(1 + k^2)$
**11.** $[4t^3 - 9t^2 + 24t - 18]/6$    **13.** $(4 - 7t + 4t^2)e^{-t}$    **15.** $t - t^2/2$

## Section 7.6, page 294

**1.** $(\sin t - \cos t + e^{-2t}[\sin t + \cos t])/8$    **3.** $e^{-t}[\sin t + \cos t]$
**5.** $e^t[4 + (t - 1)u_1(t) - (t - 3)u_3(t)]$    **9.** $[e^{-2t} - 8e^{-t} + 7 - 6t + 2t^2]/4$
**13. (a)** $x_0 \cos \omega t + [v_0/\omega] \sin \omega t + E_0[r(t - 1)u_1(t) + r(t - 2)u_2(t)], \; r(t) = [1 - \cos \omega t]/\omega^2$

## Section 7.7, page 299

**13.** $\dfrac{1}{3}\begin{bmatrix} e^t + 2e^{4t} \\ e^t - e^{4t} \\ e^t - e^{4t} \end{bmatrix}$    **15.** $\begin{bmatrix} 5[e^{3t} - e^{2t}] \\ e^{3t} \\ e^{3t} \end{bmatrix}$    **17.** $x_1(t) = [2e^{2t} - 2e^{-3t} + 5e^{3t}]/5$
$\qquad\qquad\qquad\qquad\qquad\qquad\qquad\qquad\qquad\quad x_2(t) = [4e^{2t} + e^{-3t} - 5e^{3t}]/10$
**19.** $x_1(t) = [28e^{2t} - 28e^{-3t} - 15te^{2t}]/25$    **23.** $x_1(t) = [-1 - 4e^{3t} + 5e^{2t}]/2$
$\quad\; x_2(t) = [18e^{2t} + 7e^{-3t} - 15te^{2t}]/25$    $\qquad\; x_2(t) = [-1 + 16e^{3t} - 15e^{2t}]/6$
**25.** $x_1(t) = -2e^{-3t} + 4e^{2t}$    **27. (a)** $x_1(t) = -[3\sqrt{2} \sin \sqrt{2}\,t + \sqrt{6} \sin(t\sqrt{2}/3)]/8$
$\quad\; x_2(t) = 2e^{-3t} + e^{2t}$    $\qquad\qquad\; x_2(t) = [3\sqrt{2} \sin \sqrt{2}\,t - \sqrt{6} \sin(t\sqrt{2}/3)]/4$

# CHAPTER 9

## Section 9.1, page 317

**2. (a)** $y^2 = -\cos^2 x + C$   **(c)** $y = Cx$

## Section 9.2, page 319

**1.** $(0, 0)$    **3.** $(0, 0)$    **5.** $(-3/11, -9/11)$    **7.** $(1/\sqrt{3}, 1/\sqrt{3}), (-1/\sqrt{3}, -1/\sqrt{3})$
**9.** $(0, 0)$

## Section 9.3, page 329

**1.** 1a    **3.** 1a    **5.** 3a    **7.** 3a    **9.** 3b

## Section 9.4, page 336

**1.** $(0, 0)$ 1a    **3.** $(0, 0)$ 1a    **5. (c)** $(0, 0), (2^{1/3}, 2^{2/3})$

## Section 9.5, page 339

**2. (a)** asymptotically stable   **(c)** within a circle of radius $\frac{1}{2}$, stable, but not asymptotically stable

## CHAPTER 10

### Section 10.1, page 349

1. $\|\sin nx\| = \sqrt{\pi/2}$, $n = 1, 2, 3, \ldots$     3. $\|1\| = \sqrt{\pi}$, $\|\cos nx\| = \sqrt{\pi/2}$, $n = 1, 2, 3, \ldots$
5. $\|\sin(n\pi x/L)\| = \sqrt{L/2}$, $n = 1, 2, 3, \ldots$
7. $\|1\| = \sqrt{2L}$, $\|\sin(n\pi x/L)\| = \|\cos(n\pi x/L)\| = \sqrt{L}$, $n = 1, 2, 3, \ldots$
9. (a) $\dfrac{2}{\pi} \displaystyle\sum_{n=1}^{\infty} \dfrac{1 - (-1)^n(1 + \pi)}{n} \sin nx$

11. (a) $\displaystyle\sum_{n=1}^{\infty} [P_{n-1}(0) - P_{n+1}(0)]P_n(x)$   (c) $[2P_2(x) + 3P_1(x) + 4P_0(x)]/3$

### Section 10.2, page 355

1. period $\pi$     3. not periodic     5. period $2\pi$     9. period 2     11. period 1
13. $\dfrac{2L}{\pi} \displaystyle\sum_{n=1}^{\infty} \dfrac{(-1)^{n+1} \sin(n\pi x/L)}{n}$, converges to 0 at $x = \pm L$

15. $\dfrac{1}{2} + \dfrac{1}{\pi} \displaystyle\sum_{n=1}^{\infty} \dfrac{(-1)^n - 1}{n} \sin nx$, converges to $\dfrac{1}{2}$ at $x = 0, \pm\pi$

19. $-1 + \dfrac{4}{\pi} \displaystyle\sum_{n=1}^{\infty} \dfrac{1 - (-1)^n}{n} \sin nx$, converges to $-1$ at $x = 0, \pm\pi$

### Section 10.3, page 364

1. $1, \dfrac{4}{\pi} \displaystyle\sum_{n=1}^{\infty} \dfrac{\sin(2n - 1)\pi x/L}{2n - 1}$

3. $\dfrac{\pi^2}{3} + 4 \displaystyle\sum_{n=1}^{\infty} \dfrac{(-1)^n \cos nx}{n^2}, \dfrac{2}{\pi} \displaystyle\sum_{n=1}^{\infty} \dfrac{(2 - \pi^2 n^2)(-1)^n - 2}{n^3} \sin nx$     5. $\cos x, \dfrac{8}{\pi} \displaystyle\sum_{n=1}^{\infty} \dfrac{n \sin 2nx}{4n^2 - 1}$

7. $\dfrac{3}{2} + \dfrac{2}{\pi} \displaystyle\sum_{n=1}^{\infty} \dfrac{(-1)^n}{2n - 1} \cos(2n - 1)x, \dfrac{2}{\pi} \displaystyle\sum_{n=1}^{\infty} \dfrac{1 - 2(-1)^n + \cos n\pi/2}{n} \sin nx$

11. $2 + \dfrac{4}{\pi^2} \displaystyle\sum_{n=1}^{\infty} \dfrac{(-1)^n - 1}{n^2} \cos n\pi x/2, \dfrac{2}{\pi} \displaystyle\sum_{n=1}^{\infty} \dfrac{1 - 3(-1)^n}{n} \sin n\pi x/2$

15. $\dfrac{1}{2} + \dfrac{\cos 2x}{2}, \dfrac{2}{3} \sin x + \dfrac{1}{2} \displaystyle\sum_{n=3}^{\infty} \left[ \dfrac{1 - (-1)^n}{n} - \dfrac{n(-1)^n}{n^2 - 4} \right] \sin nx$

25. $y(x) = \displaystyle\sum_{n=1}^{\infty} \dfrac{b_n \sin nx}{\omega^2 - n^2} - \dfrac{1}{\omega} \displaystyle\sum_{n=1}^{\infty} \dfrac{nb_n}{\omega^2 - n^2} \sin \omega x$

27. $y(x) = C_1 e^{-3x} + C_2 e^{-5x} + \displaystyle\sum_{n=1}^{\infty} a_n \cos(4n - 2)x + b_n \sin(4n - 2)x$,

$C_1 = \displaystyle\sum_{n=1}^{\infty} -\dfrac{5}{2} a_n - (2n - 1)b_n, C_2 = \displaystyle\sum_{n=1}^{\infty} \dfrac{3}{2} a_n + (2n - 1)b_n$,

$a_n = \dfrac{-16(2n - 1)b_n}{15 - (4n - 2)^2}, b_n = \dfrac{4[15 - (4n - 2)^2]}{64(4n - 2)^2 + [15 - (4n - 2)^2]^2}$

## Section 10.4, page 373

1. (a) $\lambda_n = (2n - 1)^2 \pi^2/4$, $\phi_n(x) = \sin(2n - 1)\pi x/2$, $\|\phi_n\| = 1/\sqrt{2}$, $n = 1, 2, 3, \ldots$
   (c) $\lambda_n = \xi_n^2$, where $\tan \xi_n = -\xi_n$, $n = 1, 2, 3, \ldots$
   $\phi_n(x) = \sin \xi_n x$, $\|\phi_n\| = \sqrt{2\xi_n - \sin 2\xi_n}/(2\sqrt{\xi_n})$, $n = 1, 2, 3, \ldots$
   (e) $\lambda_n = n^2 \pi^2 - 1$, $\phi_n(x) = \sin n\pi x$, $\|\phi_n\| = 1/\sqrt{2}$, $n = 1, 2, 3, \ldots$

2. (a) $\dfrac{4}{\pi} \displaystyle\sum_{n=1}^{\infty} \dfrac{\sin(2n-1)\pi x/2}{2n-1}$  (c) $2 \displaystyle\sum_{n=1}^{\infty} \dfrac{1 - \cos \xi_n}{\xi_n(1 + \cos^2 \xi_n)} \sin \xi_n x$  (e) $\dfrac{4}{\pi} \displaystyle\sum_{n=1}^{\infty} \dfrac{\sin(2n-1)\pi x}{2n-1}$

7. (a) $\lambda_n = n\pi/L$, $\phi_n(x) = \sin n\pi x/L$, $n = 1, 2, 3, \ldots$
   (c) $\lambda_n$ roots of $\cosh \lambda_n L \cos \lambda_n L = -1$, $n = 1, 2, 3, \ldots$
   $$\phi_n(x) = \frac{\cos \lambda_n x - \cosh \lambda_n x}{\cos \lambda_n L + \cosh \lambda_n L} + \frac{\sinh \lambda_n x - \sin \lambda_n x}{\sinh \lambda_n L + \sin \lambda_n L}, \quad n = 1, 2, 3, \ldots$$

# CHAPTER 11

## Section 11.1, page 383

1. $u = yx^2/2 + xf(y) + g(y)$   3. $u = \frac{1}{2} \sin x + f(x)e^{-y^2}$

5. $u = f(y) \cos 2yx + g(y) \sin 2yx + \dfrac{1-x}{4y^2} + \dfrac{1}{4}$   7. $u = -y^2/3 + f(x) + g(y)e^{3x}$

9. $u = xe^x + \sin y$   11. $u = x^3 \cosh \sqrt{2}\, y + \dfrac{e^{-x}}{\sqrt{2}} \sinh \sqrt{2}\, y$

## Section 11.2, page 397

1. $u = Ce^{\lambda(x-y/3)}$   3. $u = Cx^{\lambda/2}e^{-\lambda y}$   5. $u = Cx^\lambda y^{-\lambda/3}$   7. $u = Ce^{\lambda(x+y)+x^2/2-y^2}$

9. $u = Ce^{\lambda(x^3-y)}$   11. $u = Ce^{\lambda x^3 - y^3/\lambda}$

13. $\dfrac{20}{\pi} \displaystyle\sum_{n=1}^{\infty} \dfrac{\cos(n\pi/2) - (-1)^n}{n} \sin(n\pi x/50)e^{-n^2\pi^2 t/10,000}$

15. (a) $u_0 L^2 \displaystyle\sum_{n=1}^{\infty} \dfrac{1 - e^{-(1/2+n^2\pi^2/L^2)t}}{2n^2\pi^2 + L^2} \sin n\pi x/L + \dfrac{L}{2\pi} \displaystyle\sum_{n=1}^{\infty} \dfrac{(-1)^{n+1}}{n} e^{-(1/2+n^2\pi^2/L^2)t} \sin n\pi x/L$

16. (a) $6 - 2x + \dfrac{6}{\pi} \displaystyle\sum_{n=1}^{\infty} \dfrac{2 + (-1)^n}{n} \sin n\pi x\, e^{-n^2\pi^2 kt}$

21. $w(x, t) = 20x - 2x^2 - \dfrac{2}{\pi} \displaystyle\sum_{n=1}^{\infty} \dfrac{199(-1)^{n+1} - 1}{n} e^{-n^2\pi^2 kt/100} \sin n\pi x/10$

27. $u(x, t) = \dfrac{Cxe^{-\alpha t}}{2\sqrt{\pi k}} \displaystyle\int_0^t e^{\alpha \tau} e^{-x^2/(4k\tau)} \tau^{-3/2}\, d\tau$

   Note: $\mathcal{L}\{t^{-3/2}e^{-x^2/(4t)}\} = \dfrac{2\sqrt{\pi}}{x} e^{-x\sqrt{s}}$

## Section 11.3, page 408

1 (a) $0.1 \cos t \sin x$   (c) $0.1 \cos t \sin x + 0.01 \cos 2t \sin 2x$
   (e) $\dfrac{4}{\pi} \displaystyle\sum_{n=1}^{\infty} \dfrac{(-1)^{n+1}}{(2n-1)^2} \cos(2n-1)t \sin(2n-1)x$

2. (a) $0.1 \sin t \sin x$   (c) $0.1 \sin t \sin x + 0.005 \sin 2t \sin 2x$
   (e) $\dfrac{4}{\pi} \displaystyle\sum_{n=1}^{\infty} \dfrac{(-1)^{n+1}}{(2n-1)^3} \sin(2n-1)t \sin(2n-1)x$

**3.** $u(x, t) = v(x, t) + 2x/L + 1$

**7.** $u(x, t) = \sum_{n=1}^{\infty} [b_{2n-1} \cos(2n-1)c\pi t/L - \alpha_{2n-1}] \sin(2n-1)\pi x/L,$

$$b_{2n-1} = \alpha_{2n-1} + \frac{4L(-1)^{n+1}}{\pi^2(2n-1)^2}, \ \alpha_{2n-1} = \frac{4gL^2}{\rho\pi^3 c^2(2n-1)^3}$$

## Section 11.4, page 419

**1. (a)** $u(x, y) = \dfrac{200}{\pi} \sum_{n=1}^{\infty} \dfrac{\sinh(2n-1)\pi x/b}{\sinh(2n-1)\pi a/b} \dfrac{\sin(2n-1)\pi y/b}{2n-1}$

**(c)** $u(x, y) = \dfrac{200}{\pi} \sum_{n=1}^{\infty} \dfrac{\sinh(2n-1)\pi y/a}{\sinh(2n-1)\pi b/a} \dfrac{\sin(2n-1)\pi x/a}{2n-1}$

**(e)** $u(x, y) = 10x/a + \sum_{n=1}^{\infty} \left(a_n \cosh n\pi y/a + b_n \dfrac{\sinh n\pi y/a}{\sinh n\pi b/a}\right) \sin n\pi x/a$

$a_n = \dfrac{10}{n\pi}[1 + (-1)^n], \ b_n = \dfrac{20}{n\pi}[(-1)^{n+1} + 2]$

**3.** $u = R(r)\Theta(\theta)Z(z), \ Z'' - \lambda Z = 0, \ \Theta'' + \mu\Theta = 0, \ r^2R'' + rR' + (\lambda^2 r^2 - \mu)R = 0$

**5.** $A_0/2 + \sum_{n=1}^{\infty} [A_n \cos n\theta + B_n \sin n\theta](r/a)^n$

$A = \dfrac{1}{\pi} \int_{-\pi}^{\pi} f(\theta) \cos n\theta \, d\theta, \ B_n = \dfrac{1}{\pi} \int_{-\pi}^{\pi} f(\theta) \sin n\theta \, d\theta, \ n = 0, 1, 2, \ldots$

**7. (a)** $u(\rho, \phi) = 4a/\rho$

**9.** $u(\rho, \phi) = \dfrac{1}{2} \sum_{n=0}^{\infty} (2n+1) \left[\int_0^{\pi} f(\phi)P_n(\cos\phi) \sin\phi \, d\phi\right] (\rho/a)^n P_n(\cos\phi)$

# APPENDIX

## Section A.1, page 435

**1. (a)** $-2$ **(c)** $2$     **2. (a)** $-x^2$ **(c)** $e^{3x}$     **3. (a)** $20$ **(c)** $-12$     **4. (a)** $2x^3$ **(c)** $-1$

**5. (a)** $x_1 = 17, x_2 = -\frac{21}{2}$ **(b)** $x_1 = \frac{23}{2}, x_2 = -15$ **(c)** $x_1 = \frac{3}{5}, x_2 = -\frac{11}{5}$

**7. (a)** $x_1 = -\frac{19}{16}, x_2 = \frac{17}{12}, x_3 = -\frac{19}{12}$ **(b)** $x_1 = 1, x_2 = 3, x_3 = 2$

  **(c)** $x_1 = \frac{41}{8}, x_2 = -\frac{5}{2}, x_3 = -\frac{27}{8}$

**9. (a)** yes **(b)** no **(c)** yes     **12. (a)** $abc$ **(b)** $abc$ **(c)** $-abc$ **(d)** $0$ **(e)** $0$ **(f)** $-abcd$

## Section A.2, page 446

**1. (a)** $x_1 = \frac{13}{2}, x_2 = \frac{1}{5}, x_3 = \frac{37}{10}$ **(b)** $x_1 = \frac{3}{2}, x_2 = 1, x_3 = 0$

**3. (a)** $x_1 = 2, x_2 = 3, x_3 = -1$ **(b)** $x_1 = 2, x_2 = 1, x_3 = -3$

**5. (a)** $x_1 = 4, x_2 = -2, x_3 = -1, x_4 = 1$ **(b)** $x_1 = 2x_4 + 5x_3, x_2 = -4x_3 + 3x_4$

**7. (a)** $x_1 = \frac{16}{5} + 2x_2 + 3x_4, x_3 = -\frac{2}{5} - 3x_4$

  **(b)** $x_1 = (8 - 31x_4)/11, x_2 = (26 + 16x_4)/11, x_3 = (32 + 4x_4)/11$

**9. (a)** $r = (8 \pm \sqrt{2})/2$ **(b)** no values     **11. (a)** never **(b)** $r \neq 1$

# Index